Student Solutions Manual

Multivariable Calculus

Soo T. Tan
Stonehill College

with contributions by

Tao Guo
Rock Valley College

Kevin Charlwood
Washburn University

BROOKS/COLE
CENGAGE Learning™

Australia • Brazil • Japan • Korea • Mexico • Singapore • Spain • United Kingdom • United States

ISBN-13: 978-0-534-46577-3
ISBN-10: 0-534-46577-3

Brooks/Cole
20 Channel Center Street
Boston, MA 02210
USA

Cengage Learning is a leading provider of customized learning solutions with office locations around the globe, including Singapore, the United Kingdom, Australia, Mexico, Brazil, and Japan. Locate your local office at: **www.cengage.com/global**

Cengage Learning products are represented in Canada by Nelson Education, Ltd.

To learn more about Brooks/Cole, visit
www.cengage.com/brookscole

Purchase any of our products at your local college store or at our preferred online store
www.cengagebrain.com

Printed in the United States of America
1 2 3 4 5 6 7 14 13 12 11 10

Preface

This *Student Solutions Manual* contains solutions to the odd-numbered exercises in chapters 10–15 of *Calculus, First Edition*, by Soo T. Tan. These correspond to chapters 9–14 of *Calculus, Early Transcendentals, First Edition*.

I would like to thank Tao Guo, who contributed to this manual by solving the problems independently; Kevin Charlwood, who checked all of the solutions for accuracy; and Andy Bulman-Fleming, who typeset this manual. I also want to thank Jerry Grossman for his help with the accuracy checking of the odd answers for the text. Their efforts played a big role in ensuring the accuracy of this manual, and I sincerely appreciate their contributions. I also wish to thank our development editor Jeannine Lawless and our editor Liz Covello of Cengage Learning for their help and patience in bringing this supplement to completion.

Please submit any errors in the solutions manual or suggestions for improvements to me in care of the publisher: Math Editorial, Cengage Learning, 20 Channel Center Street, Boston, MA 02210.

Soo T. Tan

Contents

v

13   Functions of Several Variables   103   ET 12

14   Multiple Integrals   157   ET 13

15   Vector Analysis   199   ET 14

10.1 Concept Questions

1. a. See page 828 (824 in ET).   **b. i.** See page 829 (825 in ET).   **ii.** See page 830 (826 in ET).

3. a. See page 836 (832 in ET).   **b. i.** See page 838 (834 in ET).   **ii.** See page 838 (834 in ET).

10.1 Conic Sections

1. $x^2 = -4y$ has the form $x^2 = 4py$ with $p = -1$, so it represents the parabola with vertex $(0, 0)$, focus $(0, -1)$, and directrix $y = 1$, labeled **h**.

3. $y^2 = 8x$ has the form $y^2 = 4px$ with $p = 2$, so it represents the parabola with vertex $(0, 0)$, focus $(2, 0)$, and directrix $x = -2$, labeled **c**.

5. $\dfrac{x^2}{9} + \dfrac{y^2}{4} = 1$ has the form $\dfrac{x^2}{a^2} + \dfrac{y^2}{b^2} = 1$ with $a = 3$, $b = 2$, and $c = \sqrt{a^2 - b^2} = \sqrt{5}$, so it represents the ellipse with vertices $(\pm 3, 0)$ and foci $\left(\pm\sqrt{5}, 0\right)$, labeled **b**. Its eccentricity is $\frac{\sqrt{5}}{3}$.

7. $\dfrac{x^2}{16} - \dfrac{y^2}{9} = 1$ has the form $\dfrac{x^2}{a^2} - \dfrac{y^2}{b^2} = 1$ with $a = 4$, $b = 3$, and $c = \sqrt{a^2 + b^2} = 5$, so it represents the hyperbola with vertices $(\pm 4, 0)$ and foci $(\pm 5, 0)$, labeled **d**. Its eccentricity is $\frac{5}{4}$.

9. $y = 2x^2 \Leftrightarrow x^2 = 4py$ with $p = \frac{1}{8}$, so the parabola has vertex $(0, 0)$, focus $\left(0, \frac{1}{8}\right)$, and directrix $y = -\frac{1}{8}$.

11. $x = 2y^2 \Leftrightarrow y^2 = \frac{1}{2}x = 4px$ with $p = \frac{1}{8}$, so the parabola has vertex $(0, 0)$, focus $\left(\frac{1}{8}, 0\right)$, and directrix $x = -\frac{1}{8}$.

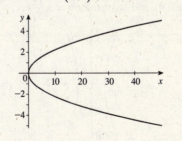

1

13. $5y^2 = 12x \Leftrightarrow y^2 = \frac{12}{5}x = 4px$ with $p = \frac{3}{5}$, so the

parabola has vertex $(0, 0)$, focus $\left(\frac{3}{5}, 0\right)$, and directrix

$x = -\frac{3}{5}$.

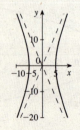

15. $\dfrac{x^2}{4} + \dfrac{y^2}{25} = 1$ has $a = 5, b = 2$, and

$c = \sqrt{a^2 - b^2} = \sqrt{21}$, so the ellipse has foci $\left(0, \pm\sqrt{21}\right)$

and vertices $(0, \pm 5)$.

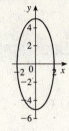

17. $4x^2 + 9y^2 = 36 \Leftrightarrow \dfrac{x^2}{9} + \dfrac{y^2}{4} = 1$ has $a = 3, b = 2$, and

$c = \sqrt{a^2 - b^2} = \sqrt{5}$, so the ellipse has foci $\left(\pm\sqrt{5}, 0\right)$

and vertices $(\pm 3, 0)$.

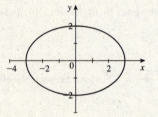

19. $x^2 + 4y^2 = 4 \Leftrightarrow \dfrac{x^2}{4} + y^2 = 1$ has $a = 2, b = 1$, and

$c = \sqrt{a^2 - b^2} = \sqrt{3}$, so the ellipse has foci $\left(\pm\sqrt{3}, 0\right)$

and vertices $(\pm 2, 0)$.

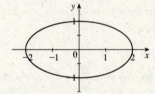

21. $\dfrac{x^2}{25} - \dfrac{y^2}{144} = 1$ has $a = 5, b = 12$, and

$c = \sqrt{a^2 + b^2} = 13$, so the hyperbola has foci $(\pm 13, 0)$,

vertices $(\pm 5, 0)$, and asymptotes $y = \pm\frac{12}{5}x$.

23. $x^2 - y^2 = 1$ has $a = 1, b = 1$, and $c = \sqrt{a^2 + b^2} = \sqrt{2}$,

so the hyperbola has foci $\left(\pm\sqrt{2}, 0\right)$, vertices $(\pm 1, 0)$, and

asymptotes $y = \pm x$.

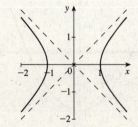

25. $y^2 - 5x^2 = 25 \Leftrightarrow \dfrac{y^2}{25} - \dfrac{x^2}{5} = 1$ has $a = 5$, $b = \sqrt{5}$, and $c = \sqrt{a^2 + b^2} = \sqrt{30}$, so the hyperbola has foci $\left(0, \pm\sqrt{30}\right)$, vertices $(0, \pm 5)$, and asymptotes $y = \pm\sqrt{5}x$.

27. The parabola with focus $(3, 0)$ and directrix $x = -3$ has $p = 3$ and axis the x-axis, so an equation is $y^2 = 4px = 12x$.

29. The parabola with focus $\left(-\frac{5}{2}, 0\right)$ and directrix $x = \frac{5}{2}$ has $p = -\frac{5}{2}$ and axis the x-axis, so an equation is $y^2 = 4px = -10x$.

31. The ellipse with foci $(\pm 1, 0)$ and vertices $(\pm 3, 0)$ has $a = 3$, $c = 1$, and $b = \sqrt{a^2 - c^2} = 2\sqrt{2}$, so an equation is $\dfrac{x^2}{9} + \dfrac{y^2}{8} = 1$.

33. The ellipse with foci $(0, \pm 1)$ and major axis of length 6 has $2a = 6 \Leftrightarrow a = 3$, $c = 1$, and $b = \sqrt{a^2 - c^2} = 2\sqrt{2}$, so an equation is $\dfrac{x^2}{8} + \dfrac{y^2}{9} = 1$.

35. The ellipse with vertices $(\pm 3, 0)$ has $a = 3$, so the ellipse has an equation of the form $\dfrac{x^2}{9} + \dfrac{y^2}{b^2} = 1$. Substituting the point $\left(1, \sqrt{2}\right)$, we find $\dfrac{1^2}{9} + \dfrac{\left(\sqrt{2}\right)^2}{b^2} = 1 \Leftrightarrow \dfrac{2}{b^2} = \dfrac{8}{9} \Leftrightarrow b^2 = \dfrac{9}{4}$, so an equation of the ellipse is $\dfrac{x^2}{9} + \dfrac{4y^2}{9} = 1$.

37. The ellipse with vertices $(0, \pm 5)$ has $a = 5$, so the ellipse has an equation of the form $\dfrac{x^2}{b^2} + \dfrac{y^2}{25} = 1$. Substituting the point $\left(2, \dfrac{3\sqrt{3}}{2}\right)$, we find $\dfrac{4}{b^2} + \dfrac{27}{4 \cdot 25} = 1 \Leftrightarrow b^2 = \dfrac{400}{73}$, so an equation is $\dfrac{73x^2}{400} + \dfrac{y^2}{25} = 1$.

39. The hyperbola with foci $(\pm 5, 0)$ and vertices $(\pm 3, 0)$ has $a = 3$, $c = 5$, and $b^2 = c^2 - a^2 = 16$, so an equation is $\dfrac{x^2}{9} - \dfrac{y^2}{16} = 1$.

41. The hyperbola with foci $(0, \pm 5)$ and conjugate axis of length 4 has $c = 5$, $2b = 4 \Leftrightarrow b = 2$, and $a^2 = c^2 - b^2 = 21$, so an equation is $\dfrac{y^2}{21} - \dfrac{x^2}{4} = 1$.

43. The hyperbola with vertices $(\pm 2, 0)$ and asymptotes $y = \pm\frac{3}{2}x$ has $a = 2$ and $\dfrac{b}{a} = \dfrac{3}{2}$, so $b = \frac{3}{2}(2) = 3$ and an equation is $\dfrac{x^2}{4} - \dfrac{y^2}{9} = 1$.

45. Referring to the table on page 840 (836 in ET) and Figure 26, we see that $(x + 3)^2 = -2(y - 4)$ is an equation of the parabola with vertex $(-3, 4)$ opening downward, labeled **b**.

47. $\dfrac{(y - 3)^2}{16} - \dfrac{(x + 1)^2}{9} = 1$ is an equation of a hyperbola centered at $(-1, 3)$, labeled **c**.

49. The parabola with focus $(3, 1)$ and directrix $x = 1$ has vertex $(2, 1)$ and opens to the right with $p = 1$, so an equation is $(y - 1)^2 = 4(1)(x - 2)$ or $(y - 1)^2 = 4(x - 2)$.

51. The parabola with vertex $(2, 2)$ and focus $\left(\frac{3}{2}, 2\right)$ opens to the left with $p = -\frac{1}{2}$, so an equation is

$(y - 2)^2 = 4\left(-\frac{1}{2}\right)(x - 2)$ or $(y - 2)^2 = -2(x - 2)$.

53. Referring to the table on page 840 (836 in ET), the parabola with axis parallel to the y-axis has equation

$(x - h)^2 = 4p(y - k)$. Substituting the three known points gives the equations $\begin{cases} (-3 - h)^2 = 4p(2 - k) \\ (0 - h)^2 = 4p\left(-\frac{5}{2} - k\right) \\ (1 - h)^2 = 4p(-6 - k) \end{cases} \Rightarrow$

$\begin{cases} h^2 + 6h + 9 = 8p - 4pk \\ h^2 = -10p - 4pk \\ h^2 - 2h + 1 = -24p - 4pk \end{cases}$ Subtracting the second equation from the first gives $6h + 9 = 18p \Leftrightarrow 2h + 3 = 6p$

(call this equation α), and subtracting the third from the second gives $2h - 1 = 14p$ (equation β). Subtracting β from α

gives $4 = -8p \Leftrightarrow p = -\frac{1}{2}$. Thus, from α, $2h + 3 = -3 \Leftrightarrow h = -3$, and finally, from the second original equation

$(-3)^2 = -10\left(-\frac{1}{2}\right) - 4\left(-\frac{1}{2}\right)k \Leftrightarrow k = 2$. An equation of the parabola is thus $(x + 3)^2 = -2(y - 2)$.

55. The ellipse with foci $(\pm 1, 3)$ and vertices $(\pm 3, 3)$ has center $(0, 3)$, $a = 3$, and $c = 1$. Thus, $b^2 = a^2 - c^2 = 8$, and an

equation is $\dfrac{x^2}{9} + \dfrac{(y - 3)^2}{8} = 1$.

57. The ellipse with foci $(\pm 1, 2)$ and major axis of length $2a = 8 \Leftrightarrow a = 4$ has vertices $(\pm 4, 2)$, center $(0, 2)$, and $c = 1$. Thus,

$b^2 = a^2 - c^2 = 15$, and an equation is $\dfrac{x^2}{16} + \dfrac{(y - 2)^2}{15} = 1$.

59. The ellipse with center $(2, 1)$, one focus at $(0, 1)$, and one vertex at $(5, 1)$ has $a = 3$ and $c = 2$, so $b^2 = a^2 - c^2 = 5$ and an

equation is $\dfrac{(x - 2)^2}{9} + \dfrac{(y - 1)^2}{5} = 1$.

61. The hyperbola with foci $(-2, 2)$ and $(8, 2)$ and vertices $(0, 2)$ and $(6, 2)$ has center $(3, 2)$, so $a = 3$, $c = 5$, and

$b^2 = c^2 - a^2 = 16$. An equation is thus $\dfrac{(x - 3)^2}{9} - \dfrac{(y - 2)^2}{16} = 1$.

63. The hyperbola with foci $(6, -3)$ and $(-4, -3)$ and asymptotes $y + 3 = \pm\frac{4}{3}(x - 1)$ has center $(1, -3)$, so $c = 5$. The

slopes of the asymptotes are $\pm\frac{b}{a} = \pm\frac{4}{3}$ and $a^2 + b^2 = c^2 = 25$, so by inspection $a = 3$ and $b = 4$. An equation is thus

$\dfrac{(x - 1)^2}{9} - \dfrac{(y + 3)^2}{16} = 1$.

65. The hyperbola with vertices $(4, -2)$ and $(4, 4)$ and asymptotes $y - 1 = \pm\frac{3}{2}(x - 4)$ has center $(4, 1)$, so $a = 3$. The slopes

of the asymptotes are $\pm\frac{a}{b} = \pm\frac{3}{2}$, so $b = 2$ and an equation is $\dfrac{(y - 1)^2}{9} - \dfrac{(x - 4)^2}{4} = 1$.

67. We complete the square and put the equation into standard form: $y^2 - 2y - 4x + 9 = 0 \Leftrightarrow (y-1)^2 - 1 - 4x + 9 = 0$ $\Leftrightarrow (y-1)^2 = 4x - 8 = 4(x-2)$. This equation has $p = 1$, so it represents a parabola with vertex $(2, 1)$, focus $(3, 1)$, and directrix $x = 1$.

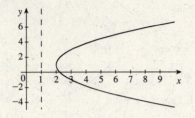

69. $x^2 + 6x - y + 11 = 0 \Leftrightarrow (x+3)^2 - 9 - y + 11 = 0 \Leftrightarrow$ $(x+3)^2 = y - 2$ represents a parabola with vertex $(-3, 2)$, focus $\left(-3, \frac{9}{4}\right)$, and directrix $y = \frac{7}{4}$.

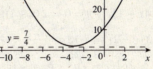

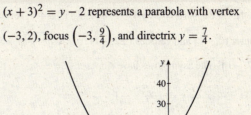

71. $4y^2 - 4y - 32x - 31 = 0 \Leftrightarrow 4\left(y - \frac{1}{2}\right)^2 - 1 - 32x - 31 = 0 \Leftrightarrow$

$\left(y - \frac{1}{2}\right)^2 = 8(x+1)$ represents a parabola with vertex $\left(-1, \frac{1}{2}\right)$, focus $\left(1, \frac{1}{2}\right)$, and directrix $x = -3$.

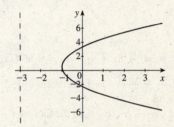

73. $(x-1)^2 + 4(y+2)^2 = 1$ has center $(1, -2)$ and vertices $(0, -2)$ and $(2, -2)$.

$c = \sqrt{a^2 - b^2} = \sqrt{1 - \frac{1}{4}} = \frac{\sqrt{3}}{2}$, so the foci are $\left(1 \pm \frac{\sqrt{3}}{2}, -2\right)$.

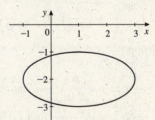

75. We complete the squares and put the equation into standard form:

$x^2 + 4y^2 - 2x + 16y + 13 = 0 \Leftrightarrow (x-1)^2 - 1 + 4(y+2)^2 - 16 + 13 = 0 \Leftrightarrow$

$(x-1)^2 + 4(y+2)^2 = 4 \Leftrightarrow \frac{(x-1)^2}{4} + (y+2)^2 = 1$ has center $(1, -2)$ and

vertices $(-1, -2)$ and $(3, -2)$. $c = \sqrt{a^2 - b^2} = \sqrt{3}$, so the foci are

$\left(1 \pm \sqrt{3}, -2\right)$.

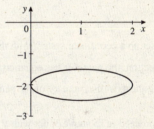

77. $4x^2 + 9y^2 - 18x - 27 = 0 \Leftrightarrow 4\left(x - \frac{9}{4}\right)^2 - \frac{81}{4} + 9y^2 - 27 = 0 \Leftrightarrow$

$\frac{16\left(x - \frac{9}{4}\right)^2}{189} + \frac{4y^2}{21} = 1$ has center $\left(\frac{9}{4}, 0\right)$ and vertices $\left(\frac{9}{4} \pm \frac{3\sqrt{21}}{4}, 0\right)$.

$c = \sqrt{a^2 - b^2} = \sqrt{\frac{189}{16} - \frac{21}{4}} = \frac{\sqrt{105}}{4}$, so the foci are $\left(\frac{9}{4} \pm \frac{\sqrt{105}}{4}, 0\right)$.

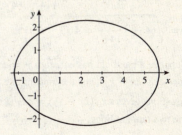

79. We complete the square in y and put the equation into standard form:

$$3x^2 - 4y^2 - 8y - 16 = 0 \Leftrightarrow 3x^2 - 4(y+1)^2 + 4 - 16 = 0 \Leftrightarrow \frac{x^2}{4} - \frac{(y+1)^2}{3} = 1$$

has center $(0, -1)$ and vertices $(\pm 2, -1)$. $c = \sqrt{a^2 + b^2} = \sqrt{7}$, so the foci are

$\left(\pm \sqrt{7}, -1\right)$. The asymptotes have slopes $\pm \frac{b}{a} = \pm \frac{\sqrt{3}}{2}$, so their equations are

$y + 1 = \pm \frac{\sqrt{3}}{2} x \Leftrightarrow y = \pm \frac{\sqrt{3}}{2} x - 1$.

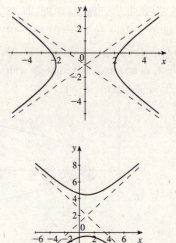

81. $2x^2 - 3y^2 - 4x + 12y + 8 = 0 \Leftrightarrow 2(x-1)^2 - 2 - 3(y-2)^2 + 12 + 8 = 0 \Leftrightarrow$

$2(x-1)^2 - 3(y-2)^2 = -18 \Leftrightarrow \frac{(y-2)^2}{6} - \frac{(x-1)^2}{9} = 1$ has center $(1, 2)$ and

vertices $\left(1, 2 \pm \sqrt{6}\right)$. $c = \sqrt{a^2 + b^2} = \sqrt{15}$, so the foci are $\left(1, 2 \pm \sqrt{15}\right)$. The

asymptotes have slopes $\pm \frac{\sqrt{6}}{3}$, so their equations are $y - 2 = \pm \frac{\sqrt{6}}{3} (x - 1)$.

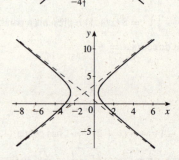

83. $4x^2 - 2y^2 + 8x + 8y - 12 = 0 \Leftrightarrow 4(x+1)^2 - 4 - 2(y-2)^2 + 8 - 12 = 0 \Leftrightarrow$

$\frac{(x+1)^2}{2} - \frac{(y-2)^2}{4} = 1$ has center $(-1, 2)$ and vertices $\left(-1 \pm \sqrt{2}, 2\right)$.

$c = \sqrt{a^2 + b^2} = \sqrt{6}$, so the foci are $\left(-1 \pm \sqrt{6}, 2\right)$. The asymptotes are

$y - 2 = \pm \sqrt{2} (x + 1)$.

85. We set up a coordinate system with the origin at the vertex of the parabola and the positive x-axis pointing to the right. By inspection, the parabola has equation $y^2 = x$, so its focus is $\left(\frac{1}{4}, 0\right)$. The light source should be placed $\frac{1}{4}$ ft from the vertex along the axis of symmetry of the parabola.

87. We proceed in Example 3, but with $y = \dfrac{60x^2}{300^2} = \dfrac{x^2}{1500} \Rightarrow y' = \dfrac{x}{750}$. The length is thus

$$s = 2 \int_0^{300} \sqrt{1 + \left(\frac{x}{750}\right)^2}\, dx = \frac{2}{750} \int_0^{300} \sqrt{750^2 + x^2}\, dx = \frac{1}{375} \left[\frac{x}{2} \sqrt{750^2 + x^2} + \frac{750^2}{2} \ln\left(x + \sqrt{750^2 + x^2}\right) \right]_0^{300}$$

≈ 616 ft

89. $\dfrac{dy}{dx} = \dfrac{W}{H} x \Rightarrow y = \dfrac{W}{2H} x^2 + C$ and $y(0) = 0 \Rightarrow y = \dfrac{W}{2H} x^2$, a parabola.

91. If the origin is at the center of the ellipse, then one arch has major axis of length $2a = 24$ and minor axis of length

$2b = 16$, so an equation is $\dfrac{x^2}{144} + \dfrac{y^2}{64} = 1$. So 6 units from the center of the base, we have $\dfrac{6^2}{144} + \dfrac{y^2}{64} = 1 \Leftrightarrow$

$y = \sqrt{64 - \dfrac{36 \cdot 64}{144}} = 4\sqrt{3} \approx 6.93$ ft.

93. Suppose a parabola has equation $y = ax^2$. Then the tangent line where $x = x_0$ has slope $2ax_0$ and equation $y = ax_0(2x - x_0)$, and the tangent line where $x = x_1$ has slope $2ax_1$ and equation $y = ax_1(2x - x_1)$. Solving these two equations gives $ax_0(2x - x_0) = ax_1(2x - x_1) \Rightarrow x = \frac{1}{2}(x_0 + x_1)$, the x-coordinate of the unique point of intersection of the two tangent lines.

95. Differentiating implicitly, we get $\dfrac{x^2}{a^2} + \dfrac{y^2}{b^2} = 1 \Rightarrow b^2x^2 + a^2y^2 = a^2b^2 \Rightarrow 2b^2x + 2a^2yy' = 0 \Rightarrow y' = -\dfrac{b^2}{a^2}\dfrac{x}{y}$.

An equation of the tangent line at (x_0, y_0) is $y - y_0 = -\dfrac{b^2}{a^2}\dfrac{x_0}{y_0}(x - x_0) \Rightarrow \dfrac{yy_0}{b^2} - \dfrac{y_0^2}{b^2} = -\dfrac{x_0}{a^2}(x - x_0) \Rightarrow$

$\dfrac{xx_0}{a^2} + \dfrac{yy_0}{b^2} = \dfrac{x_0^2}{a^2} + \dfrac{y_0^2}{b^2} = 1$.

97. Differentiating implicitly, we get $\dfrac{x^2}{a^2} - \dfrac{y^2}{b^2} = 1 \Rightarrow b^2x^2 - a^2y^2 = a^2b^2 \Rightarrow 2b^2x - 2a^2yy' = 0 \Rightarrow y' = \dfrac{b^2}{a^2}\dfrac{x}{y}$. An equation

of the tangent line at (x_0, y_0) is $y - y_0 = \dfrac{b^2}{a^2}\dfrac{x_0}{y_0}(x - x_0) \Rightarrow \dfrac{yy_0}{b^2} - \dfrac{y_0^2}{b^2} = \dfrac{x_0}{a^2}(x - x_0) \Rightarrow \dfrac{xx_0}{a^2} - \dfrac{yy_0}{b^2} = \dfrac{x_0^2}{a^2} - \dfrac{y_0^2}{b^2} = 1$.

99. We differentiate each equation implicitly: $\dfrac{x^2}{a^2} + \dfrac{2y^2}{b^2} = 1 \Rightarrow b^2x^2 + 2a^2y^2 = 1 \Rightarrow 2b^2x + 4a^2yy' = 0 \Rightarrow$

$y' = -\dfrac{b^2}{2a^2}\dfrac{x}{y}$, and $\dfrac{x^2}{a^2 - b^2} - \dfrac{2y^2}{b^2} = 1 \Rightarrow b^2x^2 - 2\left(a^2 - b^2\right)y^2 = b^2\left(a^2 - b^2\right) \Rightarrow 2b^2x - 4\left(a^2 - b^2\right)yy' = 0 \Rightarrow$

$y' = \dfrac{b^2}{2\left(a^2 - b^2\right)}\dfrac{x}{y}$. The two curves are perpendicular at a shared point (x_0, y_0) if the slopes of their tangents are negative

reciprocals. The product of these slopes is $-\dfrac{b^2}{2a^2}\dfrac{x_0}{y_0} \cdot \dfrac{b^2}{2\left(a^2 - b^2\right)}\dfrac{x_0}{y_0} = \dfrac{b^4x_0^2}{4a^2\left(b^2 - a^2\right)y_0^2} = \dfrac{b^2}{4y_0^2}\dfrac{b^2x_0^2}{\left(b^2 - a^2\right)a^2}$, but

(x_0, y_0) lies on both curves, so $\dfrac{x_0^2}{a^2} + \dfrac{2y_0^2}{b^2} = \dfrac{x_0^2}{a^2 - b^2} - \dfrac{2y_0^2}{b^2} \Leftrightarrow \dfrac{4y_0^2}{b^2} = x_0^2\left(\dfrac{1}{a^2 - b^2} - \dfrac{1}{a^2}\right) = -\dfrac{b^2x_0^2}{\left(b^2 - a^2\right)a^2}$, so our

expression for the product of the slopes becomes $\dfrac{b^2}{4y_0^2}\left(-\dfrac{4y_0^2}{b^2}\right) = -1$. Thus, the two curves are perpendicular at every

point of intersection.

101. $\dfrac{x^2}{a^2} - \dfrac{y^2}{b^2} = 1 \Rightarrow y = \pm\dfrac{b}{a}\sqrt{x^2 - a^2}$. Consider

$$\lim_{x\to\infty}\left[\pm\dfrac{b}{a}x - \left(\pm\dfrac{b}{a}\right)\sqrt{x^2 - a^2}\right] = \pm\dfrac{b}{a}\lim_{x\to\infty}\left(x - \sqrt{x^2 - a^2}\right) = \pm\dfrac{b}{a}\lim_{x\to\infty}\left[\left(x - \sqrt{x^2 - a^2}\right)\dfrac{x + \sqrt{x^2 - a^2}}{x + \sqrt{x^2 - a^2}}\right]$$

$$= \pm\dfrac{b}{a}\lim_{x\to\infty}\dfrac{a^2}{x + \sqrt{x^2 - a^2}} = 0$$

This shows that $y = \pm\dfrac{b}{a}x$ are asymptotes of $\dfrac{x^2}{a^2} - \dfrac{y^2}{b^2} = 1$.

103. Differentiating implicitly, we find $4x^2 + 25y^2 = 100 \Rightarrow 8x + 50yy' = 0 \Rightarrow y' = -\dfrac{4}{25}\dfrac{x}{y} = -\dfrac{2}{5}\dfrac{x}{\sqrt{25 - x^2}}$

for $y > 0$. So using the arc length formula and the symmetry of the ellipse, we can write

$$C = 4\int_0^5 \sqrt{1 + (y')^2}\,dx = 4\int_0^5 \sqrt{1 + \left(\dfrac{2}{5}\dfrac{x}{\sqrt{25 - x^2}}\right)^2}\,dx \approx 23.013, \text{ using a computer algebra system.}$$

105. a. To calculate $\tan\beta$, we first calculate the y-intercept of ℓ. Note that $y^2 = 4px \Rightarrow 2yy' = 4p$, so at P, $y' = \dfrac{2p}{y_0}$

and an equation of the tangent is $y - y_0 = \dfrac{2p}{y_0}(x - x_0)$. Substituting $x = 0$ gives $y = y_0 - \dfrac{2px_0}{y_0}$, so

$$\tan\beta = \dfrac{y_0 - \left(y_0 - \dfrac{2px_0}{y_0}\right)}{x_0} = \dfrac{2px_0}{x_0y_0} = \dfrac{2p}{y_0}.$$

b. $\tan\phi = \dfrac{y_0}{x_0 - p}$ by inspection.

c. ϕ and $\alpha + \beta$ are alternate angles, so $\phi = \alpha + \beta \Rightarrow \alpha = \phi - \beta$ and

$$\tan\alpha = \frac{\tan\phi - \tan\beta}{1 + \tan\phi\tan\beta} = \frac{\dfrac{y_0}{x_0 - p} - \dfrac{2p}{y_0}}{1 + \dfrac{y_0}{x_0 - p}\cdot\dfrac{2p}{y_0}} = \frac{y_0^2 - 2p(x_0 - p)}{(x_0 - p)y_0 + 2py_0}.$$ Substituting $y_0^2 = 4px_0$, we find that this is equal

to $\dfrac{4px_0 - 2p(x_0 - p)}{(x_0 - p)y_0 + 2py_0} = \dfrac{2p(x_0 + p)}{y_0(p + x_0)} = \dfrac{2p}{y_0}$. In other words, $\alpha = \beta$.

107. Using the result of Exercise 97, the slope of ℓ_T at (x_0, y_0) is

$m_T = \dfrac{b^2 x_0}{a^2 y_0}$, so the slope of ℓ_N is $m_N = -\dfrac{a^2 y_0}{b^2 x_0}$. Note that

to prove that $\alpha = \beta$, it suffices to show that $\theta_1 = \theta_2$, since $\theta_1 + \beta = \theta_2 + \alpha = \frac{\pi}{2}$. Also note that $\theta_1 = \theta_3 + \theta_4$, $\theta_2 = \theta_5 - \alpha$, and $\theta_5 = \pi - \theta_6$. An equation of the normal

line ℓ_N is $y - y_0 = -\dfrac{a^2 y_0}{b^2 x_0}(x - x_0)$. Setting $y = 0$ gives

the x-intercept of ℓ_N as $x_1 = x_0 + \dfrac{b^2 x_0}{a^2}$.

Now $\tan\theta_4 = \dfrac{y_0}{x_1 - x_0} = \dfrac{a^2 y_0}{b^2 x_0}$, $\tan\theta_3 = \dfrac{y_0}{x_0 + c}$, and $\tan\theta_6 = \dfrac{y_0}{x_0 - c}$, so

$\tan\theta_5 = \tan(\pi - \theta_6) = \dfrac{\tan\pi - \tan\theta_6}{1 + \tan\pi\tan\theta_6} = -\tan\theta_6 = \dfrac{y_0}{c - x_0}$. Using the relationships $b^2 x_0^2 - a^2 y_0^2 = a^2 b^2$ and

$c^2 = a^2 + b^2$, we can show that $\tan\theta_1 = \tan(\theta_3 + \theta_4) = \dfrac{\tan\theta_3 + \tan\theta_4}{1 - \tan\theta_3\tan\theta_4} = \dfrac{\dfrac{y_0}{x_0 + c} + \dfrac{a^2 y_0}{b^2 x_0}}{1 - \dfrac{y_0}{x_0 + c}\cdot\dfrac{a^2 y_0}{b^2 x_0}} = \dfrac{cy_0}{b}$ and

$\tan\theta_2 = \tan(\theta_5 - \theta_4) = \dfrac{\tan\theta_5 - \tan\theta_4}{1 + \tan\theta_5\tan\theta_4} = \dfrac{\dfrac{y_0}{c - x_0} - \dfrac{a^2 y_0}{b^2 x_0}}{1 + \dfrac{y_0}{c - x_0}\cdot\dfrac{a^2 y_0}{b^2 x_0}} = \dfrac{cy_0}{b}$. Therefore, $\tan\theta_1 = \tan\theta_2$, showing that

$\theta_1 = \theta_2$ and thus $\alpha = \beta$.

109. True. If $F > 0$, then $2x^2 - y^2 + F = 0 \Leftrightarrow \dfrac{y^2}{F} - \dfrac{x^2}{\left(\frac{F}{2}\right)} = 1$ is a hyperbola with transverse axis the y-axis. If $F < 0$, then

$2x^2 - y^2 + F = 0 \Leftrightarrow \dfrac{x^2}{\frac{|F|}{2}} - \dfrac{y^2}{|F|} = 1$ is a hyperbola with transverse axis the x-axis.

111. True. The ellipse $b^2x^2 + a^2y^2 = a^2b^2$ $(a > b > 0)$ is contained in the circle $x^2 + y^2 = a^2$ and contains the circle $x^2 + y^2 = b^2$.

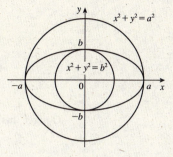

113. False. Take $A = 1$, $C = 4$, $D = 0$, $E = 0$, and $F = 5$. Then we have $x^2 + 4y^2 + 5 = 0 \Leftrightarrow x^2 + 4y^2 = -5$ which has no solution.

10.2 Concept Questions   ET 9.2

1. For the definition of a plane curve, see page 849 (845 in ET). The plane curve $C : x = \cos\theta$, $y = \sin\theta$, $0 \le \theta \le 2\pi$ is a plane curve (circle) that is not the graph of a function.

3. C_1 and C_2 have the same graph, but C_1 has initial point $P_0\,(f\,(0)\,,g\,(0))$ and terminal point $P_1\,(f\,(1)\,,g\,(1))$, whereas C_2 begins at the point P_1 and traces the curve in the opposite direction, ending at P_0.

10.2 Plane Curves and Parametric Equations   ET 9.2

1. a. $\left.\begin{matrix} x = 2t + 1 \\ y + 3 = t \end{matrix}\right\} \Rightarrow x = 2\,(y + 3) + 1$ or

$x - 2y - 7 = 0$

b.

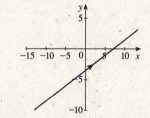

The orientation is found by observing that as t increases, so does x.

3. a. $\left.\begin{matrix} x = \sqrt{t} \\ y = 9 - t \end{matrix}\right\} \Rightarrow y = 9 - x^2, x \ge 0$

b.

The orientation is found by observing that as t increases, so does x.

5. a. $\left.\begin{array}{l} x = t^2 + 1 \\ y = 2t^2 - 1 \end{array}\right\}$ for $-2 \le t \le 2 \Rightarrow$

$y = 2(x - 1) - 1 \Rightarrow y = 2x - 3, 1 \le x \le 5$

b.

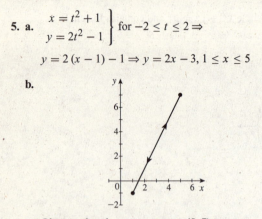

Observe that the curve starts at $(5, 7)$, moves
along the line segment to $(1, -1)$, then returns to
$(5, 7)$.

7. a. $\left.\begin{array}{l} x = t^2 \\ y = t^3 \end{array}\right\} \Rightarrow x = y^{2/3}, -8 \le y \le 8$

b.

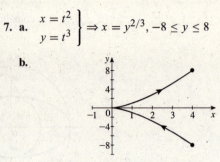

Observe that y increases as t increases.

9. a. $\left.\begin{array}{l} x = 2\sin\theta \\ y = 2\cos\theta \end{array}\right\} \Rightarrow \left\{\begin{array}{l} \sin\theta = \frac{1}{2}x \\ \cos\theta = \frac{1}{2}y \end{array}\right\} \Rightarrow$

$\left(\frac{1}{2}x\right)^2 + \left(\frac{1}{2}y\right)^2 = \sin^2\theta + \cos^2\theta = 1$, so

$x^2 + y^2 = 4$.

b.

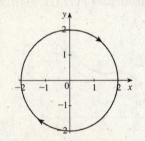

Observe that as θ increases from 0 to 2π, the
curve C is traced once in a clockwise direction
starting from the point $(0, 2)$.

11. a. $\left.\begin{array}{l} x = 2\sin\theta \\ y = 3\cos\theta \end{array}\right\} \Rightarrow \left\{\begin{array}{l} \sin\theta = \frac{1}{2}x \\ \cos\theta = \frac{1}{3}y \end{array}\right\} \Rightarrow$

$\left(\frac{1}{2}x\right)^2 + \left(\frac{1}{3}y\right)^2 = \sin^2\theta + \cos^2\theta = 1 \Rightarrow$

$\frac{1}{4}x^2 + \frac{1}{9}y^2 = 1$

b.

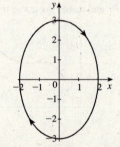

13. a. $\left.\begin{array}{l} x = 2\cos\theta + 2 \\ y = 3\sin\theta - 1 \end{array}\right\} \Rightarrow \left\{\begin{array}{l} \cos\theta = \frac{1}{2}(x - 2) \\ \sin\theta = \frac{1}{3}(y + 1) \end{array}\right\}$

$\Rightarrow \left[\frac{1}{2}(x - 2)\right]^2 + \left[\frac{1}{3}(y + 1)\right]^2 = \sin^2\theta +$

$\cos^2\theta = 1 \Rightarrow \frac{1}{4}(x - 2)^2 + \frac{1}{9}(y + 1)^2 = 1$

b.

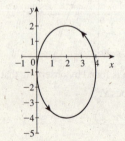

15. a. $\left.\begin{array}{l} x = \cos\theta \\ y = \cos 2\theta \end{array}\right\} \Rightarrow \left\{\begin{array}{l} x = \cos\theta \\ y = 2\cos^2\theta - 1 \end{array}\right\} \Rightarrow$

$y = 2x^2 - 1, -1 \le x \le 1$

b.

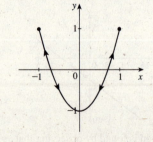

17. a. $\left.\begin{matrix} x = \sec\theta \\ y = \tan\theta \end{matrix}\right\} \Rightarrow \left\{\begin{matrix} x^2 = \sec^2\theta \\ y^2 = \tan^2\theta \end{matrix}\right\}$ From the

identity $\sec^2\theta = 1 + \tan^2\theta$, we obtain

$x^2 - y^2 = 1, x \geq 1$.

b.

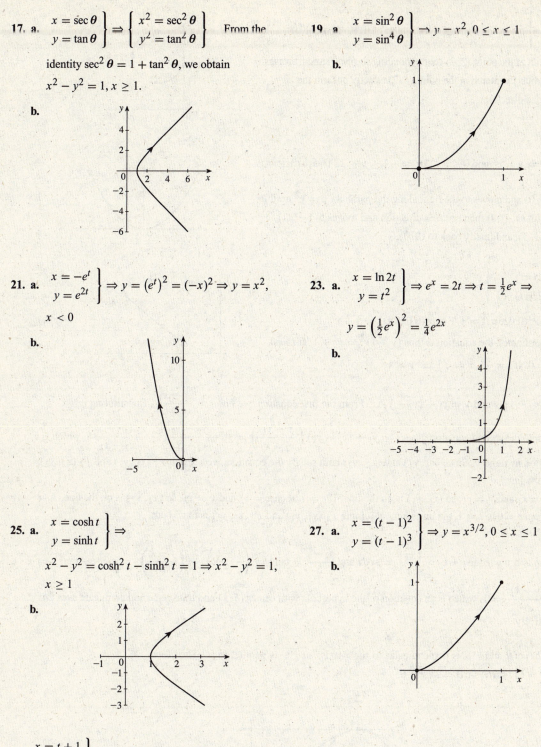

19. a. $\left.\begin{matrix} x = \sin^2\theta \\ y = \sin^4\theta \end{matrix}\right\} \rightarrow y - x^2, 0 \leq x \leq 1$

b.

21. a. $\left.\begin{matrix} x = -e^t \\ y = e^{2t} \end{matrix}\right\} \Rightarrow y = (e^t)^2 = (-x)^2 \Rightarrow y = x^2,$

$x < 0$

b.

23. a. $\left.\begin{matrix} x = \ln 2t \\ y = t^2 \end{matrix}\right\} \Rightarrow e^x = 2t \Rightarrow t = \tfrac{1}{2}e^x \Rightarrow$

$y = \left(\tfrac{1}{2}e^x\right)^2 = \tfrac{1}{4}e^{2x}$

b.

25. a. $\left.\begin{matrix} x = \cosh t \\ y = \sinh t \end{matrix}\right\} \Rightarrow$

$x^2 - y^2 = \cosh^2 t - \sinh^2 t = 1 \Rightarrow x^2 - y^2 = 1,$

$x \geq 1$

b.

27. a. $\left.\begin{matrix} x = (t-1)^2 \\ y = (t-1)^3 \end{matrix}\right\} \Rightarrow y = x^{3/2}, 0 \leq x \leq 1$

b.

29. $\left.\begin{matrix} x = t+1 \\ y = \sqrt{t} \end{matrix}\right\} \Rightarrow y = \sqrt{x-1}$ and $0 \leq t \leq 4 \Rightarrow 1 \leq x \leq 5$. At $t = 0$, the

particle is located at the point $(1, 0)$. As t increases, the particle moves

along the parabola toward the point $(5, 2)$.

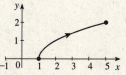

31. $\left.\begin{array}{l} x = 1 + \cos t \\ y = 2 + \sin t \end{array}\right\} \Rightarrow (x-1)^2 + (y-2)^2 = \cos^2 t + \sin^2 t = 1.$ The

particle starts out at the point $(2, 2)$ (corresponding to $t = 0$) and traverses
the circle of radius 1 centered at the point $(1, 2)$ exactly once in the
counterclockwise direction.

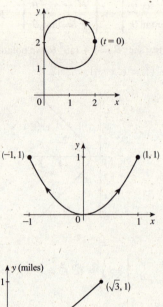

33. $\left.\begin{array}{l} x = \sin t \\ y = \sin^2 t \end{array}\right\} \Rightarrow y = x^2$ and $0 \le t \le 3\pi \Rightarrow -1 \le x \le 1.$ The particle

starts out at $(0, 0)$ and moves to the right along the parabola $y = x^2$ until it
reaches the point $(1, 1)$. It then reverses direction and moves to $(-1, 1)$,
then again to $(1, 1)$, and finally back to $(0, 0)$.

35. $\left.\begin{array}{l} x = \tan(0.025\pi t) \\ y = \sec(0.025\pi t) - 1 \end{array}\right\} \Rightarrow$

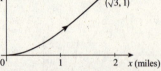

$(y+1)^2 = \sec^2(0.025\pi t) = 1 + \tan^2(0.025\pi t) = 1 + x^2.$ In

rectangular coordinates, the equation is thus $(y+1)^2 = x^2 + 1.$ If $t = 0,$

$x = 0$ and $y = 0.$ If $t = \frac{40}{3}, x = \sqrt{3}$ and $y = 1.$

37. a. $x = x_1 + (x_2 - x_1)t$ and $y = y_1 + (y_2 - y_1)t.$ From the first equation, we find $t = \dfrac{x - x_1}{x_2 - x_1}.$ Substituting this

expression into the second equation gives $y = y_1 + (y_2 - y_1)\left(\dfrac{x - x_1}{x_2 - x_1}\right) = y_1 + \dfrac{y_2 - y_1}{x_2 - x_1}x - \dfrac{x_1(y_2 - y_1)}{x_2 - x_1},$ which is a

linear equation in x and y. Since (x_1, y_1) and (x_2, y_2) both satisfy the equation, we see that $P_1(x_1, y_1)$ and $P_2(x_2, y_2)$
both lie on the line.

b. If $t = 0,$ $x = x_1$ and $y = y_1,$ so (x_1, y_1) is on the line. If $t = 1,$ then $x = x_2$ and $y = y_2,$ so (x_2, y_2) is on the line. As t
increases from $t = 0$ to $t = 1,$ the line segment joining $P_1(x_1, y_1)$ and $P_2(x_2, y_2)$ is traced out.

39. $x = a \sec t + h$ and $y = b \tan t + k \Rightarrow \dfrac{x - h}{a} = \sec t$ and $\dfrac{y - k}{b} = \tan t \Rightarrow \left(\dfrac{x - h}{a}\right)^2 - \left(\dfrac{y - k}{b}\right)^2 = \sec^2 t - \tan^2 t \Leftrightarrow$

$\dfrac{(x - h)^2}{a^2} - \dfrac{(y - k)^2}{b^2} = 1,$ which is an equation of the hyperbola with center (h, k) and transverse and conjugate axes $2a$
and $2b$ respectively.

41. $\overline{AB} = x = 2a \cot\theta$ and PB is perpendicular to the x-axis, so $\overline{PC}$ is parallel to it. Therefore,
$y = \overline{OC} \sin\theta = (2a \sin\theta) \sin\theta = 2a \sin^2\theta.$

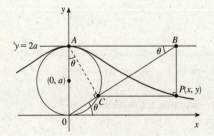

43.

45.

47.

49.

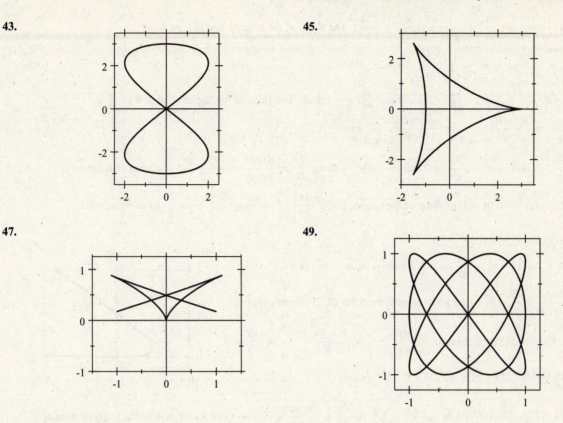

51. False. The curve with parametric representation $x = \cos^2 t$, $y = \sin^2 t$, $-\infty < t < \infty$ is contained in the line segment joining $(0, 1)$ and $(1, 0)$.

53. True. If (x', y') is a point on the curve with parametric equations $x = f(t) + a$ and $y = g(t) + b$, then $x' = f(t_0) + a$ and $y' = g(t_0) + b$ for some t_0. But the point $(f(t_0) + a, g(t_0) + b)$ is obtained from the point $(f(t_0), g(t_0))$ by shifting it a units horizontally and b units vertically.

10.3   Concept Questions

1. $m = \dfrac{g'(t_0)}{f'(t_0)}$, $f'(t_0) \neq 0$

3. a. $S = 2\pi \displaystyle\int_a^b y \, ds = 2\pi \int_a^b g(t) \sqrt{\left(\frac{dx}{dt}\right)^2 + \left(\frac{dy}{dt}\right)^2} \, dt$

b. $S = 2\pi \displaystyle\int_a^b x \, ds = 2\pi \int_a^b f(t) \sqrt{\left(\frac{dx}{dt}\right)^2 + \left(\frac{dy}{dt}\right)^2} \, dt$

10.3   The Calculus of Parametric Equations   ET 9.3

1. $x = t^2 + 1$, $y = t^2 - t \Rightarrow \frac{dx}{dt} = 2t$ and $\frac{dy}{dt} = 2t - 1$. The slope of the tangent line at $t = 1$ is

$$\left.\frac{dy}{dx}\right|_{t=1} = \left.\frac{dy/dt}{dx/dt}\right|_{t=1} = \left.\frac{2t-1}{2t}\right|_{t=1} = \frac{1}{2}.$$

3. $x = \sqrt{t}$, $y = \frac{1}{t} \Rightarrow \frac{dx}{dt} = \frac{1}{2\sqrt{t}}$ and $\frac{dy}{dt} = -\frac{1}{t^2}$. The slope of the tangent line at $t = 1$ is

$$\frac{dy}{dx}\bigg|_{t=1} = \frac{dy/dt}{dx/dt}\bigg|_{t=1} = \frac{-1/t^2}{1/(2\sqrt{t})}\bigg|_{t=1} = -2.$$

5. $x = 2\sin\theta$, $y = 3\cos\theta \Rightarrow \frac{dx}{d\theta} = 2\cos\theta$ and $\frac{dy}{d\theta} = -3\sin\theta$. The slope of the tangent line at $\theta = \frac{\pi}{4}$ is

$$\frac{dy}{dx}\bigg|_{\theta=\pi/4} = \frac{dy/d\theta}{dx/d\theta}\bigg|_{\theta=\pi/4} = \frac{-3\sin\theta}{2\cos\theta}\bigg|_{\theta=\pi/4} = -\frac{3}{2}.$$

7. $x = 2t - 1$, $y = t^3 - t^2 \Rightarrow \frac{dx}{dt} = 2$ and $\frac{dy}{dt} = 3t^2 - 2t \Rightarrow \frac{dy}{dx} = \frac{dy/dt}{dx/dt} = \frac{3t^2 - 2t}{2}$. The point of tangency

is $(x(1), y(1)) = (1, 0)$ and the slope of the tangent line is $\frac{dy}{dx}\bigg|_{t=1} = \frac{3t^2 - 2t}{2}\bigg|_{t=1} = \frac{1}{2}$, so an equation is

$y - 0 = \frac{1}{2}(x - 1)$ or $y = \frac{1}{2}x - \frac{1}{2}$.

9. $x = t^2 + t$, $y = t^2 - t^3 \Rightarrow \frac{dx}{dt} = 2t + 1$ and $\frac{dy}{dt} = 2t - 3t^2$, so

$\frac{dy}{dx} = \frac{dy/dt}{dx/dt} = \frac{2t - 3t^2}{2t + 1}$. The given point of tangency $(0, 2)$ corresponds to

$t = -1$. The slope of the tangent line is $\frac{dy}{dx}\bigg|_{t=-1} = \frac{2t - 3t^2}{2t + 1}\bigg|_{t=-1} = \frac{-5}{-1} = 5$,

so an equation is $y - 2 = 5(x - 0)$ or $y = 5x + 2$.

11. $x = 2t^2 - 1$, $y = t^3 \Rightarrow \frac{dx}{dt} = 4t$ and $\frac{dy}{dt} = 3t^2$, so $\frac{dy}{dx} = \frac{dy/dt}{dx/dt} = \frac{3t^2}{4t} = \frac{3t}{4} = 3 \Rightarrow t = 4$, so the required point is

$(x(4), y(4)) = (31, 64)$.

13. $x = t^2 - 4$, $y = t^3 - 3t \Rightarrow \frac{dx}{dt} = 2t$ and $\frac{dy}{dt} = 3t^2 - 3$. To find the point(s) where

the tangent line is horizontal, set $\frac{dy}{dt} = 0 \Rightarrow 3t^2 - 3 = 3(t + 1)(t - 1) = 0 \Rightarrow$

$t = \pm 1$. Since $\frac{dx}{dt} \neq 0$ at either of these t-values, the required points are

$(x(-1), y(-1)) = (-3, 2)$ and $(x(1), y(1)) = (-3, -2)$. To find the point(s)

where the tangent line is vertical, set $\frac{dx}{dt} = 0 \Rightarrow 2t = 0 \Rightarrow t = 0$. Since $\frac{dy}{dt} \neq 0$ at

this value of t, we see that the required point is $(x(0), y(0)) = (-4, 0)$.

15. $x = 1 + 3\cos t$, $y = 2 - 2\sin t \Rightarrow \frac{dx}{dt} = -3\sin t$ and $\frac{dy}{dt} = -2\cos t$. To find the

point(s) where the tangent line is horizontal, set $\frac{dy}{dt} = 0 \Rightarrow -2\cos t = 0 \Rightarrow$

$t = \frac{\pi}{2} + n\pi$, n an integer. Since $\frac{dx}{dt} \neq 0$ at these t-values, the required points are

$(1, 0)$ and $(1, 4)$. To find the point(s) where the tangent line is vertical, set $\frac{dx}{dt} = 0$

$\Rightarrow -3\sin t = 0 \Rightarrow t = n\pi$, π an integer. Since $\frac{dy}{dt} \neq 0$ at these t-values, the

required points are $(4, 2)$ and $(-2, 2)$.

17. $x = 3t^2 + 1$, $y = 2t^3 \Rightarrow \frac{dx}{dt} = 6t$ and $\frac{dy}{dt} = 6t^2$, so $\frac{dy}{dx} = \frac{\frac{dy}{dt}}{\frac{dx}{dt}} = \frac{6t^2}{6t} = t$ and $\frac{d^2y}{dx^2} = \frac{\frac{d}{dt}\left(\frac{dy}{dx}\right)}{\frac{dx}{dt}} = \frac{1}{6t}$.

19. $x = \sqrt{t}$, $y = \dfrac{1}{t}$ $\Rightarrow$ $\dfrac{dx}{dt} = \dfrac{1}{2\sqrt{t}}$ and $\dfrac{dy}{dt} = -\dfrac{1}{t^2}$, so $\dfrac{dy}{dx} = \dfrac{\frac{dy}{dt}}{\frac{dx}{dt}} = \dfrac{-1/t^2}{1/\left(2t^{1/2}\right)} = -2t^{-3/2} = -\dfrac{2}{t^{3/2}}$ and

$$\dfrac{d^2y}{dx^2} = \dfrac{\frac{d}{dt}\left(\frac{dy}{dx}\right)}{\frac{dx}{dt}} = \dfrac{\frac{d}{dt}\left(-2t^{-3/2}\right)}{1/\left(2t^{1/2}\right)} = 3t^{-5/2} \cdot 2t^{1/2} = \dfrac{6}{t^2}.$$

21. $x = \theta + \cos\theta$, $y = \theta - \sin\theta$ $\Rightarrow$ $\dfrac{dx}{d\theta} = 1 - \sin\theta$ and $\dfrac{dy}{d\theta} = 1 - \cos\theta$, so $\dfrac{dy}{dx} = \dfrac{\frac{dy}{d\theta}}{\frac{dx}{d\theta}} = \dfrac{1 - \cos\theta}{1 - \sin\theta}$ and

$$\dfrac{d^2y}{dx^2} = \dfrac{\frac{d}{d\theta}\left(\frac{dy}{dx}\right)}{\frac{dx}{d\theta}} = \dfrac{1}{1 - \sin\theta} \cdot \dfrac{d}{d\theta}\left(\dfrac{1 - \cos\theta}{1 - \sin\theta}\right) = \dfrac{1}{1 - \sin\theta} \cdot \dfrac{(1 - \sin\theta)(\sin\theta) - (1 - \cos\theta)(-\cos\theta)}{(1 - \sin\theta)^2}$$

$$= \dfrac{\sin\theta + \cos\theta - 1}{(1 - \sin\theta)^3}$$

23. $x = \cosh t$, $y = \sinh t$ $\Rightarrow$ $\dfrac{dx}{dt} = \sinh t$ and $\dfrac{dy}{dt} = \cosh t$, so $\dfrac{dy}{dx} = \dfrac{\frac{dy}{dt}}{\frac{dx}{dt}} = \dfrac{\cosh t}{\sinh t} = \coth t$ and

$$\dfrac{d^2y}{dx^2} = \dfrac{\frac{d}{dt}\left(\frac{dy}{dx}\right)}{\frac{dx}{dt}} = \dfrac{\frac{d}{dt}\left(\coth t\right)}{\sinh t} = \dfrac{-\operatorname{csch}^2 t}{\sinh t} = -\dfrac{1}{\sinh^3 t}.$$

25. $x = t^2$, $y = t^3 - 3t$ $\Rightarrow$ $\dfrac{dx}{dt} = 2t$ and $\dfrac{dy}{dt} = 3t^2 - 3$, so $\dfrac{dy}{dx} = \dfrac{\frac{dy}{dt}}{\frac{dx}{dt}} = \dfrac{3\left(t^2 - 1\right)}{2t}$ and

$$\dfrac{d^2y}{dx^2} = \dfrac{\frac{d}{dt}\left(\frac{dy}{dx}\right)}{\frac{dx}{dt}} = \dfrac{1}{2t} \cdot \dfrac{d}{dt}\left[\dfrac{3\left(t^2 - 1\right)}{2t}\right] = \dfrac{3}{4t}\dfrac{d}{dt}\left(t - \dfrac{1}{t}\right) = \dfrac{3}{4t}\left(1 + \dfrac{1}{t^2}\right) = \dfrac{3\left(t^2 + 1\right)}{4t^3}.$$ We see that $\dfrac{d^2y}{dx^2} < 0$ if

$t < 0$ and $\dfrac{d^2y}{dx^2} > 0$ if $t > 0$, and so the curve is concave downward on $(-\infty, 0)$ for $t < 0$ and concave upward on $(0, \infty)$ for $t > 0$.

27. $x = a\cos^3 t$, $y = a\sin^3 t$ $\Rightarrow$ $\cos t = \left(\dfrac{x}{a}\right)^{1/3}$ and $\sin t = \left(\dfrac{y}{a}\right)^{1/3}$, so $\left(\dfrac{x}{a}\right)^{2/3} + \left(\dfrac{y}{a}\right)^{2/3} = \cos^2 t + \sin^2 t = 1 \Rightarrow$
$x^{2/3} + y^{2/3} = a^{2/3}$ and so the parametric equations indeed represent the astroid $x^{2/3} + y^{2/3} = a^{2/3}$. The slope of the

tangent line to the astroid is $\dfrac{dy}{dx} = \dfrac{\frac{dy}{dt}}{\frac{dx}{dt}} = \dfrac{\frac{d}{dt}\left(a\sin^3 t\right)}{\frac{d}{dt}\left(a\cos^3 t\right)} = \dfrac{3a\sin^2 t\cos t}{-3a\cos^2 t\sin t} = -\tan t$. Setting $\dfrac{dy}{dx} = -1$ gives $\tan t = 1 \Rightarrow$

$t = \dfrac{\pi}{4} + n\pi$, n an integer, and setting $\dfrac{dy}{dx} = 1$ gives $\tan t = -1 \Rightarrow t = \dfrac{3\pi}{4} + n\pi$, n an integer. These values of t give the

points $\left(\dfrac{\sqrt{2}}{4}a, \dfrac{\sqrt{2}}{4}a\right)$ and $\left(-\dfrac{\sqrt{2}}{4}a, -\dfrac{\sqrt{2}}{4}a\right)$ where the slope of the tangent line is -1 and the points $\left(-\dfrac{\sqrt{2}}{4}a, \dfrac{\sqrt{2}}{4}a\right)$ and
$\left(\dfrac{\sqrt{2}}{4}a, -\dfrac{\sqrt{2}}{4}a\right)$ where the slope of the tangent line is 1.

29. $x = t^5 + 5t^3 + 10t + 2$, $y = 2t^3 - 3t^2 - 12t + 1$, $-2 \le t \le 2 \Rightarrow$

$\dfrac{dy}{dx} = \dfrac{\frac{dy}{dt}}{\frac{dx}{dt}} = \dfrac{\frac{d}{dt}\left(2t^3 - 3t^2 - 12t + 1\right)}{\frac{d}{dt}\left(t^5 + 5t^3 + 10t + 2\right)} = \dfrac{6t^2 - 6t - 12}{5t^4 + 15t^2 + 10} = \dfrac{6(t-2)(t+1)}{5\left(t^2+1\right)\left(t^2+2\right)}$. Setting $\dfrac{dy}{dx} = 0$ gives $t = -1$ or

$t = 2$. Since the domain is $-2 \le t \le 2$, we evaluate $(x(-2), y(-2)) = (-90, -3)$, $(x(-1), y(-1)) = (-14, 8)$, and
$(x(2), y(2)) = (94, -19)$, and see that f attains an absolute maximum value of 8 at $x = -14$ and an absolute minimum
value of -19 at $x = 94$.

31. $x = 2t^2$, $y = 3t^3$, $0 \le t \le 1 \Rightarrow \dfrac{dx}{dt} = 4t$ and $\dfrac{dy}{dt} = 9t^2$, so $ds = \sqrt{(4t)^2 + \left(9t^2\right)^2}\, dt$. Thus,

$$L = \int_0^1 ds = \int_0^1 \sqrt{16t^2 + 81t^4}\, dt = \int_0^1 t\sqrt{16 + 81t^2}\, dt = \dfrac{1}{162} \cdot \dfrac{2}{3}\left(16 + 81t^2\right)^{3/2}\Big|_0^1 = \dfrac{1}{243}\left(97^{3/2} - 64\right) \approx 3.67.$$

33. $x = \sin^2 t$, $y = \cos 2t$, $0 \le t \le \pi \Rightarrow \frac{dx}{dt} = 2\sin t \cos t = \sin 2t$ and $\frac{dy}{dt} = -2\sin 2t$, so $ds = \sqrt{\sin^2 2t + 4\sin^2 2t}\, dt$. Thus,

$L = \int_0^\pi \sqrt{5\sin^2 2t}\, dt = \sqrt{5}\int_0^\pi |\sin 2t|\, dt = 2\sqrt{5}\int_0^{\pi/2} \sin 2t\, dt = 2\sqrt{5}\left(-\frac{1}{2}\right)\cos 2t\Big|_0^{\pi/2} = -\sqrt{5}\,(-1-1) = 2\sqrt{5}$.

35. $x = a(\cos t + t\sin t)$, $y = a(\sin t - t\cos t)$, $0 \le t \le \frac{\pi}{2} \Rightarrow \frac{dx}{dt} = a(-\sin t + \sin t + t\cos t) = at\cos t$

and $\frac{dy}{dt} = a(\cos t - \cos t + t\sin t) = at\sin t$, so $ds = \sqrt{(at\cos t)^2 + (at\sin t)^2} = at\, dt$. Thus,

$L = \int_0^{\pi/2} ds = a\int_0^{\pi/2} t\, dt = \frac{1}{2}at^2\Big|_0^{\pi/2} = \frac{1}{8}a\pi^2$.

37. $x = a(2\cos t - \cos 2t)$, $y = a(2\sin t - \sin 2t) \Rightarrow \frac{dx}{dt} = a(-2\sin t + 2\sin 2t)$ and $\frac{dy}{dt} = a(2\cos t - 2\cos 2t)$, so

$ds = \sqrt{a^2(-2\sin t + 2\sin 2t)^2 + a^2(2\cos t - 2\cos 2t)^2}\, dt$. Thus,

$L = \int_0^{2\pi} ds = 2\sqrt{2}a\int_0^{2\pi} \sqrt{(\sin 2t - \sin t)^2 + (\cos t - \cos 2t)^2}\, dt = 2\sqrt{2}a\int_0^{2\pi} \sqrt{1 - \cos t}\, dt$

$= 2\sqrt{2}a\int_0^{2\pi} \sqrt{2\sin^2\frac{1}{2}t}\, dt = 4a\int_0^{2\pi} \sin\frac{1}{2}t = -8a\cos\frac{1}{2}t\Big|_0^{2\pi} = -8a(-1-1) = 16a$

39. $x = \cos^2 t$, $y = \sin^2 t$, $0 \le t \le 2\pi \Rightarrow \frac{dx}{dt} = -2\cos t\sin t$ and $\frac{dy}{dt} = 2\sin t\cos t$, so the distance covered is

$D = \int_0^{2\pi} ds = \int_0^{2\pi} \sqrt{\left(\frac{dx}{dt}\right)^2 + \left(\frac{dy}{dt}\right)^2}\, dt = \int_0^{2\pi} \sqrt{(-2\cos t\sin t)^2 + (2\sin t\cos t)^2}\, dt = 2\sqrt{2}\int_0^{2\pi} |\cos t\sin t|\, dt$

$= 2\sqrt{2}\,(4)\int_0^{\pi/2} \cos t\sin t\, dt = 8\sqrt{2}\left(\frac{1}{2}\sin^2 t\right)\Big|_0^{\pi/2} = 4\sqrt{2}$

41. $x = 800\left(t^{7/9} - t^{11/9}\right)$, $y = 1600t \Rightarrow \frac{dx}{dt} = 800\left(\frac{7}{9}t^{-2/9} - \frac{11}{9}t^{2/9}\right) = \frac{800\left(7 - 11t^{4/9}\right)}{9t^{2/9}}$ and

$\frac{dy}{dt} = 1600$. The boat is at A when $t = 0$ and at B when $t = 1$, so the distance travelled by the boat is

$D = \int_0^1 \sqrt{\left(\frac{dx}{dt}\right)^2 + \left(\frac{dy}{dt}\right)^2} = \int_0^1 \sqrt{\left[\frac{800\left(7 - 11t^{4/9}\right)}{9t^{2/9}}\right]^2 + 1600^2}\, dt \approx 1639 \text{ ft (using a CAS)}.$

43. From the result of Exercise 42, we have $y = -\frac{1}{2}\left(\frac{eE}{mv_0^2}\right)x^2$. At $x = a$, we find $y = -\frac{1}{2}\left(\frac{eE}{mv_0^2}\right)a^2$, so the electron hints

the screen at the point $\left(a, -\frac{1}{2}\left(\frac{eE}{mv_0^2}\right)a^2\right)$.

45. $x = C(t) = \int_0^t \cos\left(\frac{\pi}{2}u^2\right) du$, $y = S(t) = \int_0^t \sin\left(\frac{\pi}{2}u^2\right) du$

a.

As $t \to \infty$, the curve spirals about and converges to the point $\left(\frac{1}{2}, \frac{1}{2}\right)$. As $t \to -\infty$, the curve spirals about and converges to the point $\left(-\frac{1}{2}, -\frac{1}{2}\right)$.

b. $\frac{dx}{dt} = \frac{d}{dt}\int_0^t \cos\left(\frac{\pi}{2}u^2\right) du = \cos\left(\frac{\pi}{2}t^2\right)$ and

$\frac{dy}{dt} = \frac{d}{dt}\int_0^t \sin\left(\frac{\pi}{2}u^2\right) du = \sin\left(\frac{\pi}{2}t^2\right)$, so

$L = \int_0^a \sqrt{\left(\frac{dx}{dt}\right)^2 + \left(\frac{dy}{dt}\right)^2} = \int_0^a \sqrt{\cos^2\left(\frac{\pi}{2}t^2\right) + \sin^2\left(\frac{\pi}{2}t^2\right)}\, dt$

$= \int_0^a 1\, dt = t\big|_0^a = a$

47. $x = a(\theta - \sin\theta)$, $y = a(1 - \cos\theta)$. Using the result of Exercise 46, we find

$$A = \int_0^{2\pi} y\, dx = \int_0^{2\pi} a(1 - \cos\theta)\, a(1 - \cos\theta)\, d\theta = a^2 \int_0^{2\pi} \left(1 - 2\cos\theta + \cos^2\theta\right) d\theta$$

$$= a^2 \int_0^{2\pi} \left[1 - 2\cos\theta + \tfrac{1}{2}(1 + \cos 2\theta)\right] d\theta = a^2 \left(\tfrac{3}{2}\theta - 2\sin\theta + \tfrac{1}{4}\sin 2\theta\right)\Big|_0^{2\pi} = 3\pi a^2$$

49. $x = a\cos^3\theta$, $y = a\sin^3\theta \Rightarrow$

$$A = 4\int_{\pi/2}^0 y\, dx = 4\int_{\pi/2}^0 \left(a\sin^3\theta\right)\left(-3a\cos^2\theta\sin\theta\, d\theta\right) \text{ [since } x(\theta) \text{ is decreasing on } (0, \tfrac{\pi}{2})]$$

$$= 12a^2 \int_0^{\pi/2} \sin^4\theta\cos^2\theta\, d\theta = 12a^2 \int_0^{\pi/2} \sin^4\theta\left(1 - \sin^2\theta\right) d\theta = 12a^2 \int_0^{\pi/2} \left(\sin^4\theta - \sin^6\theta\right) d\theta$$

Using Formula 79 from the Table of Integrals repeatedly, we obtain $A = \tfrac{3}{8}\pi a^2$.

51. $x = 4\sqrt{2}\sin t$, $y = \sin 2t \Rightarrow$

$$A = 4\int_0^{\pi/2} y\, dx = 4\int_0^{\pi/2} (\sin 2t)\left(4\sqrt{2}\cos t\, dt\right) = 16\sqrt{2}\int_0^{\pi/2} (2\sin t\cos t)(\cos t)\, dt = 32\sqrt{2}\int_0^{\pi/2} \sin t\cos^2 t\, dt$$

$$= -\tfrac{32\sqrt{2}}{3}\cos^3 t\Big|_0^{\pi/2} = \tfrac{32\sqrt{2}}{3}$$

53. $x = t^3$, $y = t^2$, $0 \le t \le 1 \Rightarrow \frac{dx}{dt} = 3t^2$ and $\frac{dy}{dt} = 2t$, so

$$S = 2\pi \int_0^1 y\sqrt{\left(3t^2\right)^2 + (2t)^2}\, dt = 2\pi \int_0^1 t^2 \sqrt{9t^4 + 4t^2}\, dt$$

$$= 2\pi \int_0^1 t^3 \sqrt{9t^2 + 4}\, dt$$

$$= 2\pi \int_4^{13} \tfrac{1}{9}(u - 4)\, u^{1/2}\left(\tfrac{1}{18}\, du\right) \quad \text{(where } u = 9t^2 + 4)$$

$$= \tfrac{\pi}{81} \int_4^{13} \left(u^{3/2} - 4u^{1/2}\right) du = \tfrac{\pi}{81}\left(\tfrac{2}{5}u^{5/2} - \tfrac{8}{3}u^{3/2}\right)\Big|_4^{13}$$

$$= \frac{2\left(247\sqrt{13} + 64\right)\pi}{1215}$$

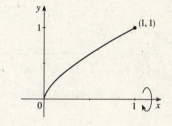

55. $x = \tfrac{1}{3}t^3$, $y = 4 - \tfrac{1}{2}t^2$, $0 \le t \le 2\sqrt{2} \Rightarrow \frac{dx}{dt} = t^2$ and $\frac{dy}{dt} = -t$, so

$$S = 2\pi \int_0^{2\sqrt{2}} y\, ds = 2\pi \int_0^{2\sqrt{2}} \left(4 - \tfrac{1}{2}t^2\right)\sqrt{t^4 + t^2}\, dt$$

$$= 2\pi \int_0^{2\sqrt{2}} \left(4t\sqrt{t^2 + 1} - \tfrac{1}{2}t^3 \sqrt{t^2 + 1}\right) dt$$

$$= 8\pi\left(\tfrac{1}{2}\right)\left(\tfrac{2}{3}\right)\left(t^2 + 1\right)^{3/2}\Big|_0^{2\sqrt{2}} - \pi \int_0^{2\sqrt{2}} t^3 \sqrt{t^2 + 1}\, dt$$

$$= \tfrac{8\pi}{3}(27 - 1) - \pi \int_1^9 (u - 1)\left(\tfrac{1}{2}\right)u^{1/2}\, du \quad \text{(where } u = t^2 + 1)$$

$$= \tfrac{208\pi}{3} - \tfrac{\pi}{2} \int_1^9 \left(u^{3/2} - u^{1/2}\right) du = \tfrac{208\pi}{3} - \left[\tfrac{\pi}{2}\left(\tfrac{2}{5}u^{5/2} - \tfrac{2}{3}u^{3/2}\right)\right]_1^9 = \tfrac{148\pi}{5}$$

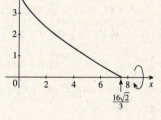

57. $x = t - \sin t$, $y = 1 - \cos t \Rightarrow \frac{dx}{dt} = 1 - \cos t$ and $\frac{dy}{dt} = \sin t$. Thus,

$$S = 2\pi \int_0^{2\pi} y\, ds = 2\pi \int_0^{2\pi} (1 - \cos t)\sqrt{(1 - \cos t)^2 + \sin^2 t}\, dt$$

$$= 2\pi \int_0^{2\pi} (1 - \cos t)\sqrt{2}\sqrt{1 - \cos t}\, dt = 2\sqrt{2}\pi \int_0^{2\pi} (1 - \cos t)^{3/2}\, dt$$

$$= 8\pi \int_0^{2\pi} \sin^3 \tfrac{1}{2}t\, dt = 8\pi \int_0^{2\pi} \left(1 - \cos^2 \tfrac{1}{2}t\right)\sin \tfrac{1}{2}t\, dt$$

$$= \left(-16\pi \cos \tfrac{1}{2}t + \tfrac{16}{3}\pi \cos^3 \tfrac{1}{2}t\right)\Big|_0^{2\pi}$$

$$= 8\pi\left(-2\cos \tfrac{1}{2}t + \tfrac{2}{3}\cos^3 \tfrac{1}{2}t\right)\Big|_0^{2\pi} = \tfrac{64\pi}{3}$$

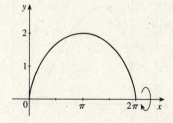

59. $x = 3t^2$, $y = 2t^3$, $0 \le t \le 1 \Rightarrow$

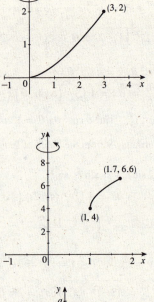

$$S = 2\pi \int_0^1 x\,ds = 2\pi \int_0^1 3t^2 \sqrt{(6t)^2 + (6t^2)^2}\,dt = 36\pi \int_0^1 t^3 \sqrt{1 + t^2}\,dt$$

$$= 36\pi \int_1^2 (u - 1)\left(\tfrac{1}{2}\right)u^{1/2}\,du \quad \text{(where } u = 1 + t^2\text{)}$$

$$= 18\pi \int_1^2 \left(u^{3/2} - u^{1/2}\right)du = 18\pi \left(\tfrac{2}{5}u^{5/2} - \tfrac{2}{3}u^{3/2}\right)\Big|_1^2$$

$$= 18\pi \left(\tfrac{8\sqrt{2}}{5} - \tfrac{4\sqrt{2}}{5} - \tfrac{2}{5} + \tfrac{2}{3}\right) = \tfrac{24\left(\sqrt{2}+1\right)\pi}{5}$$

61. $x = e^t - t$, $y = 4e^{t/2}$, $0 \le t \le 1 \Rightarrow$

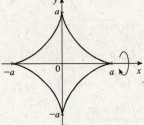

$$S = 2\pi \int_0^1 x\,ds = 2\pi \int_0^1 \left(e^t - t\right)\sqrt{\left(e^t - 1\right)^2 + \left(2e^{t/2}\right)^2}\,dt$$

$$= 2\pi \int_0^1 \left(e^t - t\right)\left(e^t + 1\right)dt = 2\pi \int_0^1 \left(e^{2t} - te^t + e^t - t\right)dt$$

$$= 2\pi \left(\tfrac{1}{2}e^{2t} - (t - 1)e^t + e^t - \tfrac{1}{2}t^2\right)\Big|_0^1 \quad \text{[since } \int te^t\,dt = (t - 1)e^t + C\text{]}$$

$$= \pi \left(e^2 + 2e - 6\right)$$

63. $x = a\cos^3 t$, $y = a\sin^3 t \Rightarrow$

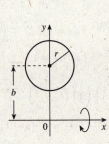

$$\sqrt{\left(\tfrac{dx}{dt}\right)^2 + \left(\tfrac{dy}{dt}\right)^2} = \sqrt{\left(-3a\cos^2 t \sin t\right)^2 + \left(3a\sin^2 t \cos t\right)^2}$$

$$= 3a\sin t \cos t \text{ for } 0 \le t \le \tfrac{\pi}{2}$$

Thus,

$$S = 2(2\pi)\int_0^{\pi/2} y\,ds = 4\pi \int_0^{\pi/2} \left(a\sin^3 t\right)(3a\sin t \cos t)\,dt$$

$$= 12\pi a^2 \int_0^{\pi/2} \sin^4 t \cos t\,dt = \frac{12\pi a^2}{5}\sin^5 t\Big|_0^{\pi/2} = \frac{12\pi a^2}{5}$$

65. $x^2 + (y - b)^2 = r^2$, $0 < r < b$. The circle can be written in the parametric form

$x = r\cos t$, $y = b + r\sin t$, $0 \le t \le 2\pi$, so $\tfrac{dx}{dt} = -r\sin t$ and $\tfrac{dy}{dt} = r\cos t$. Thus,

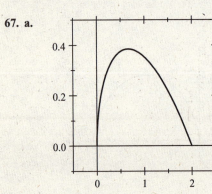

$$S = 2\pi \int_0^{2\pi} y\,ds = 2\pi \int_0^{2\pi} (b + r\sin t)\sqrt{(-r\sin t)^2 + (r\cos t)^2}\,dt$$

$$= 2\pi r \int_0^{2\pi} (b + r\sin t)\,dt = 2\pi r (bt - r\cos t)\big|_0^{2\pi} = 4\pi^2 rb$$

67. a.

b. $x = 2t^2$, $y = t - t^3$, $0 \le t \le 1 \Rightarrow$

$$L = \int_0^1 \sqrt{\left(\tfrac{dx}{dt}\right)^2 + \left(\tfrac{dy}{dt}\right)^2}\,dt$$

$$= \int_0^1 \sqrt{(4t)^2 + \left(1 - 3t^2\right)^2}\,dt$$

$$= \int_0^1 \sqrt{9t^4 + 10t^2 + 1}\,dt \approx 2.2469 \text{ (using a CAS)}$$

69. a.

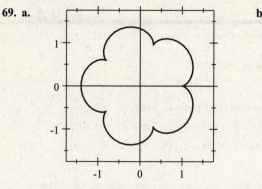

b. $x = 0.2 (6 \cos t - \cos 6t) \Rightarrow$

$\frac{dx}{dt} = 0.2 (-6 \sin t + 6 \sin 6t) = -1.2 (\sin t - \sin 6t)$ and

$y = 0.2 (6 \sin t - \sin 6t) \Rightarrow \frac{dy}{dt} = 1.2 (\cos t - \cos 6t)$, so

$\left(\frac{dx}{dt}\right)^2 + \left(\frac{dy}{dt}\right)^2 = (1.2)^2 \left[(\sin t - \sin 6t)^2 + (\cos t - \cos 6t)^2\right]$

$= (1.2)^2 \left(\sin^2 t - 2 \sin t \sin 6t + \sin^2 6t \right.$

$\left. + \cos^2 t - 2 \cos t \cos 6t + \cos^2 6t \right)$

$= 2 (1.2)^2 (1 - \sin t \sin 6t - \cos t \cos 6t) \Rightarrow$

$L = 2 \int_0^\pi \sqrt{\left(\frac{dx}{dt}\right)^2 + \left(\frac{dy}{dt}\right)^2}\, dt = (2.4) \sqrt{2} \int_0^\pi \sqrt{1 - \sin t \sin 6t - \cos t \cos 6t}\, dt \approx 9.6000$ (using a CAS).

71. $x = 4 \sin 2t,\ y = 2 \cos 3t,\ 0 \le t \le \frac{\pi}{6} \Rightarrow$

$S = 2\pi \int_0^{\pi/6} y\, ds = 2\pi \int_0^{\pi/6} 2 \cos 3t \sqrt{\left(\frac{dx}{dt}\right)^2 + \left(\frac{dy}{dt}\right)^2}\, dt$

$= 4\pi \int_0^{\pi/6} \cos 3t \sqrt{(8 \cos 2t)^2 + (-6 \sin 3t)^2}\, dt$

$= 4\pi \int_0^{\pi/6} \cos 3t \sqrt{64 \cos^2 2t + 36 \sin^2 3t}\, dt \approx 33.66$ (using a CAS)

73. $x^2 + y^2 = \left(\dfrac{2at}{1+t^2}\right)^2 + \left[\dfrac{a\left(1-t^2\right)}{1+t^2}\right]^2 = \dfrac{4a^2 t^2 + a^2 - 2a^2 t^2 + a^2 t^4}{(1+t^2)^2} = \dfrac{a^2\left(t^4 + 2t^2 + 1\right)}{(1+t^2)^2} = \dfrac{a^2\left(t^2+1\right)^2}{(1+t^2)^2} = a^2,$

an equation of the circle with radius a centered at the origin.

75. $t = \dfrac{y}{x} \Rightarrow y = tx$, and substituting this into the equation of the folium gives $x^3 + (tx)^3 = 3ax\,(tx) \Leftrightarrow x^3 + t^3 x^3 = 3ax^2 t$

$\Leftrightarrow x^3 \left(1 + t^3\right) = 3ax^2 t \Rightarrow x = \dfrac{3at}{t^3 + 1}$, so $y = tx = \dfrac{3at^2}{t^3 + 1}$.

77. $x = a \cos t \Rightarrow \frac{dx}{dt} = -a \sin t$ and $y = b \sin t \Rightarrow \frac{dy}{dt} = b \cos t$, so

$\sqrt{\left(\frac{dx}{dt}\right)^2 + \left(\frac{dy}{dt}\right)^2} = \sqrt{a^2 \sin^2 t + b^2 \cos^2 t} = \sqrt{a^2 \left(1 - \cos^2 t\right) + b^2 \cos^2 t} = \sqrt{a^2 - \left(a^2 - b^2\right) \cos^2 t} = a\sqrt{1 - e^2 \cos^2 t}$

since $e = \dfrac{\sqrt{a^2 - b^2}}{a}$. Therefore, by symmetry, $L = \int_0^{2\pi} \sqrt{\left(\frac{dx}{dt}\right)^2 + \left(\frac{dy}{dt}\right)^2}\, dt = 4a \int_0^{\pi/2} \sqrt{1 - e^2 \cos^2 t}\, dt.$

Now we substitute $t = \frac{\pi}{2} - u$, so $dt = -du$, $t = 0 \Rightarrow u = \frac{\pi}{2}$, and $t = \frac{\pi}{2} \Rightarrow u = 0$, and so

$L = 4a \int_{\pi/2}^0 \sqrt{1 - e^2 \left[\cos\left(\frac{\pi}{2} - u\right)\right]^2}\, (-du) = 4a \int_0^{\pi/2} \sqrt{1 - e^2 \sin^2 u}\, du$. The desired result follows by replacing u with t.

79. False.   In fact, $\dfrac{dy}{dx} = \dfrac{g'(t)}{f'(t)}$, so

$\dfrac{d^2 y}{dx^2} = \dfrac{d}{dx}\left[\dfrac{g'(t)}{f'(t)}\right] = \dfrac{\dfrac{d}{dt}\left[\dfrac{g'(t)}{f'(t)}\right]}{\dfrac{dx}{dt}} = \dfrac{\dfrac{f'(t)\, g''(t) - g'(t)\, f''(t)}{[f'(t)]^2}}{f'(t)} = \dfrac{f'(t)\, g''(t) - g'(t)\, f''(t)}{[f'(t)]^3}.$

10.4   Concept Questions   ET 9.4

1. Besides $P(r, \theta)$, we have $P(r, \theta + 2k\pi)$ and $P(-r, \theta + k\pi)$, where k is an integer.

3. a. The graph of $r = f(\theta)$ is symmetric with respect to the polar axis if and only if $f(-\theta) = f(\theta)$.

   b. The graph of $r = f(\theta)$ is symmetric with respect to the vertical line $\theta = \frac{\pi}{2}$ if and only if $f(\pi - \theta) = f(\theta)$.

   c. The graph of $r = f(\theta)$ is symmetric with respect to the pole if and only if $(-r, \theta) = (r, \theta)$.

10.4   Polar Coordinates   ET 9.4

1. $x = 4\cos\frac{\pi}{4} = 4\left(\frac{\sqrt{2}}{2}\right) = 2\sqrt{2}$ and

$y = 4\sin\frac{\pi}{4} = 4\left(\frac{\sqrt{2}}{2}\right) = 2\sqrt{2}$, so the point is

$\left(2\sqrt{2}, 2\sqrt{2}\right)$.

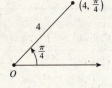

3. $x = 4\cos\frac{3\pi}{2} = 0$ and $y = 4\sin\frac{3\pi}{2} = 4(-1) = -4$, so the point is $(0, -4)$.

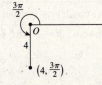

5. $x = -\sqrt{2}\cos\frac{\pi}{4} = -\sqrt{2}\left(\frac{\sqrt{2}}{2}\right) = -1$ and

$y = -\sqrt{2}\sin\frac{\pi}{4} = -\sqrt{2}\left(\frac{\sqrt{2}}{2}\right) = -1$, so the point is

$(-1, -1)$.

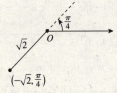

7. $x = (-4)\cos\left(-\frac{3\pi}{4}\right) = (-4)\left(-\frac{\sqrt{2}}{2}\right) = 2\sqrt{2}$ and

$y = (-4)\sin\left(-\frac{3\pi}{4}\right) = (-4)\left(-\frac{\sqrt{2}}{2}\right) = 2\sqrt{2}$, so the

point is $\left(2\sqrt{2}, 2\sqrt{2}\right)$.

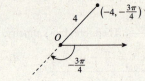

9. $r = \sqrt{2^2 + 2^2} = 2\sqrt{2}$ and $\tan\theta = \frac{2}{2} = 1 \Rightarrow$

$\theta = \tan^{-1}1 = \frac{\pi}{4}$, so the point is $\left(2\sqrt{2}, \frac{\pi}{4}\right)$.

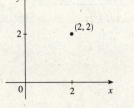

11. $r = \sqrt{0^2 + 5^2} = 5$ and $\tan\theta$ is undefined, so $\theta = \frac{\pi}{2}$. The point is $\left(5, \frac{\pi}{2}\right)$.

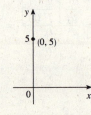

13. $r = \sqrt{\left(-\sqrt{3}\right)^2 + \left(-\sqrt{3}\right)^2} = \sqrt{6}$ and $\tan\theta = \frac{-\sqrt{3}}{-\sqrt{3}} \Rightarrow$

$\theta = \pi + \tan^{-1} 1 = \frac{5\pi}{4}$, so the point is $\left(\sqrt{6}, \frac{5\pi}{4}\right)$.

15. $r = \sqrt{5^2 + (-12)^2} = 13$ and $\tan\theta = \frac{-12}{5} \Rightarrow$

$\theta = \tan^{-1}\left(-\frac{12}{5}\right) + 2\pi$, so the point is

$\left(13, \tan^{-1}\left(-\frac{12}{5}\right) + 2\pi\right)$.

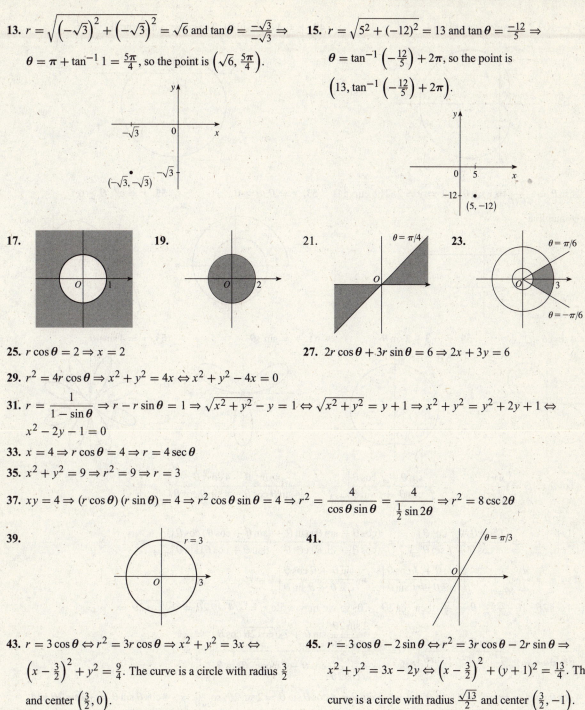

17.

19.

21.

23.

25. $r\cos\theta = 2 \Rightarrow x = 2$

27. $2r\cos\theta + 3r\sin\theta = 6 \Rightarrow 2x + 3y = 6$

29. $r^2 = 4r\cos\theta \Rightarrow x^2 + y^2 = 4x \Leftrightarrow x^2 + y^2 - 4x = 0$

31. $r = \dfrac{1}{1 - \sin\theta} \Rightarrow r - r\sin\theta = 1 \Rightarrow \sqrt{x^2 + y^2} - y = 1 \Leftrightarrow \sqrt{x^2 + y^2} = y + 1 \Rightarrow x^2 + y^2 = y^2 + 2y + 1 \Leftrightarrow$

$x^2 - 2y - 1 = 0$

33. $x = 4 \Rightarrow r\cos\theta = 4 \Rightarrow r = 4\sec\theta$

35. $x^2 + y^2 = 9 \Rightarrow r^2 = 9 \Rightarrow r = 3$

37. $xy = 4 \Rightarrow (r\cos\theta)(r\sin\theta) = 4 \Rightarrow r^2 \cos\theta \sin\theta = 4 \Rightarrow r^2 = \dfrac{4}{\cos\theta\sin\theta} = \dfrac{4}{\frac{1}{2}\sin 2\theta} \Rightarrow r^2 = 8\csc 2\theta$

39.

41.

43. $r = 3\cos\theta \Leftrightarrow r^2 = 3r\cos\theta \Rightarrow x^2 + y^2 = 3x \Leftrightarrow$

$\left(x - \frac{3}{2}\right)^2 + y^2 = \frac{9}{4}$. The curve is a circle with radius $\frac{3}{2}$

and center $\left(\frac{3}{2}, 0\right)$.

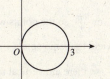

45. $r = 3\cos\theta - 2\sin\theta \Leftrightarrow r^2 = 3r\cos\theta - 2r\sin\theta \Rightarrow$

$x^2 + y^2 = 3x - 2y \Leftrightarrow \left(x - \frac{3}{2}\right)^2 + (y + 1)^2 = \frac{13}{4}$. The

curve is a circle with radius $\frac{\sqrt{13}}{2}$ and center $\left(\frac{3}{2}, -1\right)$.

47. $r = 1 + \cos\theta$. The curve is a cardioid.

49. $r = 4(1 - \sin\theta)$. The curve is a cardioid.

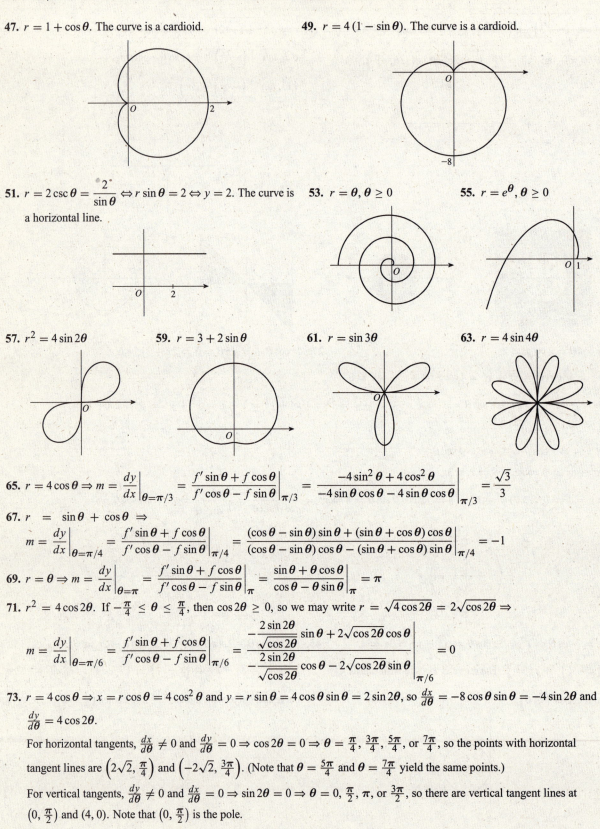

51. $r = 2\csc\theta = \dfrac{2}{\sin\theta} \Leftrightarrow r\sin\theta = 2 \Leftrightarrow y = 2$. The curve is a horizontal line.

53. $r = \theta,\ \theta \geq 0$

55. $r = e^\theta,\ \theta \geq 0$

57. $r^2 = 4\sin 2\theta$

59. $r = 3 + 2\sin\theta$

61. $r = \sin 3\theta$

63. $r = 4\sin 4\theta$

65. $r = 4\cos\theta \Rightarrow m = \left.\dfrac{dy}{dx}\right|_{\theta=\pi/3} = \left.\dfrac{f'\sin\theta + f\cos\theta}{f'\cos\theta - f\sin\theta}\right|_{\pi/3} = \left.\dfrac{-4\sin^2\theta + 4\cos^2\theta}{-4\sin\theta\cos\theta - 4\sin\theta\cos\theta}\right|_{\pi/3} = \dfrac{\sqrt{3}}{3}$

67. $r = \sin\theta + \cos\theta \Rightarrow$

$m = \left.\dfrac{dy}{dx}\right|_{\theta=\pi/4} = \left.\dfrac{f'\sin\theta + f\cos\theta}{f'\cos\theta - f\sin\theta}\right|_{\pi/4} = \left.\dfrac{(\cos\theta - \sin\theta)\sin\theta + (\sin\theta + \cos\theta)\cos\theta}{(\cos\theta - \sin\theta)\cos\theta - (\sin\theta + \cos\theta)\sin\theta}\right|_{\pi/4} = -1$

69. $r = \theta \Rightarrow m = \left.\dfrac{dy}{dx}\right|_{\theta=\pi} = \left.\dfrac{f'\sin\theta + f\cos\theta}{f'\cos\theta - f\sin\theta}\right|_{\pi} = \left.\dfrac{\sin\theta + \theta\cos\theta}{\cos\theta - \theta\sin\theta}\right|_{\pi} = \pi$

71. $r^2 = 4\cos 2\theta$. If $-\dfrac{\pi}{4} \leq \theta \leq \dfrac{\pi}{4}$, then $\cos 2\theta \geq 0$, so we may write $r = \sqrt{4\cos 2\theta} = 2\sqrt{\cos 2\theta} \Rightarrow$

$m = \left.\dfrac{dy}{dx}\right|_{\theta=\pi/6} = \left.\dfrac{f'\sin\theta + f\cos\theta}{f'\cos\theta - f\sin\theta}\right|_{\pi/6} = \left.\dfrac{-\dfrac{2\sin 2\theta}{\sqrt{\cos 2\theta}}\sin\theta + 2\sqrt{\cos 2\theta}\cos\theta}{-\dfrac{2\sin 2\theta}{\sqrt{\cos 2\theta}}\cos\theta - 2\sqrt{\cos 2\theta}\sin\theta}\right|_{\pi/6} = 0$

73. $r = 4\cos\theta \Rightarrow x = r\cos\theta = 4\cos^2\theta$ and $y = r\sin\theta = 4\cos\theta\sin\theta = 2\sin 2\theta$, so $\dfrac{dx}{d\theta} = -8\cos\theta\sin\theta = -4\sin 2\theta$ and $\dfrac{dy}{d\theta} = 4\cos 2\theta$.

For horizontal tangents, $\dfrac{dx}{d\theta} \neq 0$ and $\dfrac{dy}{d\theta} = 0 \Rightarrow \cos 2\theta = 0 \Rightarrow \theta = \dfrac{\pi}{4}, \dfrac{3\pi}{4}, \dfrac{5\pi}{4}$, or $\dfrac{7\pi}{4}$, so the points with horizontal tangent lines are $\left(2\sqrt{2}, \dfrac{\pi}{4}\right)$ and $\left(-2\sqrt{2}, \dfrac{3\pi}{4}\right)$. (Note that $\theta = \dfrac{5\pi}{4}$ and $\theta = \dfrac{7\pi}{4}$ yield the same points.)

For vertical tangents, $\dfrac{dy}{d\theta} \neq 0$ and $\dfrac{dx}{d\theta} = 0 \Rightarrow \sin 2\theta = 0 \Rightarrow \theta = 0, \dfrac{\pi}{2}, \pi$, or $\dfrac{3\pi}{2}$, so there are vertical tangent lines at $\left(0, \dfrac{\pi}{2}\right)$ and $(4, 0)$. Note that $\left(0, \dfrac{\pi}{2}\right)$ is the pole.

75. $r = \sin 2\theta \Rightarrow x = r\cos\theta = \sin 2\theta\cos\theta = 2\sin\theta\cos^2\theta$ and $y = r\sin\theta = \sin 2\theta\sin\theta = 2\sin^2\theta\cos\theta$,

so $\frac{dx}{d\theta} = 2\cos^3\theta - 4\sin^2\theta\cos\theta = 2\cos\theta\left(\cos^2\theta - 2\sin^2\theta\right) = 2\cos\theta\left(1 - 3\sin^2\theta\right)$ and

$\frac{dy}{d\theta} = -2\sin^3\theta + 4\sin\theta\cos^2\theta = 2\sin\theta\left(2\cos^2\theta - \sin^2\theta\right)$. Now $\frac{dy}{d\theta} = 0 \Rightarrow \theta = 0$,

$\tan^{-1}\sqrt{2}$, $\pi + \tan^{-1}\left(-\sqrt{2}\right)$, π, $\pi + \tan^{-1}\sqrt{2}$, or $2\pi + \tan^{-1}\left(-\sqrt{2}\right)$; and $\frac{dx}{d\theta} = 0 \Rightarrow$

$\theta = \sin^{-1}\frac{\sqrt{3}}{3}, \frac{\pi}{2}, \pi + \sin^{-1}\left(-\frac{\sqrt{3}}{3}\right), \pi + \sin^{-1}\frac{\sqrt{3}}{3}, \frac{3\pi}{2}$, or $2\pi + \sin^{-1}\left(-\frac{\sqrt{3}}{3}\right)$. The lists

are mutually exclusive, so the tangent line is horizontal at $(0,0)$, $\left(\sin\left(2\tan^{-1}\sqrt{2}\right), \tan^{-1}\sqrt{2}\right)$,

$\left(\sin 2\left(\pi + \tan^{-1}\left(-\sqrt{2}\right)\right), \pi + \tan^{-1}\left(-\sqrt{2}\right)\right) = \left(\sin\left(2\tan^{-1}\left(-\sqrt{2}\right)\right), \pi + \tan^{-1}\left(-\sqrt{2}\right)\right)$,

$(0, \pi)$, $\left(\sin\left(2\left(\pi + \tan^{-1}\sqrt{2}\right)\right), \pi + \tan^{-1}\sqrt{2}\right) = \left(\sin\left(2\tan^{-1}\sqrt{2}\right), \pi + \tan^{-1}\sqrt{2}\right)$, and

$\left(\sin\left(2\left(2\pi + \tan^{-1}\left(-\sqrt{2}\right)\right)\right), 2\pi + \tan^{-1}\left(-\sqrt{2}\right)\right) = \left(\sin\left(2\tan^{-1}\left(-\sqrt{2}\right)\right), 2\pi + \tan^{-1}\left(-\sqrt{2}\right)\right)$;

and the tangent line is vertical at $\left(\sin\left(2\sin^{-1}\frac{\sqrt{3}}{3}\right), \sin^{-1}\frac{\sqrt{3}}{3}\right)$, $\left(0, \frac{\pi}{2}\right)$,

$\left(\sin\left(2\left(\pi + \sin^{-1}\left(-\frac{\sqrt{3}}{3}\right)\right)\right), \pi + \sin^{-1}\left(-\frac{\sqrt{3}}{3}\right)\right) = \left(\sin\left(2\sin^{-1}\left(-\frac{\sqrt{3}}{3}\right)\right), \pi + \sin^{-1}\left(-\frac{\sqrt{3}}{3}\right)\right)$,

$\left(\sin\left(2\left(\pi + \sin^{-1}\frac{\sqrt{3}}{3}\right)\right), \pi + \sin^{-1}\frac{\sqrt{3}}{3}\right) = \left(\sin\left(2\sin^{-1}\frac{\sqrt{3}}{3}\right), \pi + \sin^{-1}\frac{\sqrt{3}}{3}\right)$, $\left(0, \frac{3\pi}{2}\right)$, and

$\left(\sin\left(2\left(2\pi + \sin^{-1}\left(-\frac{\sqrt{3}}{3}\right)\right)\right), 2\pi + \sin^{-1}\left(-\frac{\sqrt{3}}{3}\right)\right) = \left(\sin\left(2\sin^{-1}\left(-\frac{\sqrt{3}}{3}\right)\right), 2\pi + \sin^{-1}\left(-\frac{\sqrt{3}}{3}\right)\right)$.

77. $r = 1 + 2\cos\theta \Rightarrow x = r\cos\theta = \cos\theta + 2\cos^2\theta$ and $y = r\sin\theta = \sin\theta + 2\sin\theta\cos\theta = \sin\theta + \sin 2\theta$,

so $\frac{dx}{d\theta} = -\sin\theta - 4\cos\theta\sin\theta = -\sin\theta(4\cos\theta + 1)$ and

$\frac{dy}{d\theta} = \cos\theta + 2\cos 2\theta = \cos\theta + 2\left(2\cos^2\theta - 1\right) = 4\cos^2\theta + \cos\theta - 2$. $\frac{dy}{d\theta} = 0 \Rightarrow \theta = \cos^{-1}\frac{\sqrt{33}-1}{8}$,

$\cos^{-1}\left(\frac{-1-\sqrt{33}}{8}\right)$, $2\pi - \cos^{-1}\left(\frac{-1-\sqrt{33}}{8}\right)$, or $2\pi - \cos^{-1}\frac{\sqrt{33}-1}{8}$, and $\frac{dx}{d\theta} = 0 \Rightarrow \theta = 0, \cos^{-1}\left(-\frac{1}{4}\right), \pi$, or

$2\pi - \cos^{-1}\left(-\frac{1}{4}\right)$. From this, we see that the points where the tangent line is horizontal are $\left(1 + \frac{\sqrt{33}-1}{4}, \cos^{-1}\frac{\sqrt{33}-1}{8}\right)$,

$\left(1 - \frac{\sqrt{33}+1}{4}, \cos^{-1}\left(\frac{-\sqrt{33}-1}{8}\right)\right)$, $\left(1 - \frac{\sqrt{33}+1}{4}, 2\pi - \cos^{-1}\left(\frac{-1-\sqrt{33}}{8}\right)\right)$, and $\left(1 + \frac{\sqrt{33}-1}{4}, 2\pi - \cos^{-1}\frac{\sqrt{33}-1}{8}\right)$; and

the points where the tangent line is vertical are $(3, 0)$, $\left(\frac{1}{2}, \cos^{-1}\left(-\frac{1}{4}\right)\right)$, $(-1, \pi)$, and $\left(\frac{1}{2}, 2\pi - \cos^{-1}\left(-\frac{1}{4}\right)\right)$.

79. $x^4 - 2x^3 + 2x^2y^2 - 2xy^2 - y^2 + y^4 = 0$. Letting $x = r\cos\theta$ and $y = r\sin\theta$, we find

$r^4\cos^4\theta - 2r^3\cos^3\theta + 2r^4\cos^2\theta\sin^2\theta - 2r^3\cos\theta\sin^2\theta - r^2\sin^2\theta + r^4\sin^4\theta = 0$

$\Leftrightarrow r^4\left(\cos^4\theta + 2\cos^2\theta\sin^2\theta + \sin^4\theta\right) - 2r^3\cos\theta\left(\cos^2\theta + \sin^2\theta\right) - r^2\sin^2\theta = 0 \Leftrightarrow$

$r^4\left(\cos^2\theta + \sin^2\theta\right)^2 - 2r^3\cos\theta - r^2\sin^2\theta = 0 \Leftrightarrow r^2\left[r^2 - 2r\cos\theta - \left(1 - \cos^2\theta\right)\right] = 0 \Leftrightarrow r = 0$ or

$r^2 - 2r\cos\theta + \cos^2\theta - 1 = 0 \Leftrightarrow r = 0$ or $(r - \cos\theta)^2 = 1 \Leftrightarrow r = 0$ or $r = \cos\theta \pm 1$. This shows that $r = 1 + \cos\theta$ has

the given equation as its rectangular equation.

81. a. $d^2 = (x_2 - x_1)^2 + (y_2 - y_1)^2$. Here $x_1 = r_1\cos\theta_1$, $y_1 = r_1\sin\theta_1$, $x_2 = r_2\cos\theta_2$, and $y_2 = r_2\sin\theta_2$, so

$d^2 = (r_2\cos\theta_2 - r_1\cos\theta_1)^2 + (r_2\sin\theta_2 - r_1\sin\theta_1)^2$

$= r_2^2\cos^2\theta_2 - 2r_1r_2\cos\theta_1\cos\theta_2 + r_1^2\cos^2\theta_1 + r_2^2\sin^2\theta_2 - 2r_1r_2\sin\theta_1\sin\theta_2 + r_1^2\sin^2\theta_1$

$= r_2^2\left(\cos^2\theta_2 + \sin^2\theta_2\right) - 2r_1r_2\left(\cos\theta_1\cos\theta_2 + \sin\theta_1\sin\theta_2\right) + r_1^2\left(\cos^2\theta_1 + \sin^2\theta_1\right)$

$= r_1^2 + r_2^2 - 2r_1r_2\cos(\theta_1 - \theta_2)$

so $d = \sqrt{r_1^2 + r_2^2 - 2r_1r_2\cos(\theta_1 - \theta_2)}$.

b. Letting $r_1 = 4$, $\theta_1 = \frac{2\pi}{3}$, $r_2 = 2$, and $\theta_2 = \frac{\pi}{3}$, we find the distance between $\left(4, \frac{2\pi}{3}\right)$ and $\left(2, \frac{\pi}{3}\right)$ to be

$$d = \sqrt{4^2 + 2^2 - 2\,(4)\,(2) \cos\left(\frac{2\pi}{3} - \frac{\pi}{3}\right)} = \sqrt{20 - 16 \cos \frac{\pi}{3}} = 2\sqrt{3}.$$

83. a.

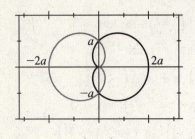

b. Let $C_1: r_1 = a\,(1 + \cos\theta)$ and $C_2: r_2 = a\,(1 - \cos\theta)$. To find the points of intersection of C_1 and C_2, we solve

$a\,(1 + \cos\theta) = a\,(1 - \cos\theta) \Leftrightarrow \cos\theta = 0 \Rightarrow \theta = \frac{\pi}{2}$ or $\frac{3\pi}{2}$. They

also intersect at the pole. The slope of the tangent line to C_1 is

$$m_1 = \frac{r_1' \sin\theta + r_1 \cos\theta}{r_1' \cos\theta - r_1 \sin\theta} = \frac{-a\sin^2\theta + a\cos^2\theta + a\cos\theta}{-a\sin\theta\cos\theta - a\sin\theta - a\sin\theta\cos\theta}$$

$$= -\frac{2\cos^2\theta + \cos\theta - 1}{\sin\theta\,(2\cos\theta + 1)}$$

and the slope of the tangent line to C_2 is $m_2 = \dfrac{r_2' \sin\theta + r_2 \cos\theta}{r_2' \cos\theta - r_2 \sin\theta} = -\dfrac{2\cos^2\theta - \cos\theta - 1}{\sin\theta\,(2\cos\theta - 1)}$, so at $\theta = \frac{\pi}{2}$, the slope

of the tangent line to C_1 is $m_1 = -\dfrac{2\cos^2\theta + \cos\theta - 1}{\sin\theta\,(2\cos\theta + 1)}\bigg|_{\pi/2} = 1$ and the slope of the tangent line to C_2 is

$m_2 = -\dfrac{2\cos^2\theta - \cos\theta - 1}{\sin\theta\,(2\cos\theta - 1)}\bigg|_{\pi/2} = -1$. Therefore, C_1 and C_2 intersect at right angles at $\left(a, \frac{\pi}{2}\right)$. Similarly, at

$\theta = \frac{3\pi}{2}$, we find $m_1 = -1$ and $m_2 = 1$, so C_1 and C_2 intersect at right angles at $\left(a, \frac{3\pi}{2}\right)$.

85.

87.

89.

91.

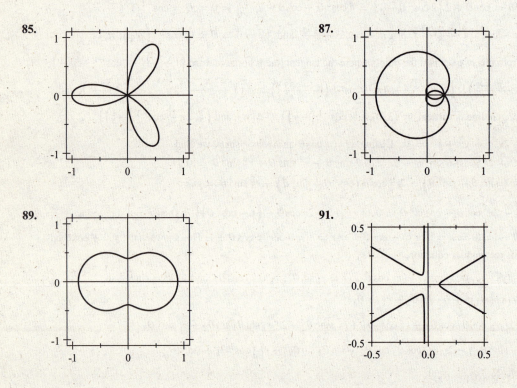

93. False. $P\left(2, \frac{\pi}{6}\right)$ and $P\left(-2, \frac{7\pi}{6}\right)$ represent the same point in polar coordinates, but $r_1 = 2 \neq -2 = r_2$.

95. False. Take $r = 1 + \sin\theta$. Then $\frac{dy}{d\theta} = 0$ at $\theta = \frac{3\pi}{2}$, but the graph has a

vertical tangent line at the point $\left(0, \frac{3\pi}{2}\right)$ (see the figure). Note that

$\frac{dx}{d\theta} = 0$ at $\theta = \frac{3\pi}{2}$.

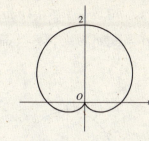

10.5 Concept Questions   ET 9.5

1. a. $A = \int_\alpha^\beta \frac{1}{2}\left[f(\theta)\right]^2 d\theta = \int_\alpha^\beta \frac{1}{2}r^2\, d\theta$. See Figure 2a.

b. $A = \int_\alpha^\beta \frac{1}{2}\left\{\left[f(\theta)\right]^2 - \left[g(\theta)\right]^2\right\} d\theta$. See Figure 6.

3. a. $S = 2\pi \int_\alpha^\beta r\sin\theta\sqrt{(r')^2 + r^2}\, d\theta$ **b.** $S = 2\pi \int_\alpha^\beta r\cos\theta\sqrt{(r')^2 + r^2}\, d\theta$

10.5 Areas and Arc Length in Polar Coordinates   ET 9.5

1. a. $r = 4\cos\theta \Rightarrow r^2 = 4r\cos\theta \Rightarrow x^2 + y^2 = 4x \Rightarrow x^2 - 4x + y^2 = 0 \Rightarrow (x-2)^2 + y^2 = 4$, an equation of the circle
with center $(2, 0)$ and radius 2. Its area is $\pi(2)^2 = 4\pi$.

b. $A = \frac{1}{2}\int_\alpha^\beta \left[f(\theta)\right]^2 d\theta = \frac{1}{2}\int_0^\pi (4\cos\theta)^2\, d\theta = 8\int_0^\pi \cos^2\theta\, d\theta = 4\int_0^\pi (1 + \cos 2\theta)\, d\theta = 4\left(\theta + \frac{1}{2}\sin 2\theta\right)\Big|_0^\pi = 4\pi$

3. $A = \frac{1}{2}\int_0^\pi \theta^2\, d\theta = \frac{1}{6}\theta^3\Big|_0^\pi = \frac{1}{6}\pi^3$

5. $A = \frac{1}{2}\int_{-\pi/2}^0 \left(e^\theta\right)^2 d\theta = \frac{1}{2}\int_{-\pi/2}^0 e^{2\theta}\, d\theta = \frac{1}{4}e^{2\theta}\Big|_{-\pi/2}^0 = \frac{1 - e^{-\pi}}{4} = \frac{e^\pi - 1}{4e^\pi}$

7. $A = \frac{1}{2}\int_0^{\pi/2} \left(\sqrt{\cos\theta}\right)^2 d\theta = \frac{1}{2}\int_0^{\pi/2} \cos\theta\, d\theta = \frac{1}{2}\sin\theta\Big|_0^{\pi/2} = \frac{1}{2}$

9. $A = \frac{1}{2}\int_0^{3\pi/2} \theta^2\, d\theta = \frac{1}{6}\theta^3\Big|_0^{3\pi/2} = \frac{9\pi^3}{16}$

11. $A = \frac{1}{2}\int_0^\pi (1 - \cos\theta)^2\, d\theta = \frac{1}{2}\int_0^\pi \left(1 - 2\cos\theta + \cos^2\theta\right) d\theta = \frac{1}{2}\int_0^\pi \left[1 - 2\cos\theta + \frac{1}{2}(1 + \cos 2\theta)\right] d\theta$

$= \frac{1}{2}\int_0^\pi \left(\frac{3}{2} - 2\cos\theta + \frac{1}{2}\cos 2\theta\right) d\theta = \frac{1}{2}\left(\frac{3}{2}\theta - 2\sin\theta + \frac{1}{4}\sin 2\theta\right)\Big|_0^\pi = \frac{3\pi}{4}$

13. $A = \frac{1}{2}\int_0^\pi (3\sin\theta)^2\,d\theta = \frac{9}{2}\int_0^\pi \sin^2\theta\,d\theta$

$= \frac{9}{4}\int_0^\pi (1 - \cos 2\theta)\,d\theta = \frac{9}{4}\left(\theta - \frac{1}{2}\sin 2\theta\right)\Big|_0^\pi$

$= \frac{9\pi}{4}$

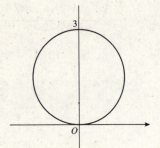

15. $r^2 = \sin\theta \Rightarrow r = \sqrt{\sin\theta}$ for $0 \le \theta \le \pi$, so

$A = 2\int_0^\pi \frac{1}{2}r^2\,d\theta = \int_0^\pi \sin\theta\,d\theta = -\cos\theta\big|_0^\pi = 2.$

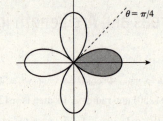

17. $r = 2\sin 2\theta \Rightarrow$

$A = 4\int_0^{\pi/2} \frac{1}{2}r^2\,d\theta = 2\int_0^{\pi/2} (2\sin 2\theta)^2\,d\theta$

$= 8\int_0^{\pi/2} \sin^2 2\theta\,d\theta = 4\int_0^{\pi/2} (1 - \cos 4\theta)\,d\theta$

$= 4\left(\theta - \frac{1}{4}\sin 4\theta\right)\Big|_0^{\pi/2} = 2\pi$

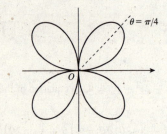

19. $r = \cos 2\theta \Rightarrow$

$A = 2\int_0^{\pi/4} \frac{1}{2}r^2\,d\theta = \int_0^{\pi/4} (\cos 2\theta)^2\,d\theta$

$= \frac{1}{2}\int_0^{\pi/4} (1 + \cos 4\theta)\,d\theta$

$= \frac{1}{2}\left(\theta + \frac{1}{4}\sin 4\theta\right)\Big|_0^{\pi/4} = \frac{\pi}{8}$

21. $r = \sin 4\theta \Rightarrow$

$A = \frac{1}{2}\int_0^{\pi/4} r^2\,d\theta = \frac{1}{2}\int_0^{\pi/4} (\sin 4\theta)^2\,d\theta$

$= \frac{1}{4}\int_0^{\pi/4} (1 - \cos 8\theta)\,d\theta$

$= \frac{1}{4}\left(\theta - \frac{1}{8}\sin 8\theta\right)\Big|_0^{\pi/4} = \frac{\pi}{16}$

23. $r = 1 + 2\cos\theta \Rightarrow$

$A = 2\int_{2\pi/3}^\pi \frac{1}{2}r^2\,d\theta = \int_{2\pi/3}^\pi (1 + 2\cos\theta)^2\,d\theta$

$= \int_{2\pi/3}^\pi \left(1 + 4\cos\theta + 4\cos^2\theta\right)d\theta$

$= \int_{2\pi/3}^\pi [1 + 4\cos\theta + 2(1 + \cos 2\theta)]\,d\theta$

$= (3\theta + 4\sin\theta + \sin 2\theta)\big|_{2\pi/3}^\pi = \pi - \frac{3\sqrt{3}}{2}$

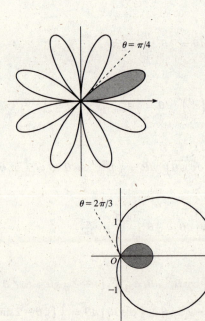

25. The two circles intersect along the line $\theta = \frac{\pi}{4}$ because setting $\sin\theta = \cos\theta$ gives $\tan\theta = 1$ or $\theta = \frac{\pi}{4}$. Therefore,

$$A = 2\int_0^{\pi/4} \frac{1}{2} r^2 \, d\theta = \int_0^{\pi/4} (\sin\theta)^2 \, d\theta = \frac{1}{2}\int_0^{\pi/4} (1 - \cos 2\theta) \, d\theta = \frac{1}{2}\left(\theta - \frac{1}{2}\sin 2\theta\right)\Big|_0^{\pi/4} = \frac{\pi - 2}{8}.$$

27. The area of the region under the outer curve and above the polar axis is

$$A_1 = \frac{1}{2}\int_0^\pi r^2 \, d\theta = \frac{1}{2}\int_0^\pi (1 + \cos\theta)^2 \, d\theta = \frac{1}{2}\int_0^\pi \left(1 + 2\cos\theta + \cos^2\theta\right) d\theta = \frac{1}{2}\int_0^\pi \left(1 + 2\cos\theta + \frac{1}{2} + \frac{1}{2}\cos 2\theta\right) d\theta$$

$$= \frac{1}{2}\left(\frac{3}{2}\theta + 2\sin\theta + \frac{1}{4}\sin 2\theta\right)\Big|_0^\pi = \frac{3\pi}{4}$$

The area of the region under the inner curve and above the polar axis is

$$A_2 = \frac{1}{2}\int_0^{\pi/4} r^2 \, d\theta = \frac{1}{2}\int_0^{\pi/4} \left(\sqrt{\cos 2\theta}\right)^2 d\theta = \frac{1}{2}\int_0^{\pi/4} \cos 2\theta \, d\theta = \frac{1}{4}\sin 2\theta\Big|_0^{\pi/4} = \frac{1}{4}.$$ Therefore, the area of the

shaded region is $A = A_1 - A_2 = \dfrac{3\pi}{4} - \dfrac{1}{4} = \dfrac{3\pi - 1}{4}.$

29. $r = 1$ and $r = 1 + \cos\theta$. Equating the two, we find $1 + \cos\theta = 1 \Rightarrow$
$\cos\theta = 0$, so $\theta = \frac{\pi}{2}$ or $\frac{3\pi}{2}$ and the points of intersection are $\left(1, \frac{\pi}{2}\right)$ and
$\left(1, \frac{3\pi}{2}\right)$.

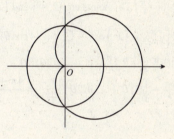

31. $r = 2$ and $r = 4\cos 2\theta$. Equating, we find $4\cos 2\theta = 2 \Rightarrow \cos 2\theta = \frac{1}{2}$, so
$2\theta = \frac{\pi}{3}, \frac{7\pi}{3}, \frac{5\pi}{3}$, or $\frac{11\pi}{3}$, implying $\theta = \frac{\pi}{6}, \frac{5\pi}{6}, \frac{7\pi}{6}$, or $\frac{11\pi}{6}$. By
symmetry, the points of intersection are $\left(2, \frac{\pi}{6}\right), \left(2, \frac{\pi}{3}\right), \left(2, \frac{2\pi}{3}\right),$
$\left(2, \frac{5\pi}{6}\right), \left(2, \frac{7\pi}{6}\right), \left(2, \frac{4\pi}{3}\right), \left(2, \frac{5\pi}{3}\right),$ and $\left(2, \frac{11\pi}{6}\right).$

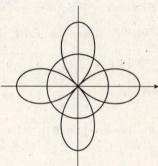

33. $r = \sin\theta$ and $r = \sin 2\theta$. Equating, we find $2\sin\theta\cos\theta = \sin\theta \Rightarrow$
$\sin\theta(1 - 2\cos\theta) = 0 \Rightarrow \sin\theta = 0$ or $\cos\theta = \frac{1}{2} \Rightarrow \theta = 0, \frac{\pi}{3}, \pi$, or $\frac{5\pi}{3}$,
and so the points of intersection are $\left(\frac{\sqrt{3}}{2}, \frac{\pi}{3}\right), \left(-\frac{\sqrt{3}}{2}, \frac{5\pi}{3}\right)$, and the pole.

35. $r = 1 + \cos\theta$ and $r = 3\cos\theta$. Equating, we find $1 + \cos\theta = 3\cos\theta \Leftrightarrow$

$\cos\theta = \frac{1}{2} \Rightarrow \theta = \frac{\pi}{3}$ or $\frac{5\pi}{3}$, and so the points of intersection are $\left(\frac{3}{2}, \frac{\pi}{3}\right)$

and $\left(\frac{3}{2}, \frac{5\pi}{3}\right)$. Thus,

$A = 2\int_0^{\pi/3} \frac{1}{2}\left[(3\cos\theta)^2 - (1 + \cos\theta)^2\right]d\theta$

$\quad = \int_0^{\pi/3}\left(8\cos^2\theta - 2\cos\theta - 1\right)d\theta$

$\quad = \int_0^{\pi/3}(3 + 4\cos 2\theta - 2\cos\theta)\,d\theta$

$\quad = (3\theta + 2\sin 2\theta - 2\sin\theta)\big|_0^{\pi/3} = \pi$

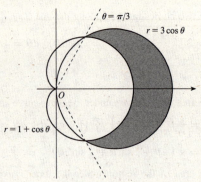

37. $r = 4\cos\theta$ and $r = 2$. Equating, we find $4\cos\theta = 2 \Leftrightarrow \cos\theta = \frac{1}{2} \Rightarrow$

$\theta = \frac{\pi}{3}$ or $\frac{5\pi}{3}$. The points are $\left(2, \frac{\pi}{3}\right)$ and $\left(2, \frac{5\pi}{3}\right)$. Thus,

$A = 2\left[\int_{\pi/3}^{\pi} \frac{1}{2}\left(2^2\right)d\theta - \int_{\pi/3}^{\pi/2} \frac{1}{2}(4\cos\theta)^2\,d\theta\right]$

$\quad = 4\theta\big|_{\pi/3}^{\pi} - 8\int_{\pi/3}^{\pi/2}(1 + \cos 2\theta)\,d\theta$

$\quad = \frac{8\pi}{3} - \left[8\left(\theta + \frac{1}{2}\sin 2\theta\right)\right]_{\pi/3}^{\pi/2} = \frac{4\pi + 6\sqrt{3}}{3}$

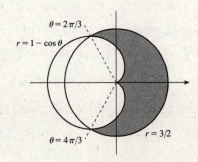

39. $r = 1 - \cos\theta$ and $r = \frac{3}{2}$. Equating, we find $1 - \cos\theta = \frac{3}{2} \Leftrightarrow \cos\theta = -\frac{1}{2}$

$\Rightarrow \theta = \frac{2\pi}{3}$ or $\frac{4\pi}{3}$, so the points are $\left(\frac{3}{2}, \frac{2\pi}{3}\right)$ and $\left(\frac{3}{2}, \frac{4\pi}{3}\right)$. Thus,

$A = 2\int_0^{2\pi/3}\left[\frac{1}{2}\left(\frac{3}{2}\right)^2 - \frac{1}{2}(1 - \cos\theta)^2\right]d\theta$

$\quad = \int_0^{2\pi/3}\left[\frac{5}{4} + 2\cos\theta - \frac{1}{2}(1 + \cos 2\theta)\right]d\theta$

$\quad = \left(\frac{3}{4}\theta + 2\sin\theta - \frac{1}{4}\sin 2\theta\right)\Big|_0^{2\pi/3} = \frac{4\pi + 9\sqrt{3}}{8}$

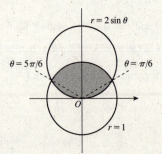

41. $r = 1$ and $r = 2\sin\theta$. To find the points of intersection, we solve

$2\sin\theta = 1 \Leftrightarrow \sin\theta = \frac{1}{2} \Rightarrow \theta = \frac{\pi}{6}$ or $\frac{5\pi}{6}$, giving the points $\left(1, \frac{\pi}{6}\right)$ and

$\left(1, \frac{5\pi}{6}\right)$. The area of the required region is

$A = 2\left[\frac{1}{2}\int_0^{\pi/6}(2\sin\theta)^2\,d\theta + \frac{1}{2}\int_{\pi/6}^{\pi/2}1^2\,d\theta\right]$

$\quad = 2\left(2\int_0^{\pi/6}\sin^2\theta\,d\theta + \left[\frac{1}{2}\theta\right]_{\pi/6}^{\pi/2}\right) = 2\left[\int_0^{\pi/6}(1 - \cos 2\theta)\,d\theta + \frac{\pi}{6}\right]$

$\quad = 2\left[\left(\theta - \frac{1}{2}\sin 2\theta\right)\Big|_0^{\pi/6} + \frac{\pi}{6}\right] = \frac{4\pi - 3\sqrt{3}}{6}$

43. $r = \sin\theta$ and $r = 1 - \sin\theta$. Equating, we find $\sin\theta = 1 - \sin\theta \Rightarrow$

$\sin\theta = \frac{1}{2} \Rightarrow \theta = \frac{\pi}{6}$ or $\frac{5\pi}{6}$, so the points of intersection are $\left(\frac{1}{2}, \frac{\pi}{6}\right)$,

$\left(\frac{1}{2}, \frac{5\pi}{6}\right)$, and the pole. Thus,

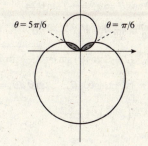

$A = 2\int_0^{\pi/6} \frac{1}{2}(\sin\theta)^2\, d\theta + 2\int_{\pi/6}^{\pi/2} \frac{1}{2}(1-\sin\theta)^2\, d\theta$

$\quad = \frac{1}{2}\int_0^{\pi/6}(1-\cos 2\theta)\, d\theta + \int_{\pi/6}^{\pi/2}\left(1 - 2\sin\theta + \sin^2\theta\right)d\theta$

$\quad = \frac{1}{2}\left(\theta - \frac{1}{2}\sin 2\theta\right)\Big|_0^{\pi/6} + \int_{\pi/6}^{\pi/2}\left(\frac{3}{2} - 2\sin\theta - \frac{1}{2}\cos 2\theta\right)d\theta$

$\quad = \frac{\pi}{12} - \frac{\sqrt{3}}{8} + \left(\frac{3}{2}\theta + 2\cos\theta - \frac{1}{4}\sin 2\theta\right)\Big|_{\pi/6}^{\pi/2} = \dfrac{7\pi - 12\sqrt{3}}{12}$

45. $r^2 = 4\cos 2\theta$ and $r = \sqrt{2}$. To find the points of intersection, we solve

$4\cos 2\theta = 2 \Rightarrow \cos 2\theta = \frac{1}{2} \Rightarrow \theta = \frac{\pi}{6}, \frac{5\pi}{6}, \frac{7\pi}{6}$, or $\frac{11\pi}{6}$, so the points of

intersection are $\left(\sqrt{2}, \frac{\pi}{6}\right)$, $\left(\sqrt{2}, \frac{5\pi}{6}\right)$, $\left(\sqrt{2}, \frac{7\pi}{6}\right)$, and $\left(\sqrt{2}, \frac{11\pi}{6}\right)$. Thus,

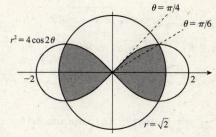

$A = 4\left[\frac{1}{2}\int_0^{\pi/6}\left(\sqrt{2}\right)^2 d\theta + \frac{1}{2}\int_{\pi/6}^{\pi/4} 4\cos 2\theta\, d\theta\right] = [2(2\theta)]_0^{\pi/6} +$

$\left[8\left(\frac{1}{2}\sin 2\theta\right)\right]_{\pi/6}^{\pi/4} = \dfrac{2\pi + 12 - 6\sqrt{3}}{3}$.

47. $r = 5\sin\theta \Rightarrow L = \int_0^\pi \sqrt{(r')^2 + r^2}\, d\theta = \int_0^\pi \sqrt{(5\cos\theta)^2 + (5\sin\theta)^2}\, d\theta = 5\int_0^\pi d\theta = 5\pi$

49. $r = e^{-\theta}$, $0 \le \theta \le 4\pi \Rightarrow$

$L = \int_0^{4\pi}\sqrt{(r')^2 + r^2}\, d\theta = \int_0^{4\pi}\sqrt{\left(-e^{-\theta}\right)^2 + \left(e^{-\theta}\right)^2}\, d\theta = \sqrt{2}\int_0^{4\pi}e^{-\theta}\, d\theta = -\sqrt{2}e^{-\theta}\Big|_0^{4\pi} = \sqrt{2}\left(1 - e^{-4\pi}\right)$

51. $r = \sin^3\frac{\theta}{3}$, $0 \le \theta \le \pi \Rightarrow$

$L = \int_0^\pi\sqrt{(r')^2 + r^2}\, d\theta = \int_0^\pi\sqrt{\left(\sin^2\frac{\theta}{3}\cos\frac{\theta}{3}\right)^2 + \left(\sin^3\frac{\theta}{3}\right)^2}\, d\theta = \int_0^\pi \sin^2\frac{\theta}{3}\, d\theta = \frac{1}{2}\int_0^\pi\left(1 - \cos\frac{2\theta}{3}\right)d\theta$

$\quad = \frac{1}{2}\left(\theta - \frac{3}{2}\sin\frac{2\theta}{3}\right)\Big|_0^\pi = \dfrac{4\pi - 3\sqrt{3}}{8}$

53. $r = a\sin^4\frac{\theta}{4} \Rightarrow r' = a\sin^3\frac{\theta}{4}\cos\frac{\theta}{4} \Rightarrow$

$(r')^2 + r^2 = a^2\sin^6\frac{\theta}{4}\cos^2\frac{\theta}{4} + a^2\sin^8\frac{\theta}{4} = a^2\sin^6\frac{\theta}{4}$, so

$L = \int_0^{4\pi}\sqrt{a^2\sin^6\frac{\theta}{4}}\, d\theta = 4a\int_0^\pi\sin^3 u\, du$

$\quad = 4a\int_0^\pi\left(1 - \cos^2 u\right)\sin u\, du$

$\quad = 4a\left(-\cos u + \frac{1}{3}\cos^3 u\right)\Big|_0^\pi = \frac{16}{3}a$

55. $r = 4\cos\theta \Rightarrow r' = -4\sin\theta \Rightarrow$

$S = 2\pi\int_0^{\pi/2} r\sin\theta\sqrt{(r')^2 + r^2}\, d\theta$

$\quad = 2\pi\int_0^{\pi/2} 4\cos\theta\sin\theta\sqrt{16\sin^2\theta + 16\cos^2\theta}\, d\theta$

$\quad = 32\pi\int_0^{\pi/2}\sin\theta\cos\theta\, d\theta = 16\pi\sin^2\theta\Big|_0^{\pi/2}$

$\quad = 16\pi$

57. $r = 2 + 2\cos\theta \Rightarrow r' = -2\sin\theta \Rightarrow$

$\quad S = 2\pi \int_0^\pi r\sin\theta \sqrt{(r')^2 + r^2}\, d\theta$

$\quad\quad = 2\pi \int_0^\pi (2 + 2\cos\theta)\sin\theta \sqrt{(-2\sin\theta)^2 + (2 + 2\cos\theta)^2}\, d\theta$

$\quad\quad = 8\sqrt{2}\pi \int_0^\pi (1 + \cos\theta)^{3/2}\sin\theta\, d\theta = -8\sqrt{2}\pi \cdot \frac{2}{5}(1 + \cos\theta)^{5/2}\Big|_0^\pi$

$\quad\quad = \frac{16\sqrt{2}\pi}{5}\left(2^{5/2}\right) = \frac{128\pi}{5}$

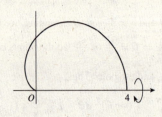

59. $r^2 = \cos 2\theta$. We can take $r = \sqrt{\cos 2\theta}$ for $-\frac{\pi}{4} \le \theta \le \frac{\pi}{4}$, so

$\quad r' = -\dfrac{\sin 2\theta}{\sqrt{\cos 2\theta}}$. Thus, $(r')^2 + r^2 = \dfrac{\sin^2 2\theta}{\cos 2\theta} + \cos 2\theta = \dfrac{1}{\cos 2\theta}$, and so

$\quad S = 2\pi \int_{-\pi/4}^{\pi/4} r\cos\theta \sqrt{(r')^2 + r^2}\, d\theta$

$\quad\quad = 2\pi \int_{-\pi/4}^{\pi/4} \sqrt{\cos 2\theta}\,(\cos\theta)\dfrac{1}{\sqrt{\cos 2\theta}}\, d\theta$

$\quad\quad = 2\pi \int_{-\pi/4}^{\pi/4} \cos\theta\, d\theta = 2\pi \sin\theta\Big|_{-\pi/4}^{\pi/4} = 2\sqrt{2}\pi$

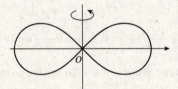

61. $\left(x^2 + y^2\right)^3 = 16x^2y^2$. The polar equation is

$\quad \left(r^2\right)^3 = 16(r\cos\theta)^2(r\sin\theta)^2 \Leftrightarrow$

$\quad r^6 = 16r^4\sin^2\theta\cos^2\theta = 16r^4\left(\dfrac{\sin 2\theta}{2}\right)^2 = 4r^4\sin^2 2\theta \Leftrightarrow$

$\quad r^2 = 4\sin^2 2\theta$. Using symmetry, we find

$\quad A = 4\int_0^{\pi/2} \frac{1}{2}r^2\, d\theta = 2\int_0^{\pi/2} 4\sin^2 2\theta\, d\theta = 4\int_0^{\pi/2}(1 - \cos 4\theta)\, d\theta$

$\quad\quad = 4\left(\theta - \frac{1}{4}\sin 4\theta\right)\Big|_0^{\pi/2} = 2\pi$

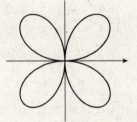

63. a. $r = e^{m\theta} \Rightarrow \dfrac{dr}{d\theta} = me^{m\theta}$, so $\tan\psi = \dfrac{r}{\frac{dr}{d\theta}} = \dfrac{e^{m\theta}}{me^{m\theta}} = \dfrac{1}{m}$, showing that ψ is constant.

b. Suppose that $\psi = k$, where k is a constant. Then, by part a, $\tan\psi = \tan k = \dfrac{r}{\frac{dr}{d\theta}}$, and this means that, if $m = \dfrac{1}{\tan k}$,

$\quad \dfrac{dr}{d\theta} = mr \Rightarrow \displaystyle\int \dfrac{dr}{r} = \int m\, d\theta \Rightarrow \ln|r| = m\theta + C_1 \Rightarrow |r| = C_2 e^{m\theta}$, where $C_2 = e^{C_1} \Rightarrow r = Ce^{m\theta}$, where $C = \pm C_2$.

65. For the parabola $y = \dfrac{1}{2p}x^2$, $y' = \dfrac{x}{p}$, and so its length is

$\quad L_1 = \displaystyle\int_0^a \sqrt{1 + (y')^2}\, dx = \int_0^a \sqrt{1 + \left(\dfrac{x}{p}\right)^2}\, dx = \dfrac{1}{|p|}\int_0^a \sqrt{p^2 + x^2}\, dx$.

For the spiral $r = p\theta$, $0 \le r \le a$, we find $0 \le p\theta \le a \Leftrightarrow 0 \le \theta \le \frac{a}{p}$, and its length is

$\quad L_2 = \displaystyle\int_0^{a/p} \sqrt{(r')^2 + r^2}\, d\theta = \int_0^{a/p} \sqrt{p^2 + p^2\theta^2}\, d\theta = p\int_0^{a/p} \sqrt{1 + \theta^2}\, d\theta$. If we put $\theta = \dfrac{x}{p}$, then $d\theta = \dfrac{dx}{p}$ and

$\quad L_2 = \displaystyle\int_0^a p\sqrt{1 + \dfrac{x^2}{p^2}} \cdot \dfrac{dx}{p} = \dfrac{1}{|p|}\int_0^a \sqrt{p^2 + x^2}\, dx = L_1$.

67. a.

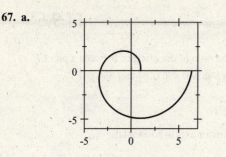

b. $r = \sqrt{1 + \theta^2} \Rightarrow r' = \dfrac{\theta}{\sqrt{\theta^2 + 1}} \Rightarrow$

$$(r')^2 + r^2 = \frac{\theta^2}{\theta^2 + 1} + \left(\sqrt{1 + \theta^2}\right)^2 = \frac{\theta^4 + 3\theta^2 + 1}{\theta^2 + 1}.$$

Using a CAS to integrate, we find

$$L = \int_0^{2\pi} \sqrt{(r')^2 + r^2}\, d\theta = \int_0^{2\pi} \sqrt{\frac{\theta^4 + 3\theta^2 + 1}{\theta^2 + 1}}\, d\theta \approx 22.01.$$

69.

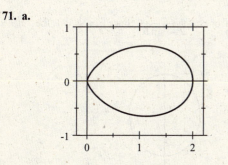

b. $r = 3 \sin\theta \cos^2\theta \Rightarrow$

$$r' = 3\cos^3\theta - 6\sin^2\theta\cos\theta = 3\cos\theta\left(\cos^2\theta - 2\sin^2\theta\right) \Rightarrow$$

$$(r')^2 + r^2 = 9\cos^2\theta\left(\cos^2\theta - 2\sin^2\theta\right)^2 + 9\sin^2\theta\cos^4\theta$$

$$= 9\cos^2\theta\left(\cos^4\theta - 3\sin^2\theta\cos^2\theta + 4\sin^4\theta\right) \Rightarrow$$

$$L = \int_0^\pi \sqrt{(r')^2 + r^2}\, d\theta$$

$$= 2 \cdot 3 \int_0^{\pi/2} \cos\theta \sqrt{\cos^4\theta - 3\sin^2\theta\cos^2\theta + 4\sin^4\theta}\, d\theta$$

$$\approx 5.37 \text{ (using a CAS)}$$

71. a.

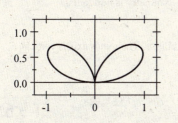

b. The area of the region is

$$A = 2\int_0^{\pi/2} \tfrac{1}{2}r^2\, d\theta = \int_0^{\pi/2} \left(2\cos^3\theta\right)^2 d\theta = 4\int_0^{\pi/2} \cos^6\theta\, d\theta.$$

Using a CAS or Formula 70 from the table of integrals, we find

$A = \dfrac{5\pi}{8}$. Next, we superimpose a Cartesian coordinate system so that the origin coincides with the pole and the x-axis lies along the polar axis. Then we have $x = r\cos\theta = 2\cos^4\theta$ and $y = r\sin\theta = 2\cos^3\theta\sin\theta$. By symmetry with respect to the x-axis, we see that $\overline{y} = 0$. Next, we calculate

$$\overline{x} = \frac{1}{A}(2)\int_0^2 xy\, dx = 2\left(\frac{8}{5\pi}\right)\int_0^{\pi/2}\left(2\cos^4\theta\right)\left(2\cos^3\theta\sin\theta\right)\left(-8\cos^3\theta\sin\theta\right)d\theta = \frac{512}{5\pi}\int_0^{\pi/2}\cos^{10}\theta\sin^2\theta\, d\theta$$

$$= \frac{512}{5\pi}\int_0^{\pi/2}\left(\cos^{10}\theta - \cos^{12}\theta\right)d\theta$$

and using a CAS or Formula 70 again, we find $\overline{x} = \dfrac{21}{20}$. Thus, the centroid is $\left(\dfrac{21}{20}, 0\right)$.

73. a. Put $x = r\cos\theta$, $y = r\sin\theta$. Then

$$(r\cos\theta)^3 + (r\sin\theta)^3 = 3(r\cos\theta)(r\sin\theta) \Leftrightarrow$$

$$r^3\left(\cos^3\theta + \sin^3\theta\right) = 3r^2\cos\theta\sin\theta \Leftrightarrow$$

$$r = \frac{3\cos\theta\sin\theta}{\cos^3\theta + \sin^3\theta}, \quad -\frac{\pi}{4} < \theta < \frac{3\pi}{4}.$$

b.

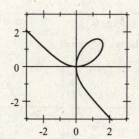

c. Using symmetry and a CAS, we find

$$A = 2\int_0^{\pi/4}\frac{r^2}{2}\, d\theta = \int_0^{\pi/4}\left[\frac{3\sin\theta\cos\theta}{\cos^3\theta + \sin^3\theta}\right]^2 d\theta = 9\int_0^{\pi/4}\frac{\sin^2\theta\cos^2\theta}{\left(\cos^3\theta + \sin^3\theta\right)^2}\, d\theta = \frac{3}{2}.$$

75. True. If $f(\theta_0) = g(\theta_0)$ for some θ_0, then the points $(\theta_0, f(\theta_0))$ and $(\theta_0, g(\theta_0))$ are the same.

10.6   Concept Questions   ET 9.6

1. The number d gives the distance between the pole and the directrix and e gives the ratio of the distance between a point P on the conic and its focus and the distance between P and its directrix. See pages 888 and 890 (884 and 886 in ET).

3. a. Since $e = 2 > 1$, the conic is a hyperbola. The transverse axis is vertical.

b. $r = \dfrac{6}{3 + \cos\theta} = \dfrac{\frac{1}{3}(6)}{1 + \frac{1}{3}\cos\theta}$, so $e = \frac{1}{3} < 1$ and the conic is an ellipse. The major axis is horizontal.

c. $r = \dfrac{2}{3(1 + \cos\theta)} = \dfrac{\frac{2}{3}}{1 + \cos\theta}$, so $e = 1$ and the conic is a parabola. The axis of symmetry is horizontal.

d. $r = \dfrac{5}{3 - 2\sin\theta} = \dfrac{\frac{5}{3}}{1 - \frac{2}{3}\sin\theta} = \dfrac{\frac{2}{3}\left(\frac{5}{2}\right)}{1 - \frac{2}{3}\sin\theta}$. Since $e = \frac{2}{3} < 1$, the conic is an ellipse. The major axis is vertical.

10.6   Conic Sections in Polar Coordinates   ET 9.6

1. Here $d = 2$ and $e = 1$, so an equation of the conic is

$r = \dfrac{2}{1 - \cos\theta}$. Since $e = 1$, the conic is a parabola.

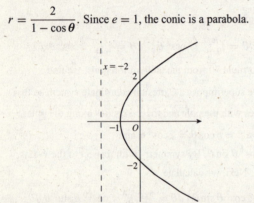

3. Here $d = 2$ and $e = \frac{1}{2}$, so an equation of the conic is

$r = \dfrac{\frac{1}{2}(2)}{1 - \frac{1}{2}\sin\theta}$ or $r = \dfrac{2}{2 - \sin\theta}$. Since $e = \frac{1}{2} < 1$, the

conic is an ellipse.

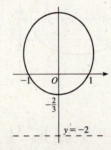

5. Here $d = 1$ and $e = \frac{3}{2}$, so an equation of the conic is

$r = \dfrac{\frac{3}{2}}{1 + \frac{3}{2}\cos\theta}$ or $r = \dfrac{3}{2 + 3\cos\theta}$. Since $e = \frac{3}{2} > 1$,

the conic is a hyperbola.

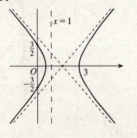

7. Here $d = 0.4 = \frac{2}{5}$ and $e = 0.4 = \frac{2}{5}$, so an equation of the

conic is $r = \dfrac{\frac{2}{5}\left(\frac{2}{5}\right)}{1 + \frac{2}{5}\sin\theta}$ or $r = \dfrac{4}{25 + 10\sin\theta}$. Since

$e = \frac{2}{5} < 1$, the conic is an ellipse.

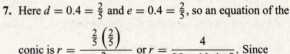
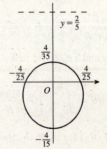

9. $r = \dfrac{8}{6 + 2\sin\theta} = \dfrac{\frac{1}{3}(4)}{1 + \frac{1}{3}\sin\theta}$, so $d = 4$ and $e = \dfrac{1}{3}$.

a. The eccentricity is $\frac{1}{3}$ and an equation of the directrix is $y = 4$.

b. The conic is an ellipse because $e = \frac{1}{3} < 1$.

c.

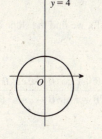

11. $r = \dfrac{10}{4 + 6\cos\theta} = \dfrac{\frac{3}{2}\left(\frac{5}{3}\right)}{1 + \frac{3}{2}\cos\theta}$, so $d = \dfrac{5}{3}$ and $e = \dfrac{3}{2}$.

a. The eccentricity is $\frac{3}{2}$ and an equation of the directrix is $x = \frac{5}{3}$.

b. The conic is a hyperbola because $e = \frac{3}{2} > 1$.

c.

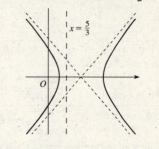

13. $r = \dfrac{5}{2 + 2\cos\theta} = \dfrac{\frac{5}{2}}{1 + \cos\theta}$, so $d = \dfrac{5}{2}$ and $e = 1$.

a. The eccentricity is 1 and an equation of the directrix is $x = \frac{5}{2}$.

b. The conic is a parabola because $e = 1$.

c.

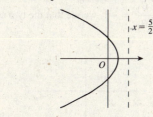

15. $r = \dfrac{1}{3 - 2\cos\theta} = \dfrac{\frac{2}{3}\left(\frac{1}{2}\right)}{1 - \frac{2}{3}\cos\theta}$, so $d = \dfrac{1}{2}$ and $e = \dfrac{2}{3}$.

a. The eccentricity is $\frac{2}{3}$ and an equation of the directrix is $x = -\frac{1}{2}$.

b. The conic is an ellipse because $e = \frac{2}{3} < 1$.

c.

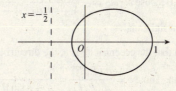

17. $r = \dfrac{1}{1 - \sin\theta}$, so $d = 1$ and $e = 1$.

a. The eccentricity is 1 and an equation of the directrix is $y = -1$.

b. The conic is a parabola because $e = 1$.

c.

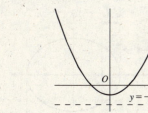

19. $r = -\dfrac{6}{\sin\theta - 2} = \dfrac{3}{1 - \frac{1}{2}\sin\theta} = \dfrac{\frac{1}{2}(6)}{1 - \frac{1}{2}\sin\theta}$, so $d = 6$ and $e = \dfrac{1}{2}$.

a. The eccentricity is $\frac{1}{2}$ and an equation of the directrix is $y = -6$.

b. The conic is an ellipse because $e = \frac{1}{2} < 1$.

c.

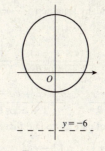

21. $\dfrac{x^2}{9} + \dfrac{y^2}{16} = 1$. Here $a = 4$ and $b = 3$, so $c = \sqrt{16 - 9} = \sqrt{7}$. Therefore, $e = \dfrac{c}{a} = \dfrac{\sqrt{7}}{4}$.

23. $x^2 - y^2 = 1$. Here $a = 1$ and $b = 1$, so $c = \sqrt{1^2 + 1^2} = \sqrt{2}$. Therefore, $e = \dfrac{c}{a} = \dfrac{\sqrt{2}}{1} = \sqrt{2}$.

25. $x^2 - 9y^2 + 2x - 54y = 105$. Completing the squares in x and y gives $\left(x^2 + 2x + 1\right) - 9\left(y^2 + 6y + 9\right) = 105 + 1 - 81 = 25$

$\Leftrightarrow \dfrac{(x+1)^2}{25} - \dfrac{(y+3)^2}{\frac{25}{9}} = 1$. Here $a = 5$ and $b = \frac{5}{3}$, so $c = \sqrt{5^2 + \left(\dfrac{5}{3}\right)^2} = \dfrac{5\sqrt{10}}{3}$. Therefore,

$e = \dfrac{c}{a} = \dfrac{5\sqrt{10}}{3} \cdot \dfrac{1}{5} = \dfrac{\sqrt{10}}{3}$.

27. Let $C_1 : r = \dfrac{c}{1 + \sin\theta}$ and $C_2 : r = \dfrac{d}{1 - \sin\theta}$. Using Equation 10.4.3 (9.4.3 in ET), we find the slope of the tangent line to C_1 at (r, θ):

$\dfrac{dy}{dx} = \dfrac{\frac{dr}{d\theta}\sin\theta + r\cos\theta}{\frac{dr}{d\theta}\cos\theta - r\sin\theta} = \dfrac{\dfrac{-c\cos\theta}{(1 + \sin\theta)^2}\sin\theta + \dfrac{c}{1 + \sin\theta}\cos\theta}{\dfrac{-c\cos\theta}{(1 + \sin\theta)^2}\cos\theta - \dfrac{c}{1 + \sin\theta}\sin\theta} = \dfrac{-c\cos\theta\sin\theta + c\cos\theta + c\cos\theta\sin\theta}{-c\cos^2\theta - c\sin\theta - c\sin^2\theta} = -\dfrac{\cos\theta}{1 + \sin\theta}$.

Similarly, the slope of the tangent line to C_2 at (r, θ) is

$\dfrac{dy}{dx} = \dfrac{\dfrac{d\cos\theta}{(1 - \sin\theta)^2}\cdot\sin\theta + \dfrac{d}{1 - \sin\theta}\cdot\cos\theta}{\dfrac{d\cos\theta}{(1 - \sin\theta)^2}\cdot\cos\theta - \dfrac{d}{1 - \sin\theta}\cdot\sin\theta} = \dfrac{d\cos\theta\sin\theta + d\cos\theta - d\cos\theta\sin\theta}{d\cos^2\theta - d\sin\theta + d\sin^2\theta} = \dfrac{\cos\theta}{1 - \sin\theta}$. So at a point of

intersection (r_1, θ_1), we find the slope of the tangent lines to C_1 and C_2 to be $m_1 = -\dfrac{\cos\theta_1}{1 + \sin\theta_1}$ and $m_2 = \dfrac{\cos\theta_1}{1 - \sin\theta_1}$

respectively. Because $m_1 m_2 = -\dfrac{\cos\theta_1}{1 + \sin\theta_1} \cdot \dfrac{\cos\theta_1}{1 - \sin\theta_1} = -\dfrac{\cos^2\theta_1}{1 - \sin^2\theta_1} = -\dfrac{\cos^2\theta_1}{\cos^2\theta_1} = -1$, we see that the two curves

do intersect at right angles.

29. We are given that the eccentricity is e, the directrix is $y = d$, and the focus

is at the origin. Thus, $e = \dfrac{d(P, F)}{d(P, \ell)} = \dfrac{r}{d - r\sin\theta} \Leftrightarrow ed - er\sin\theta = r$

$\Leftrightarrow ed = r(1 + e\sin\theta) \Leftrightarrow r = \dfrac{ed}{1 + e\sin\theta}$.

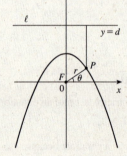

31. a. By the definition of eccentricity, $\dfrac{a - c}{k} = e \Leftrightarrow k = \dfrac{a - c}{e}$. Also,

$d = k + a - c$ and, by Equation 5, $\dfrac{c}{a} = e \Leftrightarrow c = ea$. Thus,

$r = \dfrac{ed}{1 - e\cos\theta} = \dfrac{e(k + a - c)}{1 - e\cos\theta} = \dfrac{e\left[\left(\dfrac{a - c}{e}\right) + a - c\right]}{1 - e\cos\theta}$

$= \dfrac{a - c + ea - ec}{1 - e\cos\theta} = \dfrac{a - ea + ea - e^2 a}{1 - e\cos\theta} = \dfrac{a\left(1 - e^2\right)}{1 - e\cos\theta}$

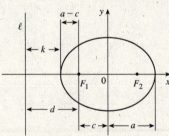

b. Using the result of part a, we find that r is minimized when $\theta = \pi$, so the perihelion distance is

$r(\pi) = \dfrac{a\left(1 - e^2\right)}{1 - e\cos\pi} = \dfrac{a\left(1 - e^2\right)}{1 + e} = a(1 - e)$.

33. $e = 0.056$ and $a = 1.427 \times 10^9$, so $r = \dfrac{a\left(1 - e^2\right)}{1 - e\cos\theta} = \dfrac{1.423 \times 10^9}{1 - 0.056\cos\theta}$. The perihelion distance is obtained by

setting $\theta = \pi$, giving $\dfrac{1.423 \times 10^9}{1 + 0.056} \approx 1.347 \times 10^9$ km. The aphelion distance is obtained by setting $\theta = 0$, giving

$\dfrac{1.423 \times 10^9}{1 - 0.056} \approx 1.507 \times 10^9$ km.

35. We are given that $a(1 - e) = 4.6 \times 10^7$ and $a(1 + e) = 7 \times 10^7$, so $\dfrac{1 + e}{1 - e} = \dfrac{7}{4.6} \Rightarrow 4.6 + 4.6e = 7 - 7e \Rightarrow e \approx 0.207$.

Chapter 10   Review   ET 9

Concept Review

1. a. equidistant; point; line; point; focus; line; directrix   **b.** vertex; focus; directrix

3. a. sum; foci; constant   **b.** foci; major axis; center; minor axis

5. a. difference; foci; constant   **b.** vertices; transverse; transverse; center; two separate

7. $x = f(t)$, $y = g(t)$; parameter

9. a. $f'(t)$; $g'(t)$; simultaneously zero; endpoints   **b.** $\int_a^b \sqrt{[f'(t)]^2 + [g'(t)]^2}\, dt = \int_a^b \sqrt{\left(\frac{dx}{dt}\right)^2 + \left(\frac{dy}{dt}\right)^2}\, dt$

11. a. $r\cos\theta$; $r\sin\theta$ **b.** $x^2 + y^2$; $\dfrac{y}{x}\ (x \neq 0)$

13. a. $A = \frac{1}{2}\int_\alpha^\beta r^2\, d\theta = \frac{1}{2}\int_\alpha^\beta [f(\theta)]^2\, d\theta$ **b.** $A = \frac{1}{2}\int_\alpha^\beta \left\{[f(\theta)]^2 - [g(\theta)]^2\right\} d\theta$

15. $\dfrac{d(P,F)}{d(P,\ell)} = e$; $0 < e < 1$; $e = 1$; $e > 1$

Review Exercises

1. $\dfrac{x^2}{4} + \dfrac{y^2}{9} = 1$. Here $a = 3$ and $b = 2$, so

$c = \sqrt{a^2 - b^2} = \sqrt{3^2 - 2^2} = \sqrt{5}$. The center of the

ellipse is $(0,0)$, so the vertices are $(0, \pm 3)$ and the foci are

$\left(0, \pm\sqrt{5}\right)$.

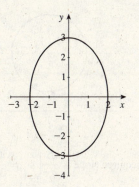

3. $x^2 - 9y^2 = 9 \Rightarrow \dfrac{x^2}{9} - \dfrac{y^2}{1} = 1$. Here $a = 3$ and $b = 1$, so

$c = \sqrt{a^2 + b^2} = \sqrt{3^2 + 1} = \sqrt{10}$. The center of the

hyperbola is $(0,0)$, so the vertices are $(\pm 3, 0)$ and the foci

are $\left(\pm\sqrt{10}, 0\right)$. The asymptotes are $y = \pm\frac{b}{a}x = \pm\frac{1}{3}x$.

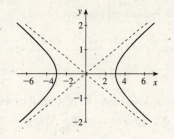

5. $y^2 - 9x^2 + 8y + 7 = 0 \Rightarrow y^2 + 8y + 16 - 9x^2 = 9 \Rightarrow$

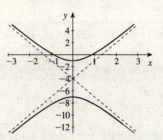

$\dfrac{(y+4)^2}{9} - \dfrac{x^2}{1} = 1$, so $a = 3$, $b = 1$, and

$c = \sqrt{a^2 + b^2} = \sqrt{3^2 + 1} = \sqrt{10}$. The center of the hyperbola is $(0, -4)$,

the vertices are $(0, -1)$ and $(0, -7)$, the foci are $\left(0, -4 \pm \sqrt{10}\right)$, and the

asymptotes are $y = \pm 3x - 4$.

7. If the focus of a parabola is $(-2, 0)$ and its directrix is $x = 2$, then its equation can be written $(y - k)^2 = 4p(x - h)$.

Because its vertex is $(0, 0)$ and $p = -2$, an equation is $y^2 = -8x$.

9. An equation of the ellipse with vertices $(\pm 7, 0)$ and foci $(\pm 2, 0)$ has the form $\dfrac{(x - h)^2}{a^2} + \dfrac{\left(y - k^2\right)}{b^2} = 1$. Its center is

$(0, 0)$, $a = 7$, and $c = 2 \Rightarrow b^2 = a^2 - c^2 = 7^2 - 2^2 = 45$. Thus, an equation is $\dfrac{x^2}{49} + \dfrac{y^2}{45} = 1$.

11. An equation of the hyperbola with foci $\left(0, \pm \frac{3}{2}\sqrt{5}\right)$ and vertices $(0, \pm 3)$ has the form $\dfrac{(y - k)^2}{a^2} - \dfrac{(x - h)^2}{b^2} = 1$. Its center

is $(0, 0)$, and so $h = k = 0$. Since $a = 3$ and $c = \frac{3}{2}\sqrt{5}$, we have $b^2 = c^2 - a^2 = \left(\frac{3}{2}\sqrt{5}\right)^2 - 3^2 = \frac{9}{4}$, so $b = \frac{3}{2}$. An

equation is thus $\dfrac{y^2}{9} - \dfrac{x^2}{\frac{9}{4}} = 1$ or $y^2 - 4x^2 = 9$.

13. Let m be a real number and let (x_0, y_0) be a point on the parabola $x^2 = 4py$. Differentiating, we find $2x = 4py'$

$\Rightarrow y' = \dfrac{x}{2p}$, so the slope of the tangent line to the parabola at (x_0, y_0) is $\dfrac{x_0}{2p}$. This is equal to m, so $\dfrac{x_0}{2p} = m$.

An equation of the tangent line is thus $y - y_0 = m(x - x_0) \Rightarrow y = y_0 + mx - mx_0$. But $x_0^2 = 4py_0$, so

$y = \dfrac{x_0^2}{4p} + mx - mx_0 = \dfrac{4p^2m^2}{4p} + mx - m(2pm) = pm^2 + mx - 2pm^2 \Rightarrow y = mx - pm^2$.

15. a. $x = 1 + 2t \Rightarrow t = \frac{1}{2}(x - 1)$, so

$y = 3 - 2t = 3 - 2\left(\frac{1}{2}\right)(x - 1) = 4 - x \Rightarrow y = 4 - x$.

b. As t increases, x increases and y decreases, as shown in the diagram.

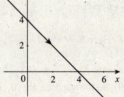

17. a. $\begin{cases} x = 1 + 2\sin t \\ y = 3 + 2\cos t \end{cases} \Rightarrow \begin{cases} \sin t = \frac{1}{2}(x - 1) \\ \cos t = \frac{1}{2}(y - 3) \end{cases}$ Thus, by the identity

$\sin^2 t + \cos^2 t = 1$, $\dfrac{(x - 1)^2}{4} + \dfrac{(y - 3)^2}{4} = 1 \Rightarrow$

$(x - 1)^2 + (y - 3)^2 = 4$.

b. If $t = 0$, then $x = 1$ and $y = 5$. A small increase in t results in a small

increase in x, but a decrease in y. This gives the direction of the path.

19. $\begin{cases} x = t^3 + 1 \\ y = 2t^2 - 1 \end{cases} \Rightarrow \begin{cases} \dfrac{dx}{dt} = 3t^2 \\ \dfrac{dy}{dt} = 4t \end{cases}$ Thus, $\dfrac{dy}{dx} = \dfrac{\frac{dy}{dt}}{\frac{dx}{dt}} = \dfrac{4t}{3t^2} = \dfrac{4}{3t}$ and the required slope is $\dfrac{dy}{dx}\bigg|_{t=1} = \dfrac{4}{3}$.

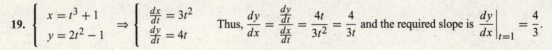

21. $\begin{cases} x = te^{-t} \\ y = \dfrac{1}{t^2+1} \end{cases} \Rightarrow \begin{cases} \dfrac{dx}{dt} = e^{-t} - te^{-t} = (1-t)\,e^{-t} \\ \dfrac{dy}{dt} = -\dfrac{2t}{\left(t^2+1\right)^2} \end{cases}$ Thus, the required slope is

$\dfrac{dy}{dx}\bigg|_{t=0} = \dfrac{\frac{dy}{dt}}{\frac{dx}{dt}}\bigg|_{t=0} = -\dfrac{2t}{\left(t^2+1\right)^2} \cdot \dfrac{1}{(1-t)\,e^{-t}}\bigg|_{t=0} = 0.$

23. $\begin{cases} x = t^3 + 1 \\ y = t^4 + 2t^2 \end{cases} \Rightarrow \begin{cases} \dfrac{dx}{dt} = 3t^2 \\ \dfrac{dy}{dt} = 4t^3 + 4t \end{cases} \Rightarrow \dfrac{dy}{dx} = \dfrac{\frac{dy}{dt}}{\frac{dx}{dt}} = \dfrac{4t\left(t^2+1\right)}{3t^2} = \dfrac{4\left(t^2+1\right)}{3t} \Rightarrow$

$\dfrac{d^2y}{dx^2} = \dfrac{\frac{d}{dt}\left(\frac{dy}{dx}\right)}{\frac{dx}{dt}} = \dfrac{\dfrac{d}{dt}\left[\dfrac{4\left(t^2+1\right)}{3t}\right]}{3t^2} = \dfrac{\frac{4}{3}\frac{d}{dt}\left(t + t^{-1}\right)}{9t^2} = \dfrac{4\left(1 - t^{-2}\right)}{9t^2} = \dfrac{4\left(t^2 - 1\right)}{9t^4}$

25. $\begin{cases} x = t^3 - 4t \\ y = t^2 + 2 \end{cases} \Rightarrow \begin{cases} \dfrac{dx}{dt} = 3t^2 - 4 \\ \dfrac{dy}{dt} = 2t \end{cases}$ If the tangent line is vertical, then $\frac{dx}{dt} = 0$ and $\frac{dy}{dt} \neq 0$. Since $3t^2 - 4 = 0 \Leftrightarrow$

$t = \pm\frac{2\sqrt{3}}{3}$ and $\frac{dy}{dt} \neq 0$ for these values of t, the curve has vertical tangent lines at $\left(\pm\frac{16\sqrt{3}}{9}, \frac{10}{3}\right)$.

If the tangent line is horizontal, then $\frac{dy}{dt} = 0$ and $\frac{dx}{dt} \neq 0$. Now $\frac{dy}{dt} = 0 \Rightarrow t = 0$ and $\frac{dx}{dt}\big|_{t=0} = -4 \neq 0$. Thus, the curve

has a horizontal tangent line at $(0, 2)$.

27. $\begin{cases} x = \frac{1}{6}t^6 \\ y = 2 - \frac{1}{4}t^4 \end{cases} \Rightarrow \begin{cases} \dfrac{dx}{dt} = t^5 \\ \dfrac{dy}{dt} = -t^3 \end{cases}$ Thus,

$L = \int_a^b \sqrt{\left(\dfrac{dx}{dt}\right)^2 + \left(\dfrac{dy}{dt}\right)^2}\,dt = \int_0^{\sqrt[4]{8}} \sqrt{\left(t^5\right)^2 + \left(-t^3\right)^2}\,dt = \int_0^{\sqrt[4]{8}} t^3\sqrt{t^4+1}\,dt = \frac{1}{4} \cdot \frac{2}{3}\left(t^4+1\right)^{3/2}\bigg|_0^{\sqrt[4]{8}} = \frac{13}{3}.$

29. $\begin{cases} x = e^{-t}\cos t \\ y = e^{-t}\sin t \end{cases} \Rightarrow \begin{cases} \dfrac{dx}{dt} = -e^{-t}\left(\cos t + \sin t\right) \\ \dfrac{dy}{dt} = e^{-t}\left(\cos t - \sin t\right) \end{cases}$ Thus,

$L = \int_a^b \sqrt{\left(\dfrac{dx}{dt}\right)^2 + \left(\dfrac{dy}{dt}\right)^2}\,dt = \int_0^{\pi/2} \sqrt{\left[-e^{-t}\left(\cos t + \sin t\right)\right]^2 + \left[e^{-t}\left(\cos t - \sin t\right)\right]^2}\,dt = \int_0^{\pi/2} \sqrt{2e^{-2t}}\,dt$

$= \sqrt{2} \int_0^{\pi/2} e^{-t}\,dt = \sqrt{2}\left(-e^{-t}\right)\big|_0^{\pi/2} = \sqrt{2}\left(1 - e^{-\pi/2}\right)$

31. $\begin{cases} x = t^2 \\ y = \frac{1}{3}t\left(3 - t^2\right) \end{cases} \Rightarrow \begin{cases} \dfrac{dx}{dt} = 2t \\ \dfrac{dy}{dt} = 1 - t^2 \end{cases}$ Thus,

$S = 2\pi \int_a^b y\,ds = 2\pi \int_0^{\sqrt{3}} \frac{1}{3}t\left(3 - t^2\right)\sqrt{(2t)^2 + \left(1 - t^2\right)^2}\,dt = \frac{2\pi}{3}\int_0^{\sqrt{3}} t\left(3 - t^2\right)\left(t^2 + 1\right)dt$

$= \frac{2\pi}{3}\int_0^{\sqrt{3}}\left(-t^5 + 2t^3 + 3t\right)dt = \frac{2\pi}{3}\left(-\frac{1}{6}t^6 + \frac{1}{2}t^4 + \frac{3}{2}t^2\right)\bigg|_0^{\sqrt{3}} = 3\pi$

33. The graph of $r = 2 \sin \theta$ is the circle with radius 1 and center $(0, 1)$.

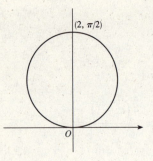

35. $r = 2 \cos 5\theta$

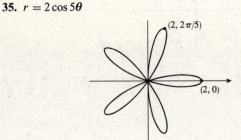

37. $r^2 = \cos 2\theta$

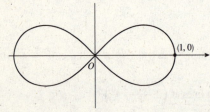

39. $r = f(\theta) = e^{2\theta} \Rightarrow f'(\theta) = 2e^{2\theta} \Rightarrow \dfrac{dy}{dx} = \dfrac{f(\theta)\cos\theta + f'(\theta)\sin\theta}{-f(\theta)\sin\theta + f'(\theta)\cos\theta} = \dfrac{e^{2\theta}\cos\theta + 2e^{2\theta}\sin\theta}{-e^{2\theta}\sin\theta + 2e^{2\theta}\cos\theta} = \dfrac{2\sin\theta + \cos\theta}{2\cos\theta - \sin\theta}$,

so the required slope is $\dfrac{dy}{dx}\Big|_{\theta=\pi/2} = \dfrac{2\sin\theta + \cos\theta}{2\cos\theta - \sin\theta}\Big|_{\theta=\pi/2} = -2.$

41. $\begin{cases} r = \sin\theta \\ r = 1 - \sin\theta \end{cases} \Rightarrow \sin\theta = 1 - \sin\theta \Leftrightarrow \sin\theta = \frac{1}{2} \Rightarrow \theta = \frac{\pi}{6} \text{ or } \frac{5\pi}{6}.$ If $\theta = \frac{\pi}{6}$, then $r = \frac{1}{2}$; if $\theta = \frac{5\pi}{6}$, then $r = \frac{1}{2}$.

Thus, the points of intersection are $\left(\frac{1}{2}, \frac{\pi}{6}\right)$, $\left(\frac{1}{2}, \frac{5\pi}{6}\right)$, and the pole.

43. $A = \dfrac{1}{2}\displaystyle\int_0^{2\pi} r^2 \, d\theta = \dfrac{1}{2}\int_0^{2\pi} (2 + \cos\theta)^2 \, d\theta$

$= \dfrac{1}{2}\displaystyle\int_0^{2\pi} \left(4 + 4\cos\theta + \cos^2\theta\right) d\theta$

$= \dfrac{1}{2}\displaystyle\int_0^{2\pi} \left[4 + 4\cos\theta + \dfrac{1 + \cos 2\theta}{2}\right] d\theta$

$= \dfrac{1}{2}\left(4\theta + 4\sin\theta + \dfrac{1}{2}\theta + \dfrac{1}{4}\sin 2\theta\right)\Big|_0^{2\pi} = \dfrac{9\pi}{2}$

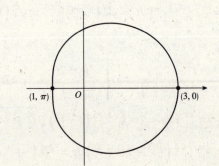

45. The required area is 8 times the area of the shaded region R. By symmetry,

$$R = \int_0^{\pi/8} \tfrac{1}{2} (2 \sin 2\theta)^2 \, d\theta + \int_{\pi/8}^{\pi/4} \tfrac{1}{2} (2 \cos 2\theta)^2 \, d\theta$$

$$= 2\left(\tfrac{1}{2}\right) \int_0^{\pi/8} (2 \sin 2\theta)^2 \, d\theta = 4 \int_0^{\pi/8} \sin^2 2\theta \, d\theta$$

$$= 2 \int_0^{\pi/8} (1 - \cos 4\theta) \, d\theta = 2\left(\theta - \tfrac{1}{4} \sin 4\theta\right)\Big|_0^{\pi/8}$$

$$= 2\left(\tfrac{\pi}{8} - \tfrac{1}{4}\right) = \tfrac{1}{4}(\pi - 2)$$

so the required area is $8 \cdot \tfrac{1}{4}(\pi - 2) = 2(\pi - 2)$.

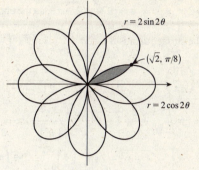

$r = 2 \sin 2\theta$

$(\sqrt{2}, \pi/8)$

$r = 2 \cos 2\theta$

47. $r = \theta^2 \Rightarrow \dfrac{dr}{d\theta} = 2\theta$, so

$$L = \int_0^{2\pi} \sqrt{r^2 + \left(\dfrac{dr}{d\theta}\right)^2} \, d\theta = \int_0^{2\pi} \sqrt{\left(\theta^2\right)^2 + (2\theta)^2} \, d\theta = \int_0^{2\pi} \theta \sqrt{\theta^2 + 4} \, d\theta = \tfrac{1}{2} \cdot \tfrac{2}{3} \left(\theta^2 + 4\right)^{3/2}\Big|_0^{2\pi}$$

$$= \tfrac{8}{3}\left[\left(\pi^2 + 1\right)^{3/2} - 1\right]$$

49. $r = f(\theta) = 2 \sin \theta \Rightarrow f'(\theta) = 2 \cos \theta$, so

$$S = 2\pi \int_\alpha^\beta f(\theta) \sin \theta \sqrt{[f(\theta)]^2 + [f'(\theta)]^2} \, d\theta$$

$$= 2\pi \int_0^\pi 2 \sin^2 \theta \sqrt{(2 \sin \theta)^2 + (2 \cos \theta)^2} \, d\theta$$

$$= 8\pi \int_0^\pi \sin^2 \theta \, d\theta = 4\pi \int_0^\pi (1 - \cos 2\theta) \, d\theta$$

$$= 4\pi \left(\theta - \tfrac{1}{2} \sin 2\theta\right)\Big|_0^\pi = 4\pi^2$$

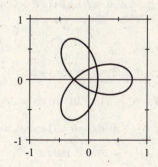

$(2, \pi/2)$

O

51. $r = \dfrac{1}{1 + \sin \theta}$. Here $e = 1$ and the curve is a parabola.   **53.**

$d = 1$, so the directrix is $y = 1$.

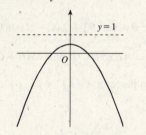

$y = 1$

O

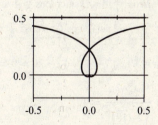

55.

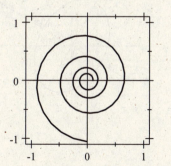

57.

59. $x^2 + 2y^2 = 2 \Rightarrow y = \pm\frac{\sqrt{2}}{2}\sqrt{2-x^2}$, so the upper half of the ellipse has equation $y = \frac{\sqrt{2}}{2}\sqrt{2-x^2}$. If $y = 0$, then $x = \pm\sqrt{2}$,

so the surface area is $S = 2\pi \int_{-\sqrt{2}}^{\sqrt{2}} y\sqrt{1 + \left(\frac{dy}{dx}\right)^2}\, dx$. But $\frac{dy}{dx} = \frac{\sqrt{2}}{2} \cdot \frac{1}{2}\left(2 - x^2\right)^{-1/2}(-2x) = -\frac{x}{\sqrt{2}\sqrt{2-x^2}}$, so

$1 + \left(\frac{dy}{dx}\right)^2 = 1 + \frac{x^2}{2\left(2-x^2\right)} = \frac{4-x^2}{2\left(2-x^2\right)}$, and therefore,

$L = 2\pi \int_{-\sqrt{2}}^{\sqrt{2}} \frac{\sqrt{2}}{2}\sqrt{2-x^2}\,\frac{\sqrt{4-x^2}}{\sqrt{2}\sqrt{2-x^2}}\, dx = \pi \int_{-\sqrt{2}}^{\sqrt{2}} \sqrt{4-x^2}\, dx = 2\pi \int_0^{\sqrt{2}} \sqrt{4-x^2}\, dx$

$= 2\pi\left[\frac{x}{2}\sqrt{4-x^2} + 2\sin^{-1}\frac{x}{2}\right]_0^{\sqrt{2}} = 2\pi\left(\frac{\sqrt{2}}{2}\cdot\sqrt{2} + 2\sin^{-1}\frac{\sqrt{2}}{2}\right) = 2\pi\left(1 + \frac{\pi}{2}\right) = \pi\left(\pi + 2\right)$

Challenge Problems

1. a. $x = \overline{ON} - \overline{MN} = \overline{OK}\cos\theta - \overline{JK} = x'\cos\theta - y'\sin\theta$ and

$y = \overline{MJ} + \overline{JP} = \overline{NK} + \overline{JP} = \overline{OK}\sin\theta + \overline{KP}\cos\theta = x'\sin\theta +$
$y'\cos\theta$.

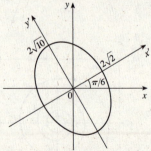

b. Using the result of part a, we require that

$Ax^2 + Bxy + Cy^2 + F$

$\qquad = A\left(x'\cos\theta - y'\sin\theta\right)^2 + B\left(x'\cos\theta - y'\sin\theta\right)\left(x'\sin\theta + y'\cos\theta\right) + C\left(x'\sin\theta + y'\cos\theta\right)^2 + F$

$\qquad = \left(A\cos^2\theta + B\sin\theta\cos\theta + C\sin^2\theta\right)\left(x'\right)^2 + \left[B\left(\cos^2\theta - \sin^2\theta\right) + 2\left(C - A\right)\sin\theta\cos\theta\right]x'y'$

$\qquad\qquad\qquad\qquad\qquad + \left(A\sin^2\theta - B\sin\theta\cos\theta + C\cos^2\theta\right)\left(y'\right)^2 + F$

$\qquad = 0$

Let us choose θ such that the cross term $x'y' = 0$. To do this, we set $B\left(\cos^2\theta - \sin^2\theta\right) + 2\left(C - A\right)\sin\theta\cos\theta = 0$

$\Leftrightarrow B\cos 2\theta + (C - A)\sin 2\theta = 0 \Leftrightarrow \cot 2\theta = \frac{A - C}{B}$. Finally, writing $A' = A\cos^2\theta + B\sin\theta\cos\theta + C\sin^2\theta$ and

$C' = A\sin^2\theta - B\sin\theta\cos\theta + C\cos^2\theta$, we can write the given equation in the form $A'\left(x'\right)^2 + C'\left(y'\right)^2 + F = 0$.

c. $2x^2 + \sqrt{3}xy + y^2 - 20 = 0$. Here $A = 2$, $B = \sqrt{3}$, $C = 1$, and

$F = -20$, so $\cot 2\theta = \frac{2-1}{\sqrt{3}} = \frac{\sqrt{3}}{3} \Rightarrow 2\theta = \cot^{-1}\frac{\sqrt{3}}{3} = \frac{\pi}{6}$.

Thus, $A' = 2\cos^2\frac{\pi}{6} + \sqrt{3}\sin\frac{\pi}{6}\cos\frac{\pi}{6} + \sin^2\frac{\pi}{6} = \frac{5}{2}$ and

$C' = 2\sin^2\frac{\pi}{6} - \sqrt{3}\sin\frac{\pi}{6}\cos\frac{\pi}{6} + \cos^2\frac{\pi}{6} = \frac{1}{2}$, so in the $x'y'$-plane,

an equation is $\frac{5}{2}\left(x'\right)^2 + \frac{1}{2}\left(y'\right)^2 - 20 = 0 \Leftrightarrow \frac{\left(x'\right)^2}{8} + \frac{\left(y'\right)^2}{40} = 1 \Leftrightarrow$

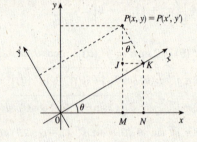

$\frac{\left(x'\right)^2}{\left(2\sqrt{2}\right)^2} + \frac{\left(y'\right)^2}{\left(2\sqrt{10}\right)^2} = 1$.

3. $x = \int_1^t \frac{\cos u}{u}\, du$ and $y = \int_1^t \frac{\sin u}{u}\, du$. $\frac{dx}{dt} = \frac{\cos t}{t}$ and $\frac{dy}{dt} = \frac{\sin t}{t}$. At the point where

the tangent is vertical, $\frac{dx}{dt} = 0$ and $\frac{dy}{dt} \neq 0$. We see that this is the case when $t = \frac{\pi}{2}$.

Also, the origin corresponds to $t = 1$, since $\int_1^1 \frac{\cos u}{u}\, du = 0 = \int_1^1 \frac{\sin u}{u}\, du$. Therefore,

$$L = \int_1^{\pi/2} \sqrt{\left(\frac{dx}{dt}\right)^2 + \left(\frac{dy}{dt}\right)^2}\, dt = \int_1^{\pi/2} \sqrt{\left(\frac{\cos t}{t}\right)^2 + \left(\frac{\sin t}{t}\right)^2}\, dt = \int_1^{\pi/2} \frac{dt}{t} = \ln t\big|_1^{\pi/2} = \ln \frac{\pi}{2}.$$

5. $x^3 + y^3 = 3axy$

a. $x = r\cos\theta$ and $y = r\sin\theta$, so we have $(r\cos\theta)^3 + (r\sin\theta)^3 = 3a(r\cos\theta)(r\sin\theta) \Leftrightarrow$

$$r^3\left(\cos^3\theta + \sin^3\theta\right) = 3ar^2\cos\theta\sin\theta \Leftrightarrow r = \frac{3a\cos\theta\sin\theta}{\cos^3\theta + \sin^3\theta} = \frac{\dfrac{3a\cos\theta\sin\theta}{\cos^3\theta}}{\dfrac{\cos^3\theta + \sin^3\theta}{\cos^3\theta}} = \frac{3a\sec\theta\tan\theta}{1 + \tan^3\theta}.$$

b. $A = \int_0^{\pi/2} \frac{r^2}{2}\, d\theta = \frac{1}{2}\int_0^{\pi/2} \left(\frac{3a\sec\theta\tan\theta}{1 + \tan^3\theta}\right)^2 d\theta = \frac{9a^2}{2}\int_0^{\pi/2} \frac{\sec^2\theta\tan^2\theta}{(1 + \tan^3\theta)^2}\, d\theta$. Let $u = 1 + \tan^3\theta$.

Then $du = 3\tan^2\theta\sec^2\theta\, d\theta$, so $\int \frac{\sec^2\theta\tan^2\theta}{(1 + \tan^3\theta)^2}\, d\theta = \frac{1}{3}\int \frac{du}{u^2} = -\frac{1}{3u} + C$. Therefore,

$$A = -\frac{3a^2}{2}\left(\frac{1}{1 + \tan^3\theta}\right)\Bigg|_0^{\pi/2} = \frac{3a^2}{2}.$$

7. a. Let us find a polar equation describing the path taken by the ant that starts
out at the northeast corner. Let $P\,(r, \theta)$ denote the position of that ant and
let Q denote the position of the ant toward which it is crawling. Since the
ants are the same distance from the pole, $Q = \left(r, \theta + \frac{\pi}{2}\right)$. In terms of
rectangular coordinates, $P = (r\cos\theta, r\sin\theta)$ and
$Q = \left(r\cos\left(\theta + \frac{\pi}{2}\right), r\sin\left(\theta + \frac{\pi}{2}\right)\right) = (-r\sin\theta, r\cos\theta)$. So the slope
of the line passing through P and Q is

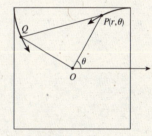

$\dfrac{r\cos\theta - r\sin\theta}{-r\sin\theta - r\cos\theta} = \dfrac{\sin\theta - \cos\theta}{\sin\theta + \cos\theta}$. But this slope must be equal to the tangent line to the

curve at P, which is given by $\dfrac{r'\sin\theta + r\cos\theta}{r'\cos\theta - r\sin\theta}$. Thus, $\dfrac{r'\sin\theta + r\cos\theta}{r'\cos\theta - r\sin\theta} = \dfrac{\sin\theta - \cos\theta}{\sin\theta + \cos\theta} \Leftrightarrow$

$r'\sin^2\theta + r'\sin\theta\cos\theta + r\cos\theta\sin\theta + r\cos^2\theta = r'\cos\theta\sin\theta - r'\cos^2\theta - r\sin^2\theta + r\cos\theta\sin\theta \Leftrightarrow$

$r'\left(\cos^2\theta + \sin^2\theta\right) + r\left(\cos^2\theta + \sin^2\theta\right) = 0 \Leftrightarrow r' + r = 0$. The equation $\dfrac{dr}{d\theta} = -r$ is separable, and we find

$\displaystyle\int \frac{dr}{r} = -\int d\theta \Rightarrow \ln r = -\theta + C_1 \Rightarrow r = Ce^{-\theta}$, where $C = \pm e^{C_1}$. To find C, we note that $r\left(\frac{\pi}{4}\right) = \frac{\sqrt{2}}{2}a$ (remember

that the ant starts at the northeast corner), giving $r\left(\frac{\pi}{4}\right) = Ce^{-\pi/4} = \frac{\sqrt{2}}{2}a \Rightarrow C = \frac{\sqrt{2}}{2}ae^{\pi/4}$. Therefore, the required

equation is $r = \frac{\sqrt{2}}{2}ae^{\pi/4}e^{-\theta} = \frac{\sqrt{2}}{2}ae^{(\pi/4)-\theta}$.

b. $r' = -\frac{\sqrt{2}}{2}ae^{(\pi/4)-\theta}$, so $r^2 + (r')^2 = \frac{1}{2}a^2 e^{(\pi/2)-2\theta} + \frac{1}{2}a^2 e^{(\pi/2)-2\theta} = a^2 e^{(\pi/2)-2\theta}$. Therefore, the distance
traveled by the ant is

$L = \int_{\pi/4}^{\infty} \sqrt{r^2 + (r')^2}\, d\theta = \int_{\pi/4}^{\infty} ae^{(\pi/4)-\theta}\, d\theta = ae^{\pi/4}\lim_{b\to\infty}\int_{\pi/4}^{b} e^{-\theta}\, d\theta = ae^{\pi/4}\lim_{b\to\infty}\left(-e^{-\theta}\right)\Big|_{\pi/4}^{b}$

$= ae^{\pi/4}\lim_{b\to\infty}\left(-e^{-b} + e^{-\pi/4}\right) = ae^{\pi/4}e^{-\pi/4} = a$

11 Vectors and the Geometry of Space   ET 10

11.1 Concept Questions   ET 10.1

1. **a.** See page 902 (898 in ET). Force and velocity are vectors.

   b. See page 902 (898 in ET).

   c. See page 903 (899 in ET).

   d. See page 903 (899 in ET).

3. See page 905 (901 in ET).

11.1 Vectors in the Plane   ET 10.1

1. **a.** The amount of water in a swimming pool is a scalar. It has no direction.

   b. It is a vector because it has both magnitude and direction.

   c. It is a scalar because it has magnitude but no direction.

   d. It is a vector because it has both magnitude and direction.

3. **a.** No. A vector and a scalar cannot be added.

   b. Yes. Since $|\mathbf{a}|$ is a scalar, $|\mathbf{a}|\,\mathbf{b}$ is obtained by taking the scalar product of the vector $\mathbf{b}$ and the scalar $|\mathbf{a}|$.

   c. No. $2\mathbf{a} + \mathbf{b}$ is a vector and 0 is a scalar, and the two cannot be added.

   d. No. Division of one vector by another is not defined.

   e. Yes. $\dfrac{1}{|\mathbf{b}|}$ is a scalar, so $\dfrac{\mathbf{a}}{|\mathbf{b}|}$ is obtained by taking the scalar product of the vector $\mathbf{a}$ and the scalar $\dfrac{1}{|\mathbf{b}|}$.

   f. Yes. $\dfrac{|\mathbf{a}|\,\mathbf{b} - |\mathbf{b}|\,\mathbf{a}}{|\mathbf{a}|^2} = \left(\dfrac{1}{|\mathbf{a}|}\right)\mathbf{b} - \left(\dfrac{|\mathbf{b}|}{|\mathbf{a}|^2}\right)\mathbf{a}.$

5. $\mathbf{a} = \langle -2 - 3, 4 - 1 \rangle = \langle -5, 3 \rangle = \mathbf{b}$

7. $\mathbf{a} = \langle 2 - (-2), -3 - 1 \rangle = \langle 4, -4 \rangle = \mathbf{b}$

9. $\overrightarrow{AB} = \langle -3 - 2, 3 - 3 \rangle = \langle -5, 0 \rangle$

11. $\overrightarrow{AB} = \langle 0 - (-3), 0 - 4 \rangle = \langle 3, -4 \rangle$

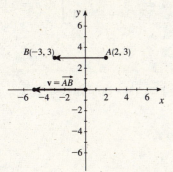

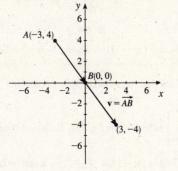

13. $2\mathbf{a} - 3\mathbf{b}$

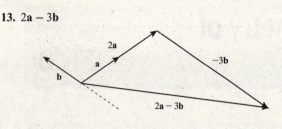

15. $(\mathbf{a} + 2\mathbf{b}) + \mathbf{c}$

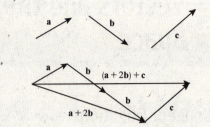

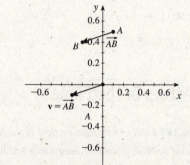

17. $\mathbf{b} + \mathbf{a} = -\mathbf{v}$ or $\mathbf{v} = -(\mathbf{a} + \mathbf{b})$.

19. $\frac{1}{3}\mathbf{v} + \mathbf{a} = \mathbf{b}$ or $\mathbf{v} = 3(\mathbf{b} - \mathbf{a})$

21. $\overrightarrow{AB} = \langle 1 - 3, 3 - 4 \rangle = \langle -2, -1 \rangle$

23. $\overrightarrow{AB} = \langle -0.2 - 0.1, 0.4 - 0.5 \rangle = \langle -0.3, -0.1 \rangle$

25. Let $A(a_1, a_2)$. Then $\overrightarrow{AB} = \langle 0 - a_1, 2 - a_2 \rangle = \langle -a_1, 2 - a_2 \rangle = \langle -1, 4 \rangle \Rightarrow -a_1 = -1$ and $2 - a_2 = 4$, so $a_1 = 1$ and $a_2 = -2$. Thus, the point is $A(1, -2)$.

27. $2\mathbf{a} = 2\langle -1, 2 \rangle = \langle -2, 4 \rangle$, $\mathbf{a} + \mathbf{b} = \langle -1, 2 \rangle + \langle 3, 1 \rangle = \langle 2, 3 \rangle$, $\mathbf{a} - \mathbf{b} = \langle -1, 2 \rangle - \langle 3, 1 \rangle = \langle -4, 1 \rangle$,
$2\mathbf{a} + \mathbf{b} = 2\langle -1, 2 \rangle + \langle 3, 1 \rangle = \langle 1, 5 \rangle$, and $|2\mathbf{a} + \mathbf{b}| = |\langle 1, 5 \rangle| = \sqrt{1^2 + 5^2} = \sqrt{26}$.

29. $2\mathbf{a} = 2(3\mathbf{i} - 2\mathbf{j}) = 6\mathbf{i} - 4\mathbf{j}$, $\mathbf{a} + \mathbf{b} = (3\mathbf{i} - 2\mathbf{j}) + 2\mathbf{i} = 5\mathbf{i} - 2\mathbf{j}$, $\mathbf{a} - \mathbf{b} = (3\mathbf{i} - 2\mathbf{j}) - 2\mathbf{i} = \mathbf{i} - 2\mathbf{j}$,
$2\mathbf{a} + \mathbf{b} = 2(3\mathbf{i} - 2\mathbf{j}) + 2\mathbf{i} = 8\mathbf{i} - 4\mathbf{j}$, and $|2\mathbf{a} + \mathbf{b}| = |8\mathbf{i} - 4\mathbf{j}| = \sqrt{8^2 + 4^2} = 4\sqrt{5}$.

31. $2\mathbf{a} = 2\left(\frac{1}{2}\mathbf{i} + \frac{3}{2}\mathbf{j}\right) = \mathbf{i} + 3\mathbf{j}$, $\mathbf{a} + \mathbf{b} = \left(\frac{1}{2}\mathbf{i} + \frac{3}{2}\mathbf{j}\right) + \left(\frac{3}{4}\mathbf{i} - \frac{1}{4}\mathbf{j}\right) = \frac{5}{4}\mathbf{i} + \frac{5}{4}\mathbf{j}$, $\mathbf{a} - \mathbf{b} = \left(\frac{1}{2}\mathbf{i} + \frac{3}{2}\mathbf{j}\right) - \left(\frac{3}{4}\mathbf{i} - \frac{1}{4}\mathbf{j}\right) = -\frac{1}{4}\mathbf{i} + \frac{7}{4}\mathbf{j}$,
$2\mathbf{a} + \mathbf{b} = 2\left(\frac{1}{2}\mathbf{i} + \frac{3}{2}\mathbf{j}\right) + \left(\frac{3}{4}\mathbf{i} - \frac{1}{4}\mathbf{j}\right) = \frac{7}{4}\mathbf{i} + \frac{11}{4}\mathbf{j}$, and $|2\mathbf{a} + \mathbf{b}| = \left|\frac{7}{4}\mathbf{i} + \frac{11}{4}\mathbf{j}\right| = \sqrt{\left(\frac{7}{4}\right)^2 + \left(\frac{11}{4}\right)^2} = \frac{\sqrt{170}}{4}$.

33. $2\mathbf{a} - 3\mathbf{b} = 2(2\mathbf{i}) - 3(-6\mathbf{j}) = 4\mathbf{i} + 18\mathbf{j}$, $\frac{1}{2}\mathbf{a} + \frac{1}{3}\mathbf{b} = \frac{1}{2}(2\mathbf{i}) + \frac{1}{3}(-6\mathbf{j}) = \mathbf{i} - 2\mathbf{j}$

35. $a\mathbf{u} + b\mathbf{v} = \mathbf{w} \Rightarrow a\langle -1, 3 \rangle + b\langle 2, 4 \rangle = \langle 6, 4 \rangle \Leftrightarrow \langle -a + 2b, 3a + 4b \rangle = \langle 6, 4 \rangle \Leftrightarrow \left.\begin{array}{r} -a + 2b = 6 \\ 3a + 4b = 4 \end{array}\right\} \Leftrightarrow a = -1.6$ and
$b = 2.2$.

37. $\mathbf{b} = 9\mathbf{i} - 6\mathbf{j} = 3(3\mathbf{i} - 2\mathbf{j}) = 3\mathbf{a}$ and so $\mathbf{b}$ is parallel to $\mathbf{a}$.

39. If $\mathbf{b}$ is parallel to $\mathbf{a}$, then there exists a constant c such that $\mathbf{b} = c\mathbf{a}$; that is, $\mathbf{i} - \mathbf{j} = c(3\mathbf{i} - 2\mathbf{j}) = 3c\mathbf{i} - 2c\mathbf{j} \Leftrightarrow \left.\begin{array}{r} 3c = 1 \\ -2c = -1 \end{array}\right\}$
$\Rightarrow \left.\begin{array}{l} c = \frac{1}{3} \\ c = \frac{1}{2} \end{array}\right\}$ which is impossible. So no such c exists, and $\mathbf{b}$ is not parallel to $\mathbf{a}$.

41. $\mathbf{b} = \langle -1, 2 \rangle + 4\langle 1, -1 \rangle = \langle 3, -2 \rangle = \mathbf{a}$ and so $\mathbf{b}$ is parallel to $\mathbf{a}$.

43. Since **a** and **b** are parallel, there exists a constant k such that $\mathbf{b} = k\mathbf{a}$; that is, $k(c\mathbf{i} - 2\mathbf{j}) = -4\mathbf{i} + 3\mathbf{j}$. This implies that

$$\left.\begin{array}{r} kc = -4 \\ -2k = 3 \end{array}\right\} \Leftrightarrow k = -\tfrac{3}{2} \text{ and } c = \tfrac{8}{3}.$$

45. a. The required vector is $\mathbf{u} = \dfrac{\mathbf{a}}{|\mathbf{a}|} = \dfrac{\langle 2, 1\rangle}{\sqrt{2^2 + 1^2}} = \left\langle \dfrac{2\sqrt{5}}{5}, \dfrac{\sqrt{5}}{5} \right\rangle.$

 b. The required vector is $\mathbf{v} = -\mathbf{u} = \left\langle -\dfrac{2\sqrt{5}}{5}, -\dfrac{\sqrt{5}}{5} \right\rangle.$

47. a. The required vector is $\mathbf{u} = \dfrac{\mathbf{a}}{|\mathbf{a}|} = \dfrac{\langle -\sqrt{3}, 1\rangle}{\sqrt{\left(-\sqrt{3}\right)^2 + 1}} = \left\langle -\dfrac{\sqrt{3}}{2}, \dfrac{1}{2} \right\rangle.$

 b. The required vector is $\mathbf{v} = -\mathbf{u} = \left\langle \dfrac{\sqrt{3}}{2}, -\dfrac{1}{2} \right\rangle.$

49. A unit vector in the direction of **b** is $\mathbf{u} = \dfrac{\mathbf{b}}{|\mathbf{b}|} = \left\langle \dfrac{\sqrt{2}}{2}, \dfrac{\sqrt{2}}{2} \right\rangle$, and so $\mathbf{a} = 5\mathbf{u} = \left\langle \dfrac{5\sqrt{2}}{2}, \dfrac{5\sqrt{2}}{2} \right\rangle.$

51. A unit vector in the direction of **b** is $\mathbf{u} = \dfrac{\mathbf{b}}{|\mathbf{b}|} = \dfrac{3\mathbf{i}}{3} = \mathbf{i}$, and so $\mathbf{a} = \sqrt{3}\mathbf{u} = \sqrt{3}\mathbf{i}.$

53. $2\mathbf{a} - 3\mathbf{b} = 2\langle -3, 4\rangle - 3\langle 1, 2\rangle = \langle -9, 2\rangle$ and a unit vector in the direction of this vector is

$\mathbf{u} = \dfrac{2\mathbf{a} - 3\mathbf{b}}{|2\mathbf{a} - 3\mathbf{b}|} = \dfrac{\langle -9, 2\rangle}{\sqrt{(-9)^2 + 2^2}} = \left\langle -\dfrac{9\sqrt{85}}{85}, \dfrac{2\sqrt{85}}{85} \right\rangle$. Thus, the required vector is $3\mathbf{u} = \left\langle -\dfrac{27\sqrt{85}}{85}, \dfrac{6\sqrt{85}}{85} \right\rangle.$

55. $|\mathbf{F}| = \sqrt{3^2 + 1^2} = \sqrt{10}$. A unit vector in the same direction as **F** is $\mathbf{u} = \dfrac{\mathbf{F}}{|\mathbf{F}|} = \dfrac{\langle 3, 1\rangle}{\sqrt{10}} = \left\langle \dfrac{3\sqrt{10}}{10}, \dfrac{\sqrt{10}}{10} \right\rangle$, so

$\mathbf{F} = |\mathbf{F}|\,\mathbf{u} = \sqrt{10}\left\langle \dfrac{3\sqrt{10}}{10}, \dfrac{\sqrt{10}}{10} \right\rangle.$ The horizontal and vertical components of **F** are $F_1 = 3$ and $F_2 = 1$, respectively.

57. $|\mathbf{F}| = \sqrt{\left(\sqrt{3}\right)^2 + 6^2} = \sqrt{39}$. A unit vector in the same direction as **F** is $\mathbf{u} = \dfrac{\mathbf{F}}{|\mathbf{F}|} = \dfrac{\sqrt{39}}{39}\left(\sqrt{3}\mathbf{i} + 6\mathbf{j}\right) = \dfrac{\sqrt{13}}{13}\mathbf{i} + \dfrac{2\sqrt{39}}{13}\mathbf{j}$, so

$\mathbf{F} = |\mathbf{F}|\,\mathbf{u} = \sqrt{39}\left(\dfrac{\sqrt{13}}{13}\mathbf{i} + \dfrac{2\sqrt{39}}{13}\mathbf{j}\right).$ The horizontal and vertical components of **F** are $F_1 = \sqrt{3}$ and $F_2 = 6$, respectively.

59. $\mathbf{a} = |\mathbf{a}|\cos\theta\,\mathbf{i} + |\mathbf{a}|\sin\theta\,\mathbf{j} = 2(\cos 0)\mathbf{i} + 2(\sin 0)\mathbf{j} = 2\mathbf{i}$

61. $\mathbf{a} = |\mathbf{a}|\cos\theta\,\mathbf{i} + |\mathbf{a}|\sin\theta\,\mathbf{j} = 3\cos\dfrac{5\pi}{3}\mathbf{i} + 3\sin\dfrac{5\pi}{3}\mathbf{j} = \dfrac{3}{2}\mathbf{i} - \dfrac{3\sqrt{3}}{2}\mathbf{j}$

63. a. $\mathbf{v}_1 + \mathbf{v}_2 = \langle a_1, b_1\rangle + \langle a_2, b_2\rangle = \langle a_1 + a_2, b_1 + b_2\rangle$. The company produced $a_1 + a_2$ model A systems and $b_1 + b_2$ model B systems last year.

 b. The required vector is $\mathbf{v} = 1.1\langle a_1 + a_2, b_1 + b_2\rangle = \langle 1.1(a_1 + a_2), 1.1(b_1 + b_2)\rangle.$

65. The horizontal component of **v** is $\mathbf{v}_1 = (|\mathbf{v}|\cos 45°)\mathbf{i} = 800\left(\dfrac{\sqrt{2}}{2}\right)\mathbf{i} = 400\sqrt{2}\mathbf{i}$ ft/s and its vertical component is

$\mathbf{v}_2 = (|\mathbf{v}|\sin 45°)\mathbf{j} = 800\left(\dfrac{\sqrt{2}}{2}\right)\mathbf{j} = 400\sqrt{2}\mathbf{j}$ ft/s.

67. $\mathbf{F} = \mathbf{F}_1 + \mathbf{F}_2 + \mathbf{F}_3 = (30\cos 45°\,\mathbf{i} + 30\sin 45°\,\mathbf{j}) + (-15\cos 30°\,\mathbf{i} + 15\sin 30°\,\mathbf{j}) - (|\mathbf{F}_3|\cos\theta\,\mathbf{i} + |\mathbf{F}_3|\sin\theta\,\mathbf{j})$

$= \left[30\left(\dfrac{\sqrt{2}}{2}\right) - 15\left(\dfrac{\sqrt{3}}{2}\right) - |\mathbf{F}_3|\cos\theta\right]\mathbf{i} + \left[30\left(\dfrac{\sqrt{2}}{2}\right) + 15\left(\dfrac{1}{2}\right) - |\mathbf{F}_3|\sin\theta\right]\mathbf{j}$

If the object is in equilibrium, then $\mathbf{F} = 0$, so $\left.\begin{array}{l} 15\sqrt{2} - \dfrac{15\sqrt{3}}{2} - |\mathbf{F}_3|\cos\theta = 0 \\ 15\sqrt{2} + \dfrac{15}{2} - |\mathbf{F}_3|\sin\theta = 0 \end{array}\right\} \Leftrightarrow \left.\begin{array}{l} |\mathbf{F}_3|\cos\theta = \dfrac{30\sqrt{2} - 15\sqrt{3}}{2} \\ |\mathbf{F}_3|\sin\theta = \dfrac{30\sqrt{2} + 15}{2} \end{array}\right\}$

Therefore, $\tan\theta = \dfrac{30\sqrt{2} + 15}{30\sqrt{2} - 15\sqrt{3}} \Rightarrow \theta \approx 74°$. Also, $|\mathbf{F}_3| \approx \dfrac{30\sqrt{2} + 15}{2} \cdot \dfrac{1}{\sin 74°} \approx 30$ lb.

69. Let $\mathbf{v} = a\mathbf{j}$ be the required velocity of the boat and let $\mathbf{w} = 5\mathbf{i}$ be the

velocity of the river. Referring to the figure, we see that

$(-10\cos\theta)\,\mathbf{i} + (10\sin\theta)\,\mathbf{j} + 5\mathbf{i} = a\mathbf{j}$. This implies that $-10\cos\theta + 5 = 0$

and $a = 10\sin\theta$. We find $\cos\theta = \frac{1}{2} \Rightarrow \theta = 60°$, so

$a = 10\left(\frac{\sqrt{3}}{2}\right) = 5\sqrt{3}$. Thus, it takes $\frac{1/2}{5\sqrt{3}} = \frac{\sqrt{3}}{30}$ hr or about 3.5 minutes to

cross the river.

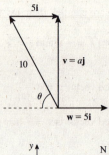
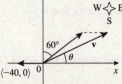

71. Let $|\mathbf{v}|$ denote the airspeed of the airplane. Then $\mathbf{v} = |\mathbf{v}|\,\langle\cos\theta, \sin\theta\rangle$.

From the figure, we see that

$\langle|\mathbf{v}|\cos\theta, |\mathbf{v}|\sin\theta\rangle + \langle-40, 0\rangle = \langle240\sin 60°, 240\cos 60°\rangle$, so

$|\mathbf{v}|\cos\theta - 40 = 120\sqrt{3}$ and $|\mathbf{v}|\sin\theta = 120 \Rightarrow \tan\theta = \frac{120}{120\sqrt{3}+40} \Rightarrow$

$\theta \approx 25.8°$, and $|\mathbf{v}| \approx 275.4$. Thus, the airspeed is approximately

275.4 mi/h and the heading is E 25.8° N.

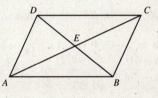

73. Let E be the point of intersection of the diagonals AC and BD. Now

$\overrightarrow{AC} = \overrightarrow{AB} + \overrightarrow{BC} = \overrightarrow{AB} + \overrightarrow{AD}$ and $\overrightarrow{DB} = \overrightarrow{AB} - \overrightarrow{AD}$, so

$\overrightarrow{AE} = a\overrightarrow{AC} = a\left(\overrightarrow{AB} + \overrightarrow{AD}\right)$ for some constant a and

$\overrightarrow{DE} = b\overrightarrow{DB} = b\left(\overrightarrow{AB} - \overrightarrow{AD}\right)$ for some constant b. But then

$\overrightarrow{AD} = \overrightarrow{AE} - \overrightarrow{DE} = a\left(\overrightarrow{AB} + \overrightarrow{AD}\right) - b\left(\overrightarrow{AB} - \overrightarrow{AD}\right) = (a-b)\,\overrightarrow{AB} + (a+b)\,\overrightarrow{AD}$. Therefore, $\left.\begin{array}{c} a - b = 0 \\ a + b = 1 \end{array}\right\} \Rightarrow$

$a = b = \frac{1}{2}$, and the desired result follows.

75. $(\mathbf{a} + \mathbf{b}) + \mathbf{c} = (\langle a_1, a_2\rangle + \langle b_1, b_2\rangle) + \langle c_1, c_2\rangle = \langle a_1 + b_1, a_2 + b_2\rangle + \langle c_1, c_2\rangle = \langle(a_1 + b_1) + c_1, (a_2 + b_2) + c_2\rangle$

$\qquad = \langle a_1 + b_1 + c_1, a_2 + b_2 + c_2\rangle$

and

$\mathbf{a} + (\mathbf{b} + \mathbf{c}) = \langle a_1, a_2\rangle + (\langle b_1, b_2\rangle + \langle c_1, c_2\rangle) = \langle a_1, a_2\rangle + \langle b_1 + c_1, b_2 + c_2\rangle = \langle a_1 + (b_1 + c_1), a_2 + (b_2 + c_2)\rangle$

$\qquad = \langle a_1 + b_1 + c_1, a_2 + b_2 + c_2\rangle$

so $(\mathbf{a} + \mathbf{b}) + \mathbf{c} = \mathbf{a} + (\mathbf{b} + \mathbf{c})$.

77. $\mathbf{a} + (-\mathbf{a}) = \langle a_1, a_2\rangle + (-\langle a_1, a_2\rangle) = \langle a_1, a_2\rangle - \langle a_1, a_2\rangle = \langle a_1 - a_1, a_2 - a_2\rangle = \langle 0, 0\rangle = \mathbf{0}$

79. $c\,(d\mathbf{a}) = c\,(d\,\langle a_1, a_2\rangle) = c\,\langle da_1, da_2\rangle = \langle cda_1, cda_2\rangle$ and $(cd)\,\mathbf{a} = cd\,\langle a_1, a_2\rangle = \langle cda_1, cda_2\rangle$, so $c\,(d\mathbf{a}) = (cd)\,\mathbf{a}$.

81. $1\mathbf{a} = 1\,\langle a_1, a_2\rangle = \langle 1 \cdot a_1, 1 \cdot a_2\rangle = \langle a_1, a_2\rangle = \mathbf{a}$

83. False. $\mathbf{v} - \mathbf{v} = \mathbf{0}$, the zero vector.

85. True. $\mathbf{v} = \dfrac{\mathbf{v}}{|\mathbf{v}|}\,|\mathbf{v}| = |\mathbf{v}|\,\mathbf{u}$ provided $\mathbf{v} \neq \mathbf{0}$.

87. False. $\mathbf{u}$ and $-\mathbf{u}$ point in opposite directions.

11.2 Concept Questions   ET 10.2

1. a. See page 914 (910 in ET).

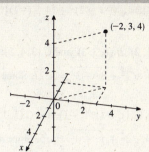

  b. $d\left(P_1, P_2\right) = \sqrt{\left(a_2 - a_1\right)^2 + \left(b_2 - b_1\right)^2 + \left(c_2 - c_1\right)^2}$

  c. $\left(\dfrac{a_1 + a_2}{2}, \dfrac{b_1 + b_2}{2}, \dfrac{c_1 + c_2}{2}\right)$

3. a. $\mathbf{a} + \mathbf{b} = \langle a_1, a_2, a_3 \rangle + \langle b_1, b_2, b_3 \rangle = \langle a_1 + b_1, a_2 + b_2, a_3 + c_3 \rangle$, $c\mathbf{a} = c\langle a_1, a_2, a_3 \rangle = \langle ca_1, ca_2, ca_3 \rangle$, and

  $\|\mathbf{a}\| = \sqrt{a_1^2 + a_2^2 + a_3^2}$.

  b. $\overrightarrow{P_1 P_2} = \langle b_1 - a_1, b_2 - a_2, b_3 - a_3 \rangle$ and $\overrightarrow{P_2 P_1} = \langle a_1 - b_1, a_2 - b_2, a_3 - b_3 \rangle$, so $\overrightarrow{P_1 P_2} = -\overrightarrow{P_2 P_1}$.

11.2 Coordinate Systems and Vectors in Three-Space   ET 10.2

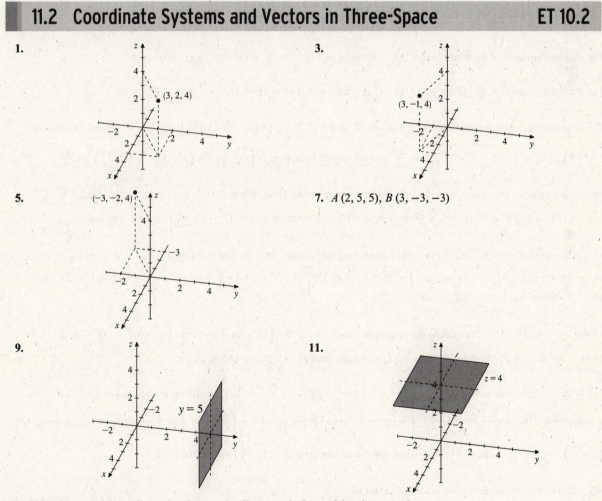

1.

3.

5.

7. $A\,(2, 5, 5)$, $B\,(3, -3, -3)$

9.

11.

13. The subspace of three-dimensional space that lies on or in front of the plane $x = 3$

15. The subspace of three-dimensional space that lies strictly above the plane $z = 3$

17. a. $A\,(2,-1,3)$ and $B\,(-1,4,1)$, in units of 1000 ft.

b. $d\,(A,B) = \sqrt{(-1-2)^2 + [4-(-1)]^2 + (1-3)^2} = \sqrt{38}$ or approximately 6164 feet.

19. For $A\,(0,1,2)$, $B\,(4,3,3)$, and $C\,(3,4,5)$: $d\,(A,B) = \sqrt{4^2 + 2^2 + 1^2} = \sqrt{21}$, $d\,(B,C) = \sqrt{(-1)^2 + 1^2 + 2^2} = \sqrt{6}$, and $d\,(A,C) = \sqrt{3^2 + 3^2 + 3^2} = \sqrt{27}$. Since $d^2\,(A,C) = d^2\,(B,C) + d^2\,(A,B)$, the triangle is a right triangle.

21. For $A\,(-1,0,1)$, $B\,(1,1,-1)$, and $C\,(1,1,1)$: $d\,(A,B) = \sqrt{2^2 + 1^2 + (-2)^2} = 3$, $d\,(B,C) = \sqrt{0^2 + 0^2 + 2^2} = 2$, and $d\,(A,C) = \sqrt{2^2 + 1^2 + 0^2} = \sqrt{5}$. Since $d^2\,(A,B) = d^2\,(B,C) + d^2\,(A,C)$, the triangle is a right triangle.

23. For $A\,(2,3,2)$, $B\,(-4,0,5)$, and $C\,(4,4,1)$: $d\,(A,B) = \sqrt{(-6)^2 + (-3)^2 + 3^2} = \sqrt{54} = 3\sqrt{6}$, $d\,(B,C) = \sqrt{8^2 + 4^2 + (-4)^2} = \sqrt{96} = 4\sqrt{6}$, and $d\,(A,C) = \sqrt{2^2 + 1^2 + (-1)^2} = \sqrt{6}$. Since $d\,(A,B) + d\,(A,C) = d\,(B,C)$, the three points are collinear.

25. For $A\,(2,4,-6)$ and $B\,(-4,2,4)$, the midpoint is $\left(\dfrac{2-4}{2}, \dfrac{4+2}{2}, \dfrac{-6+4}{2}\right) = (-1,3,-1)$.

27. The standard equation of the sphere with $C\,(2,1,3)$ and $r = 3$ is $(x-2)^2 + (y-1)^2 + (z-3)^2 = 9$.

29. The standard equation of the sphere with $C\,(3,-1,2)$ and $r = 4$ is $(x-3)^2 + (y+1)^2 + (z-2)^2 = 16$.

31. The center of the sphere is the midpoint of $A\,(2,-3,4)$ and $B\,(3,2,1)$: $\left(\frac{2+3}{2}, \frac{-3+2}{2}, \frac{4+1}{2}\right) = \left(\frac{5}{2}, -\frac{1}{2}, \frac{5}{2}\right)$ and its radius is $r = \frac{1}{2}d\,(A,B) = \frac{1}{2}\sqrt{1^2 + 5^2 + (-3)^2} = \frac{\sqrt{35}}{2}$, so the required equation is $\left(x - \frac{5}{2}\right)^2 + \left(y + \frac{1}{2}\right)^2 + \left(z - \frac{5}{2}\right)^2 = \frac{35}{4}$.

33. Denote the point on the sphere by $A\,(1,3,5)$. Then its radius is $d\,(A,C)$, where $C\,(-1,2,4)$ is its center. Thus, $r = \sqrt{(-2)^2 + (-1)^2 + (-1)^2} = \sqrt{6}$. Therefore, the required equation is $(x+1)^2 + (y-2)^2 + (z-4)^2 = 6$.

35. $x^2 + y^2 + z^2 - 2x - 4y - 6z + 10 = 0$. Completing the squares in x, y, and z, we obtain $\left[x^2 - 2x + (-1)^2\right] + \left[y^2 - 4y + (-2)^2\right] + \left[z^2 - 6z + (-3)^2\right] = -10 + 1 + 4 + 9 \Leftrightarrow (x-1)^2 + (y-2)^2 + (z-3)^2 = 4$. Thus, the sphere has center $(1,2,3)$ and radius 2.

37. $x^2 + y^2 + z^2 - 4x + 6y = 0$. Completing the squares in x and y, we obtain $\left[x^2 - 4x + (-2)^2\right] + \left[y^2 + 6y + (3)^2\right] + z^2 = 4 + 9$ $\Leftrightarrow (x-2)^2 + (y+3)^2 + z^2 = 13$. Thus, the sphere has center $(2,-3,0)$ and radius $\sqrt{13}$.

39. $2x^2 + 2y^2 + 2z^2 - 6x - 4y + 2z = 1 \Leftrightarrow x^2 + y^2 + z^2 - 3x - 2y + z = \frac{1}{2}$. Completing the squares in x, y, and z, we obtain $\left[x^2 - 3x + \left(-\frac{3}{2}\right)^2\right] + \left[y^2 - 2y + (-1)^2\right] + \left[z^2 + z + \left(\frac{1}{2}\right)^2\right] = \frac{1}{2} + \frac{9}{4} + 1 + \frac{1}{4} \Leftrightarrow$ $\left(x - \frac{3}{2}\right)^2 + (y-1)^2 + \left(z + \frac{1}{2}\right)^2 = 4$. Thus, the sphere has center $\left(\frac{3}{2}, 1, -\frac{1}{2}\right)$ and radius 2.

41. All points inside the sphere with radius 2 and center $(0,0,0)$.

43. All points lying on or between two concentric spheres with radii 1 and 3 and center $(0,0,0)$

45. For $A\,(3, 2, 1)$ and $B\,(1, 4, 5)$,

$$\overrightarrow{AB} = \langle 1 - 3, 4 - 2, 5 - 1 \rangle = \langle -2, 2, 4 \rangle.$$

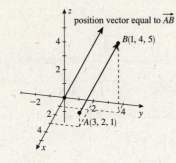

47. For $A\,(-2, 3, 0)$ and $B\,(2, -2, 5)$,

$$\overrightarrow{AB} = \langle 2 - (-2), -2 - 3, 5 - 0 \rangle = \langle 4, -5, 5 \rangle.$$

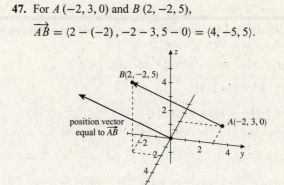

49. For $A\,(2, 1, 0)$ and $B\,(1, 4, 5)$,

$$\overrightarrow{AB} = \langle 1 - 2, 4 - 1, 5 - 0 \rangle = \langle -1, 3, 5 \rangle.$$

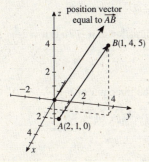

51. $\overrightarrow{AB} = \langle -1, 3, 4 \rangle$ and $B\,(2, -3, 1)$. Let $A\,(a_1, a_2, a_3)$. Then $\overrightarrow{AB} = \langle 2 - a_1, -3 - a_2, 1 - a_3 \rangle = \langle -1, 3, 4 \rangle$. Then
$2 - a_1 = -1 \Rightarrow a_1 = 3$, $-3 - a_2 = 3 \Rightarrow a_2 = -6$, and $1 - a_3 = 4 \Rightarrow a_3 = -3$, so the required point is $A\,(3, -6, -3)$.

53. $\mathbf{a} = \langle -1, 2, 0 \rangle$ and $\mathbf{b} = \langle 2, 3, -1 \rangle$, so $\mathbf{a} + \mathbf{b} = \langle 1, 5, -1 \rangle$,

$2\mathbf{a} - 3\mathbf{b} = 2\langle -1, 2, 0 \rangle - 3\langle 2, 3, -1 \rangle = \langle -2, 4, 0 \rangle - \langle 6, 9, -3 \rangle = \langle -8, -5, 3 \rangle$,

$|3\mathbf{a}| = |3\langle -1, 2, 0 \rangle| = |\langle -3, 6, 0 \rangle| = \sqrt{(-3)^2 + 6^2 + 0^2} = 3\sqrt{5}$,

$|-2\mathbf{b}| = |-2\langle 2, 3, -1 \rangle| = |\langle -4, -6, 2 \rangle| = \sqrt{(-4)^2 + (-6)^2 + 2^2} = 2\sqrt{14}$, and

$|\mathbf{a} - \mathbf{b}| = |\langle -1, 2, 0 \rangle - \langle 2, 3, -1 \rangle| = |\langle -3, -1, 1 \rangle| = \sqrt{(-3)^2 + (-1)^2 + 1^2} = \sqrt{11}$.

55. $\mathbf{a} = \langle 0, 2.1, 3.4 \rangle$ and $\mathbf{b} = \langle 1, 4.1, -5.6 \rangle$, so $\mathbf{a} + \mathbf{b} = \langle 1, 6.2, -2.2 \rangle$,

$2\mathbf{a} - 3\mathbf{b} = \langle 0, 4.2, 6.8 \rangle - \langle 3, 12.3, -16.8 \rangle = \langle -3, -8.1, 23.6 \rangle$,

$|3\mathbf{a}| = |3\langle 0, 2.1, 3.4 \rangle| = |\langle 0, 6.3, 10.2 \rangle| = \sqrt{(6.3)^2 + (10.2)^2} \approx 11.99$,

$|-2\mathbf{b}| = |-2\langle 1, 4.1, -5.6 \rangle| = |\langle -2, -8.2, 11.2 \rangle| = \sqrt{(-2)^2 + (-8.2)^2 + (11.2)^2} \approx 14.02$, and

$|\mathbf{a} - \mathbf{b}| = |\langle 0, 2.1, 3.4 \rangle - \langle 1, 4.1, -5.6 \rangle| = |\langle -1, -2, 9 \rangle| = \sqrt{(-1)^2 + (-2)^2 + 9^2} \approx 9.27$.

57. $\mathbf{u} = \langle -1, 3, -2 \rangle$, $\mathbf{v} = \langle 2, 1, 4 \rangle$, and $\mathbf{w} = \langle 3, -2, 1 \rangle$.

$a\mathbf{u} + b\mathbf{v} + c\mathbf{w} = \langle -a, 3a, -2a \rangle + \langle 2b, b, 4b \rangle + \langle 3c, -2c, c \rangle = \langle -a + 2b + 3c, 3a + b - 2c, -2a + 4b + c \rangle = \langle 2, 0, 2 \rangle$

$$\Rightarrow \left. \begin{array}{r} -a + 2b + 3c = 2 \\ 3a + b - 2c = 0 \\ -2a + 4b + c = 2 \end{array} \right\} \quad \text{Solving the system, we find } a = \frac{4}{35}, b = \frac{16}{35}, \text{ and } c = \frac{2}{5}.$$

59. $\mathbf{b} = \langle 3, -6, 15 \rangle = 3\langle 1, -2, 5 \rangle = 3\mathbf{a}$, and so $\mathbf{b}$ is parallel to $\mathbf{a}$.

61. Suppose $\mathbf{b}$ is parallel to $\mathbf{a}$. Then there exists a real number $c \neq 0$ such that $\mathbf{b} = c\mathbf{a} \Rightarrow$
$2\mathbf{i} - 3\mathbf{j} + 10\mathbf{k} = c\,(\mathbf{i} - 2\mathbf{j} + 5\mathbf{k}) = c\mathbf{i} - 2c\mathbf{j} + 5c\mathbf{k} \Leftrightarrow 2 = c,\ -3 = -2c,$ and $10 = 5c$. This system is not consistent, and so no such c exists. Thus, $\mathbf{b}$ is not parallel to $\mathbf{a}$.

63. $\mathbf{a} = \langle 1, 2, 2 \rangle$, so $|\mathbf{a}| = \sqrt{1^2 + 2^2 + 2^2} = 3$.

   a. $\mathbf{u}_1 = \frac{1}{3}\mathbf{a} = \left\langle \frac{1}{3}, \frac{2}{3}, \frac{2}{3} \right\rangle$ **b.** $\mathbf{u}_2 = \left\langle -\frac{1}{3}, -\frac{2}{3}, -\frac{2}{3} \right\rangle$

65. $\mathbf{a} = -\mathbf{i} + 3\mathbf{j} - \mathbf{k}$, so $|\mathbf{a}| = \sqrt{(-1)^2 + 3^2 + (-1)^2} = \sqrt{11}$.

   a. $\mathbf{u}_1 = \frac{1}{\sqrt{11}}\,(-\mathbf{i} + 3\mathbf{j} - \mathbf{k}) = -\frac{\sqrt{11}}{11}\mathbf{i} + \frac{3\sqrt{11}}{11}\mathbf{j} - \frac{\sqrt{11}}{11}\mathbf{k}$ **b.** $\mathbf{u}_2 = \frac{\sqrt{11}}{11}\mathbf{i} - \frac{3\sqrt{11}}{11}\mathbf{j} + \frac{\sqrt{11}}{11}\mathbf{k}$

67. $|\mathbf{a}| = 10$, $\mathbf{b} = \langle 1, 1, 1 \rangle$, and $|\mathbf{b}| = \sqrt{1^2 + 1^2 + 1^2} = \sqrt{3}$, so a unit vector in the direction of $\mathbf{b}$ is $\mathbf{u} = \frac{1}{\sqrt{3}} \langle 1, 1, 1 \rangle$.

Therefore, the required vector is $\mathbf{a} = 10\mathbf{u} = \frac{10}{\sqrt{3}} \langle 1, 1, 1 \rangle = \left\langle \frac{10\sqrt{3}}{3}, \frac{10\sqrt{3}}{3}, \frac{10\sqrt{3}}{3} \right\rangle$.

69. $|\mathbf{a}| = 3$, $\mathbf{b} = 2\mathbf{i} + 4\mathbf{j}$, and $|\mathbf{b}| = \sqrt{2^2 + 4^2} = 2\sqrt{5}$, so a unit vector in the direction of $\mathbf{b}$ is $\mathbf{u} = \frac{1}{2\sqrt{5}}\,(2\mathbf{i} + 4\mathbf{j})$. Therefore, the

required vector is $\mathbf{a} = 3\mathbf{u} = \frac{3}{2\sqrt{5}}\,(2\mathbf{i} + 4\mathbf{j}) = \frac{3\sqrt{5}}{5}\mathbf{i} + \frac{6\sqrt{5}}{5}\mathbf{j}$.

71. $\mathbf{a} = \langle 3, -1, 2 \rangle$ and $\mathbf{b} = \langle 1, 0, -1 \rangle$, so $\mathbf{a} - 2\mathbf{b} = \langle 3, -1, 2 \rangle - 2 \langle 1, 0, -1 \rangle = \langle 1, -1, 4 \rangle$ and

$|\mathbf{a} - 2\mathbf{b}| = \sqrt{1^2 + (-1)^2 + 4^2} = \sqrt{18} = 3\sqrt{2}$. Therefore, a unit vector in the direction of $\mathbf{a} - 2\mathbf{b}$ is $\mathbf{u} = \frac{1}{3\sqrt{2}} \langle 1, -1, 4 \rangle$,

and the required vector is $2\mathbf{u} = \frac{2}{3\sqrt{2}} \langle 1, -1, 4 \rangle = \left\langle \frac{\sqrt{2}}{3}, -\frac{\sqrt{2}}{3}, \frac{4\sqrt{2}}{3} \right\rangle$.

73. $\mathbf{F} = \langle -3, 4, 5 \rangle$ and $|\mathbf{F}| = \sqrt{(-3)^2 + 4^2 + 5^2} = 5\sqrt{2}$, so the unit vector in the direction of $\mathbf{F}$ is

$\mathbf{u} = \frac{\mathbf{F}}{|\mathbf{F}|} = \frac{1}{5\sqrt{2}} \langle -3, 4, 5 \rangle = \left\langle -\frac{3\sqrt{2}}{10}, \frac{2\sqrt{2}}{5}, \frac{\sqrt{2}}{2} \right\rangle$. Therefore, $\mathbf{F} = 5\sqrt{2}\mathbf{u} = 5\sqrt{2} \left\langle -\frac{3\sqrt{2}}{10}, \frac{2\sqrt{2}}{5}, \frac{\sqrt{2}}{2} \right\rangle$.

75. By the Law of Cosines, $|\mathbf{a} - \mathbf{b}|^2 = |\mathbf{a}|^2 + |\mathbf{b}|^2 - 2\,|\mathbf{a}|\,|\mathbf{b}| \cos \theta$.

77. Let $M\,(a, b, c)$ be the required midpoint. Then $\overrightarrow{P_1 M} = \frac{1}{2} \overrightarrow{P_1 P_2} \Rightarrow$

$\langle a - x_1, b - y_1, c - z_1 \rangle = \frac{1}{2} \langle x_2 - x_1, y_2 - y_1, z_2 - z_1 \rangle$

$\qquad\qquad\qquad\qquad = \left\langle \frac{1}{2}\,(x_2 - x_1), \frac{1}{2}\,(y_2 - y_1), \frac{1}{2}\,(z_2 - z_1) \right\rangle$

Therefore, $a - x_1 = \frac{1}{2}\,(x_2 - x_1) \Leftrightarrow a = \frac{1}{2}\,(x_1 + x_2)$,

$b - y_1 = \frac{1}{2}\,(y_2 - y_1) \Leftrightarrow b = \frac{1}{2}\,(y_1 + y_2)$, and $c - z_1 = \frac{1}{2}\,(z_2 - z_1) \Leftrightarrow$

$c = \frac{1}{2}\,(z_1 + z_2)$. Therefore, the midpoint is

$\left(\frac{1}{2}\,(x_1 + x_2), \frac{1}{2}\,(y_1 + y_2), \frac{1}{2}\,(z_1 + z_2) \right)$.

79. By Newton's Law of Gravitation, the magnitude of the force is $|\mathbf{F}| = \dfrac{Gm_1 m_2}{\left| \overrightarrow{BA} \right|^2}$ in the direction of $\mathbf{u} = \dfrac{\overrightarrow{BA}}{\left| \overrightarrow{BA} \right|}$. Therefore,

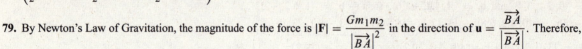

$\mathbf{F} = |\mathbf{F}|\,\mathbf{u} = \left(\dfrac{Gm_1 m_2}{\left| \overrightarrow{BA} \right|^2} \right) \mathbf{u} = \dfrac{Gm_1 m_2}{\left| \overrightarrow{BA} \right|^3} \overrightarrow{BA}$.

81. The force $\mathbf{F}_1$ exerted by the charge at A on the charge at B has direction $\dfrac{\overrightarrow{AB}}{\left| \overrightarrow{AB} \right|}$. For like charges $q_1 q_2 > 0$ and

the force is repulsive, whereas for unlike charges $q_1 q_2 < 0$ and the force is attractive. Thus, by Coulomb's Law,

$\mathbf{F}_1 = \dfrac{kq_1 q_2}{\left| \overrightarrow{AB} \right|^2} \left(\dfrac{\overrightarrow{AB}}{\left| \overrightarrow{AB} \right|} \right) = \dfrac{kq_1 q_2}{\left| \overrightarrow{AB} \right|^3} \overrightarrow{AB}$ where k is the constant of proportionality. Similarly, we see that $\mathbf{F}_2 = \dfrac{kq_1 q_2}{\left| \overrightarrow{BA} \right|^3} \overrightarrow{BA}$.

83. True. In an equation of a sphere, the coefficients of x^2, y^2, and z^2 must be the same.

85. False. Let $P_1(1, 2, 3)$, $P_2(4, 5, 6)$, $Q_1(0, 0, 0)$, and $Q_2(3, 3, 3)$. Then $P_1 \neq Q_1$ and $P_2 \neq Q_2$, but $\overrightarrow{P_1 P_2} = \langle 3, 3, 3 \rangle = \overrightarrow{Q_1 Q_2}$.

11.3 Concept Questions   ET 10.3

1. a. $\mathbf{a} \cdot \mathbf{b} = \langle a_1, a_2, a_3 \rangle \cdot \langle b_1, b_2, b_3 \rangle = a_1 b_1 + a_2 b_2 + a_3 b_3$

b. See page 924 (920 in ET).

c. $|\mathbf{a}| = \sqrt{\mathbf{a} \cdot \mathbf{a}}$

d. $\theta = \cos^{-1} \left(\dfrac{\mathbf{a} \cdot \mathbf{b}}{|\mathbf{a}| \, |\mathbf{b}|} \right), \ 0 \le \theta \le \pi$

3. a. The direction cosines of a vector $\mathbf{a}$ in 3-space are the cosines of the angles the vector makes with the coordinate axes. If $\mathbf{a} = \langle a_1, a_2, a_3 \rangle$, then $\cos \alpha = \dfrac{a_1}{|\mathbf{a}|}$, $\cos \beta = \dfrac{a_2}{|\mathbf{a}|}$, and $\cos \gamma = \dfrac{a_3}{|\mathbf{a}|}$.

b. $\mathbf{a} = |\mathbf{a}| \langle \cos \alpha, \cos \beta, \cos \gamma \rangle$

5. $W = \mathbf{F} \cdot \overrightarrow{PQ}$

11.3 The Dot Product   ET 10.3

1. a. Yes. $\mathbf{a} \cdot \mathbf{b}$ is a scalar and so $(\mathbf{a} \cdot \mathbf{b}) \, \mathbf{c}$ is a vector.

b. No. $\mathbf{b} \cdot \mathbf{c}$ is a scalar and it does not make sense to take the dot product of a vector and a scalar.

c. No. $\mathbf{b} \cdot \mathbf{c}$ is a scalar and $\mathbf{a}$ is a vector and addition of a vector and a scalar is undefined.

d. Yes. $\mathbf{a} \cdot \mathbf{b}$ is a scalar and so is $|\mathbf{a}| \, |\mathbf{b}|$, so they can be subtracted.

e. Yes. Both $|\mathbf{a}| \, \mathbf{b}$ and $(\mathbf{a} \cdot \mathbf{b}) \, \mathbf{a}$ are vectors.

f. Yes. $\dfrac{\mathbf{b}}{|\mathbf{b}|}$ is a vector and can be dotted with $\mathbf{a}$.

3. $\mathbf{a} \cdot \mathbf{b} = \langle 1, 3 \rangle \cdot \langle 2, -1 \rangle = 2 - 3 = -1$

5. $\mathbf{a} \cdot \mathbf{b} = (2\mathbf{i} + 3\mathbf{j}) \cdot (\mathbf{i} - 2\mathbf{j}) = 2 - 6 = -4$

7. $\mathbf{a} \cdot \mathbf{b} = \langle 0, 1, -3 \rangle \cdot \langle 10, \pi, -\pi \rangle = \pi + 3\pi = 4\pi$

9. $\mathbf{a} \cdot (\mathbf{b} + \mathbf{c}) = \langle 1, -3, 2 \rangle \cdot [\langle -2, 4, 1 \rangle + \langle 2, -4, 1 \rangle] = \langle 1, -3, 2 \rangle \cdot \langle 0, 0, 2 \rangle = 4$

11. $(2\mathbf{a} + 3\mathbf{b}) \cdot (3\mathbf{c}) = [2 \langle 1, -3, 2 \rangle + 3 \langle -2, 4, 1 \rangle] \cdot [3 \langle 2, -4, 1 \rangle] = [\langle 2, -6, 4 \rangle + \langle -6, 12, 3 \rangle] \cdot \langle 6, -12, 3 \rangle$

$$= \langle -4, 6, 7 \rangle \cdot \langle 6, -12, 3 \rangle = -24 - 72 + 21 = -75$$

13. $(\mathbf{a} \cdot \mathbf{b}) \, \mathbf{c} = [\langle 1, -3, 2 \rangle \cdot \langle -2, 4, 1 \rangle] \langle 2, -4, 1 \rangle = (-2 - 12 + 2) \langle 2, -4, 1 \rangle = -12 \langle 2, -4, 1 \rangle = \langle -24, 48, -12 \rangle$

15. $|\mathbf{a} - \mathbf{b}|^2 + |\mathbf{a} + \mathbf{b}|^2 = |\langle 1, -3, 2 \rangle - \langle -2, 4, 1 \rangle|^2 + |\langle 1, -3, 2 \rangle + \langle -2, 4, 1 \rangle|^2 = |\langle 3, -7, 1 \rangle|^2 + |\langle -1, 1, 3 \rangle|^2$

$$= (9 + 49 + 1) + (1 + 1 + 9) = 70$$

17. $\theta = \cos^{-1} \left(\dfrac{\langle 2, 1 \rangle \cdot \langle 3, 4 \rangle}{\sqrt{4 + 1} \sqrt{9 + 16}} \right) = \cos^{-1} \dfrac{10}{5\sqrt{5}} \approx 26.6°$

19. $\theta = \cos^{-1} \left(\dfrac{\langle 1, 1, 1 \rangle \cdot \langle 2, 3, -6 \rangle}{\sqrt{1 + 1 + 1} \sqrt{4 + 9 + 36}} \right) = \cos^{-1} \left(\dfrac{-1}{7\sqrt{3}} \right) \approx 94.7°$

21. $\theta = \cos^{-1} \left(\dfrac{(-2\mathbf{j} + 3\mathbf{k}) \cdot (\mathbf{i} + \mathbf{j} + 2\mathbf{k})}{\sqrt{4 + 9} \sqrt{1 + 1 + 4}} \right) = \cos^{-1} \dfrac{4}{\sqrt{13}\sqrt{6}} \approx 63.1°$

23. $\cos 45° = \dfrac{\langle 1, c \rangle \cdot \langle 1, 2 \rangle}{\sqrt{1 + c^2}\sqrt{1 + 4^2}} \Rightarrow \dfrac{\sqrt{2}}{2} = \dfrac{1 + 2c}{\sqrt{1 + c^2}\sqrt{5}} \Rightarrow \sqrt{10}\sqrt{1 + c^2} = 2 + 4c \Rightarrow 10\left(1 + c^2\right) = 4 + 16c + 16c^2 \Rightarrow$

$6c^2 + 16c - 6 = 0 \Rightarrow (3c - 1)(c + 3) = 0 \Rightarrow c = -3$ or $\frac{1}{3}$. If $c = \frac{1}{3}$, then $\theta = 45°$, and if $c = -3$, then $\theta = 135°$, so the

desired value of c is $\frac{1}{3}$.

25. $\mathbf{a} = \langle 1, 2 \rangle$ and $\mathbf{b} = \langle 3, 0 \rangle$. $\mathbf{a} \neq c\mathbf{b}$ for any scalar c, so $\mathbf{a}$ and $\mathbf{b}$ are not parallel. $\mathbf{a} \cdot \mathbf{b} = \langle 1, 2 \rangle \cdot \langle 3, 0 \rangle = 3 \neq 0$, and so $\mathbf{a}$ and $\mathbf{b}$ are not orthogonal. Thus, $\mathbf{a}$ and $\mathbf{b}$ are neither orthogonal nor parallel.

27. $\mathbf{a} = \mathbf{i} - 2\mathbf{j} + \mathbf{k}$ and $\mathbf{b} = 3\mathbf{i} + 2\mathbf{j} - 2\mathbf{k}$. $\mathbf{a} \neq c\mathbf{b}$ for any scalar c, so $\mathbf{a}$ and $\mathbf{b}$ are not parallel.

$\mathbf{a} \cdot \mathbf{b} = (\mathbf{i} - 2\mathbf{j} + \mathbf{k}) \cdot (3\mathbf{i} + 2\mathbf{j} - 2\mathbf{k}) = 3 - 4 - 2 = -3 \neq 0$, so $\mathbf{a}$ and $\mathbf{b}$ are not orthogonal either.

29. $\mathbf{a} = \langle 2, 3, -1 \rangle$ and $\mathbf{b} = \langle 2, -1, 1 \rangle$. $\mathbf{a} \neq c\mathbf{b}$ for any scalar c, so $\mathbf{a}$ and $\mathbf{b}$ are not parallel.

$\mathbf{a} \cdot \mathbf{b} = \langle 2, 3, -1 \rangle \cdot \langle 2, -1, 1 \rangle = 4 - 3 - 1 = 0$, so $\mathbf{a}$ and $\mathbf{b}$ are orthogonal.

31. If $\langle c, 2, -1 \rangle$ and $\langle 2, 3, c \rangle$ are orthogonal, then $\langle c, 2, -1 \rangle \cdot \langle 2, 3, c \rangle = 0$; that is, $2c + 6 - c = 0 \Leftrightarrow c = -6$.

33. $\mathbf{a} = \langle 1, 2, 3 \rangle$, so $|\mathbf{a}| = \sqrt{1^2 + 2^2 + 3^2} = \sqrt{14}$. Thus, $\cos \alpha = \frac{1}{\sqrt{14}} = \frac{\sqrt{14}}{14} \Rightarrow \alpha \approx 74.5°$, $\cos \beta = \frac{2}{\sqrt{14}} = \frac{\sqrt{14}}{7} \Rightarrow$

$\beta \approx 57.7°$, and $\cos \gamma = \frac{3}{\sqrt{14}} = \frac{3\sqrt{14}}{14} \Rightarrow \gamma \approx 36.7°$.

35. $\mathbf{a} = -\mathbf{i} + 3\mathbf{j} + 5\mathbf{k}$, so $|\mathbf{a}| = \sqrt{1 + 9 + 25} = \sqrt{35}$. Thus, $\cos \alpha = -\frac{\sqrt{35}}{35} \Rightarrow \alpha \approx 99.7°$, $\cos \beta = \frac{3\sqrt{35}}{35} \Rightarrow \beta \approx 59.5°$, and

$\cos \gamma = \frac{\sqrt{35}}{7} \Rightarrow \gamma \approx 32.3°$.

37. $\alpha = \frac{\pi}{3}$ and $\gamma = \frac{\pi}{4}$. Since $\cos^2 \alpha + \cos^2 \beta + \cos^2 \gamma = 1$, we have $\cos^2\left(\frac{\pi}{3}\right) + \cos^2 \beta + \cos^2\left(\frac{\pi}{4}\right) = 1 \Rightarrow$

$\cos^2 \beta = 1 - \left(\frac{1}{2}\right)^2 - \left(\frac{\sqrt{2}}{2}\right)^2 = \frac{1}{4}$. Thus, $\cos \beta = \pm\frac{1}{2}$, and so $\beta = \frac{\pi}{3}$ or $\frac{2\pi}{3}$.

39. $\mathbf{a} = \langle 2, 3 \rangle$ and $\mathbf{b} = \langle 1, 4 \rangle$.

    a. $\text{proj}_{\mathbf{a}} \mathbf{b} = \left(\dfrac{\mathbf{a} \cdot \mathbf{b}}{|\mathbf{a}|^2}\right)\mathbf{a} = \left(\dfrac{\langle 2, 3 \rangle \cdot \langle 1, 4 \rangle}{4 + 9}\right)\langle 2, 3 \rangle = \frac{2 + 12}{13}\langle 2, 3 \rangle = \left\langle \frac{28}{13}, \frac{42}{13} \right\rangle$

    b. $\text{proj}_{\mathbf{b}} \mathbf{a} = \left(\dfrac{\mathbf{a} \cdot \mathbf{b}}{|\mathbf{b}|^2}\right)\mathbf{b} = \left(\dfrac{\langle 2, 3 \rangle \cdot \langle 1, 4 \rangle}{1 + 16}\right)\langle 1, 4 \rangle = \frac{2 + 12}{17}\langle 1, 4 \rangle = \left\langle \frac{14}{17}, \frac{56}{17} \right\rangle$

41. $\mathbf{a} = 2\mathbf{i} + \mathbf{j} + 4\mathbf{k}$ and $\mathbf{b} = 3\mathbf{i} + \mathbf{k}$.

    a. $\text{proj}_{\mathbf{a}} \mathbf{b} = \left(\dfrac{\mathbf{a} \cdot \mathbf{b}}{|\mathbf{a}|^2}\right)\mathbf{a} = \left[\dfrac{(2\mathbf{i} + \mathbf{j} + 4\mathbf{k}) \cdot (3\mathbf{i} + \mathbf{k})}{4 + 1 + 16}\right](2\mathbf{i} + \mathbf{j} + 4\mathbf{k}) = \frac{6 + 4}{21}(2\mathbf{i} + \mathbf{j} + 4\mathbf{k}) = \frac{20}{21}\mathbf{i} + \frac{10}{21}\mathbf{j} + \frac{40}{21}\mathbf{k}$

    b. $\text{proj}_{\mathbf{b}} \mathbf{a} = \left(\dfrac{\mathbf{a} \cdot \mathbf{b}}{|\mathbf{b}|^2}\right)\mathbf{b} = \left[\dfrac{(2\mathbf{i} + \mathbf{j} + 4\mathbf{k}) \cdot (3\mathbf{i} + \mathbf{k})}{9 + 1}\right](3\mathbf{i} + \mathbf{k}) = \frac{6 + 4}{10}(3\mathbf{i} + \mathbf{k}) = 3\mathbf{i} + \mathbf{k}$

43. $\mathbf{a} = \langle -3, 4, -2 \rangle$ and $\mathbf{b} = \langle 0, 1, 0 \rangle$.

    a. $\text{proj}_{\mathbf{a}} \mathbf{b} = \left(\dfrac{\mathbf{a} \cdot \mathbf{b}}{|\mathbf{a}|^2}\right)\mathbf{a} = \left(\dfrac{\langle -3, 4, -2 \rangle \cdot \langle 0, 1, 0 \rangle}{9 + 16 + 4}\right)\langle -3, 4, -2 \rangle = \frac{4}{29}\langle -3, 4, -2 \rangle = \left\langle -\frac{12}{29}, \frac{16}{29}, -\frac{8}{29} \right\rangle$

    b. $\text{proj}_{\mathbf{b}} \mathbf{a} = \left(\dfrac{\mathbf{a} \cdot \mathbf{b}}{|\mathbf{b}|^2}\right)\mathbf{b} = \left(\dfrac{\langle -3, 4, -2 \rangle \cdot \langle 0, 1, 0 \rangle}{1}\right)\langle 0, 1, 0 \rangle = 4\langle 0, 1, 0 \rangle = \langle 0, 4, 0 \rangle$

45. $\mathbf{a} = \langle 1, 3 \rangle$ and $\mathbf{b} = \langle 2, 4 \rangle$. Because $\text{proj}_{\mathbf{a}} \mathbf{b} = \left(\dfrac{\mathbf{a} \cdot \mathbf{b}}{|\mathbf{a}|^2}\right)\mathbf{a} = \left(\dfrac{\langle 1, 3 \rangle \cdot \langle 2, 4 \rangle}{1 + 9}\right)\langle 1, 3 \rangle = \frac{2 + 12}{10}\langle 1, 3 \rangle = \left\langle \frac{7}{5}, \frac{21}{5} \right\rangle$ is parallel to

$\mathbf{a}$, we can write $\mathbf{b} = \text{proj}_{\mathbf{a}} \mathbf{b} + (\mathbf{b} - \text{proj}_{\mathbf{a}} \mathbf{b}) = \left\langle \frac{7}{5}, \frac{21}{5} \right\rangle + \left\langle \frac{3}{5}, -\frac{1}{5} \right\rangle$.

47. $\mathbf{a} = \mathbf{i} + 2\mathbf{j} + 3\mathbf{k}$ and $\mathbf{b} = 2\mathbf{i} - \mathbf{j} + \mathbf{k}$. Because

$\text{proj}_{\mathbf{a}} \mathbf{b} = \left(\dfrac{\mathbf{a} \cdot \mathbf{b}}{|\mathbf{a}|^2}\right)\mathbf{a} = \left[\dfrac{(\mathbf{i} + 2\mathbf{j} + 3\mathbf{k}) \cdot (2\mathbf{i} - \mathbf{j} + \mathbf{k})}{1 + 4 + 9}\right](\mathbf{i} + 2\mathbf{j} + 3\mathbf{k}) = \left(\dfrac{2 - 2 + 3}{14}\right)(\mathbf{i} + 2\mathbf{j} + 3\mathbf{k}) = \frac{3}{14}\mathbf{i} + \frac{3}{7}\mathbf{j} + \frac{9}{14}\mathbf{k}$ is

parallel to $\mathbf{a}$, we can write

$\mathbf{b} = \text{proj}_{\mathbf{a}} \mathbf{b} + (\mathbf{b} - \text{proj}_{\mathbf{a}} \mathbf{b}) = \left(\frac{3}{14}\mathbf{i} + \frac{3}{7}\mathbf{j} + \frac{9}{14}\mathbf{k}\right) + \left[(2\mathbf{i} - \mathbf{j} + \mathbf{k}) - \left(\frac{3}{14}\mathbf{i} + \frac{3}{7}\mathbf{j} + \frac{9}{14}\mathbf{k}\right)\right]$

$= \left(\frac{3}{14}\mathbf{i} + \frac{3}{7}\mathbf{j} + \frac{9}{14}\mathbf{k}\right) + \left(\frac{25}{14}\mathbf{i} - \frac{10}{7}\mathbf{j} + \frac{5}{14}\mathbf{k}\right)$

49. $\overrightarrow{PQ} = [2 - (-1)]\mathbf{i} + [1 - (-2)]\mathbf{j} + (5 - 2)\mathbf{k} = 3\mathbf{i} + 3\mathbf{j} + 3\mathbf{k}$, so $W = \mathbf{F} \cdot \overrightarrow{PQ} = (2\mathbf{i} + 3\mathbf{j} - \mathbf{k}) \cdot (3\mathbf{i} + 3\mathbf{j} + 3\mathbf{k}) = 12$.

51. From the figure,

$$\cos\theta = \frac{\overrightarrow{OC}\cdot\mathbf{i}}{\left|\overrightarrow{OC}\right||\mathbf{i}|} = \frac{(a\mathbf{i} + a\mathbf{j} + a\mathbf{k})\cdot\mathbf{i}}{\sqrt{a^2 + a^2 + a^2}}\,(1)$$

$$= \frac{a}{\sqrt{3}\,a} = \frac{\sqrt{3}}{3}$$

so $\theta = \cos^{-1}\left(\frac{\sqrt{3}}{3}\right) \approx 54.7°$.

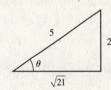

53. Let us label the points $A\,(3, 0, 0)$, $B\,(3, 6, 4)$, $C\,(0, 6, 4)$ and $O\,(0, 0, 0)$. Then $\overrightarrow{OA} = \langle 3, 0, 0\rangle$, $\overrightarrow{OB} = \langle 3, 6, 4\rangle$, and

$\overrightarrow{OC} = \langle 0, 6, 4\rangle$, so $\cos\theta = \left(\dfrac{\overrightarrow{OA}\cdot\overrightarrow{OB}}{\left|\overrightarrow{OA}\right|\left|\overrightarrow{OB}\right|}\right) = \dfrac{\langle 3, 0, 0\rangle\cdot\langle 3, 6, 4\rangle}{3\sqrt{9 + 36 + 16}} = \dfrac{9}{3\sqrt{61}} = \dfrac{3\sqrt{61}}{61} \Rightarrow \theta = \cos^{-1}\left(\dfrac{3\sqrt{61}}{61}\right) \approx 67.4°$ and

$\cos\psi = \left(\dfrac{\overrightarrow{OB}\cdot\overrightarrow{OC}}{\left|\overrightarrow{OB}\right|\left|\overrightarrow{OC}\right|}\right) = \dfrac{\langle 3, 6, 4\rangle\cdot\langle 0, 6, 4\rangle}{\sqrt{9 + 36 + 16}\sqrt{36 + 16}} = \dfrac{52}{\sqrt{61}\sqrt{52}} = \dfrac{2\sqrt{13}\sqrt{61}}{61} \Rightarrow \psi = \cos^{-1}\left(\dfrac{2\sqrt{13}\sqrt{61}}{61}\right) \approx 22.6°$.

55. $W = (24\cos 30°)(50) = 600\sqrt{3} \approx 1039.2$ ft-lb

57. Using the result of Exercise 11.1.63 (10.1.63 in ET), we have

$\mathbf{F}_1 = \langle 3000\cos\theta, 3000\sin\theta\rangle$ and $\mathbf{F}_2 = \left\langle 1200\sqrt{3}, 3000\sin\theta - 1200\right\rangle$.

The displacement of the ship is $\overrightarrow{OA} = \langle 100, 0\rangle$, where A denotes the point

$(100, 0)$, so the work done by Tugboat I is

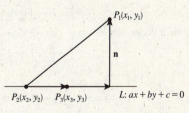

$W_1 = \mathbf{F}_1\cdot\overrightarrow{OA} = \langle 3000\cos\theta, 3000\sin\theta\rangle\cdot\langle 100, 0\rangle = (3000\cos\theta)\,100 = 3000\left(\frac{\sqrt{21}}{5}\right)100 \approx 274{,}955$ ft-lb and the work

done by Tugboat II is $W_2 = \mathbf{F}_2\cdot\overrightarrow{OA} = \left\langle 1200\sqrt{3}, 3000\sin\theta - 1200\right\rangle\cdot\langle 100, 0\rangle = \left(1200\sqrt{3}\right)(100) \approx 207{,}846$ ft-lb.

59. Let $\mathbf{a} = \langle a_1, a_2, a_3\rangle$ and $\mathbf{b} = \langle b_1, b_2, b_3\rangle$. Then $(c\mathbf{a})\cdot\mathbf{b} = \left[c\,\langle a_1, a_2, a_3\rangle\right]\cdot$
$\langle b_1, b_2, b_3\rangle = \langle ca_1, ca_2, ca_3\rangle\cdot\langle b_1, b_2, b_3\rangle = ca_1b_1 + ca_2b_2 + ca_3b_3 = c\,(a_1b_1 + a_2b_2 + a_3b_3)$ (1). Next,
$c\,(\mathbf{a}\cdot\mathbf{b}) = c\left[\langle a_1, a_2, a_3\rangle\cdot\langle b_1, b_2, b_3\rangle\right] = c\,(a_1b_1 + a_2b_2 + a_3b_3)$ (2). Finally,
$\mathbf{a}\cdot(c\mathbf{b}) = \langle a_1, a_2, a_3\rangle\cdot\left[c\,\langle b_1, b_2, b_3\rangle\right] = \langle a_1, a_2, a_3\rangle\cdot\langle cb_1, cb_2, cb_3\rangle = a_1\,(cb_1) + a_2\,(cb_2) + a_3\,(cb_3)$

$= c\,(a_1b_1 + a_2b_2 + a_3b_3)$ (3)

Comparing (1), (2), and (3), we see that $(c\mathbf{a})\cdot\mathbf{b} = c\,(\mathbf{a}\cdot\mathbf{b}) = \mathbf{a}\cdot(c\mathbf{b})$.

61. a.

b. $\mathbf{a}\cdot\left(\mathbf{b} - \text{proj}_{\mathbf{a}}\,\mathbf{b}\right) = \mathbf{a}\cdot\mathbf{b} - \mathbf{a}\cdot\text{proj}_{\mathbf{a}}\,\mathbf{b} = \mathbf{a}\cdot\mathbf{b} - \mathbf{a}\cdot\left[\left(\dfrac{\mathbf{a}\cdot\mathbf{b}}{|\mathbf{a}|^2}\right)\mathbf{a}\right]$

$= \mathbf{a}\cdot\mathbf{b} - \dfrac{\mathbf{a}\cdot\mathbf{b}}{|\mathbf{a}|^2}\,(\mathbf{a}\cdot\mathbf{a}) = \mathbf{a}\cdot\mathbf{b} - \mathbf{a}\cdot\mathbf{b}$

$= 0$

Therefore, $\mathbf{b} - \text{proj}_{\mathbf{a}}\,\mathbf{b}$ is orthogonal to $\mathbf{a}$.

63. a. Pick $P_3\,(x_3, y_3)$ to be a point on L distinct from $P_2\,(x_2, y_2)$ and consider

$\mathbf{n}\cdot\overrightarrow{P_2P_3} = \langle a, b\rangle\cdot\langle x_3 - x_2, y_3 - y_2\rangle = a\,(x_3 - x_2) + b\,(y_3 - y_2)$

$= (ax_3 + by_3) - (ax_2 + by_2)$ (1)

Since P_2 and P_3 both lie on L, we have $ax_2 + by_2 = -c$ and

$ax_3 + by_3 = -c$. Substituting into (1) gives $\mathbf{n}\cdot\overrightarrow{P_2P_3} = -c - (-c) = 0$,

so $\mathbf{n}$ is orthogonal to L.

b. $d = \dfrac{\left|\overrightarrow{P_1 P_2} \cdot \mathbf{n}\right|}{|\mathbf{n}|} = \dfrac{|\langle x_2 - x_1, y_2 - y_1 \rangle \cdot \langle a, b \rangle|}{\sqrt{a^2 + b^2}} = \dfrac{|a(x_2 - x_1) + b(y_2 - y_1)|}{\sqrt{a^2 + b^2}}$. Since $P_2(x_2, y_2)$ lies on L,

$ax_2 + by_2 = -c$, so $d = \dfrac{|ax_1 + by_1 + c|}{\sqrt{a^2 + b^2}}$.

c. Here $P_1(1, -3)$ and $L : 2x + 3y - 6 = 0$, so $d = \dfrac{|2(1) + 3(-3) - 6|}{\sqrt{4 + 9}} = \dfrac{13}{\sqrt{13}} = \sqrt{13}$.

65. False. Take $\mathbf{a} = \mathbf{i}$ and $\mathbf{b} = \mathbf{j}$. Then $(\mathbf{a} \cdot \mathbf{b})^2 = (\mathbf{i} \cdot \mathbf{j})^2 = 0^2 = 0$, but $|\mathbf{a}|^2 |\mathbf{b}|^2 = |\mathbf{i}|^2 |\mathbf{j}|^2 = (1)(1) = 1$.

67. True. $(2\mathbf{u} + 3\mathbf{v}) \cdot \mathbf{w} = 2\mathbf{u} \cdot \mathbf{w} + 3\mathbf{v} \cdot \mathbf{w} = 2(0) + 3(0) = 0$ since $\mathbf{u} \cdot \mathbf{w} = 0$ and $\mathbf{v} \cdot \mathbf{w} = 0$. Therefore, $2\mathbf{u} + 3\mathbf{v}$ is orthogonal to $\mathbf{w}$.

11.4   Concept Questions   ET 10.4

1. a. See page 934 (930 in ET).

  b. See page 937 (933 in ET).

  c. It is normal to the plane determined by $\mathbf{a}$ and $\mathbf{b}$.

  d. $\mathbf{a} \times \mathbf{b} = (|\mathbf{a}|\, |\mathbf{b}| \sin \theta)\, \mathbf{n}$, where $\mathbf{n}$ is a unit vector in the direction of $\mathbf{a} \times \mathbf{b}$.

11.4   The Cross Product   ET 10.4

1. a. Yes. $\mathbf{a} \times \mathbf{b}$ is a vector and so we can take its dot product with the vector $\mathbf{c}$.

  b. No. $\mathbf{b} \cdot \mathbf{c}$ is a scalar and so $\mathbf{a} \times (\mathbf{b} \cdot \mathbf{c})$ is not defined.

  c. Yes. $\mathbf{a} + \mathbf{b}$ and $\mathbf{c} \times \mathbf{d}$ are both vectors and so we may take their dot product.

  d. Yes. $\mathbf{a} \times \mathbf{b}$ and $\mathbf{c} \times \mathbf{d}$ are both vectors and so we may take their dot product.

  e. Yes. $\mathbf{b} \cdot \mathbf{c}$ is a scalar and so $(\mathbf{b} \cdot \mathbf{c})\, \mathbf{d}$ is a vector. Therefore, we can take the cross product of $\mathbf{a}$ and $(\mathbf{b} \cdot \mathbf{c})\, \mathbf{d}$.

  f. Yes. $\mathbf{a} \times \mathbf{b}$ and $\mathbf{c} \times \mathbf{d}$ are both vectors and so we may take their cross product.

3. $\mathbf{a} \times \mathbf{b} = \begin{vmatrix} \mathbf{i} & \mathbf{j} & \mathbf{k} \\ 1 & 1 & 0 \\ 0 & 2 & 3 \end{vmatrix} = 3\mathbf{i} - 3\mathbf{j} + 2\mathbf{k}$ **5.** $\mathbf{a} \times \mathbf{b} = \begin{vmatrix} \mathbf{i} & \mathbf{j} & \mathbf{k} \\ 1 & -2 & 1 \\ 3 & 1 & -2 \end{vmatrix} = 3\mathbf{i} + 5\mathbf{j} + 7\mathbf{k}$

7. $\mathbf{a} \times \mathbf{b} = \begin{vmatrix} \mathbf{i} & \mathbf{j} & \mathbf{k} \\ 2 & 0 & 3 \\ -3 & 2 & -1 \end{vmatrix} = -6\mathbf{i} - 7\mathbf{j} + 4\mathbf{k}$ **9.** $\mathbf{a} \times \mathbf{b} = \begin{vmatrix} \mathbf{i} & \mathbf{j} & \mathbf{k} \\ 2 & 1 & -3 \\ \frac{2}{3} & \frac{1}{3} & -1 \end{vmatrix} = 0\mathbf{i} + 0\mathbf{j} + 0\mathbf{k} = \mathbf{0}$

11. $\mathbf{a} \times \mathbf{b} = \begin{vmatrix} \mathbf{i} & \mathbf{j} & \mathbf{k} \\ 1 & 2 & 3 \\ 2 & -1 & -1 \end{vmatrix} = \mathbf{i} + 7\mathbf{j} - 5\mathbf{k}$ and $\mathbf{b} \times \mathbf{a} = \begin{vmatrix} \mathbf{i} & \mathbf{j} & \mathbf{k} \\ 2 & -1 & -1 \\ 1 & 2 & 3 \end{vmatrix} = -\mathbf{i} - 7\mathbf{j} + 5\mathbf{k}$.

13. One such vector is $\mathbf{a} \times \mathbf{b} = \begin{vmatrix} \mathbf{i} & \mathbf{j} & \mathbf{k} \\ 2 & -3 & 4 \\ -1 & 1 & -2 \end{vmatrix} = 2\mathbf{i} - \mathbf{k}$; another is $\mathbf{b} \times \mathbf{a} = -2\mathbf{i} + \mathbf{k}$.

15. $\mathbf{a} \times \mathbf{b} = \begin{vmatrix} \mathbf{i} & \mathbf{j} & \mathbf{k} \\ -3 & 1 & -2 \\ 1 & 1 & 1 \end{vmatrix} = 3\mathbf{i} + \mathbf{j} - 4\mathbf{k}$, so the required unit vectors are $\pm \left(\dfrac{3\mathbf{i} + \mathbf{j} - 4\mathbf{k}}{\sqrt{9 + 1 + 16}} \right)$; that is, $\pm \dfrac{\sqrt{26}}{26}(3\mathbf{i} + \mathbf{j} - 4\mathbf{k})$.

17. For $P(1, 0, 0)$, $Q(0, 1, 0)$, and $R(0, 0, 1)$, $\overrightarrow{PQ} = -\mathbf{i} + \mathbf{j}$ and $\overrightarrow{PR} = -\mathbf{i} + \mathbf{k}$, so $\overrightarrow{PQ} \times \overrightarrow{PR} = \begin{vmatrix} \mathbf{i} & \mathbf{j} & \mathbf{k} \\ -1 & 1 & 0 \\ -1 & 0 & 1 \end{vmatrix} = \mathbf{i} + \mathbf{j} + \mathbf{k}$, so

the area of $\triangle PQR$ is $\frac{1}{2}\left|\overrightarrow{PQ} \times \overrightarrow{PR}\right| = \frac{1}{2}\left|\mathbf{i} + \mathbf{j} + \mathbf{k}\right| = \frac{1}{2}\sqrt{1 + 1 + 1} = \frac{\sqrt{3}}{2}$.

19. For $P(1, -1, 2)$, $Q(2, 3, 1)$, and $R(-2, 3, 4)$, $\overrightarrow{PQ} = \mathbf{i} + 4\mathbf{j} - \mathbf{k}$ and $\overrightarrow{PR} = -3\mathbf{i} + 4\mathbf{j} + 2\mathbf{k}$,

so $\overrightarrow{PQ} \times \overrightarrow{PR} = \begin{vmatrix} \mathbf{i} & \mathbf{j} & \mathbf{k} \\ 1 & 4 & -1 \\ -3 & 4 & 2 \end{vmatrix} = 12\mathbf{i} + \mathbf{j} + 16\mathbf{k}$, so the area of $\triangle PQR$ is

$\frac{1}{2}\left|\overrightarrow{PQ} \times \overrightarrow{PR}\right| = \frac{1}{2}\left|12\mathbf{i} + \mathbf{j} + 16\mathbf{k}\right| = \frac{1}{2}\sqrt{144 + 1 + 256} = \frac{\sqrt{401}}{2}$.

21. $(2\mathbf{a}) \times \mathbf{b} = 2(\mathbf{a} \times \mathbf{b}) = 2\begin{vmatrix} \mathbf{i} & \mathbf{j} & \mathbf{k} \\ 1 & -1 & 1 \\ 2 & 3 & -1 \end{vmatrix} = 2(-2\mathbf{i} + 3\mathbf{j} + 5\mathbf{k}) = -4\mathbf{i} + 6\mathbf{j} + 10\mathbf{k}$

23. $\mathbf{a} \times \mathbf{b} = \begin{vmatrix} \mathbf{i} & \mathbf{j} & \mathbf{k} \\ 1 & -1 & 1 \\ 2 & 3 & -1 \end{vmatrix} = -2\mathbf{i} + 3\mathbf{j} + 5\mathbf{k} \Rightarrow (\mathbf{a} \times \mathbf{b}) \times \mathbf{c} = \begin{vmatrix} \mathbf{i} & \mathbf{j} & \mathbf{k} \\ -2 & 3 & 5 \\ -1 & 1 & 2 \end{vmatrix} = \mathbf{i} - \mathbf{j} + \mathbf{k}$

25. $\mathbf{a} \times \mathbf{b} = \begin{vmatrix} \mathbf{i} & \mathbf{j} & \mathbf{k} \\ 1 & -1 & 1 \\ 2 & 3 & -1 \end{vmatrix} = -2\mathbf{i} + 3\mathbf{j} + 5\mathbf{k} \Rightarrow (\mathbf{a} \times \mathbf{b}) \cdot \mathbf{c} = (-2\mathbf{i} + 3\mathbf{j} + 5\mathbf{k}) \cdot (-\mathbf{i} + \mathbf{j} + 2\mathbf{k}) = 15$

27. $\mathbf{a} \cdot (\mathbf{b} \times \mathbf{c}) = \begin{vmatrix} 1 & 1 & 0 \\ 0 & 1 & -2 \\ 1 & 2 & 3 \end{vmatrix} = 5 \Rightarrow V = |\mathbf{a} \cdot (\mathbf{b} \times \mathbf{c})| = |5| = 5$

29. For $P(0, 0, 0)$, $Q(3, -2, 1)$, $R(1, 2, 2)$, and $S(1, 1, 4)$, $\overrightarrow{PQ} = \langle 3, -2, 1 \rangle$, $\overrightarrow{PR} = \langle 1, 2, 2 \rangle$, and $\overrightarrow{PS} = \langle 1, 1, 4 \rangle$, so

$\overrightarrow{PQ} \cdot \left(\overrightarrow{PR} \times \overrightarrow{PS}\right) = \begin{vmatrix} 3 & -2 & 1 \\ 1 & 2 & 2 \\ 1 & 1 & 4 \end{vmatrix} = 21$. Thus, $V = \left|\overrightarrow{PQ} \cdot \left(\overrightarrow{PR} \times \overrightarrow{PS}\right)\right| = |21| = 21$.

31. $h = \left|\text{proj}_{\mathbf{a} \times \mathbf{b}}\,\mathbf{c}\right|$, but $\mathbf{a} \times \mathbf{b} = \begin{vmatrix} \mathbf{i} & \mathbf{j} & \mathbf{k} \\ 1 & 2 & 1 \\ 2 & 1 & -1 \end{vmatrix} = -3\mathbf{i} + 3\mathbf{j} - 3\mathbf{k}$, so

$h = \frac{|(\mathbf{a} \times \mathbf{b}) \cdot \mathbf{c}|}{|\mathbf{a} \times \mathbf{b}|} = \frac{|(-3\mathbf{i} + 3\mathbf{j} - 3\mathbf{k}) \cdot (\mathbf{i} + \mathbf{j} + 3\mathbf{k})|}{\sqrt{9 + 9 + 9}} = \frac{9}{\sqrt{27}} = \sqrt{3}$.

33. $\mathbf{a} \cdot (\mathbf{b} \times \mathbf{c}) = \begin{vmatrix} 1 & 2 & 4 \\ -2 & 3 & -1 \\ 0 & 1 & 1 \end{vmatrix} = 0$, so $\mathbf{a}$, $\mathbf{b}$, and $\mathbf{c}$ are coplanar.

35. For $P(1, 0, 1)$, $Q(2, 3, 1)$, $R(-1, 2, -3)$, and $S\left(\frac{2}{3}, -1, 1\right)$, $\overrightarrow{PQ} = \langle 1, 3, 0 \rangle$, $\overrightarrow{PR} = \langle -2, 2, -4 \rangle$, and $\overrightarrow{PS} = \left\langle -\frac{1}{3}, -1, 0 \right\rangle$,

so $\overrightarrow{PQ} \cdot \left(\overrightarrow{PR} \times \overrightarrow{PS}\right) = \begin{vmatrix} 1 & 3 & 0 \\ -2 & 2 & -4 \\ -\frac{1}{3} & -1 & 0 \end{vmatrix} = 1(0 - 4) - 3\left(0 - \frac{4}{3}\right) = 0$. Thus, the points are coplanar.

37. $\left|\overrightarrow{\tau}\right| = |\mathbf{r}|\,|\mathbf{F}|\sin\theta = \frac{3}{4}\,(50)\sin 60° \approx 32.5$ ft-lb

39. $\mathbf{B} = \langle 0, 0, 2\rangle$ and $\mathbf{v} = \left\langle \frac{3}{2}\times 10^5, 0, \frac{3\sqrt{3}}{2}\times 10^5\right\rangle$, so

$$\mathbf{F} = q\mathbf{v}\times\mathbf{B} = \left(1.6\times 10^{-19}\right)\left(10^5\right)\left\langle\frac{3}{2}, 0, \frac{3\sqrt{3}}{2}\right\rangle\times\langle 0, 0, 2\rangle = 1.6\times 10^{-14}\begin{vmatrix} \mathbf{i} & \mathbf{j} & \mathbf{k} \\ \frac{3}{2} & 0 & \frac{3\sqrt{3}}{2} \\ 0 & 0 & 2 \end{vmatrix}$$

$$= 1.6\times 10^{-14}\,(-3\mathbf{j}) = -4.8\times 10^{-14}\mathbf{j} \text{ newtons}$$

41. $\mathbf{b}\times\mathbf{c} = \begin{vmatrix} \mathbf{i} & \mathbf{j} & \mathbf{k} \\ b_1 & b_2 & b_3 \\ c_1 & c_2 & c_3 \end{vmatrix} = (b_2c_3 - c_2b_3)\,\mathbf{i} - (b_1c_3 - c_1b_3)\,\mathbf{j} + (b_1c_2 - c_1b_2)\,\mathbf{k}$, so

$\mathbf{a}\cdot(\mathbf{b}\times\mathbf{c}) = (a_1\mathbf{i} + a_2\mathbf{j} + a_3\mathbf{k})\cdot\left[(b_2c_3 - c_2b_3)\,\mathbf{i} - (b_1c_3 - c_1b_3)\,\mathbf{j} + (b_1c_2 - c_1b_2)\,\mathbf{k}\right]$

$\qquad = a_1b_2c_3 - a_1b_3c_2 - a_2b_1c_3 + a_2b_3c_1 + a_3b_1c_2 - a_3b_2c_1$

$\mathbf{a}\times\mathbf{b} = \begin{vmatrix} \mathbf{i} & \mathbf{j} & \mathbf{k} \\ a_1 & a_2 & a_3 \\ b_1 & b_2 & b_3 \end{vmatrix} = (a_2b_3 - b_2a_3)\,\mathbf{i} - (a_1b_3 - b_1a_3)\,\mathbf{j} + (a_1b_2 - b_1a_2)\,\mathbf{k}$, so

$(\mathbf{a}\times\mathbf{b})\cdot\mathbf{c} = \left[(a_2b_3 - b_2a_3)\,\mathbf{i} - (a_1b_3 - b_1a_3)\,\mathbf{j} + (a_1b_2 - b_1a_2)\,\mathbf{k}\right]\cdot(c_1\mathbf{i} + c_2\mathbf{j} + c_3\mathbf{k})$

$\qquad = a_2b_3c_1 - a_3b_2c_1 - a_1b_3c_2 + a_3b_1c_2 + a_1b_2c_3 - a_2b_1c_3$

and $\begin{vmatrix} a_1 & a_2 & a_3 \\ b_1 & b_2 & b_3 \\ c_1 & c_2 & c_3 \end{vmatrix} = a_1\begin{vmatrix} b_2 & b_3 \\ c_2 & c_3 \end{vmatrix} - a_2\begin{vmatrix} b_1 & b_3 \\ c_1 & c_3 \end{vmatrix} + a_3\begin{vmatrix} b_1 & b_2 \\ c_1 & c_2 \end{vmatrix} = a_1b_2c_3 - a_1b_3c_2 - a_2b_1c_3 + a_2b_3c_1 + a_3b_1c_2 - a_3b_2c_1.$

These three expressions are equal, which is the desired result.

43. $\mathbf{a}\times(\mathbf{b}\times\mathbf{c}) + \mathbf{b}\times(\mathbf{c}\times\mathbf{a}) + \mathbf{c}\times(\mathbf{a}\times\mathbf{b}) = (\mathbf{a}\cdot\mathbf{c})\mathbf{b} - (\mathbf{a}\cdot\mathbf{b})\mathbf{c} + (\mathbf{b}\cdot\mathbf{a})\mathbf{c} - (\mathbf{b}\cdot\mathbf{c})\mathbf{a} + (\mathbf{c}\cdot\mathbf{b})\mathbf{a} - (\mathbf{c}\cdot\mathbf{a})\mathbf{b}$

$\qquad = (\mathbf{a}\cdot\mathbf{c})\mathbf{b} - (\mathbf{a}\cdot\mathbf{b})\mathbf{c} + (\mathbf{a}\cdot\mathbf{b})\mathbf{c} - (\mathbf{b}\cdot\mathbf{c})\mathbf{a} + (\mathbf{b}\cdot\mathbf{c})\mathbf{a} - (\mathbf{a}\cdot\mathbf{c})\mathbf{b} = \mathbf{0}$

45. $|\mathbf{a}\times\mathbf{b}|^2 = (|\mathbf{a}|\,|\mathbf{b}|\sin\theta)^2 = |\mathbf{a}|^2\,|\mathbf{b}|^2\sin^2\theta = |\mathbf{a}|^2\,|\mathbf{b}|^2\left(1 - \cos^2\theta\right) = |\mathbf{a}|^2\,|\mathbf{b}|^2 - |\mathbf{a}|^2\,|\mathbf{b}|^2\cos^2\theta = |\mathbf{a}|^2\,|\mathbf{b}|^2 -$

$(|\mathbf{a}|\,|\mathbf{b}|\cos\theta)^2 = |\mathbf{a}|^2\,|\mathbf{b}|^2 - (\mathbf{a}\cdot\mathbf{b})^2$

47. Let $\mathbf{a} = \langle a_1, a_2, a_3 \rangle$, $\mathbf{b} = \langle b_1, b_2, b_3 \rangle$, and $\mathbf{c} = \langle c_1, c_2, c_3 \rangle$. $\mathbf{b} + \mathbf{c} = \langle b_1 + c_1, b_2 + c_2, b_3 + c_3 \rangle$, so

$$\mathbf{a} \times (\mathbf{b} + \mathbf{c}) = \begin{vmatrix} \mathbf{i} & \mathbf{j} & \mathbf{k} \\ a_1 & a_2 & a_3 \\ b_1 + c_1 & b_2 + c_2 & b_3 + c_3 \end{vmatrix}$$

$$= \left[a_2 (b_3 + c_3) - a_3 (b_2 + c_2) \right] \mathbf{i} - \left[a_1 (b_3 + c_3) - a_3 (b_1 + c_1) \right] \mathbf{j} + \left[a_1 (b_2 + c_2) - a_2 (b_1 + c_1) \right] \mathbf{k}$$

$$= (a_2 b_3 + a_2 c_3 - a_3 b_2 - a_3 c_2) \mathbf{i} - (a_1 b_3 + a_1 c_3 - a_3 b_1 - a_3 c_1) \mathbf{j} + (a_1 b_2 + a_1 c_2 - a_2 b_1 - a_2 c_1) \mathbf{k}$$

On the other hand,

$$\mathbf{a} \times \mathbf{b} + \mathbf{a} \times \mathbf{c} = \begin{vmatrix} \mathbf{i} & \mathbf{j} & \mathbf{k} \\ a_1 & a_2 & a_3 \\ b_1 & b_2 & b_3 \end{vmatrix} + \begin{vmatrix} \mathbf{i} & \mathbf{j} & \mathbf{k} \\ a_1 & a_2 & a_3 \\ c_1 & c_2 & c_3 \end{vmatrix}$$

$$= (a_2 b_3 - a_3 b_2) \mathbf{i} - (a_1 b_3 - a_3 b_1) \mathbf{j} + (a_1 b_2 - a_2 b_1) \mathbf{k} + (a_2 c_3 - a_3 c_2) \mathbf{i} - (a_1 c_3 - a_3 c_1) \mathbf{j} + (a_1 c_2 - a_2 c_1) \mathbf{k}$$

$$= (a_2 b_3 - a_3 b_2 + a_2 c_3 - a_3 c_2) \mathbf{i} - (a_1 b_3 - a_3 b_1 + a_1 c_3 - a_3 c_1) \mathbf{j} + (a_1 b_2 + a_1 c_2 - a_2 b_1 - a_2 c_1) \mathbf{k}$$

The result follows when we compare the two expressions.

49. $\mathbf{a} \times \mathbf{0} = \begin{vmatrix} \mathbf{i} & \mathbf{j} & \mathbf{k} \\ a_1 & a_2 & a_3 \\ 0 & 0 & 0 \end{vmatrix} = 0\mathbf{i} + 0\mathbf{j} + 0\mathbf{k} = \mathbf{0}$ and $\mathbf{0} \times \mathbf{a} = \begin{vmatrix} \mathbf{i} & \mathbf{j} & \mathbf{k} \\ 0 & 0 & 0 \\ a_1 & a_2 & a_3 \end{vmatrix} = 0\mathbf{i} + 0\mathbf{j} + 0\mathbf{k} = \mathbf{0}$

51. Let $\mathbf{a} = a_1\mathbf{i} + a_2\mathbf{j} + a_3\mathbf{k}$ and $\mathbf{b} = b_1\mathbf{i} + b_2\mathbf{j} + b_3\mathbf{k}$. Then

$$(\mathbf{a} + \mathbf{b}) \times \mathbf{c} = \begin{vmatrix} \mathbf{i} & \mathbf{j} & \mathbf{k} \\ a_1 + b_1 & a_2 + b_2 & a_3 + b_3 \\ c_1 & c_2 & c_3 \end{vmatrix}$$

$$= \left[(a_2 + b_2) c_3 - (a_3 + b_3) c_2 \right] \mathbf{i} - \left[(a_1 + b_1) c_3 - (a_3 + b_3) c_1 \right] \mathbf{j} + \left[(a_1 + b_1) c_2 - (a_2 + b_2) c_1 \right] \mathbf{k}$$

$$= (a_2 c_3 + b_2 c_3 - a_3 c_2 - b_3 c_2) \mathbf{i} - (a_1 c_3 + b_1 c_3 - a_3 c_1 - b_3 c_1) \mathbf{j} + (a_1 c_2 + b_1 c_2 - a_2 c_1 - b_2 c_1) \mathbf{k} \quad (1)$$

On the other hand,

$$\mathbf{a} \times \mathbf{c} + \mathbf{b} \times \mathbf{c} = \begin{vmatrix} \mathbf{i} & \mathbf{j} & \mathbf{k} \\ a_1 & a_2 & a_3 \\ c_1 & c_2 & c_3 \end{vmatrix} + \begin{vmatrix} \mathbf{i} & \mathbf{j} & \mathbf{k} \\ b_1 & b_2 & b_3 \\ c_1 & c_2 & c_3 \end{vmatrix}$$

$$= (a_2 c_3 - c_2 a_3) \mathbf{i} - (a_1 c_3 - c_1 a_3) \mathbf{j} + (a_1 c_2 - c_1 a_2) \mathbf{k} + (b_2 c_3 - c_2 b_3) \mathbf{i} - (b_1 c_3 - b_3 c_1) \mathbf{j} + (b_1 c_2 - c_1 b_2) \mathbf{k}$$

$$= (a_2 c_3 - c_2 a_3 + b_2 c_3 - c_2 b_3) \mathbf{i} - (a_1 c_3 - c_1 a_3 + b_1 c_3 - b_3 c_1) \mathbf{j} + (a_1 c_2 - c_1 a_2 + b_1 c_2 - c_1 b_2) \mathbf{k} \quad (2)$$

Comparing (1) and (2), we see that $(\mathbf{a} + \mathbf{b}) \times \mathbf{c} = \mathbf{a} \times \mathbf{c} + \mathbf{b} \times \mathbf{c}$

53. The unit vector parellel to the axis of rotation is $\dfrac{2\mathbf{i} + 2\mathbf{j} + \mathbf{k}}{\sqrt{4 + 4 + 1}} = \frac{2}{3}\mathbf{i} + \frac{2}{3}\mathbf{j} + \frac{1}{3}\mathbf{k}$. Therefore, $\vec{\omega} = \pm \left(\frac{2}{3}\mathbf{i} + \frac{2}{3}\mathbf{j} + \frac{1}{3}\mathbf{k} \right) \omega$.

The velocity is $\mathbf{v} = \omega \times \mathbf{R} = \begin{vmatrix} \mathbf{i} & \mathbf{j} & \mathbf{k} \\ \frac{2}{3} & \frac{2}{3} & \frac{1}{3} \\ 3 & 5 & 2 \end{vmatrix} = \pm \frac{1}{3}\omega \left(-\mathbf{i} - \mathbf{j} + 4\mathbf{k} \right)$ and the speed of the particle is

$$|\mathbf{v}| = \tfrac{1}{3}\omega \sqrt{(-1)^2 + (-1)^2 + 4^2} = \tfrac{1}{3}\omega \cdot 3\sqrt{2} = \sqrt{2}\,\omega.$$

55. If $\mathbf{a} = \langle -1, a, 3 \rangle$ and $\mathbf{b} = \langle 2, 3, b \rangle$ are parallel, then $\mathbf{a} \times \mathbf{b} = \mathbf{0} \Leftrightarrow$

$$\mathbf{a} \times \mathbf{b} = \begin{vmatrix} \mathbf{i} & \mathbf{j} & \mathbf{k} \\ -1 & a & 3 \\ 2 & 3 & b \end{vmatrix} = (ab - 9)\,\mathbf{i} - (-b - 6)\,\mathbf{j} + (-3 - 2a)\,\mathbf{k} = 0\mathbf{i} + 0\mathbf{j} + 0\mathbf{k} \Leftrightarrow ab - 9 = 0, \; b + 6 = 0, \text{ and}$$

$-3 - 2a = 0$. Solving, we find $a = -\frac{3}{2}$ and $b = -6$.

57. True. $\mathbf{b} \times \mathbf{a} = -\mathbf{a} \times \mathbf{b}$, so $\mathbf{a} \times \mathbf{b} + \mathbf{b} \times \mathbf{a} = \mathbf{0}$.

59. True. If $\mathbf{a} = \langle a_1, a_2, a_3 \rangle$ and $\mathbf{b} = \langle b_1, b_2, b_3 \rangle$, then $\mathbf{a} \cdot (\mathbf{a} \times \mathbf{b}) = \begin{vmatrix} a_1 & a_2 & a_3 \\ a_1 & a_2 & a_3 \\ b_1 & b_2 & b_3 \end{vmatrix} = 0$.

61. True.

$$(\mathbf{a} - \mathbf{b}) \times (\mathbf{a} + \mathbf{b}) = (\mathbf{a} - \mathbf{b}) \times \mathbf{a} + (\mathbf{a} - \mathbf{b}) \times \mathbf{b} = \mathbf{a} \times \mathbf{a} - \mathbf{b} \times \mathbf{a} + \mathbf{a} \times \mathbf{b} - \mathbf{b} \times \mathbf{b} = -\mathbf{b} \times \mathbf{a} + \mathbf{a} \times \mathbf{b}$$

$$= \mathbf{a} \times \mathbf{b} + \mathbf{a} \times \mathbf{b} = 2\mathbf{a} \times \mathbf{b}$$

11.5 Concept Questions   ET 10.5

1. a. $x = x_0 + at$, $y = y_0 + bt$, $z = z_0 + ct$

 b. $\dfrac{x - x_0}{a} = \dfrac{y - y_0}{b} = \dfrac{z - z_0}{c}$

 c. The parametric equations are $x = x_0 + at$, $y = y_0 + bt$, $z = z_0$ and the symmetric equations are $\dfrac{x - x_0}{a} = \dfrac{y - y_0}{b}$, $z = z_0$.

3. a. $a\,(x - x_0) + b\,(y - y_0) + c\,(z - z_0) = 0$

 b. $ax + by + cz = d$

11.5 Lines and Planes in Space   ET 10.5

1. a. The direction of the required line is the same as the direction of the given line and so it can be obtained from the parametric equations of the latter. Then use this information with the given point to write down the required equation.

 b. Obtain the vectors $\mathbf{v}_1$ and $\mathbf{v}_2$ from the two given lines. Find the vector $\mathbf{n} = \mathbf{v}_1 \times \mathbf{v}_2$. Then the required line has the direction of $\mathbf{n}$ and contains the given point.

 c. A vector parallel to the required line is $\mathbf{n}$, a vector normal to the plane. It can be obtained from the equation of the plane. Use $\mathbf{n}$ and the given point to write down the required parametric equation.

 d. Obtain $\mathbf{n}_1$ and $\mathbf{n}_2$, the normals to the two planes, from the given equations of the planes. The direction of the required line is given by $\mathbf{v} = \mathbf{n}_1 \times \mathbf{n}_2$. Find a point on the required line by setting one variable, say z, equal to 0, and solving the resulting simultaneous equations in two variables.

3. Parametric equations: $x = 1 + 2t$, $y = 3 + 4t$, $z = 2 + 5t$. Symmetric equations: $\dfrac{x - 1}{2} = \dfrac{y - 3}{4} = \dfrac{z - 2}{5}$.

5. Parametric equations: $x = 3 + 2t$, $y = -t$, $z = -2 + 3t$. Symmetric equations: $\dfrac{x - 3}{2} = \dfrac{y}{-1} = \dfrac{z + 2}{3}$.

7. Let $A\,(2, 1, 4)$ and $B\,(1, 3, 7)$. Then $\mathbf{v} = \overrightarrow{AB} = \langle -1, 2, 3 \rangle$ is parallel to the required line. Thus, parametric equations of the line are $x = 2 - t$, $y = 1 + 2t$, $z = 4 + 3t$ and symmetric equations are $\dfrac{x - 2}{-1} = \dfrac{y - 1}{2} = \dfrac{z - 4}{3}$.

9. Let $A\left(-1, -2, -\frac{1}{2}\right)$ and $B\left(1, \frac{3}{2}, -3\right)$. Then $\vec{AB} = \left\langle 2, \frac{7}{2}, -\frac{5}{2}\right\rangle$, and $\mathbf{v} = 2\vec{AB} = \langle 4, 7, -5\rangle$ is parallel to the required line. Thus, parametric equations of the line are $x = -1 + 4t$, $y = -2 + 7t$, $z = -\frac{1}{2} - 5t$ and symmetric equations are

$$\frac{x+1}{4} = \frac{y+2}{7} = \frac{z+\frac{1}{2}}{-5}.$$

11. The direction of the given line is the same as that of $\mathbf{v} = \langle 1, 2, -3\rangle$, so parametric equations of the required line are

$x = 1 + t$, $y = 2 + 2t$, $z = -1 - 3t$ and symmetric equations are $\dfrac{x-1}{1} = \dfrac{y-2}{2} = \dfrac{z+1}{-3}$. Setting $z = 0$ gives $t = -\frac{1}{3} \Rightarrow$

$x = \frac{2}{3}$ and $y = \frac{4}{3}$, so the line intersects the xy-plane at the point $\left(\frac{2}{3}, \frac{4}{3}, 0\right)$. Setting $y = 0$ gives $t = -1 \Rightarrow x = 0$ and

$z = 2$, so the line intersects the xz-plane at the point $(0, 0, 2)$. Setting $x = 0$ gives $t = -1 \Rightarrow y = 0$ and $z = 2$, so the line

intersects the yz-plane at the point $(0, 0, 2)$.

13. Parametric equations of the line L are $x = -1 - t$, $y = 4 + t$, $z = 3 - t$. Suppose the point $(-3, 6, 1)$ lies on L. Then there

exists some value t_0 such that $-1 - t_0 = -3$, $4 + t_0 = 6$, and $3 - t_0 = 1$. Solving, we find $t_0 = 2$, so the point $(-3, 6, 1)$

does lie on L.

15. A vector parallel to L_1 is $\mathbf{v}_1 = \langle 3, 3, 1\rangle$ and a vector parallel to L_2 is $\mathbf{v}_2 = \langle 4, 6, 1\rangle$. Because $\mathbf{v}_1 \neq c\mathbf{v}_2$ for any scalar c, we

see that $\mathbf{v}_1$ and $\mathbf{v}_2$ (and therefore L_1 and L_2) are not parallel. Suppose the two lines intersect at a point. Then there exist

numbers t_1 and t_2 such that $\left. \begin{array}{rcl} -1 + 3t_1 &=& 1 + 4t_2 \\ -2 + 3t_1 &=& -2 + 6t_2 \\ 3 + t_1 &=& 4 + t_2 \end{array} \right\} \Rightarrow \left. \begin{array}{rcl} 3t_1 - 4t_2 &=& 2 \\ 3t_1 - 6t_2 &=& 0 \\ t_1 - t_2 &=& 1 \end{array} \right\} \Rightarrow t_1 = 2$ and $t_2 = 1$, so the lines intersect at

$(5, 4, 5)$.

17. Parametric equations are $L_1 : x = 2 + 4t$, $y = 3$, $z = 1 - t$ and $L_2 : x = 2 + 2t$, $y = 3 + 2t$, $z = 1 + t$. A vector parallel to

L_1 is $\mathbf{v}_1 = \langle 4, 0, -1\rangle$ and a vector parallel to L_2 is $\mathbf{v}_2 = \langle 2, 2, 1\rangle$. Because $\mathbf{v}_1$ is not a scalar multiple of $\mathbf{v}_2$, we see that $\mathbf{v}_1$

and $\mathbf{v}_2$ (and therefore L_1 and L_2) are not parallel. Suppose the two lines intersect at a point. Then there exist numbers t_1

and t_2 such that $\left. \begin{array}{rcl} 2 + 4t_1 &=& 2 + 2t_2 \\ 3 &=& 3 + 2t_2 \\ 1 - t_1 &=& 1 + t_2 \end{array} \right\} \Rightarrow \left. \begin{array}{rcl} 4t_1 - 2t_2 &=& 0 \\ 2t_2 &=& 0 \\ t_1 + t_2 &=& 0 \end{array} \right\} \Rightarrow t_1 = 0$ and $t_2 = 0$, so the lines intersect at $(2, 3, 1)$.

19. A vector parallel to L_1 is $\mathbf{v}_1 = \langle -1, -2, 1\rangle$ and a vector parallel to L_2 is $\mathbf{v}_2 = \langle 3, 2, 1\rangle$. Because $\mathbf{v}_1$ is

not a scalar multiple of $\mathbf{v}_2$, we see that $\mathbf{v}_1$ and $\mathbf{v}_2$ (and therefore L_1 and L_2) are not parallel. Suppose

the two lines intersect at a point. Then there exist numbers t_1 and t_2 such that $\left. \begin{array}{rcl} 1 - t_1 &=& 2 + 3t_2 \\ 3 - 2t_1 &=& 3 + 2t_2 \\ t_1 &=& 1 + t_2 \end{array} \right\} \Rightarrow$

$\left. \begin{array}{rcl} t_1 + 3t_2 &=& -1 \\ 2t_1 + 2t_2 &=& 0 \\ t_1 - t_2 &=& 1 \end{array} \right\} \Rightarrow t_1 = \frac{1}{2}$ and $t_2 = -\frac{1}{2}$, so the lines intersect at $\left(\frac{1}{2}, 2, \frac{1}{2}\right)$. The angle between L_1 and L_2 is

$\theta = \cos^{-1}\left(\dfrac{|\mathbf{v}_1 \cdot \mathbf{v}_2|}{|\mathbf{v}_1||\mathbf{v}_2|}\right) = \cos^{-1}\left(\dfrac{|\langle -1, -2, 1\rangle \cdot \langle 3, 2, 1\rangle|}{\sqrt{1+4+1}\sqrt{9+4+1}}\right) = \cos^{-1}\dfrac{6}{\sqrt{6}\sqrt{14}} \approx 49.1°.$

21. Parametric equations are $L_1 : x = 1 - 3t, y = -2 + 2t, z = -1 + 4t$ and $L_2 : x = -2 + 2t, y = 4 + 4t, z = 3 + t$. A vector parallel to L_1 is $\mathbf{v}_1 = \langle -3, 2, 4 \rangle$ and a vector parallel to L_2 is $\mathbf{v}_2 = \langle 2, 4, 1 \rangle$. Because $\mathbf{v}_1$ is not a scalar multiple of $\mathbf{v}_2$, we see that $\mathbf{v}_1$ and $\mathbf{v}_2$ (and therefore L_1 and L_2) are not parallel. Suppose the two lines intersect at a point. Then there

$$\text{exist numbers } t_1 \text{ and } t_2 \text{ such that } \left. \begin{array}{rcl} 1 - 3t_1 &=& -2 + 2t_2 \\ -2 + 2t_1 &=& 4 + 4t_2 \\ -1 + 4t_1 &=& 3 + t_2 \end{array} \right\} \Rightarrow \left. \begin{array}{rcl} 3t_1 + 2t_2 &=& 3 \\ 2t_1 - 4t_2 &=& 6 \\ 4t_1 - t_2 &=& 4 \end{array} \right\} \quad \text{Solving the first two equations gives}$$

$t_1 = \frac{3}{2}$ and $t_2 = -\frac{3}{4}$, but substituting these into the third equation leads to $4 \left(\frac{3}{2} \right) - \left(-\frac{3}{4} \right) = \frac{27}{4} = 4$, which is false. Thus, the lines are skew and do not intersect.

23. An equation of the plane containing $P(2, 1, 5)$ with normal vector $\mathbf{n} = \langle 1, 2, 4 \rangle$ is $1(x - 2) + 2(y - 1) + 4(z - 5) = 0 \Leftrightarrow x + 2y + 4z = 24$.

25. An equation of the plane containing $P(1, 3, 0)$ with normal vector $\mathbf{n} = 2\mathbf{i} - 4\mathbf{k}$ is $2(x - 1) + 0(y - 3) - 4(z - 0) = 0 \Leftrightarrow 2x - 4z = 2 \Leftrightarrow x - 2z = 1$.

27. A normal to the given plane is $\mathbf{n} = \langle 2, 3, -1 \rangle$. Because the required plane is parallel to the given plane, $\mathbf{n}$ is also normal to the required plane, so an equation is $2(x - 3) + 3(y - 6) - 1(z + 2) = 0 \Leftrightarrow 2x + 3y - z = 26$.

29. A normal to the given plane is $\mathbf{n} = \langle 1, 0, -3 \rangle$, and this is also normal to the required plane, so an equation is $1(x + 1) + 0(y + 2) - 3(z + 3) = 0$ or $x - 3z = 8$.

31. Let $A(1, 0, -2)$, $B(1, 3, 2)$, and $C(2, 3, 0)$. Then $\overrightarrow{AB} = \langle 0, 3, 4 \rangle$ and $\overrightarrow{AC} = \langle 1, 3, 2 \rangle$. A normal to the required plane is

$$\mathbf{n} = \overrightarrow{AB} \times \overrightarrow{AC} = \begin{vmatrix} \mathbf{i} & \mathbf{j} & \mathbf{k} \\ 0 & 3 & 4 \\ 1 & 3 & 2 \end{vmatrix} = \langle -6, 4, -3 \rangle, \text{ so an equation of the plane is } -6(x - 1) + 4(y - 0) - 3(z + 2) = 0 \Leftrightarrow$$

$6x - 4y + 3z = 0$.

33. Let $P(1, 3, 2)$. Setting $t = 0$ gives the point $Q(1, -1, 3)$ on the line. Also, a vector in the same direction as the line is

$$\mathbf{v} = \langle 1, -2, 2 \rangle, \text{ so a vector normal to the required plane is } \mathbf{n} = \overrightarrow{PQ} \times \mathbf{v} = \begin{vmatrix} \mathbf{i} & \mathbf{j} & \mathbf{k} \\ 0 & -4 & 1 \\ 1 & -2 & 2 \end{vmatrix} = -6\mathbf{i} + \mathbf{j} + 4\mathbf{k}, \text{ and an equation of}$$

the plane is $-6(x - 1) + 1(y - 3) + 4(z - 2) = 0 \Leftrightarrow 6x - y - 4z = -5$.

35. Let $P(3, -4, 5)$. A point on the line is $Q(2, -1, -3)$ and a vector in the same direction as the line is $\mathbf{v} = \langle 2, -3, 5 \rangle$, so a

$$\text{vector normal to the required plane is } \mathbf{n} = \overrightarrow{PQ} \times \mathbf{v} = \begin{vmatrix} \mathbf{i} & \mathbf{j} & \mathbf{k} \\ -1 & 3 & -8 \\ 2 & -3 & 5 \end{vmatrix} = -9\mathbf{i} - 11\mathbf{j} - 3\mathbf{k} = -(9\mathbf{i} + 11\mathbf{j} + 3\mathbf{k}). \text{ Thus, an}$$

equation of the plane is $9(x - 3) + 11(y + 4) + 3(z - 5) = 0 \Leftrightarrow 9x + 11y + 3z = -2$.

37. Let $P(2, 1, 1)$ and $Q(-1, 3, 2)$, so $\overrightarrow{PQ} = \langle -3, 2, 1 \rangle$. If $\mathbf{n} = \langle 2, 3, -4 \rangle$ is normal to the given plane, then a normal to the

$$\text{required plane is } \overrightarrow{PQ} \times \mathbf{n} = \begin{vmatrix} \mathbf{i} & \mathbf{j} & \mathbf{k} \\ -3 & 2 & 1 \\ 2 & 3 & -4 \end{vmatrix} = -11\mathbf{i} - 10\mathbf{j} - 13\mathbf{k} = -(11\mathbf{i} + 10\mathbf{j} + 13\mathbf{k}). \text{ Thus, an equation of the required}$$

plane is $11(x - 2) + 10(y - 1) + 13(z - 1) = 0 \Leftrightarrow 11x + 10y + 13z = 45$.

39. A normal to the plane $x + 2y + z = 1$ is $\mathbf{n}_1 = \langle 1, 2, 1 \rangle$. A normal to the plane $2x - 3y + 4z = 3$ is $\mathbf{n}_2 = \langle 2, -3, 4 \rangle$. Because $\mathbf{n}_1$ is not a scalar multiple of $\mathbf{n}_2$, the two vectors are not parallel, showing that the two planes are not parallel. Because $\mathbf{n}_1 \cdot \mathbf{n}_2 = \langle 1, 2, 1 \rangle \cdot \langle 2, -3, 4 \rangle = 0$, the two vectors $\mathbf{n}_1$ and $\mathbf{n}_2$ are orthogonal, and thus the two given planes are orthogonal.

41. The normals to the planes $3x - y + 2z = 2$ and $2x + 3y + z = 4$ are $\mathbf{n}_1 = \langle 3, -1, 2 \rangle$ and $\mathbf{n}_2 = \langle 2, 3, 1 \rangle$, respectively. Since $\mathbf{n}_1$ is not a scalar multiple of $\mathbf{n}_2$, the two vectors are not parallel, and thus the two planes are not parallel. $\mathbf{n}_1 \cdot \mathbf{n}_2 = \langle 3, -1, 2 \rangle \cdot \langle 2, 3, 1 \rangle = 5 \neq 0$, so the two planes are not orthogonal either. The angle between the two planes is the same as the angle between their normals; that is,

$$\theta = \cos^{-1}\left(\frac{|\mathbf{n}_1 \cdot \mathbf{n}_2|}{|\mathbf{n}_1|\,|\mathbf{n}_2|}\right) = \cos^{-1}\left(\frac{|\langle 3, -1, 2 \rangle \cdot \langle 2, 3, 1 \rangle|}{\sqrt{9+1+4}\sqrt{4+9+1}}\right) = \cos^{-1}\frac{5}{\sqrt{14}\sqrt{14}} \approx 69.1°.$$

43. A normal to the plane $x + y + 2z = 6$ is $\mathbf{n} = \langle 1, 1, 2 \rangle$ and a vector parallel to the line

$L : x = 1 + t, y = 2 + t, z = -1 + t$ is $\mathbf{v} = \langle 1, 1, 1 \rangle$, so the angle between the normal to the plane and the line is

$$\theta = \cos^{-1}\left(\frac{|\mathbf{n} \cdot \mathbf{v}|}{|\mathbf{n}|\,|\mathbf{v}|}\right) = \cos^{-1}\left(\frac{|\langle 1, 1, 2 \rangle \cdot \langle 1, 1, 1 \rangle|}{\sqrt{1+1+4}\sqrt{1+1+1}}\right) = \cos^{-1}\frac{4}{\sqrt{6}\sqrt{3}} \approx 19.5°.$$ Therefore, the required angle is about $90° - 19.5° = 70.5°$.

45. A normal to the plane $2x - 3y + 4z = 3$ is $\mathbf{n}_1 = \langle 2, -3, 4 \rangle$ and a normal to the plane $x + 4y - 2z = 7$ is $\mathbf{n}_2 = \langle 1, 4, -2 \rangle$,

so a vector in the direction of the required line is $\mathbf{v} = \mathbf{n}_1 \times \mathbf{n}_2 = \begin{vmatrix} \mathbf{i} & \mathbf{j} & \mathbf{k} \\ 2 & -3 & 4 \\ 1 & 4 & -2 \end{vmatrix} = -10\mathbf{i} + 8\mathbf{j} + 11\mathbf{k}$. To find a point on

the line of intersection of the planes, we solve the system of equations $\left.\begin{array}{r} 2x - 3y + 4z = 3 \\ x + 4y - 2z = 7 \end{array}\right\}$ Choosing $y = 0$ gives

$\left.\begin{array}{r} 2x + 4z = 3 \\ x - 2z = 7 \end{array}\right\}$ Solving, we get $x = \frac{17}{4}$ and $z = -\frac{11}{8}$, so a point on the line is $\left(\frac{17}{4}, 0, -\frac{11}{8}\right)$ and parametric equations

are $x = \frac{17}{4} - 10t, y = 8t, z = -\frac{11}{8} + 11t$.

47. A normal to the plane $2x + 4y - 3z = 4$ is $\mathbf{n} = \langle 2, 4, -3 \rangle$. Since the line is perpendicular to the plane, its direction is given by $\mathbf{v} = \mathbf{n}$, so parametric equations of the line are $x = 2 + 2t, y = 3 + 4t, z = -1 - 3t$.

49. A vector parallel to the line $L_1 : x = -1 + 2t, y = 2 - 3t, z = 1 + t$ is $\mathbf{v}_1 = \langle 2, -3, 1 \rangle$ and a vector parallel to the line $L_2 : x = 2 - t, y = 1 - 2t, z = 5 - 3t$ is $\mathbf{v}_2 = \langle -1, -2, -3 \rangle$. Thus, a normal to the required plane is

$\mathbf{n} = \mathbf{v}_1 \times \mathbf{v}_2 = \begin{vmatrix} \mathbf{i} & \mathbf{j} & \mathbf{k} \\ 2 & -3 & 1 \\ -1 & -2 & -3 \end{vmatrix} = 11\mathbf{i} + 5\mathbf{j} - 7\mathbf{k}$. We can pick a point in the plane by taking $t = 0$, obtaining $(-1, 2, 1)$,

and so an equation of the plane is $11(x + 1) + 5(y - 2) - 7(z - 1) = 0 \Leftrightarrow 11x + 5y - 7z = -8$.

51. Normals to the planes $2x - 3y + z = 3$ and $x + 2y - 3z = 5$ are $\mathbf{n}_1 = \langle 2, -3, 1 \rangle$ and $\mathbf{n}_2 = \langle 1, 2, -3 \rangle$, respectively, so a

vector parallel to the line of intersection of the two planes is $\mathbf{n}_1 \times \mathbf{n}_2 = \begin{vmatrix} \mathbf{i} & \mathbf{j} & \mathbf{k} \\ 2 & -3 & 1 \\ 1 & 2 & -3 \end{vmatrix} = 7\mathbf{i} + 7\mathbf{j} + 7\mathbf{k} = 7\langle 1, 1, 1 \rangle$.

Then, since $\langle 3, 2, -4 \rangle$ is normal to the plane $3x + 2y - 4z = 7$, a vector normal to the required plane is

$\mathbf{n} = (\mathbf{i} + \mathbf{j} + \mathbf{k}) \times \langle 3, 2, -4 \rangle = \begin{vmatrix} \mathbf{i} & \mathbf{j} & \mathbf{k} \\ 1 & 1 & 1 \\ 3 & 2 & -4 \end{vmatrix} = -6\mathbf{i} + 7\mathbf{j} - \mathbf{k}$. Also, a point on the line of intersection of the two planes

$2x - 3y + z = 3$ and $x + 2y - 3z = 5$ is found by solving the system $\left.\begin{array}{r} 2x - 3y + z = 3 \\ x + 2y - 3z = 5 \end{array}\right\}$ Choosing $y = 0$ gives

$\left.\begin{array}{r} 2x + z = 3 \\ x - 3z = 5 \end{array}\right\} \Rightarrow x = 2, z = -1$, and so one such point is $(2, 0, -1)$. Therefore, an equation of the required plane is

$-6(x - 2) + 7(y - 0) - 1(z + 1) = 0 \Leftrightarrow 6x - 7y + z = 11$.

53. Suppose the line and the plane intersect at a point $P(x_0, y_0, z_0)$. Then there exists a number t_0 such that
$2(2 + 3t_0) + 3(-1 + t_0) - (3 - 2t_0) = 9 \Leftrightarrow t_0 = 1$, so the point of intersection is $(5, 0, 1)$.

55. To find the distance between $P(3, 1, 2)$ and the plane $2x - 3y + 4z = 7 \Leftrightarrow 2x - 3y + 4z - 7 = 0$, we write

$$\mathbf{n} = \langle a, b, c \rangle = \langle 2, -3, 4 \rangle \Rightarrow D = \frac{|2(3) - 3(1) + 4(2) - 7|}{\sqrt{2^2 + 3^2 + 4^2}} = \frac{4\sqrt{29}}{29}.$$

57. For the planes $x + 2y - 4z = 1$ and $x + 2y - 4z = 7$, the normal vectors are $\mathbf{n}_1 = \langle 1, 2, -4 \rangle$ and $\mathbf{n}_2 = \langle 1, 2, -4 \rangle$, which are identical. Thus, the planes are parallel. Pick a point on the first plane, say $(1, 0, 0)$ (set $y = z = 0$). Then

$$D = \frac{|1 + 0 + 0 - 7|}{\sqrt{1 + 4 + 16}} = \frac{2\sqrt{21}}{7}.$$

59. From the figure, we see that

$$D = \left|\overrightarrow{QP}\right| \sin\theta = \frac{\left|\overrightarrow{QP}\right| |\mathbf{u}| \sin\theta}{|\mathbf{u}|} = \frac{\left|\overrightarrow{QP} \times \mathbf{u}\right|}{|\mathbf{u}|}.$$

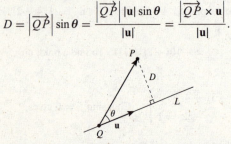

61. Here $P(1, -2, 3)$ and we can take $\mathbf{u} = \langle 3, 1, 2 \rangle$. Choosing $Q(-2, 1, -3)$, we find

$$\overrightarrow{QP} \times \mathbf{u} = \begin{vmatrix} \mathbf{i} & \mathbf{j} & \mathbf{k} \\ 3 & -3 & 6 \\ 3 & 1 & 2 \end{vmatrix} = -12\mathbf{i} + 12\mathbf{j} + 12\mathbf{k}, \text{ so } D = \frac{\left|\overrightarrow{QP} \times \mathbf{u}\right|}{|\mathbf{u}|} = \frac{|-12\mathbf{i} + 12\mathbf{j} + 12\mathbf{k}|}{\sqrt{9 + 1 + 4}} = \frac{12\sqrt{3}}{\sqrt{14}} = \frac{6\sqrt{42}}{7}.$$

63. Let us denote the planes by $\pi_1 : ax + by + cz = d_1$ and $\pi_2 : ax + by + cz = d_2$. Choose a point $P(x_0, y_0, z_0)$ on π_1 and observe that a normal to π_2 is $\mathbf{n} = \langle a, b, c \rangle$. So, using the result of Exercise 59, $D = \dfrac{|ax_0 + by_0 + cz_0 - d_2|}{\sqrt{a^2 + b^2 + c^2}}$. But since

$P(x_0, y_0, z_0)$ lies on π_1, it satisfies the equation of the plane; that is, $ax_0 + by_0 + cz_0 = d_1$. Thus, $D = \dfrac{|d_1 - d_2|}{\sqrt{a^2 + b^2 + c^2}}$.

65. Write $L_1 : \dfrac{x - 1}{-2} = \dfrac{y - 4}{-6} = \dfrac{z - 3}{-2}$ and $L_2 : x - 2 = \dfrac{y + 2}{-5} = \dfrac{z - 1}{-3}$. Since the lines are skew, each line is contained in one of two parallel planes. The vectors $\mathbf{v}_1 = \langle -2, -6, -2 \rangle$ and $\mathbf{v}_2 = \langle 1, -5, -3 \rangle$ are parallel to L_1 and L_2, respectively,

so a vector normal to the planes is $\mathbf{n} = \left(-\tfrac{1}{2}\mathbf{v}_1\right) \times \mathbf{v}_2 = \begin{vmatrix} \mathbf{i} & \mathbf{j} & \mathbf{k} \\ 1 & 3 & 1 \\ 1 & -5 & -3 \end{vmatrix} = -4\mathbf{i} + 4\mathbf{j} - 8\mathbf{k} = -4(\mathbf{i} - \mathbf{j} + 2\mathbf{k})$.

Choosing $P(1, 4, 3)$ on L_1, we see that the plane containing L_1 has equation $(x - 1) - (y - 4) + 2(z - 3) = 0$
$\Leftrightarrow x - y + 2z = 3$. Choosing $P(2, -2, 1)$ on L_2, we see that the plane containing L_2 has equation
$(x - 2) - (y + 2) + 2(z - 1) = 0 \Leftrightarrow x - y + 2z = 6$. Finally, using the result of Exercise 63 with $d_1 = 3$ and $d_2 = 6$, we

find $D = \dfrac{|d_1 - d_2|}{\sqrt{a^2 + b^2 + c^2}} = \dfrac{|3 - 6|}{\sqrt{1 + 1 + 4}} = \dfrac{\sqrt{6}}{2}$.

67. False. Let $L_1 : x = t, y = 0, z = 0$, $L_2 : x = 0, y = t, z = 0$, and $L_3 : x = 0, y = 0, z = t$. Then L_1 and L_2 are both perpendicular to L_3, but L_1 and L_2 are not parallel.

69. False. The planes $z = 1$ and $z = 2$ are both perpendicular to the plane $y = 0$, but they are not perpendicular to each other.

71. False in general. This is true if and only if the point does not line on the line.

11.6   Concept Questions

1. The trace of a surface in a plane is the intersection of the surface and the plane. The trace of $z = x^2 + y^2$ in the plane $z = 4$ is the circle with radius 2 centered on the z-axis and lying in the plane $z = 4$.

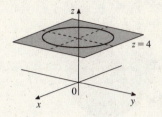

3. **a.** A quadric surface is the graph of the second-degree equation

$Ax^2 + By^2 + Cz^2 + Dxy + Exz + Fyz + Gx + Hy + Iz + J = 0$, where $A, B, \dots, J$ are constants.

  b. (1) $\dfrac{x^2}{a^2} + \dfrac{y^2}{b^2} + \dfrac{z^2}{c^2} = 1$, (2) $\dfrac{x^2}{a^2} + \dfrac{y^2}{b^2} - \dfrac{z^2}{c^2} = 1$, (3) $-\dfrac{x^2}{a^2} - \dfrac{y^2}{b^2} + \dfrac{z^2}{c^2} = 1$, (4) $\dfrac{x^2}{a^2} + \dfrac{y^2}{b^2} - \dfrac{z^2}{c^2} = 0$, (5) $\dfrac{x^2}{a^2} + \dfrac{y^2}{b^2} = cz$,

  (6) $\dfrac{x^2}{a^2} - \dfrac{y^2}{b^2} = cz$

11.6   Surfaces in Space

1.   3.   5.

7.   9.   11.

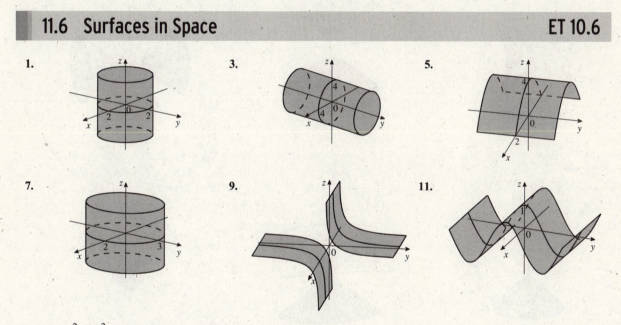

13. $x^2 + \dfrac{y^2}{16} + \dfrac{z^2}{4} = 1$ is an equation of the ellipsoid with center at the origin and axes the x-, y-, and z-axes with intercepts ± 1, ± 4, and ± 2, respectively, so it has graph a.

15. $-2x^2 - 2y^2 + z^2 = -\dfrac{x^2}{1/2} - \dfrac{y^2}{1/2} + \dfrac{z^2}{1} = 1$ is an equation of a hyperboloid of two sheets with axis the z-axis, so it has graph f.

17. $x^2 + \dfrac{z^2}{4} = y$ is an equation of an elliptic paraboloid with axis the y-axis, so it has graph e.

19. $\dfrac{x^2}{4} + \dfrac{y^2}{9} + \dfrac{z^2}{25} = 1$ is an equation of an ellipsoid with center at the origin and axes the coordinate axes with intercepts $x = \pm 2$, $y = \pm 3$, and $z = \pm 5$, so it has graph b.

21. $\dfrac{x^2}{1^2} + \dfrac{y^2}{2^2} + \dfrac{z^2}{2^2} = 1$

23. $\dfrac{x^2}{2^2} + \dfrac{y^2}{3^2} + \dfrac{z^2}{6^2} = 1$

25. $x^2 + y^2 - \dfrac{z^2}{2^2} = 1$

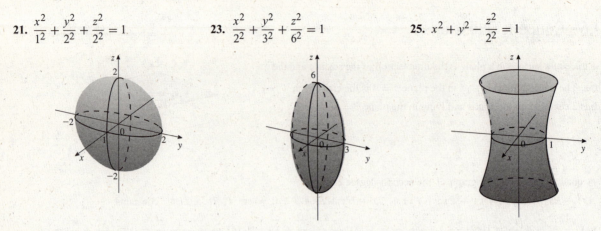

27. $\dfrac{x^2}{2^2} + \dfrac{y^2}{1^2} - \dfrac{z^2}{2^2} = 1$

29. $-x^2 - y^2 + z^2 = 1$

31. $-\dfrac{x^2}{1^2} + \dfrac{y^2}{2^2} - \dfrac{z^2}{\left(\sqrt{2}\right)^2} = 1$

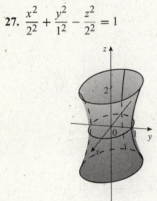

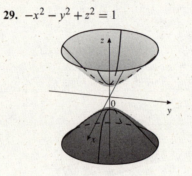

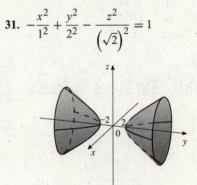

33. $x^2 + y^2 - z^2 = 0$

35. $\dfrac{x^2}{1^2} + \dfrac{y^2}{\left(\frac{3}{2}\right)^2} - \dfrac{z^2}{3^2} = 0$

37. $x^2 + y^2 = z$

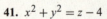

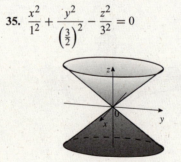

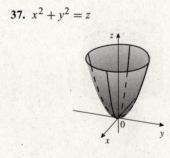

39. $\dfrac{x^2}{3^2} + \dfrac{y^2}{1^2} = \dfrac{z}{3^2}$

41. $x^2 + y^2 = z - 4$

43. $y^2 - x^2 = z$

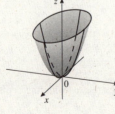

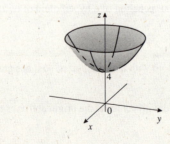

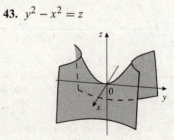

45.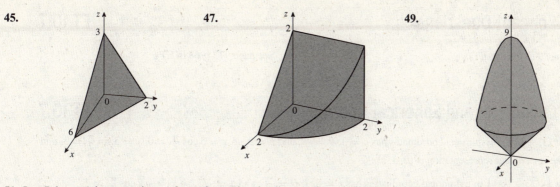

47.

49.

51. Let $P(x, y, z)$ be any point on the surface. Then its distance from the plane $x = 3$ is $D_1 = |x - 3|$. Its distance

from the point $(-3, 0, 0)$ is $D_2 = \sqrt{(x + 3)^2 + y^2 + z^2}$. Setting $D_1 = D_2$ gives $(x - 3)^2 = (x + 3)^2 + y^2 + z^2 \Leftrightarrow$

$x^2 - 6x + 9 = x^2 + 6x + 9 + y^2 + z^2 \Rightarrow y^2 + z^2 = -12x$, so the required surface is a paraboloid.

53. We have $\left.\begin{array}{l} 2x^2 + y^2 - 3z^2 \qquad\ + 2y = 6 \\ 4x^2 + 2y^2 - 6z^2 - 4x \qquad = 4 \end{array}\right\} \Rightarrow \left.\begin{array}{l} 4x^2 + 2y^2 - 6z^2 \qquad\ + 4y = 12 \\ 4x^2 + 2y^2 - 6z^2 - 4x \qquad = 4 \end{array}\right\}$ Subtracting the second

equation from the first yields $4x + 4y = 8 \Leftrightarrow x + y = 2$, an equation of a plane.

55.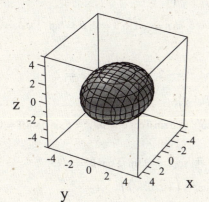

57.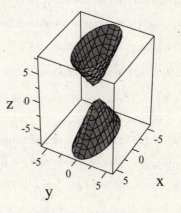

59.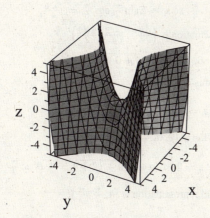

61. False. The graph of $y = x + 3$ is a plane in 3-space.

63. True. $\dfrac{x^2}{a^2} + \dfrac{y^2}{b^2} + \dfrac{z^2}{c^2} = 4 \Leftrightarrow \dfrac{x^2}{(2a)^2} + \dfrac{y^2}{(2b)^2} + \dfrac{z^2}{(2c)^2} = 1$

11.7 Concept Questions   ET 10.7

1. a. See page 971 (967 in ET).   **b.** See page 972 (968 in ET).

11.7 Cylindrical and Spherical Coordinates   ET 10.7

1. The point $\left(3, \frac{\pi}{2}, 2\right)$ in cylindrical coordinates has $r = 3$, $\theta = \frac{\pi}{2}$, and $z = 2$, so $x = 3\cos\frac{\pi}{2} = 0$, $y = 3\sin\frac{\pi}{2} = 3$, and $z = 2$. The rectangular representation is $(0, 3, 2)$.

3. The point $\left(\sqrt{2}, \frac{\pi}{4}, \sqrt{3}\right)$ in cylindrical coordinates has $r = \sqrt{2}$, $\theta = \frac{\pi}{4}$, and $z = \sqrt{3}$, so $x = \sqrt{2}\cos\frac{\pi}{4} = 1$, $y = \sqrt{2}\sin\frac{\pi}{4} = 1$, and $z = \sqrt{3}$. The rectangular representation is $\left(1, 1, \sqrt{3}\right)$.

5. The point $\left(3, -\frac{\pi}{6}, 2\right)$ in cylindrical coordinates has $r = 3$, $\theta = -\frac{\pi}{6}$, and $z = 2$, so $x = 3\cos\left(-\frac{\pi}{6}\right) = \frac{3\sqrt{3}}{2}$, $y = 3\sin\left(-\frac{\pi}{6}\right) = -\frac{3}{2}$, and $z = 2$. The rectangular representation is $\left(\frac{3\sqrt{3}}{2}, -\frac{3}{2}, 2\right)$.

7. The point $(2, 0, 3)$ in rectangular coordinates has $x = 2$, $y = 0$, and $z = 3$, so $r = \sqrt{4 + 0} = 2$, $\theta = \tan^{-1} 0 = 0$, and $z = 3$. The cylindrical representation is $(2, 0, 3)$.

9. The point $\left(1, \sqrt{3}, 5\right)$ in rectangular coordinates has $x = 1$, $y = \sqrt{3}$, and $z = 5$, so $r = \sqrt{1 + 3} = 2$, $\theta = \tan^{-1}\sqrt{3} = \frac{\pi}{3}$, and $z = 5$. The cylindrical representation is $\left(2, \frac{\pi}{3}, 5\right)$.

11. The point $\left(\sqrt{3}, 1, -2\right)$ in rectangular coordinates has $x = \sqrt{3}$, $y = 1$, and $z = -2$, so $r = \sqrt{3 + 1} = 2$, $\theta = \tan^{-1}\left(\frac{1}{\sqrt{3}}\right) = \frac{\pi}{6}$, and $z = -2$. The cylindrical representation is $\left(2, \frac{\pi}{6}, -2\right)$.

13. The point $(5, 0, 0)$ in spherical coordinates has $\rho = 5$, $\theta = 0$, and $\phi = 0$, so $x = 5\sin 0 \cos 0 = 0$, $y = 5\sin 0 \sin 0 = 0$, and $z = 5\cos 0 = 5$. The rectangular representation is $(0, 0, 5)$.

15. The point $\left(2, 0, \frac{\pi}{4}\right)$ in spherical coordinates has $\rho = 2$, $\theta = 0$, and $\phi = \frac{\pi}{4}$, so $x = 2\sin\frac{\pi}{4}\cos 0 = \sqrt{2}$, $y = 2\sin\frac{\pi}{4}\sin 0 = 0$, and $z = 2\cos\frac{\pi}{4} = \sqrt{2}$. The rectangular representation is $\left(\sqrt{2}, 0, \sqrt{2}\right)$.

17. The point $\left(5, \frac{\pi}{6}, \frac{\pi}{4}\right)$ in spherical coordinates has $\rho = 5$, $\theta = \frac{\pi}{6}$, and $\phi = \frac{\pi}{4}$, so $x = 5\sin\frac{\pi}{4}\cos\frac{\pi}{6} = \frac{5\sqrt{6}}{4}$, $y = 5\sin\frac{\pi}{4}\sin\frac{\pi}{6} = \frac{5\sqrt{2}}{4}$, and $z = 5\cos\frac{\pi}{4} = \frac{5\sqrt{2}}{2}$. The rectangular representation is $\left(\frac{5\sqrt{6}}{4}, \frac{5\sqrt{2}}{4}, \frac{5\sqrt{2}}{2}\right)$.

19. The point $(-2, 0, 0)$ in rectangular coordinates has $x = -2$ and $y = z = 0$, so $\rho = \sqrt{4 + 0 + 0} = 2$, $\theta = \tan^{-1}\left(\frac{0}{-2}\right) = \pi$, and $\phi = \cos^{-1} 0 = \frac{\pi}{2}$. The spherical representation is $\left(2, \pi, \frac{\pi}{2}\right)$.

21. The point $\left(\sqrt{3}, 0, 1\right)$ in rectangular coordinates has $x = \sqrt{3}$, $y = 0$, and $z = 1$, so $\rho = \sqrt{3 + 0 + 1} = 2$, $\theta = \tan^{-1}\left(\frac{0}{\sqrt{3}}\right) = 0$, and $\phi = \cos^{-1}\left(\frac{1}{2}\right) = \frac{\pi}{3}$. The spherical representation is $\left(2, 0, \frac{\pi}{3}\right)$.

23. The point $\left(0, 2\sqrt{3}, 2\right)$ in rectangular coordinates has $x = 0$, $y = 2\sqrt{3}$, and $z = 2$, so $\rho = \sqrt{0 + 12 + 4} = 4$, $\theta = \frac{\pi}{2}$ because $\tan^{-1}\left(\frac{2\sqrt{3}}{0}\right)$ is undefined, and $\phi = \cos^{-1}\left(\frac{2}{4}\right) = \frac{\pi}{3}$. The spherical representation is $\left(4, \frac{\pi}{2}, \frac{\pi}{3}\right)$.

25. The point $\left(2, \frac{\pi}{4}, 0\right)$ in cylindrical coordinates has $r = 2$, $\theta = \frac{\pi}{4}$, and $z = 0$, so $\rho = \sqrt{4 + 0} = 2$ and $\phi = \cos^{-1}\left(\frac{0}{2}\right) = \frac{\pi}{2}$. The spherical representation is $\left(2, \frac{\pi}{4}, \frac{\pi}{2}\right)$.

27. The point $\left(4, \frac{\pi}{3}, -4\right)$ in cylindrical coordinates has $r = 4$, $\theta = \frac{\pi}{3}$, and $z = -4$, so $\rho = \sqrt{16 + 16} = 4\sqrt{2}$ and $\phi = \cos^{-1}\left(\frac{-4}{4\sqrt{2}}\right) = \frac{3\pi}{4}$. The spherical representation is $\left(4\sqrt{2}, \frac{\pi}{3}, \frac{3\pi}{4}\right)$.

29. The point $\left(4, \frac{\pi}{6}, 6\right)$ in cylindrical coordinates has $r = 4$, $\theta = \frac{\pi}{6}$, and $z = 6$, so $\rho = \sqrt{16 + 36} = 2\sqrt{13}$ and

$\phi = \cos^{-1}\left(\frac{6}{2\sqrt{13}}\right)$. The spherical representation is $\left(2\sqrt{13}, \frac{\pi}{6}, \cos^{-1}\left(\frac{3\sqrt{13}}{13}\right)\right)$.

31. The point $(3, 0, 0)$ in spherical coordinates has $\rho = 3$, $\theta = 0$, and $\phi = 0$, so $r = 3\sin 0 = 0$ and $z = 3\cos 0 = 3$. The cylindrical representation is $(0, 0, 3)$.

33. The point $\left(2, \frac{3\pi}{2}, \frac{\pi}{2}\right)$ in spherical coordinates has $\rho = 2$, $\theta = \frac{3\pi}{2}$, and $\phi = \frac{\pi}{2}$, so $r = 2\sin\frac{\pi}{2} = 2$ and $z = 2\cos\frac{\pi}{2} = 0$.

The cylindrical representation is $\left(2, \frac{3\pi}{2}, 0\right)$.

35. The point $\left(1, \frac{\pi}{4}, \frac{\pi}{3}\right)$ in spherical coordinates has $\rho = 1$, $\theta = \frac{\pi}{4}$, and $\phi = \frac{\pi}{3}$, so $r = \sin\frac{\pi}{3} = \frac{\sqrt{3}}{2}$ and $z = \cos\frac{\pi}{3} = \frac{1}{2}$.

The cylindrical representation is $\left(\frac{\sqrt{3}}{2}, \frac{\pi}{4}, \frac{1}{2}\right)$.

37. First we find the points in rectangular coordinates. The cylindrical coordinates of $P_1\left(2, \frac{\pi}{3}, 0\right)$ are $r = 2$, $\theta = \frac{\pi}{3}$, and

$z = 0$, so $x = 2\cos\frac{\pi}{3} = 1$ and $y = 2\sin\frac{\pi}{3} = \sqrt{3}$. Therefore, the representation of P_1 in rectangular coordinates

is $P_1\left(1, \sqrt{3}, 0\right)$. The cylindrical coordinates of $P_2\left(1, \pi, 2\right)$ are $r = 1$, $\theta = \pi$, and $z = 2$, so $x = 1\cos\pi = -1$

and $y = 1\sin\pi = 0$. Therefore, the representation of P_2 in rectangular coordinates is $P_2\left(-1, 0, 2\right)$. Finally,

$$D\left(P_1, P_2\right) = \sqrt{(-1 - 1)^2 + \left(0 - \sqrt{3}\right)^2 + 2^2} = \sqrt{11}.$$

39. $r = 2$. This surface is the circular cylinder with radius 2 and central axis the z-axis.

41. $\rho = 2$. This surface is the sphere with center the origin and radius 2.

43. $\phi = \frac{\pi}{4}$. This surface is the upper half of a right circular cone with vertex the origin and axis the positive z-axis.

45. $z = 4 - r^2$. The rectangular equation is $z = 4 - x^2 - y^2$; this represents a paraboloid with vertex $(0, 0, 4)$ and axis the z-axis, opening downward.

47. $\rho\cos\phi = 3 \Leftrightarrow z = 3$ in rectangular coordinates. This surface is the plane parallel to the xy-plane and three units above it.

49. $r\sec\theta = 4 \Leftrightarrow r = 4\cos\theta \Leftrightarrow r^2 = 4r\cos\theta \Leftrightarrow x^2 + y^2 = 4x \Leftrightarrow (x - 2)^2 + y^2 = 4$ in rectangular coordinates. This surface is the circular cylinder with radius 2 and central axis the line parallel to the z-axis passing through $(2, 0, 0)$.

51. $z = r^2\sin^2\theta = (r\sin\theta)^2 = y^2$ in rectangular coordinates. This is an equation of a parabolic cylinder.

53. $r^2 + z^2 = 16 \Leftrightarrow x^2 + y^2 + z^2 = 16$ in rectangular coordinates. This surface is the sphere with radius 4 centered at the origin.

55. $\rho = 2\csc\phi\sec\theta \Rightarrow 2 = \rho\sin\phi\cos\theta \Rightarrow x = 2$ in rectangular coordinates. This surface is a plane parallel to the yz-plane passing through $(2, 0, 0)$.

57. $r^2 - 3r + 2 = 0 \Leftrightarrow (r - 2)(r - 1) = 0 \Leftrightarrow r = 1$ or $r = 2$. This surface consists of two circular cylinders with radii 1 and 2 and axis the z-axis.

Note: For Exercises 59–65, we retain only those expressions necessary to describe the entire surface.

59. a. $x^2 + y^2 + z^2 = 4 \Leftrightarrow \left(x^2 + y^2\right) + z^2 = 4 \Rightarrow r^2 + z^2 = 4$

b. $x^2 + y^2 + z^2 = 4 \Leftrightarrow \rho^2 = 4 \Rightarrow \rho = 2$

61. a. $x^2 + y^2 = 2z \Leftrightarrow r^2 - 2z = 0$

b. $x^2 + y^2 = 2z \Leftrightarrow (\rho\sin\phi\cos\theta)^2 + (\rho\sin\phi\sin\theta)^2 = 2\rho\cos\phi \Leftrightarrow$

$\rho^2\sin^2\phi\left(\cos^2\theta + \sin^2\theta\right) - 2\rho\cos\phi = \rho\left(\rho\sin^2\phi - 2\cos\phi\right) = 0 \Rightarrow \rho\sin^2\phi - 2\cos\phi = 0$

63. a. $2x + 3y - 4z = 12 \Leftrightarrow 2r\cos\theta + 3r\sin\theta - 4z = 12 \Rightarrow r\left(2\cos\theta + 3\sin\theta\right) - 4z = 12$

b. $2x + 3y - 4z = 12 \Leftrightarrow 2\rho\sin\phi\cos\theta + 3\rho\sin\phi - 4\rho\cos\phi = 12 \Rightarrow \rho\left(2\sin\phi\cos\theta + 3\sin\phi\sin\theta - 4\cos\phi\right) = 12$

65. a. $x^2 + z^2 = 4 \Leftrightarrow r^2\cos^2\theta + z^2 = 4$

b. $x^2 + z^2 = 4 \Leftrightarrow (\rho\sin\phi\cos\theta)^2 + (\rho\cos\phi)^2 = 4 \Leftrightarrow \rho^2\left(\sin^2\phi\cos^2\theta + \cos^2\phi\right) = 4$

67. $r \leq z \leq 2$. $r = z \Rightarrow x^2 + y^2 = z^2$. The region is the solid under the plane $z = 2$ and above the top half of the cone $x^2 + y^2 = z^2$.

69. $0 \leq \theta \leq 2\pi, 0 \leq \phi \leq \frac{\pi}{6}, 0 \leq \rho \leq a \sec \phi$. Observe that $\rho = a \sec \phi \Leftrightarrow \rho \cos \phi = a \Leftrightarrow z = a$. The region is the solid above the cone $\phi = \frac{\pi}{6}$ and below the plane $z = a$.

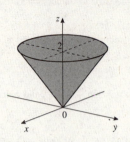

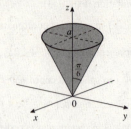

71. a. For Los Angeles, $\phi = 90 - 34.06 = 55.94°$ and $\theta = 360 - 118.25 = 241.75°$, so the location of L.A. is $L\,(3960, 241.75°, 55.94°)$.

For Paris, $\phi = 90 - 48.52 = 41.48°$ and $\theta = 2.20°$, so the location of Paris is $P\,(3960, 2.20°, 41.48°)$.

b. For Los Angeles, $x = 3960 \sin 55.94° \cos 241.75° \approx -1552.8$, $y = 3960 \sin 55.94° \sin 241.75° \approx -2889.9$, and $z = 3960 \cos 55.94° \approx 2217.8$, so the location of L.A. in rectangular coordinates is $(-1552.8, -2889.9, 2217.8)$.

For Paris, $x = 3960 \sin 41.48° \cos 2.20° \approx 2621.0$, $y = 3960 \sin 41.48° \sin 2.20° \approx 100.7$, and $z = 3960 \cos 41.48° \approx 2966.8$, so the location of Paris in rectangular coordinates is $(2621.0, 100.7, 2966.8)$.

c. Let $\mathbf{a} = \overrightarrow{OL} = \langle -0.39\rho, -0.73\rho, 0.56\rho \rangle$ and $\mathbf{b} = \overrightarrow{OP} = \langle 0.66\rho, 0.03\rho, 0.75\rho \rangle$. To find the angle ψ between $\mathbf{a}$ and $\mathbf{b}$, we write $\cos \psi = \dfrac{\mathbf{a} \cdot \mathbf{b}}{|\mathbf{a}|\,|\mathbf{b}|} = \dfrac{\rho^2 \langle -0.39, -0.73, 0.56 \rangle \cdot \langle 0.66, 0.03, 0.75 \rangle}{\rho^2 \sqrt{(-0.39)^2 + (-0.73)^2 + (0.56)^2}\sqrt{(0.66)^2 + (0.03)^2 + (0.75)^2}} \Rightarrow$

$\psi \approx 81.9° = 1.43$ radians. Therefore, the great circle distance between L.A. and Paris is about $\rho\psi \approx 3960\,(1.43) \approx 5663$ mi.

73. True. The point $\left(2, \frac{\pi}{6}\right)$ can also be represented as $\left(2, \frac{13\pi}{6}\right)$, etc.

75. True. $\theta = c$ in cylindrical coordinates represents a vertical plane, whereas $\theta = c$ in spherical coordinates represents a half-plane.

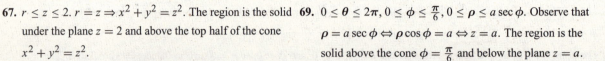

Chapter 11   Review   ET 10

Concept Review

1. a. direction; magnitude

    b. arrow; arrow; direction; length

    c. initial; A; terminal; B

    d. direction; magnitude

5. a. $(h, k, l); r$

    b. $\left(\dfrac{x_1 + x_2}{2}, \dfrac{y_1 + y_2}{2}, \dfrac{z_1 + z_2}{2}\right)$

3. a. $\langle a_1, a_2 \rangle$; a_1; a_2; scalar; $\langle 0, 0 \rangle$

    b. $\langle a_1 + b_1, a_2 + b_2 \rangle$; $\langle ca_1, ca_2 \rangle$

7. a. $a_1 b_1 + a_2 b_2 + a_3 b_3$; scalar

    b. $\sqrt{\mathbf{a} \cdot \mathbf{a}}$

    c. $\dfrac{\mathbf{a} \cdot \mathbf{b}}{|\mathbf{a}|\,|\mathbf{b}|}$

9. a. vector projection; vector component

b. scalar component

c. $\left(\dfrac{\mathbf{b} \cdot \mathbf{a}}{|\mathbf{a}|^2}\right)\mathbf{a}$

d. $\mathbf{F} \cdot \overrightarrow{PQ}$

11. a. $\mathbf{a} \cdot (\mathbf{b} \times \mathbf{c})$

b. $|\mathbf{a} \cdot (\mathbf{b} \times \mathbf{c})|$

13. a. $a\,(x - x_0) + b\,(y - y_0) + c\,(z - z_0) = 0$

b. plane; $\mathbf{n} = \langle a, b, c \rangle$; normal vectors

Review Exercises

1. $2\mathbf{a} - 3\mathbf{b} = 2\,(2\mathbf{i} - \mathbf{j} + 3\mathbf{k}) - 3\,(\mathbf{i} + 2\mathbf{j} - \mathbf{k}) = \mathbf{i} - 8\mathbf{j} + 9\mathbf{k}$

3. $|3\mathbf{a} + 2\mathbf{b}| = |3\,(2\mathbf{i} - \mathbf{j} + 3\mathbf{k}) + 2\,(\mathbf{i} + 2\mathbf{j} - \mathbf{k})| = |8\mathbf{i} + \mathbf{j} + 7\mathbf{k}| = \sqrt{8^2 + 1^2 + 7^2} = \sqrt{114}$

5. $\mathbf{a} \times \mathbf{c} = \begin{vmatrix} \mathbf{i} & \mathbf{j} & \mathbf{k} \\ 2 & -1 & 3 \\ 3 & -2 & 1 \end{vmatrix} = 5\mathbf{i} + 7\mathbf{j} - \mathbf{k}$

7. $\mathbf{a} \cdot (\mathbf{b} \times \mathbf{c}) = \begin{vmatrix} 2 & -1 & 3 \\ 1 & 2 & -1 \\ 3 & -2 & 1 \end{vmatrix} = -20$

9. $\mathbf{a} \times (\mathbf{b} + \mathbf{c}) = (2\mathbf{i} - \mathbf{j} + 3\mathbf{k}) \times \left[(\mathbf{i} + 2\mathbf{j} - \mathbf{k}) + (3\mathbf{i} - 2\mathbf{j} + \mathbf{k})\right] = (2\mathbf{i} - \mathbf{j} + 3\mathbf{k}) \times (4\mathbf{i}) = \begin{vmatrix} \mathbf{i} & \mathbf{j} & \mathbf{k} \\ 2 & -1 & 3 \\ 4 & 0 & 0 \end{vmatrix} = 12\mathbf{j} + 4\mathbf{k}$

11. $|\mathbf{c}| = \sqrt{3^2 + (-2)^2 + 1^2} = \sqrt{14}$ and the required unit vectors are $\mathbf{u}_1 = \dfrac{\mathbf{c}}{|\mathbf{c}|} = \dfrac{3\sqrt{14}}{14}\mathbf{i} - \dfrac{\sqrt{14}}{7}\mathbf{j} + \dfrac{\sqrt{14}}{14}\mathbf{k}$ and

$\mathbf{u}_2 = -\dfrac{\mathbf{c}}{|\mathbf{c}|} = -\dfrac{3\sqrt{14}}{14}\mathbf{i} + \dfrac{\sqrt{14}}{7}\mathbf{j} - \dfrac{\sqrt{14}}{14}\mathbf{k}$.

13. $|\mathbf{b}| = \sqrt{1^2 + 2^2 + (-1)^2} = \sqrt{6}$ and so $\cos\alpha = \dfrac{1}{\sqrt{6}} = \dfrac{\sqrt{6}}{6}$, $\cos\beta = \dfrac{2}{\sqrt{6}} = \dfrac{\sqrt{6}}{3}$, and $\cos\gamma = -\dfrac{1}{\sqrt{6}} = -\dfrac{\sqrt{6}}{6}$.

15. The vector projection of $\mathbf{b}$ onto $\mathbf{a}$ is

$$\operatorname{proj}_{\mathbf{a}} \mathbf{b} = \left(\dfrac{\mathbf{b} \cdot \mathbf{a}}{|\mathbf{a}|^2}\right)\mathbf{a} = \left[\dfrac{(\mathbf{i} + 2\mathbf{j} - \mathbf{k}) \cdot (2\mathbf{i} - \mathbf{j} + 3\mathbf{k})}{\left(\sqrt{2^2 + (-1)^2 + 3^2}\right)^2}\right](2\mathbf{i} - \mathbf{j} + 3\mathbf{k}) = -\dfrac{3}{14}\,(2\mathbf{i} - \mathbf{j} + 3\mathbf{k}) = -\dfrac{3}{7}\mathbf{i} + \dfrac{3}{14}\mathbf{j} - \dfrac{9}{14}\mathbf{k}.$$

17. Using the result of Exercise 7, we see that the required volume is $V = |\mathbf{a} \cdot (\mathbf{b} \times \mathbf{c})| = |-20| = 20$.

19. $\mathbf{a} \cdot \mathbf{b} = (2\mathbf{i} - 3\mathbf{j} + 4\mathbf{k}) \cdot (3\mathbf{i} + 6\mathbf{j} + 3\mathbf{k}) = 6 - 18 + 12 = 0$, and so $\mathbf{a}$ and $\mathbf{b}$ are orthogonal.

21. The vectors are coplanar provided $\mathbf{a} \cdot (\mathbf{b} \times \mathbf{c}) = \begin{vmatrix} 2 & 3 & 1 \\ 1 & 2 & 3 \\ 1 & -3 & c \end{vmatrix} = 2\,(2c + 9) - 3\,(c - 3) + 1\,(-3 - 2) = c + 22 = 0 \Leftrightarrow$

$c = -22$.

23. The figure shows a cube with side length a. The required angle is θ, where

$$\cos\theta = \frac{\overrightarrow{OC} \cdot \overrightarrow{AB}}{\left|\overrightarrow{OC}\right|\left|\overrightarrow{AB}\right|} = \frac{|\langle a, a, a\rangle \cdot \langle a, -a, a\rangle|}{\sqrt{a^2 + a^2 + a^2}\sqrt{a^2 + (-a^2) + a^2}}$$

$$= \frac{a^2}{\sqrt{3a^2}\sqrt{3a^2}} = \frac{1}{3}$$

$$\Leftrightarrow \theta = \cos^{-1}\tfrac{1}{3} \approx 70.5°$$

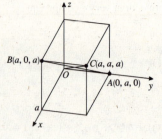

25. a. A vector perpendicular to the plane is $\mathbf{n} = \overrightarrow{PQ} \times \overrightarrow{PR} = \begin{vmatrix} \mathbf{i} & \mathbf{j} & \mathbf{k} \\ 3 & 1 & 3 \\ 4 & 0 & 3 \end{vmatrix} = 3\mathbf{i} + 3\mathbf{j} - 4\mathbf{k}.$

b. The area of the triangle is $A = \tfrac{1}{2}\left|\overrightarrow{PQ} \times \overrightarrow{PR}\right| = \tfrac{1}{2}\,|3\mathbf{i} + 3\mathbf{j} - 4\mathbf{k}| = \tfrac{1}{2}\sqrt{3^2 + 3^2 + (-4)^2} = \tfrac{1}{2}\sqrt{34}$

27. The force is $\mathbf{F} = |\mathbf{F}|\left(\dfrac{\mathbf{a}}{|\mathbf{a}|}\right) = 20\left(\dfrac{2\mathbf{i} - \mathbf{j} + 3\mathbf{k}}{\sqrt{2^2 + (-1)^2 + 3^2}}\right) = \dfrac{20\sqrt{14}}{7}\mathbf{i} - \dfrac{10\sqrt{14}}{7}\mathbf{j} + \dfrac{30\sqrt{14}}{7}\mathbf{k}$ and the work done by $\mathbf{F}$ in moving

the particle from $A\,(1, 2, 1)$ to $B\,(2, 1, 4)$ is $W = \mathbf{F} \cdot \overrightarrow{AB} = \left(\dfrac{20\sqrt{14}}{7}\mathbf{i} - \dfrac{10\sqrt{14}}{7}\mathbf{j} + \dfrac{30\sqrt{14}}{7}\mathbf{k}\right) \cdot (\mathbf{i} - \mathbf{j} + 3\mathbf{k}) = \dfrac{120\sqrt{14}}{7} \approx 64$ J.

29. Let $\mathbf{r}_1 = 7\mathbf{i}$ and $\mathbf{r}_2 = 3\mathbf{i}$. Since

$\mathbf{F}_1 = (-10\cos 30°)\,\mathbf{i} + (10\sin 30°)\,\mathbf{j} = -5\sqrt{3}\mathbf{i} + 5\mathbf{j}$ and $\mathbf{F}_2 = -8\mathbf{j}$, we

see that the torque about the point O is

$$\vec{\tau} = \mathbf{r}_1 \times \mathbf{F}_1 + \mathbf{r}_2 \times \mathbf{F}_2 = \begin{vmatrix} \mathbf{i} & \mathbf{j} & \mathbf{k} \\ 7 & 0 & 0 \\ -5\sqrt{3} & 5 & 0 \end{vmatrix} + \begin{vmatrix} \mathbf{i} & \mathbf{j} & \mathbf{k} \\ 3 & 0 & 0 \\ 0 & -8 & 0 \end{vmatrix}$$

$$= 35\mathbf{k} - 24\mathbf{k} = 11\mathbf{k}$$

31. $\mathbf{v} = \langle 2 - (-1), -1 - 2, 3 - (-4)\rangle = \langle 3, -3, 7\rangle$

a. $x = -1 + 3t,\ y = 2 - 3t,\ z = -4 + 7t$ **b.** $\dfrac{x + 1}{3} = \dfrac{y - 2}{-3} = \dfrac{z + 4}{7}$

33. The direction vector is $\mathbf{u} \times \mathbf{v} = \begin{vmatrix} \mathbf{i} & \mathbf{j} & \mathbf{k} \\ 1 & -2 & 1 \\ 3 & 2 & 5 \end{vmatrix} = -12\mathbf{i} - 2\mathbf{j} + 8\mathbf{k} = -2\,(6\mathbf{i} + \mathbf{j} - 4\mathbf{k}).$

a. $x = 1 + 6t,\ y = 2 + t,\ z = 4 - 4t$ **b.** $\dfrac{x - 1}{6} = \dfrac{y - 2}{1} = \dfrac{z - 4}{-4}$

35. A normal to the required plane is $\mathbf{n} = \langle 2, 4, -3\rangle$, so an equation of the plane is $2\,(x + 2) + 4\,(y - 4) - 3\,(z - 3) = 0 \Leftrightarrow$ $2x + 4y - 3z = 3$.

37. Here $\mathbf{n} = \mathbf{j}$, so an equation of the required plane is $0\,(x - 3) + 1\,(y - 2) + 0\,(z - 2) = 0 \Leftrightarrow y = 2$.

39. $D = \dfrac{|2\,(2) - 3\,(1) + 4\,(4) - 12|}{\sqrt{2^2 + (-3)^2 + 4^2}} = \dfrac{5}{\sqrt{29}} = \dfrac{5\sqrt{29}}{29}$

41. $L_1 : x = 1 - 2t, y = 3 - t, z = 2t$ and $L_2 : x = 2 - 3t, y = 1 + t, z = -1 + 3t$. Assuming that L_1 and

L_2 intersect, there exist numbers t_1 and t_2 such that $\left.\begin{array}{r} 1 - 2t_1 = 2 - 3t_2 \\ 3 - t_1 = 1 + t_2 \\ 2t_1 = -1 + 3t_2 \end{array}\right\}$ Solving the first two equations

simultaneously, we find $t_1 = t_2 = 1$, and substituting these values into the third equation gives $2 = -1 + 3 = 2$, so

the system is satisfied. This shows that the lines intersect; the point of intersection is $(-1, 2, 2)$. Vectors parallel

to L_1 and L_2 are $\mathbf{v}_1 = \langle -2, -1, 2 \rangle$ and $\mathbf{v}_2 = \langle -3, 1, 3 \rangle$, respectively. Thus, the required angle is θ such that

$$\cos\theta = \frac{|\mathbf{v}_1 \cdot \mathbf{v}_2|}{|\mathbf{v}_1||\mathbf{v}_2|} = \frac{|\langle -2, -1, 2 \rangle \cdot \langle -3, 1, 3 \rangle|}{\sqrt{(-2)^2 + (-1)^2 + 2^2}\sqrt{(-3)^2 + 1^2 + 3^2}} = \frac{11}{3\sqrt{19}} \Rightarrow \theta = \cos^{-1}\frac{11\sqrt{19}}{57} \approx 32.7°.$$

43. Pick a point in the plane $2x + 4y - 6z = 6$ by setting $x = y = 0$, obtaining $(0, 0, -1)$. The required distance is

$$D = \frac{|1(0) + 2(0) - 3(-1) - 2|}{\sqrt{1^2 + 2^2 + (-3)^2}} = \frac{1}{\sqrt{14}} = \frac{\sqrt{14}}{14}.$$

45. $D = \dfrac{|2(3) + 4(4) - 3(5) - 12|}{\sqrt{2^2 + 4^2 + (-3)^2}} = \dfrac{|-5|}{\sqrt{29}} = \dfrac{5\sqrt{29}}{29}$

47. The region consists of all points lying on or inside the (infinite) circular cylinder of radius 2 with axis the z-axis.

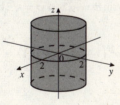

49. The region consists of all points lying on or inside the prism shown in the diagram.

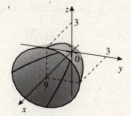

51. The surface is a plane perpendicular to the xy-plane.

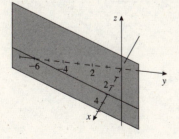

53. The surface is a paraboloid. Observe that the trace on the plane $x = 9$ is the circle $y^2 + z^2 = 9$. This helps us make an accurate sketch of the surface.

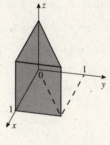

55. The surface is an elliptic cone with axis the y-axis.

$4x^2 + 9z^2 = y^2$, so the trace in the plane $y = 6$ is

$$\frac{x^2}{3^2} + \frac{z^2}{2^2} = 1.$$

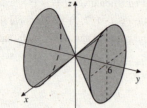

57. The surface $x^2 - z^2 = y$ is a hyperbolic paraboloid.

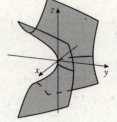

59. We are given that $x = 1$, $y = 1$, and $z = \sqrt{2}$. In cylindrical coordinates, $r = \sqrt{1^2 + 1^2} = \sqrt{2}$ and $\tan\theta = \frac{1}{1} = 1 \Rightarrow$

$\theta = \frac{\pi}{4}$, so the point is $\left(\sqrt{2}, \frac{\pi}{4}, \sqrt{2}\right)$. In spherical coordinates, $\rho = \sqrt{1^2 + 1^2 + \left(\sqrt{2}\right)^2} = 2$ and $\cos\phi = \frac{z}{\rho} = \frac{\sqrt{2}}{2} \Rightarrow$

$\phi = \frac{\pi}{4}$, so the point is $\left(2, \frac{\pi}{4}, \frac{\pi}{4}\right)$.

61. We are given $\rho = 2$, $\theta = \frac{\pi}{4}$, and $\phi = \frac{\pi}{3}$. In rectangular coordinates, $x = 2\sin\frac{\pi}{3}\cos\frac{\pi}{4} = \frac{\sqrt{6}}{2}$, $y = 2\sin\frac{\pi}{3}\sin\frac{\pi}{4} = \frac{\sqrt{6}}{2}$,

and $z = 2\cos\frac{\pi}{3} = 1$, so the point is $\left(\frac{\sqrt{6}}{2}, \frac{\sqrt{6}}{2}, 1\right)$. In cylindrical coordinates, $r = \rho\sin\phi = 2\sin\frac{\pi}{3} = \sqrt{3}$ and

$z = \rho\cos\phi = 2\cos\frac{\pi}{3} = 1$, so the point is $\left(\sqrt{3}, \frac{\pi}{4}, 1\right)$.

63. $\theta = \frac{\pi}{3}$ represents the half-plane containing the z-axis making an angle of $\frac{\pi}{3}$ with the positive x-axis.

65. $r = 2\sin\theta \Rightarrow r^2 = 2r\sin\theta \Leftrightarrow x^2 + y^2 = 2y \Leftrightarrow x^2 + (y-1)^2 = 1$ in rectangular coordinates. This describes a right circular cylinder with radius 1 and axis parallel to the z-axis passing through $(0, 1, 0)$.

67. a. $x^2 + y^2 = 2 \Rightarrow r^2 = 2 \Rightarrow r = \sqrt{2}$

     b. $x^2 + y^2 = 2 \Rightarrow (\rho\sin\phi\cos\theta)^2 + (\rho\sin\phi\sin\theta)^2 = 2 \Rightarrow \rho^2\sin^2\phi = 2 \Rightarrow \rho\sin\phi = \sqrt{2} \Rightarrow \rho = \sqrt{2}\csc\phi$

69. a. $x^2 + y^2 + 2z^2 = 1 \Rightarrow r^2 + 2z^2 = 1$

     b. $x^2 + y^2 + 2z^2 = 1 \Rightarrow \left(x^2 + y^2 + z^2\right) + z^2 = 1 \Rightarrow \rho^2 + (\rho\cos\phi)^2 = 1 \Rightarrow \rho^2\left(1 + \cos^2\phi\right) = 1$

71. $0 \le r \le z$, $0 \le \theta \le \frac{\pi}{2}$

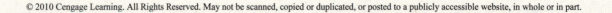

Challenge Problems

1. a. $|t\mathbf{a}+\mathbf{b}|^2 = (t\mathbf{a}+\mathbf{b})\cdot(t\mathbf{a}+\mathbf{b}) = t^2\mathbf{a}\cdot\mathbf{a}+t(\mathbf{a}\cdot\mathbf{b})+t(\mathbf{b}\cdot\mathbf{a})+\mathbf{b}\cdot\mathbf{b} = |\mathbf{a}|^2 t^2 + 2(\mathbf{a}\cdot\mathbf{b})t + |\mathbf{b}|^2 \ge 0$ for all values of t.

b. Since the quadratic function $f(t) = |\mathbf{a}|^2 t^2 + 2(\mathbf{a}\cdot\mathbf{b})t + |\mathbf{b}|^2$ is nonnegative it has either one real double root or no real root. Thus its discriminant $b^2 - 4ac \ge 0$, where $a = |\mathbf{a}|^2$, $b = 2(\mathbf{a}\cdot\mathbf{b})$, and $c = |\mathbf{b}|^2$; that is,
$[2(\mathbf{a}\cdot\mathbf{b})]^2 - 4|\mathbf{a}|^2|\mathbf{b}|^2 \le 0 \Leftrightarrow 4(\mathbf{a}\cdot\mathbf{b})^2 - 4|\mathbf{a}|^2|\mathbf{b}|^2 \le 0 \Leftrightarrow (\mathbf{a}\cdot\mathbf{b})^2 \le |\mathbf{a}|^2|\mathbf{b}|^2 \Leftrightarrow |\mathbf{a}\cdot\mathbf{b}| \le |\mathbf{a}||\mathbf{b}|$.

3. a. Since $\mathbf{c}$ is perpendicular to both $\mathbf{a}$ and $\mathbf{b}$, we have $\mathbf{c}\cdot\mathbf{a} = (x\mathbf{i}+y\mathbf{j}+z\mathbf{k})\cdot(a_1\mathbf{i}+a_2\mathbf{j}+a_3\mathbf{k}) = a_1 x + a_2 y + a_3 z = 0$ and $\mathbf{c}\cdot\mathbf{b} = (x\mathbf{i}+y\mathbf{j}+z\mathbf{k})\cdot(b_1\mathbf{i}+b_2\mathbf{j}+b_3\mathbf{k}) = b_1 x + b_2 y + b_3 z = 0$.

b. We solve the system $\left.\begin{array}{r}a_1 x + a_2 y = -a_3 z \\ b_1 x + b_2 y = -b_3 z\end{array}\right\}$ for x and y. Using Cramer's Rule, we find

$$x = \frac{\begin{vmatrix} -a_3 z & a_2 \\ -b_3 z & b_2 \end{vmatrix}}{\begin{vmatrix} a_1 & a_2 \\ b_1 & b_2 \end{vmatrix}} = \frac{a_2 b_3 - a_3 b_2}{a_1 b_2 - a_2 b_1}z \text{ and } y = \frac{\begin{vmatrix} a_1 & -a_3 z \\ b_1 & -b_3 z \end{vmatrix}}{\begin{vmatrix} a_1 & a_2 \\ b_1 & b_2 \end{vmatrix}} = \frac{a_3 b_1 - a_1 b_3}{a_1 b_2 - a_2 b_1}z.$$ Observe that $a_1 b_2 - a_2 b_1 \ne 0$ because $\mathbf{a}$

and $\mathbf{b}$ are nonparallel.

c. Since z is arbitrary, we choose it to be $a_1 b_2 - a_2 b_1$. Then $x = a_2 b_3 - a_3 b_2$ and $y = a_3 b_1 - a_1 b_3$, so $\mathbf{c} = x\mathbf{i}+y\mathbf{j}+z\mathbf{k} = (a_2 b_3 - a_3 b_2)\mathbf{i} + (a_3 b_1 - a_1 b_3)\mathbf{j} + (a_1 b_2 - a_2 b_1)\mathbf{k} = \mathbf{a}\times\mathbf{b}$.

5. a. Suppose $\mathbf{v} = \alpha\mathbf{a} + \beta\mathbf{b} + \gamma\mathbf{c}$. Then $\mathbf{v}\cdot(\mathbf{b}\times\mathbf{c}) = (\alpha\mathbf{a}+\beta\mathbf{b}+\gamma\mathbf{c})\cdot(\mathbf{b}\times\mathbf{c}) = \alpha[\mathbf{a}\cdot(\mathbf{b}\times\mathbf{c})] + \beta[\mathbf{b}\cdot(\mathbf{b}\times\mathbf{c})] + \gamma[\mathbf{c}\cdot(\mathbf{b}\times\mathbf{c})] = \alpha[\mathbf{a}\cdot(\mathbf{b}\times\mathbf{c})]$ since $\mathbf{b}\cdot(\mathbf{b}\times\mathbf{c}) = \mathbf{c}\cdot(\mathbf{b}\times\mathbf{c}) = 0$. Thus, $\alpha = \dfrac{\mathbf{v}\cdot(\mathbf{b}\times\mathbf{c})}{\mathbf{a}\cdot(\mathbf{b}\times\mathbf{c})}$. Similarly, we find
$\beta = \dfrac{\mathbf{v}\cdot(\mathbf{a}\times\mathbf{c})}{\mathbf{b}\cdot(\mathbf{a}\times\mathbf{c})}$ and $\gamma = \dfrac{\mathbf{v}\cdot(\mathbf{a}\times\mathbf{b})}{\mathbf{c}\cdot(\mathbf{a}\times\mathbf{b})}$. Since $\mathbf{a}$, $\mathbf{b}$, and $\mathbf{c}$ are noncoplanar, each of the denominators is nonzero, and so α, β, and γ are uniquely determined. Therefore, $\mathbf{v}$ can be represented as asserted.

b. Here $\alpha = \dfrac{\mathbf{v}\cdot(\mathbf{b}\times\mathbf{c})}{\mathbf{a}\cdot(\mathbf{b}\times\mathbf{c})} = \dfrac{\begin{vmatrix} 3 & 2 & 4 \\ 2 & -1 & 1 \\ 3 & 1 & 2 \end{vmatrix}}{\begin{vmatrix} 1 & 3 & 1 \\ 2 & -1 & 1 \\ 3 & 1 & 2 \end{vmatrix}} = \dfrac{9}{-1} = -9$, $\beta = \dfrac{\mathbf{v}\cdot(\mathbf{a}\times\mathbf{c})}{\mathbf{b}\cdot(\mathbf{a}\times\mathbf{c})} = \dfrac{\begin{vmatrix} 3 & 2 & 4 \\ 1 & 3 & 1 \\ 3 & 1 & 2 \end{vmatrix}}{\begin{vmatrix} 2 & -1 & 1 \\ 1 & 3 & 1 \\ 3 & 1 & 2 \end{vmatrix}} = \dfrac{-15}{1} = -15$, and

$\gamma = \dfrac{\mathbf{v}\cdot(\mathbf{a}\times\mathbf{b})}{\mathbf{c}\cdot(\mathbf{a}\times\mathbf{b})} = \dfrac{\begin{vmatrix} 3 & 2 & 4 \\ 1 & 3 & 1 \\ 2 & -1 & 1 \end{vmatrix}}{\begin{vmatrix} 3 & 1 & 2 \\ 1 & 3 & 1 \\ 2 & -1 & 1 \end{vmatrix}} = \dfrac{-14}{-1} = 14$, so $\mathbf{v} = -9\langle 1,3,1\rangle - 15\langle 2,-1,1\rangle + 14\langle 3,1,2\rangle = \langle 3,2,4\rangle$.

7. If the line intersects the paraboloid, then there exists a number t_0 such that $x = 2 + t_0$, $y = 2t_0$, $z = 24 + 16t_0$ satisfy the equation $z = 4x^2 + y^2$; that is, $24 + 16t_0 = 4(2+t_0)^2 + (2t_0)^2 = 16 + 16t_0 + 4t_0^2 + 4t_0^2 \Leftrightarrow 8t_0^2 = 8 \Leftrightarrow t_0 = \pm 1$, and the points of intersection are $(1, -2, 8)$ and $(3, 2, 40)$.

12 Vector-Valued Functions ET 11

12.1 Concept Questions ET 11.1

1. **a.** See page 985 (981 in ET).

   b. Answers will vary.

3. **a.** See page 989 (985 in ET).

   b. Let $\mathbf{r}(t) = \left\langle t, t^2, [\![t]\!] \right\rangle$. Then $\mathbf{r}(t)$ is defined on $(-1, 1)$, but fails to be continuous at 0 because $h(t) = [\![t]\!]$ is not continuous at 0.

12.1 Vector-Valued Functions ET 11.1

1. $g(t) = 1/t$ is not defined at 0, so the domain of $\mathbf{r}$ is $(-\infty, 0) \cup (0, \infty)$.

3. $f(t) = \sqrt{t}$ is defined for $t \geq 0$, $g(t) = \dfrac{1}{t-1}$ is not defined at 1, and $h(t) = \ln t$ is defined for $t > 0$, so the domain of $\mathbf{r}$ is $(0, 1) \cup (1, \infty)$.

5. $f(t) = \ln t$ is defined for $t > 0$ and both $g(t) = \cosh t$ and $h(t) = \tanh t$ are defined everywhere, so the domain of $\mathbf{r}$ is $(0, \infty)$.

7. The answer is c. The curve is the half-line $x = y = z$ that lies in the first octant and passes through the origin.

9. The answer is e. The curve lies in the plane $y = x$. The highest point on the curve is $(0, 0, 1)$, and the curve lies above the xy-plane.

11. The answer is b. The curve starts at the point $(2, 0, 1)$ and spirals upward on the elliptic cylinder $\dfrac{x^2}{4} + \dfrac{y^2}{9} = 1$ in a counterclockwise direction (viewed from above).

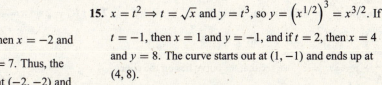

13. $x = 2t$ and $y = 3t + 1$, so $t = \frac{1}{2}x$ and $y = 3\left(\frac{1}{2}x\right) + 1 = \frac{3}{2}x + 1$. If $t = -1$, then $x = -2$ and $y = -2$, and if $t = 2$, then $x = 4$ and $y = 7$. Thus, the curve is the line segment with initial point $(-2, -2)$ and terminal point $(4, 7)$.

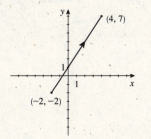

15. $x = t^2 \Rightarrow t = \sqrt{x}$ and $y = t^3$, so $y = \left(x^{1/2}\right)^3 = x^{3/2}$. If $t = -1$, then $x = 1$ and $y = -1$, and if $t = 2$, then $x = 4$ and $y = 8$. The curve starts out at $(1, -1)$ and ends up at $(4, 8)$.

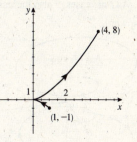

75

17. $x = e^t$ and $y = e^{2t}$, so $y = \left(e^t\right)^2 = x^2$, and the curve is contained in the parabola $y = x^2$. As $t \to -\infty$, $x \to 0$ and $y \to 0$, and as $t \to \infty$, $x \to \infty$ and $y \to \infty$, so the curve starts out from the origin (but does not include it) and extends indefinitely upward and rightward along the parabola.

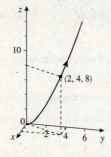

19. $x = 1 + t$, $y = 2 - t$, $z = 3 - 2t$ is the line passing through $(1, 2, 3)$ and $(0, 3, 5)$ (at $t = -1$). As t increases, z decreases, so the orientation is downward.

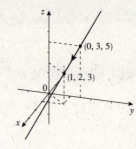

21. $x = t$, $y = t^2$, $z = t^3$. Since $y = x^2$, the curve lies on the cylinder $y = x^2$. As t increases, so does z, so the curve twists and spirals upward, passing through $(0, 0, 0)$ (at $t = 0$), $(1, 1, 1)$ (when $t = 1$), and $(2, 4, 8)$ (when $t = 2$).

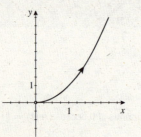

23. $x = 2 \cos t$, $y = 4 \sin t$, $z = t$. Since $\left(\frac{x}{2}\right)^2 + \left(\frac{y}{4}\right)^2 = \cos^2 t + \sin^2 t = 1$, the curve is a helix lying on the elliptical cylinder $\frac{1}{4}x^2 + \frac{1}{16}y^2 = 1$. It starts out at $(2, 0, 0)$ and makes one full turn, spiraling upward in a counterclockwise direction and ending at $(2, 0, 2\pi)$.

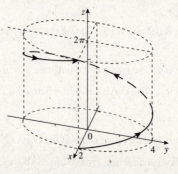

25. $x = t \cos t$, $y = t \sin t$, $z = t$. Since $\left(\frac{x}{t}\right)^2 + \left(\frac{y}{t}\right)^2 = \cos^2 t + \sin^2 t = 1 \Leftrightarrow$ $x^2 + y^2 = t^2 = z^2$, we see that the curve lies on the cone $x^2 + y^2 = z^2$. As t increases, z also increases, and so the curve spirals upward.

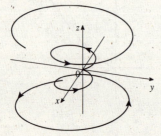

27. $\mathbf{r}(t) = 2 \sin \pi t\, \mathbf{i} + 3 \cos \pi t\, \mathbf{j} + 0.1t\, \mathbf{k}$, $0 \le t \le 10$

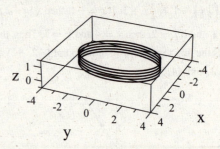

29. $\mathbf{r}(t) = \sin 3t \cos t\, \mathbf{i} + \sin 3t \sin t\, \mathbf{j} + \dfrac{t}{2\pi}\mathbf{k},\ -2\pi \le t \le 2\pi$

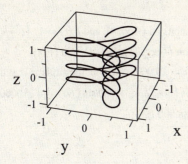

31. a. $x = \sqrt{1 - 0.09\cos^2 10t}\,\cos t,\ y = \sqrt{1 - 0.09\cos^2 10t}\,\sin t,\ z = 0.3\cos 10t.$ Thus,

$$x^2 + y^2 + z^2 = \left(1 - 0.09\cos^2 10t\right)\cos^2 t + \left(1 - 0.09\cos^2 10t\right)\sin^2 t + 0.09\cos^2 10t$$

$$= \left(1 - 0.09\cos^2 10t\right)\left(\cos^2 t + \sin^2 t\right) + 0.09\cos^2 10t = 1$$

so the curve lies on the sphere with radius 1 centered at the origin.

b.

33. $x + y + 2z = 1 \Rightarrow z = \frac{1}{2}\left(1 - x - y\right),$ but $x^2 + y^2 = 1$ can be parametrized by $x = \cos t,\ y = \sin t,\ 0 \le t \le 2\pi,$ so the

curve is described by $\mathbf{r}(t) = \left\langle \cos t,\ \sin t,\ \dfrac{1 - \cos t - \sin t}{2} \right\rangle,\ 0 \le t \le 2\pi.$

35. $x + y + z = 1 \Rightarrow z = 1 - x - y.$ Substituting this into $z = \sqrt{x^2 + y^2}$ yields $1 - x - y = \sqrt{x^2 + y^2}$

$\Rightarrow (1 - x - y)^2 = x^2 + y^2 \Rightarrow 1 + x^2 + y^2 - 2x - 2y + 2xy = x^2 + y^2 \Rightarrow 1 - 2x - 2y + 2xy = 0$

$\Rightarrow y(2x - 2) = 2x - 1 \Rightarrow y = \dfrac{2x - 1}{2x - 2}.$ Let $x = t.$ Then the required vector function is

$$\mathbf{r}(t) = \left\langle t,\ \frac{2t - 1}{2t - 2},\ 1 - t - \frac{2t - 1}{2t - 2} \right\rangle = \left\langle t,\ \frac{2t - 1}{2t - 2},\ \frac{-2t^2 + 2t - 1}{2t - 2} \right\rangle.$$

37. $\displaystyle\lim_{t \to 0}\left[\left(t^2 + 1\right)\mathbf{i} + \cos t\,\mathbf{j} - 3\mathbf{k} \right] = \mathbf{i} + \mathbf{j} - 3\mathbf{k}$

39. $\displaystyle\lim_{t \to 2}\left[\sqrt{t}\,\mathbf{i} + \frac{t^2 - 4}{t - 2}\mathbf{j} + \frac{t}{t^2 + 1}\mathbf{k} \right] = \lim_{t \to 2}\sqrt{t}\,\mathbf{i} + \lim_{t \to 2}(t + 2)\mathbf{j} + \lim_{t \to 2}\frac{t}{t^2 + 1}\mathbf{k} = \sqrt{2}\,\mathbf{i} + 4\mathbf{j} + \frac{2}{5}\mathbf{k}$

41. $\displaystyle\lim_{t \to \infty}\left\langle e^{-t},\ \frac{1}{t},\ \frac{2t^2}{t^2 + 1} \right\rangle = \left\langle \lim_{t \to \infty} e^{-t},\ \lim_{t \to \infty}\frac{1}{t},\ \lim_{t \to \infty}\frac{2}{1 + \left(1/t^2\right)} \right\rangle = \langle 0, 0, 2\rangle$

43. $\mathbf{r}(t) = \sqrt{t + 1}\,\mathbf{i} + \frac{1}{t}\mathbf{j}.$ $f(t) = \sqrt{t + 1}$ is continuous on $[-1, \infty)$ and $g(t) = \frac{1}{t}$ is continuous on $(-\infty, 0)$ and $(0, \infty),$ so $\mathbf{r}$ is continuous on $[-1, 0)$ and $(0, \infty).$

45. Since $f(t) = \dfrac{\cos t - 1}{t}$ has domain $(-\infty, 0) \cup (0, \infty)$, $g(t) = \dfrac{\sqrt{t}}{1 + 2t}$ is continuous on $[0, \infty)$, and $h(t) = te^{-1/t}$ is continuous on $(-\infty, 0)$ and $(0, \infty)$, we see that **r** is continuous on $(0, \infty)$.

47. Since $f(t) = e^{-t}$ is continuous on $(-\infty, \infty)$, $g(t) = \cos\sqrt{4 - t}$ is continuous on $(-\infty, 4]$, and $h(t) = 1/\left(t^2 - 1\right)$ is continuous on $(-\infty, -1)$, $(-1, 1)$, and $(1, \infty)$, we see that **r** is continuous on $(-\infty, -1)$, $(-1, 1)$, and $(1, 4]$.

49. The airplane is moving along a circle of radius 44,000 ft lying on the plane $z = 10{,}000$. At $t = 0$, the airplane is at the point $(44000, 0, 10000)$ and 2 minutes $(\frac{1}{30}$ hour$)$ later it is at the point $\left(44000 \cos\left(60 \cdot \frac{1}{30}\right), 44000 \sin\left(60 \cdot \frac{1}{30}\right), 10000\right)$.

Therefore, the distance covered by the airplane is $D = r\theta = 44{,}000\,(2) = 88{,}000$ ft.

51. Let $\mathbf{u}(t) = \langle u_1(t), u_2(t), u_3(t) \rangle$ and $\mathbf{v}(t) = \langle v_1(t), v_2(t), v_3(t) \rangle$. Then

$$\lim_{t \to a} [\mathbf{u}(t) + \mathbf{v}(t)] = \lim_{t \to a} \left[\langle u_1(t), u_2(t), u_3(t) \rangle + \langle v_1(t), v_2(t), v_3(t) \rangle \right]$$

$$= \lim_{t \to a} \langle u_1(t) + v_1(t), u_2(t) + v_2(t), u_3(t) + v_3(t) \rangle$$

$$= \left\langle \lim_{t \to a} [u_1(t) + v_1(t)], \lim_{t \to a} [u_2(t) + v_2(t)], \lim_{t \to a} [u_3(t) + v_3(t)] \right\rangle$$

$$= \left\langle \lim_{t \to a} u_1(t) + \lim_{t \to a} v_1(t), \lim_{t \to a} u_2(t) + \lim_{t \to a} v_2(t), \lim_{t \to a} u_3(t) + \lim_{t \to a} v_3(t) \right\rangle$$

$$= \left\langle \lim_{t \to a} u_1(t), \lim_{t \to a} u_2(t), \lim_{t \to a} u_3(t) \right\rangle + \left\langle \lim_{t \to a} v_1(t), \lim_{t \to a} v_2(t), \lim_{t \to a} v_3(t) \right\rangle$$

$$= \lim_{t \to a} \mathbf{u}(t) + \lim_{t \to a} \mathbf{v}(t)$$

53. Let $\mathbf{u}(t) = \langle u_1(t), u_2(t), u_3(t) \rangle$ and $\mathbf{v}(t) = \langle v_1(t), v_2(t), v_3(t) \rangle$. Then

$$\lim_{t \to a} [\mathbf{u}(t) \cdot \mathbf{v}(t)] = \lim_{t \to a} \left[\langle u_1(t), u_2(t), u_3(t) \rangle \cdot \langle v_1(t), v_2(t), v_3(t) \rangle \right]$$

$$= \lim_{t \to a} \left[u_1(t)\,v_1(t) + u_2(t)\,v_2(t) + u_3(t)\,v_3(t) \right]$$

$$= \left[\lim_{t \to a} u_1(t) \right]\left[\lim_{t \to a} v_1(t) \right] + \left[\lim_{t \to a} u_2(t) \right]\left[\lim_{t \to a} v_2(t) \right] + \left[\lim_{t \to a} u_3(t) \right]\left[\lim_{t \to a} v_3(t) \right]$$

$$= \lim_{t \to a} \mathbf{u}(t) \cdot \lim_{t \to a} \mathbf{v}(t)$$

55. a. Let $\mathbf{r}(t) = \langle x(t), y(t), z(t) \rangle$. Since **r** is continuous at a, so are x, y, and z. $|\mathbf{r}(t)| = \sqrt{[x(t)]^2 + [y(t)]^2 + [z(t)]^2}$.

Since x, y, and z are continuous at a, so are x^2, y^2, and z^2, and therefore $\sqrt{x^2 + y^2 + z^2} = |\mathbf{r}(t)|$ is continuous at a.

b. Consider $\mathbf{r}(t) = \langle t, t, g(t) \rangle$ on $[-1, 1]$ where $g(t) = \begin{cases} -1 & \text{if } -1 \le t \le 0 \\ 1 & \text{if } 0 < t \le 1 \end{cases}$ Then $\mathbf{r}(t)$ is discontinuous at 0, but

$|\mathbf{r}(t)| = \sqrt{t^2 + t^2 + 1} = \sqrt{2t^2 + 1}$ is continuous at 0.

57. $\displaystyle\lim_{t \to 0} \left[\frac{\sin t}{t}\mathbf{i} + \frac{1 - \cos t}{t^2}\mathbf{j} + \frac{\ln\left(1 + t^2\right)}{\cos t - e^{-t}}\mathbf{k} \right] = \left(\lim_{t \to 0} \frac{\sin t}{t} \right)\mathbf{i} + \left(\lim_{t \to 0} \frac{1 - \cos t}{t^2} \right)\mathbf{j} + \left[\lim_{t \to 0} \frac{\ln\left(1 + t^2\right)}{\cos t - e^{-t}} \right]\mathbf{k}$

$$= \mathbf{i} + \left(\lim_{t \to 0} \frac{\sin t}{2t} \right)\mathbf{j} + \left[\lim_{t \to 0} \frac{2t/\left(1 + t^2\right)}{\sin t + e^{-t}} \right]\mathbf{k} \quad \text{(using l'Hôpital's Rule)}$$

$$= \mathbf{i} + \tfrac{1}{2}\mathbf{j} + 0\mathbf{k} = \mathbf{i} + \tfrac{1}{2}\mathbf{j}$$

59. False. The function $\mathbf{r}_1$ represents a half-line in space with initial point $(0, 0, 0)$, whereas $\mathbf{r}_2$ represents the line in space that contains the half-line $\mathbf{r}_1$.

61. True. Each point on the curve has coordinates $(f(t), g(t), c)$ and hence lies in the plane $z = c$.

12.2 Concept Questions   ET 11.2

1. a. See page 992 (988 in ET).

  b. See page 993 (989 in ET).

  c. Let $\mathbf{r}(t) = \left\langle t, t^2 + 1, |t| \right\rangle$. Since $h(t) = |t|$ is not differentiable at 0, $\mathbf{r}'(0)$ does not exist.

3. a. See page 996 (992 in ET).

  b. See page 996 (992 in ET).

12.2 Differentiation and Integration of Vector-Valued Functions   ET 11.2

1. $\mathbf{r}(t) = t\mathbf{i} + t^2\mathbf{j} + t^3\mathbf{k} \Rightarrow \mathbf{r}'(t) = \mathbf{i} + 2t\mathbf{j} + 3t^2\mathbf{k} \Rightarrow \mathbf{r}''(t) = 2\mathbf{j} + 6t\mathbf{k}$

3. $\mathbf{r}(t) = \left\langle t^2 - 1, \sqrt{t^2 + 1} \right\rangle \Rightarrow \mathbf{r}'(t) = \left\langle 2t, \dfrac{t}{\sqrt{t^2 + 1}} \right\rangle$ and since

$$\frac{d}{dt} \frac{t}{\sqrt{t^2 + 1}} = \frac{d}{dt}\left[t\left(t^2 + 1\right)^{-1/2} \right] = \left(t^2 + 1\right)^{-1/2} + t\left(-\tfrac{1}{2}\right)\left(t^2 + 1\right)^{-3/2}(2t) = \frac{1}{\left(t^2 + 1\right)^{3/2}},$$

$\mathbf{r}''(t) = \left\langle 2, \dfrac{1}{\left(t^2 + 1\right)^{3/2}} \right\rangle$.

5. $\mathbf{r}(t) = \langle t\cos t - \sin t, t\sin t + \cos t \rangle \Rightarrow \mathbf{r}'(t) = \langle -t\sin t, t\cos t \rangle \Rightarrow \mathbf{r}''(t) = \langle -\sin t - t\cos t, \cos t - t\sin t \rangle$

7. $\mathbf{r}(t) = e^{-t}\sin t\,\mathbf{i} + e^{-t}\cos t\,\mathbf{j} + \tan^{-1} t\,\mathbf{k} \Rightarrow \mathbf{r}'(t) = (\cos t - \sin t)\,e^{-t}\mathbf{i} - (\cos t + \sin t)\,e^{-t}\mathbf{j} + \dfrac{1}{t^2 + 1}\mathbf{k} \Rightarrow$

$\mathbf{r}''(t) = -2e^{-t}\cos t\,\mathbf{i} + 2e^{-t}\sin t\,\mathbf{j} - \dfrac{2t}{\left(t^2 + 1\right)^2}\mathbf{k}$

9. a. $\mathbf{r}(t) = \sqrt{t}\,\mathbf{i} + (t - 4)\,\mathbf{j} \Rightarrow \mathbf{r}(2) = \sqrt{2}\,\mathbf{i} - 2\mathbf{j}; \mathbf{r}'(t) = \dfrac{1}{2\sqrt{t}}\mathbf{i} + \mathbf{j} \Rightarrow$

  $\mathbf{r}'(2) = \dfrac{\sqrt{2}}{4}\mathbf{i} + \mathbf{j}$

  b. $x = \sqrt{t}, y = t - 4 = x^2 - 4\ (x \geq 0)$, so the curve lies on the right

  portion of the parabola $y = x^2 - 4$.

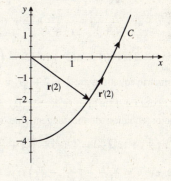

11. a. $\mathbf{r}(t) = \langle 4\cos t, 2\sin t \rangle \Rightarrow \mathbf{r}\left(\frac{\pi}{3}\right) = \left\langle 2, \sqrt{3} \right\rangle;$

  $\mathbf{r}'(t) = \langle -4\sin t, 2\cos t \rangle \Rightarrow \mathbf{r}'\left(\frac{\pi}{3}\right) = \left\langle -2\sqrt{3}, 1 \right\rangle$

  b. $x = 4\cos t$ and $y = 2\sin t$, so the curve lies on the ellipse

  $\dfrac{x^2}{16} + \dfrac{y^2}{4} = 1$.

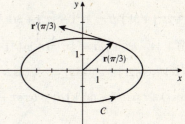

13. a. $\mathbf{r}(t) = (2 + 3t)\mathbf{i} + (1 - 2t)\mathbf{j} \Rightarrow \mathbf{r}(1) = 5\mathbf{i} - \mathbf{j}; \mathbf{r}'(t) = 3\mathbf{i} - 2\mathbf{j} \Rightarrow$

$\mathbf{r}'(1) = 3\mathbf{i} - 2\mathbf{j}$

b. $x = 2 + 3t$ and $y = 1 - 2t$, so the curve lies on the line

$y = 1 - 2\left(\dfrac{x - 2}{3}\right) \Leftrightarrow y = -\frac{2}{3}x + \frac{7}{3}$.

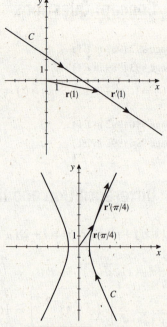

15. a. $\mathbf{r}(t) = \sec t\,\mathbf{i} + 2\tan t\,\mathbf{j} \Rightarrow \mathbf{r}\left(\frac{\pi}{4}\right) = \sqrt{2}\mathbf{i} + 2\mathbf{j};$

$\mathbf{r}'(t) = \sec t\tan t\,\mathbf{i} + 2\sec^2 t\,\mathbf{j} \Rightarrow \mathbf{r}'\left(\frac{\pi}{4}\right) = \sqrt{2}\mathbf{i} + 4\mathbf{j}$

b. $x = \sec t$ and $y = 2\tan t$, so the identity $\sec^2 t = 1 + \tan^2 t$ gives

$x^2 = 1 + \left(\frac{y}{2}\right)^2 \Leftrightarrow x^2 - \dfrac{y^2}{4} = 1$, a hyperbola that contains C.

17. $\mathbf{r}(t) = t\mathbf{i} + 2t\mathbf{j} + 3t\mathbf{k} \Rightarrow \mathbf{r}'(t) = \mathbf{i} + 2\mathbf{j} + 3\mathbf{k} \Rightarrow \mathbf{r}'(1) = \mathbf{i} + 2\mathbf{j} + 3\mathbf{k}$, so $|\mathbf{r}'(1)| = \sqrt{1^2 + 2^2 + 3^2} = \sqrt{14}$, and therefore

$\mathbf{T}(1) = \dfrac{\mathbf{r}'(1)}{|\mathbf{r}'(1)|} = \dfrac{\mathbf{i} + 2\mathbf{j} + 3\mathbf{k}}{\sqrt{14}} = \frac{\sqrt{14}}{14}\mathbf{i} + \frac{\sqrt{14}}{7}\mathbf{j} + \frac{3\sqrt{14}}{14}\mathbf{k}$.

19. $\mathbf{r}(t) = 2\sin 2t\,\mathbf{i} + 3\cos 2t\,\mathbf{j} + 3\mathbf{k} \Rightarrow \mathbf{r}'(t) = 4\cos 2t\,\mathbf{i} - 6\sin 2t\,\mathbf{j} \Rightarrow \mathbf{r}'\left(\frac{\pi}{6}\right) = 2\mathbf{i} - 3\sqrt{3}\mathbf{j}$, so

$\left|\mathbf{r}'\left(\frac{\pi}{6}\right)\right| = \sqrt{2^2 + \left(-3\sqrt{3}\right)^2} = \sqrt{31}$, and therefore $\mathbf{T}\left(\frac{\pi}{6}\right) = \dfrac{\mathbf{r}'\left(\frac{\pi}{6}\right)}{\left|\mathbf{r}'\left(\frac{\pi}{6}\right)\right|} = \frac{1}{\sqrt{31}}\left(2\mathbf{i} - 3\sqrt{3}\mathbf{j}\right) = \frac{2\sqrt{31}}{31}\mathbf{i} - \frac{3\sqrt{93}}{31}\mathbf{j}$.

21. $\mathbf{r}(t) = t\mathbf{i} + t^2\mathbf{j} + t^3\mathbf{k} \Rightarrow \mathbf{r}'(t) = \mathbf{i} + 2t\mathbf{j} + 3t^2\mathbf{k}$, so a vector equation of the tangent line when $t = 1$ is

$\mathbf{r}'(t) = \mathbf{r}(1) + t\mathbf{r}'(1) = (\mathbf{i} + \mathbf{j} + \mathbf{k}) + t(\mathbf{i} + 2\mathbf{j} + 3\mathbf{k}) = (1 + t)\mathbf{i} + (1 + 2t)\mathbf{j} + (1 + 3t)\mathbf{k}$, and parametric equations are

$x = 1 + t, y = 1 + 2t, z = 1 + 3t$.

23. $\mathbf{r}(t) = \sqrt{t + 2}\,\mathbf{i} + \dfrac{1}{t + 1}\mathbf{j} + \dfrac{2}{t^2 + 4}\mathbf{k} \Rightarrow \mathbf{r}'(t) = \dfrac{1}{2\sqrt{t + 2}}\mathbf{i} - \dfrac{1}{(t + 1)^2}\mathbf{j} - \dfrac{4t}{\left(t^2 + 4\right)^2}\mathbf{k}$, so a vector equation of the tangent

line when $t = 2$ is $\mathbf{r}'(t) = \mathbf{r}(2) + t\mathbf{r}'(2) = \left(2\mathbf{i} + \frac{1}{3}\mathbf{j} + \frac{1}{4}\mathbf{k}\right) + t\left(\frac{1}{4}\mathbf{i} - \frac{1}{9}\mathbf{j} - \frac{1}{8}\mathbf{k}\right) = \left(2 + \frac{1}{4}t\right)\mathbf{i} + \left(\frac{1}{3} - \frac{1}{9}t\right)\mathbf{j} + \left(\frac{1}{4} - \frac{1}{8}t\right)\mathbf{k}$

and parametric equations are $x = 2 + \frac{1}{4}t, y = \frac{1}{3} - \frac{1}{9}t, z = \frac{1}{4} - \frac{1}{8}t$.

25. $\mathbf{r}(t) = t\cos t\,\mathbf{i} + t\sin t\,\mathbf{j} + te^t\mathbf{k} \Rightarrow \mathbf{r}'(t) = (\cos t - t\sin t)\mathbf{i} + (\sin t + t\cos t)\mathbf{j} + (1 + t)e^t\mathbf{k}$,

so a vector equation of the tangent line when $t = \frac{\pi}{6}$ is

$\mathbf{r}'(t) = \mathbf{r}\left(\frac{\pi}{6}\right) + t\mathbf{r}'\left(\frac{\pi}{6}\right) = \left(\frac{\sqrt{3}}{12}\mathbf{i} + \frac{\pi}{12}\mathbf{j} + \frac{\pi}{6}e^{\pi/6}\mathbf{k}\right) + \left(\frac{\sqrt{3}}{2} - \frac{\pi}{12}\right)t\mathbf{i} + \left(\frac{1}{2} + \frac{\sqrt{3}\pi}{12}\right)t\mathbf{j} + \left(1 + \frac{\pi}{6}\right)e^{\pi/6}t\mathbf{k}$ and parametric

equations are $x = \frac{\sqrt{3}\pi}{12} + \left(\frac{\sqrt{3}}{2} - \frac{\pi}{12}\right)t, y = \frac{\pi}{12} + \left(\frac{1}{2} + \frac{\sqrt{3}\pi}{12}\right)t, z = \frac{\pi}{6}e^{\pi/6} + \left(1 + \frac{\pi}{6}\right)e^{\pi/6}t$.

27. $\int \left(t\mathbf{i} + 2t^2\mathbf{j} + 3\mathbf{k}\right) dt = \frac{1}{2}t^2\mathbf{i} + \frac{2}{3}t^3\mathbf{j} + 3t\mathbf{k} + \mathbf{C}$

29. $\int \left(\sqrt{t}\,\mathbf{i} + \frac{1}{t}\mathbf{j} - t^{3/2}\mathbf{k}\right) dt = \frac{2}{3}t^{3/2}\mathbf{i} + \ln|t|\,\mathbf{j} - \frac{2}{5}t^{5/2}\mathbf{k} + \mathbf{C}$

31. $\int \left(\sin 2t\,\mathbf{i} + \cos 2t\,\mathbf{j} + e^{-t}\mathbf{k}\right) dt = -\frac{1}{2}\cos 2t\,\mathbf{i} + \frac{1}{2}\sin 2t\,\mathbf{j} - e^{-t}\mathbf{k} + \mathbf{C}$

33. $\int \left(t\cos t\,\mathbf{i} + t\sin t^2\,\mathbf{j} - te^{t^2}\mathbf{k}\right) dt = (\cos t + t\sin t)\mathbf{i} - \frac{1}{2}\cos t^2\,\mathbf{j} - \frac{1}{2}e^{t^2} + \mathbf{C}$. (Note that we have integrated $\int t\cos t\,dt$ by parts.)

35. $\mathbf{r}(t) = \int \mathbf{r}'(t)\,dt = \int \left(2\mathbf{i} + 4t\mathbf{j} - 6t^2\mathbf{k}\right) dt = 2t\mathbf{i} + 2t^2\mathbf{j} - 2t^3\mathbf{k} + \mathbf{C}$ and $\mathbf{r}(0) = \mathbf{i} + \mathbf{k} \Rightarrow \mathbf{C} = \mathbf{i} + \mathbf{k}$, so

$\mathbf{r}(t) = (2t + 1)\mathbf{i} + 2t^2\mathbf{j} - \left(2t^3 - 1\right)\mathbf{k}$.

37. $\mathbf{r}(t) = \int \mathbf{r}'(t)\,dt = \int \left(2e^{2t}\mathbf{i} + 3e^{-t}\mathbf{j} + e^t\mathbf{k}\right) dt = e^{2t}\mathbf{i} - 3e^{-t}\mathbf{j} + e^t\mathbf{k} + \mathbf{C}$ and $\mathbf{r}(0) = \mathbf{i} - \mathbf{j} + \mathbf{k} \Rightarrow$

$\mathbf{i} - 3\mathbf{j} + \mathbf{k} + \mathbf{C} = \mathbf{i} - \mathbf{j} + \mathbf{k} \Rightarrow \mathbf{C} = 2\mathbf{j}$, so $\mathbf{r}(t) = e^{2t}\mathbf{i} - \left(3e^{-t} - 2\right)\mathbf{j} + e^t\mathbf{k}$.

39. $\mathbf{r}'(t) = \int \mathbf{r}''(t)\,dt = \int \left(\sqrt{t}\mathbf{i} + \sec^2 t\,\mathbf{j} + e^t\mathbf{k}\right) dt = \frac{2}{3}t^{3/2}\mathbf{i} + \tan t\,\mathbf{j} + e^t\mathbf{k} + \mathbf{C}_1$ and

$\mathbf{r}'(0) = \mathbf{i} + \mathbf{k} \Leftrightarrow \mathbf{k} + \mathbf{C}_1 = \mathbf{i} + \mathbf{k} \Rightarrow \mathbf{C}_1 = \mathbf{i}$, so $\mathbf{r}'(t) = \left(\frac{2}{3}t^{3/2} + 1\right)\mathbf{i} + \tan t\,\mathbf{j} + e^t\mathbf{k}$. Next,

$\mathbf{r}(t) = \int \left[\left(\frac{2}{3}t^{3/2} + 1\right)\mathbf{i} + \tan t\,\mathbf{j} + e^t\mathbf{k}\right] dt = \left(\frac{4}{15}t^{5/2} + t\right)\mathbf{i} - \ln|\cos t|\,\mathbf{j} + e^t\mathbf{k} + \mathbf{C}_2$ and $\mathbf{r}(0) = 2\mathbf{i} + \mathbf{j} - \mathbf{k} \Leftrightarrow$

$\mathbf{k} + \mathbf{C}_2 = 2\mathbf{i} + \mathbf{j} - \mathbf{k} \Rightarrow \mathbf{C}_2 = 2\mathbf{i} + \mathbf{j} - 2\mathbf{k}$, so $\mathbf{r}(t) = \left(\frac{4}{15}t^{5/2} + t + 2\right)\mathbf{i} - (\ln|\cos t| - 1)\mathbf{j} + \left(e^t - 2\right)\mathbf{k}$.

41. $\frac{d}{dt}\left[\mathbf{u}(t) + \mathbf{v}(t)\right] = \frac{d}{dt}\left[\left(t^2\mathbf{i} - 2t\mathbf{j} + 2\mathbf{k}\right) + \left(\cos t\,\mathbf{i} + \sin t\,\mathbf{j} + t^2\mathbf{k}\right)\right] = \frac{d}{dt}\left[\left(t^2 + \cos t\right)\mathbf{i} - (2t - \sin t)\mathbf{j} + \left(2 + t^2\right)\mathbf{k}\right]$

$= (2t - \sin t)\mathbf{i} - (2 - \cos t)\mathbf{j} + 2t\mathbf{k}$ (1)

On the other hand,

$\mathbf{u}'(t) + \mathbf{v}'(t) = \frac{d}{dt}\left(t^2\mathbf{i} - 2t\mathbf{j} + 2\mathbf{k}\right) + \frac{d}{dt}\left(\cos t\,\mathbf{i} + \sin t\,\mathbf{j} + t^2\mathbf{k}\right) = (2t\mathbf{i} - 2\mathbf{j}) + (-\sin t\,\mathbf{i} + \cos t\,\mathbf{j} + 2t\mathbf{k})$

$= (2t - \sin t)\mathbf{i} + (\cos t - 2)\mathbf{j} + 2t\mathbf{k}$ (2)

The desired result follows by comparing (1) and (2).

43. $\frac{d}{dt}\left[f(t)\mathbf{u}(t)\right] = \frac{d}{dt}\left[e^{2t}\left(t^2\mathbf{i} - 2t\mathbf{j} + 2\mathbf{k}\right)\right] = \frac{d}{dt}\left(t^2e^{2t}\mathbf{i} - 2te^{2t}\mathbf{j} + 2e^{2t}\mathbf{k}\right) = 2e^{2t}\left(t^2 + t\right)\mathbf{i} - 2e^{2t}(2t + 1)\mathbf{j} + 4e^{2t}\mathbf{k}$.

On the other hand,

$f'(t)\mathbf{u}(t) + f(t)\mathbf{u}'(t) = \left[\frac{d}{dt}\left(e^{2t}\right)\right]\left(t^2\mathbf{i} - 2t\mathbf{j} + 2\mathbf{k}\right) + e^{2t}\frac{d}{dt}\left(t^2\mathbf{i} - 2t\mathbf{j} + 2\mathbf{k}\right) = 2e^{2t}\left(t^2\mathbf{i} - 2t\mathbf{j} + 2\mathbf{k}\right) + e^{2t}(2t\mathbf{i} - 2\mathbf{j})$

$= 2e^{2t}\left(t^2 + t\right)\mathbf{i} - 2e^{2t}(2t + 1)\mathbf{j} + 4e^{2t}\mathbf{k}$

and the desired result follows.

45. $\mathbf{u}(t) \times \mathbf{v}(t) = \begin{vmatrix} \mathbf{i} & \mathbf{j} & \mathbf{k} \\ t^2 & -2t & 2 \\ \cos t & \sin t & t^2 \end{vmatrix} = \left(-2t^3 - 2\sin t\right)\mathbf{i} - \left(t^4 - 2\cos t\right)\mathbf{j} + \left(t^2\sin t + 2t\cos t\right)\mathbf{k}$, so

$\frac{d}{dt}[\mathbf{u}(t) \times \mathbf{v}(t)] = -\left(6t^2 + 2\cos t\right)\mathbf{i} - \left(4t^3 + 2\sin t\right)\mathbf{j} + \left(t^2\cos t + 2\cos t\right)\mathbf{k}$. On the other hand,

$\mathbf{u}'(t) \times \mathbf{v}(t) + \mathbf{u}(t) \times \mathbf{v}'(t) = \begin{vmatrix} \mathbf{i} & \mathbf{j} & \mathbf{k} \\ 2t & -2 & 0 \\ \cos t & \sin t & t^2 \end{vmatrix} + \begin{vmatrix} \mathbf{i} & \mathbf{j} & \mathbf{k} \\ t^2 & -2t & 2 \\ -\sin t & \cos t & 2t \end{vmatrix}$

$= -2t^2\mathbf{i} - 2t^3\mathbf{j} + (2t\sin t + 2\cos t)\mathbf{k} + \left(-4t^2 - 2\cos t\right)\mathbf{i} - \left(2t^3 + 2\sin t\right)\mathbf{j} + \left(t^2\cos t - 2t\sin t\right)\mathbf{k}$

$= -\left(6t^2 + 2\cos t\right)\mathbf{i} - \left(4t^3 + 2\sin t\right)\mathbf{j} + \left(t^2\cos t + 2\cos t\right)\mathbf{k}$

and the desired result follows.

47. Let $\mathbf{u}(t) = x_1(t)\mathbf{i} + y_1(t)\mathbf{j} + z_1(t)\mathbf{k}$ and $\mathbf{v}(t) = x_2(t)\mathbf{i} + y_2(t)\mathbf{j} + z_2(t)\mathbf{k}$. Then

$\frac{d}{dt}\left[\mathbf{u}(t) + \mathbf{v}(t)\right] = \frac{d}{dt}\left\{[x_1(t) + x_2(t)]\mathbf{i} + [y_1(t) + y_2(t)]\mathbf{j} + [z_1(t) + z_2(t)]\mathbf{k}\right\}$

$= [x_1'(t) + x_2'(t)]\mathbf{i} + [y_1'(t) + y_2'(t)]\mathbf{j} + [z_1'(t) + z_2'(t)]\mathbf{k}$

$= [x_1'(t)\mathbf{i} + y_1'(t)\mathbf{j} + z_1'(t)\mathbf{k}] + [x_2'(t)\mathbf{i} + y_2'(t)\mathbf{j} + z_2'(t)\mathbf{k}] = \mathbf{u}'(t) + \mathbf{v}'(t)$

49. Let $\mathbf{u}(t) = x_1(t)\mathbf{i} + y_1(t)\mathbf{j} + z_1(t)\mathbf{k}$. Then

$$\frac{d}{dt}[c\mathbf{u}(t)] = \frac{d}{dt}[cx_1(t)\mathbf{i} + cy_1(t)\mathbf{j} + cz_1(t)\mathbf{k}] = cx_1'(t)\mathbf{i} + cy_1'(t)\mathbf{j} + cz_1'(t)\mathbf{k} = c[x_1'(t)\mathbf{i} + y_1'(t)\mathbf{j} + z_1'(t)\mathbf{k}]$$

$$= c\mathbf{u}'(t)$$

51. Let $\mathbf{u}(t) = x_1(t)\mathbf{i} + y_1(t)\mathbf{j} + z_1(t)\mathbf{k}$ and $\mathbf{v}(t) = x_2(t)\mathbf{i} + y_2(t)\mathbf{j} + z_2(t)\mathbf{k}$. Then

$$\mathbf{u}(t) \times \mathbf{v}(t) = \begin{vmatrix} \mathbf{i} & \mathbf{j} & \mathbf{k} \\ x_1 & y_1 & z_1 \\ x_2 & y_2 & z_2 \end{vmatrix} = (y_1 z_2 - y_2 z_1)\mathbf{i} - (x_1 z_2 - x_2 z_1)\mathbf{j} + (x_1 y_2 - x_2 y_1)\mathbf{k},\ \text{so}$$

$$\frac{d}{dt}(\mathbf{u} \times \mathbf{v}) = (y_1' z_2 + y_1 z_2' - y_2' z_1 - y_2 z_1')\mathbf{i} - (x_1' z_2 + x_1 z_2' - x_2' z_1 - x_2 z_1')\mathbf{j} + (x_1' y_2 + x_1 y_2' - x_2' y_1 - x_2 y_1')\mathbf{k}.$$

On the other hand, $\mathbf{u}' \times \mathbf{v} = \begin{vmatrix} \mathbf{i} & \mathbf{j} & \mathbf{k} \\ x_1' & y_1' & z_1' \\ x_2 & y_2 & z_2 \end{vmatrix} = (y_1' z_2 - y_2 z_1')\mathbf{i} - (x_1' z_2 - x_2 z_1')\mathbf{j} + (x_1' y_2 - x_2 y_1')\mathbf{k}$

and $\mathbf{u} \times \mathbf{v}' = \begin{vmatrix} \mathbf{i} & \mathbf{j} & \mathbf{k} \\ x_1 & y_1 & z_1 \\ x_2' & y_2' & z_2' \end{vmatrix} = (y_1 z_2' - y_2' z_1)\mathbf{i} - (x_1 z_2' - z_1 x_2')\mathbf{j} + (x_1 y_2' - x_2' y_1)\mathbf{k}$, so

$\mathbf{u}' \times \mathbf{v} + \mathbf{u} \times \mathbf{v}' = (y_1' z_2 - y_2 z_1' + y_1 z_2' - y_2' z_1)\mathbf{i} - (x_1' z_2 - x_2 z_1' + x_1 z_2' - z_1 x_2')\mathbf{j} + (x_1' y_2 - x_2 y_1' + x_1 y_2' - x_2' y_1)\mathbf{k}$
and the desired result follows.

53. $\frac{d}{dt}[\mathbf{r}(t) \times \mathbf{r}'(t)] = \mathbf{r}'(t) \times \mathbf{r}'(t) + \mathbf{r}(t) \times \mathbf{r}''(t) = \mathbf{r}(t) \times \mathbf{r}''(t)$ since $\mathbf{r}'(t) \times \mathbf{r}'(t) = \mathbf{0}$.

55. $\frac{d}{dt}\left[\mathbf{r}(-t) + \mathbf{r}\left(\frac{1}{t}\right)\right] = \mathbf{r}'(-t)(-1) + \mathbf{r}'\left(\frac{1}{t}\right)\frac{d}{dt}\left(\frac{1}{t}\right) = -\mathbf{r}'(-t) - \frac{1}{t^2}\mathbf{r}'\left(\frac{1}{t}\right)$

57. $\frac{d}{dt}\{\mathbf{r}(t) \cdot [\mathbf{r}'(t) \times \mathbf{r}''(t)]\} = \mathbf{r}'(t) \cdot [\mathbf{r}'(t) \times \mathbf{r}''(t)] + \mathbf{r}(t) \cdot \frac{d}{dt}[\mathbf{r}'(t) \times \mathbf{r}''(t)]$

$$= \mathbf{r}'(t) \cdot [\mathbf{r}'(t) \times \mathbf{r}''(t)] + \mathbf{r}(t) \cdot [\mathbf{r}''(t) \times \mathbf{r}''(t) + \mathbf{r}'(t) \times \mathbf{r}'''(t)]$$

$$= \mathbf{r}'(t) \cdot [\mathbf{r}'(t) \times \mathbf{r}''(t)] + \mathbf{r}(t) \cdot [\mathbf{r}'(t) \times \mathbf{r}'''(t)]$$

since $\mathbf{r}''(t) \times \mathbf{r}''(t) = \mathbf{0}$.

59. Let $\mathbf{u}(t) = x_1(t)\mathbf{i} + y_1(t)\mathbf{j} + z_1(t)\mathbf{k}$ and $\mathbf{v}(t) = x_2(t)\mathbf{i} + y_2(t)\mathbf{j} + z_2(t)\mathbf{k}$. Then

$$\int_a^b [\mathbf{u}(t) + \mathbf{v}(t)]\,dt = \int_a^b \{[x_1(t) + x_2(t)]\mathbf{i} + [y_1(t) + y_2(t)]\mathbf{j} + [z_1(t) + z_2(t)]\mathbf{k}\}\,dt$$

$$= \int_a^b [x_1(t) + x_2(t)]\,dt\,\mathbf{i} + \int_a^b [y_1(t) + y_2(t)]\,dt\,\mathbf{j} + \int_a^b [z_1(t) + z_2(t)]\,dt\,\mathbf{k}$$

$$= \left[\int_a^b x_1(t)\,dt + \int_a^b x_2(t)\,dt\right]\mathbf{i} + \left[\int_a^b y_1(t)\,dt + \int_a^b y_2(t)\,dt\right]\mathbf{j} + \left[\int_a^b z_1(t)\,dt + \int_a^b z_2(t)\,dt\right]\mathbf{k}$$

$$= \left[\int_a^b x_1(t)\,dt\,\mathbf{i} + \int_a^b y_1(t)\,dt\,\mathbf{j} + \int_a^b z_1(t)\,dt\,\mathbf{k}\right] + \left[\int_a^b x_2(t)\,dt\,\mathbf{i} + \int_a^b y_2(t)\,dt\,\mathbf{j} + \int_a^b z_2(t)\,dt\,\mathbf{k}\right]$$

$$= \int_a^b \mathbf{u}(t)\,dt + \int_a^b \mathbf{v}(t)\,dt$$

61. a. Let $\mathbf{r}(t) = x(t)\mathbf{i} + y(t)\mathbf{j} + z(t)\mathbf{k}$ and $\mathbf{c} = c_1\mathbf{i} + c_2\mathbf{j} + c_2\mathbf{k}$. Then

$$\int_a^b \mathbf{c} \cdot \mathbf{r}(t)\,dt = \int_a^b (c_1\mathbf{i} + c_2\mathbf{j} + c_3\mathbf{k}) \cdot [x(t)\mathbf{i} + y(t)\mathbf{j} + z(t)\mathbf{k}]\,dt = \int_a^b [c_1 x(t) + c_2 y(t) + c_3 z(t)]\,dt$$

$$= c_1 \int_a^b x(t)\,dt + c_2 \int_a^b y(t)\,dt + c_3 \int_a^b z(t)\,dt$$

On the other hand,

$$\mathbf{c} \cdot \int_a^b \mathbf{r}(t)\,dt = (c_1\mathbf{i} + c_2\mathbf{j} + c_3\mathbf{k}) \cdot \left\{\int_a^b [x(t)\mathbf{i} + y(t)\mathbf{j} + z(t)\mathbf{k}]\,dt\right\}$$

$$= (c_1\mathbf{i} + c_2\mathbf{j} + c_3\mathbf{k}) \cdot \left[\int_a^b x(t)\,dt\,\mathbf{i} + \int_a^b y(t)\,dt\,\mathbf{j} + \int_a^b z(t)\,dt\,\mathbf{k}\right]$$

$$= c_1 \int_a^b x(t)\,dt + c_2 \int_a^b y(t)\,dt + c_3 \int_a^b z(t)\,dt$$

b. $\int_0^\pi \mathbf{c} \cdot \mathbf{r}(t)\, dt = \int_0^\pi (2\mathbf{i} + 3\mathbf{j} - \mathbf{k}) \cdot (\sin t\, \mathbf{i} + \cos t\, \mathbf{j} + t\mathbf{k})\, dt = \int_0^\pi (2\sin t + 3\cos t - t)\, dt$

$$= \left. \left(-2\cos t + 3\sin t - \tfrac{1}{2}t^2\right) \right|_0^\pi = 4 - \tfrac{1}{2}\pi^2$$

On the other hand,

$$\mathbf{c} \cdot \int_0^\pi \mathbf{r}(t)\, dt = (2\mathbf{i} + 3\mathbf{j} - \mathbf{k}) \cdot \int_0^\pi (\sin t\, \mathbf{i} + \cos t\, \mathbf{j} + t\mathbf{k})\, dt = (2\mathbf{i} + 3\mathbf{j} - \mathbf{k}) \cdot \left. \left[-\cos t\, \mathbf{i} + \sin t\, \mathbf{j} + \tfrac{1}{2}t^2\mathbf{k} \right] \right|_0^\pi$$

$$= (2\mathbf{i} + 3\mathbf{j} - \mathbf{k}) \cdot \left(2\mathbf{i} + \tfrac{1}{2}\pi^2\mathbf{k}\right) = 4 - \tfrac{1}{2}\pi^2$$

63. True. $\dfrac{d}{dt}\left(|\mathbf{u}|^2\right) = \dfrac{d}{dt}(\mathbf{u} \cdot \mathbf{u}) = \mathbf{u} \cdot \mathbf{u}' + \mathbf{u}' \cdot \mathbf{u} = \mathbf{u} \cdot \mathbf{u}' + \mathbf{u} \cdot \mathbf{u}' = 2\mathbf{u} \cdot \mathbf{u}'.$

65. True. Using the result of Exercise 63, we have $\dfrac{d}{dt}|\mathbf{r}(t)|^2 = 2\mathbf{r}(t) \cdot \mathbf{r}'(t)$, so if $\mathbf{r}(t) \cdot \mathbf{r}'(t) = 0$, then $\dfrac{d}{dt}|\mathbf{r}(t)|^2 = 0 \Rightarrow$
$|\mathbf{r}(t)| = c$, a constant, showing that $\mathbf{r}$ has constant length.

12.3 Concept Questions ET 11.3

1. See page 1000 (996 in ET).

3. a. See page 1001 (997 in ET).

  b. See page 1003 (999 in ET). If t is not the arc length parameter, then $\mathbf{T}(t) = \dfrac{\mathbf{r}'(t)}{|\mathbf{r}'(t)|}.$

12.3 Arc Length and Curvature ET 11.3

1. $\mathbf{r}(t) = t\mathbf{i} + 2t\mathbf{j} + 3t\mathbf{k} \Rightarrow \mathbf{r}'(t) = \mathbf{i} + 2\mathbf{j} + 3\mathbf{k} \Rightarrow |\mathbf{r}'(t)| = \sqrt{1^2 + 2^2 + 3^2} = \sqrt{14}$, so
$L = \int_a^b |\mathbf{r}'(t)|\, dt = \int_0^4 \sqrt{14}\, dt = 4\sqrt{14}.$

3. $\mathbf{r}(t) = 4\sin t\, \mathbf{i} + 3t\mathbf{j} + 4\cos t\, \mathbf{k} \Rightarrow \mathbf{r}'(t) = 4\cos t\, \mathbf{i} + 3\mathbf{j} - 4\sin t\, \mathbf{k} \Rightarrow$
$|\mathbf{r}'(t)|^2 = (4\cos t)^2 + 3^2 + (-4\sin t)^2 = 16\left(\cos^2 t + \sin^2 t\right) + 9 = 25$ and so $|\mathbf{r}'(t)| = 5$. Thus,
$L = \int_a^b |\mathbf{r}'(t)|\, dt = \int_0^{2\pi} 5\, dt = 10\pi.$

5. $\mathbf{r}(t) = \langle e^t \cos t, e^t \sin t, e^t \rangle \Rightarrow \mathbf{r}'(t) = \langle e^t (\cos t - \sin t), e^t (\sin t + \cos t), e^t \rangle \Rightarrow$
$|\mathbf{r}'(t)|^2 = \left[e^t (\cos t - \sin t)\right]^2 + \left[e^t (\cos t + \sin t)\right]^2 + \left(e^t\right)^2$

$$= e^{2t}\left(\cos^2 t - 2\cos t \sin t + \sin^2 t + \cos^2 t + 2\cos t \sin t + \sin^2 t + 1\right) = 3e^{2t}$$

so $|\mathbf{r}'(t)| = \sqrt{3e^{2t}} = \sqrt{3}e^t$. Thus, $L = \int_a^b |\mathbf{r}'(t)|\, dt = \int_0^{2\pi} \sqrt{3}e^t\, dt = \left. \sqrt{3}e^t \right|_0^{2\pi} = \sqrt{3}\left(e^{2\pi} - 1\right).$

7. $\mathbf{r}(t) = 2t\mathbf{i} + t^2\mathbf{j} + \ln t\, \mathbf{k} \Rightarrow \mathbf{r}'(t) = 2\mathbf{i} + 2t\mathbf{j} + \dfrac{1}{t}\mathbf{k} \Rightarrow$

$$|\mathbf{r}'(t)|^2 = 4 + 4t^2 + \frac{1}{t^2} = \frac{4t^4 + 4t^2 + 1}{t^2} = \frac{\left(2t^2 + 1\right)^2}{t^2} \Rightarrow |\mathbf{r}'(t)| = \frac{2t^2 + 1}{t}, \ t > 0. \ \text{Thus,}$$

$$L = \int_a^b |\mathbf{r}'(t)|\, dt = \int_1^e \frac{2t^2 + 1}{t}\, dt = \int_1^e \left(2t + \frac{1}{t}\right) dt = \left. \left(t^2 + \ln t\right) \right|_1^e = \left(e^2 + 1\right) - 1 = e^2.$$

9. $\mathbf{r}(t) = t\sin t\,\mathbf{i} + t\cos t\,\mathbf{j} + t\mathbf{k} \Rightarrow \mathbf{r}'(t) = (t\cos t + \sin t)\,\mathbf{i} + (\cos t - t\sin t)\,\mathbf{j} + \mathbf{k}$

$\Rightarrow |\mathbf{r}'(t)|^2 = (t\cos t + \sin t)^2 + (\cos t - t\sin t)^2 + 1 = t^2 + 2$. Thus,

$L = \int_a^b |\mathbf{r}'(t)|\,dt = \int_0^{2\pi} \sqrt{t^2 + 2}\,dt = \left[\frac{1}{2}t\sqrt{t^2+2} + \ln\left(t + \sqrt{t^2+2}\right)\right]_0^{2\pi}$

$= \pi\sqrt{4\pi^2 + 2} + \ln\dfrac{2\pi + \sqrt{4\pi^2 + 2}}{\sqrt{2}} = \pi\sqrt{4\pi^2 + 2} + \ln\left(\sqrt{2\pi^2 + 1} + \sqrt{2}\pi\right)$

11. $\mathbf{r}(t) = (1+t)\,\mathbf{i} + (1+2t)\,\mathbf{j} + 3t\mathbf{k} \Rightarrow \mathbf{r}'(t) = \mathbf{i} + 2\mathbf{j} + 3\mathbf{k} \Rightarrow |\mathbf{r}'(t)| = \sqrt{1^2 + 2^2 + 3^2} = \sqrt{14}$, so

$s(t) = \int_0^t |\mathbf{r}'(u)|\,du = \int_0^t \sqrt{14}\,du = \sqrt{14}\,t,\ t \geq 0$. Thus, $t = \frac{1}{\sqrt{14}}s = \frac{\sqrt{14}}{14}s$ and the required parametrization of C in terms

of s is $\mathbf{r}(t(s)) = \left(1 + \frac{\sqrt{14}}{14}s\right)\mathbf{i} + \left(1 + \frac{2\sqrt{14}}{14}s\right)\mathbf{j} + \frac{3\sqrt{14}}{14}s\mathbf{k},\ s \geq 0$.

13. $\mathbf{r}(t) = e^t\cos t\,\mathbf{i} + e^t\sin t\,\mathbf{j} + e^t\mathbf{k} \Rightarrow \mathbf{r}'(t) = e^t(\cos t - \sin t)\,\mathbf{i} + e^t(\sin t + \cos t)\,\mathbf{j} + e^t\mathbf{k}$

$\Rightarrow |\mathbf{r}'(t)|^2 = e^{2t}\left[(\cos t - \sin t)^2 + (\sin t + \cos t)^2 + 1\right] = 3e^{2t} \Rightarrow |\mathbf{r}'(t)| = \sqrt{3}e^t$, so

$s(t) = \int_0^t |\mathbf{r}'(u)|\,du = \int_0^t \sqrt{3}e^u\,du = \sqrt{3}\left(e^t - 1\right) \Rightarrow e^t - 1 = \frac{\sqrt{3}}{3}s \Leftrightarrow e^t = 1 + \frac{\sqrt{3}}{3}s$

$\Leftrightarrow t = \ln\left(1 + \frac{\sqrt{3}}{3}s\right),\ s \geq 0$, and the required parametrization of C in terms of s is

$\mathbf{r}(t(s)) = \left(1 + \frac{\sqrt{3}}{3}s\right)\cos\left(\ln\left(1 + \frac{\sqrt{3}}{3}s\right)\right)\mathbf{i} + \left(1 + \frac{\sqrt{3}}{3}s\right)\sin\left(\ln\left(1 + \frac{\sqrt{3}}{3}s\right)\right)\mathbf{j} + \left(1 + \frac{\sqrt{3}}{3}s\right)\mathbf{k},\ s \geq 0$.

15. $\mathbf{r}(t) = 2t\mathbf{i} + 2t\mathbf{j} + \mathbf{k} \Rightarrow \mathbf{r}'(t) = 2\mathbf{i} + 2\mathbf{j} \Rightarrow \mathbf{r}''(t) = 0\mathbf{i} + 0\mathbf{j}$, so $\mathbf{r}'(t) \times \mathbf{r}''(t) = \mathbf{0} \Rightarrow |\mathbf{r}'(t) \times \mathbf{r}''(t)| = 0$. Also,

$|\mathbf{r}'(t)| = \sqrt{2^2 + 2^2} = 2\sqrt{2}$, so $\kappa = \dfrac{|\mathbf{r}'(t) \times \mathbf{r}''(t)|}{|\mathbf{r}'(t)|^3} = \dfrac{0}{\left(2\sqrt{2}\right)^3} = 0$.

17. $\mathbf{r}(t) = t\mathbf{i} + \frac{1}{2}t^2\mathbf{j} + t^2\mathbf{k} \Rightarrow \mathbf{r}'(t) = \mathbf{i} + t\mathbf{j} + 2t\mathbf{k} \Rightarrow \mathbf{r}''(t) = \mathbf{j} + 2\mathbf{k}$, so $\mathbf{r}'(t) \times \mathbf{r}''(t) = \begin{vmatrix} \mathbf{i} & \mathbf{j} & \mathbf{k} \\ 1 & t & 2t \\ 0 & 1 & 2 \end{vmatrix} = -2\mathbf{j} + \mathbf{k}$

$\Rightarrow |\mathbf{r}'(t) \times \mathbf{r}''(t)| = \sqrt{(-2)^2 + 1} = \sqrt{5}$. Also, $|\mathbf{r}'(t)| = \sqrt{1 + t^2 + 4t^2} = \sqrt{1 + 5t^2}$, so

$\kappa = \dfrac{|\mathbf{r}'(t) \times \mathbf{r}''(t)|}{|\mathbf{r}'(t)|^3} = \dfrac{\sqrt{5}}{\left(\sqrt{1 + 5t^2}\right)^3} = \dfrac{\sqrt{5}}{(1 + 5t^2)^{3/2}}$.

19. $\mathbf{r}(t) = 2\sin t\,\mathbf{i} + 2\cos t\,\mathbf{j} + 2t\mathbf{k} \Rightarrow \mathbf{r}'(t) = 2\cos t\,\mathbf{i} - 2\sin t\,\mathbf{j} + 2\mathbf{k} \Rightarrow \mathbf{r}''(t) = -2\sin t\,\mathbf{i} - 2\cos t\,\mathbf{j}$,

so $\mathbf{r}'(t) \times \mathbf{r}''(t) = \begin{vmatrix} \mathbf{i} & \mathbf{j} & \mathbf{k} \\ 2\cos t & -2\sin t & 2 \\ -2\sin t & -2\cos t & 0 \end{vmatrix} = 4\cos t\,\mathbf{i} - 4\sin t\,\mathbf{j} - 4\mathbf{k} \Rightarrow$

$|\mathbf{r}'(t) \times \mathbf{r}''(t)| = \sqrt{(4\cos t)^2 + (-4\sin t)^2 + (-4)^2} = 4\sqrt{2}$. Also, $|\mathbf{r}'(t)| = \sqrt{(2\cos t)^2 + (-2\sin t)^2 + 2^2} = 2\sqrt{2}$, so

$\kappa = \dfrac{|\mathbf{r}'(t) \times \mathbf{r}''(t)|}{|\mathbf{r}'(t)|^3} = \dfrac{4\sqrt{2}}{\left(2\sqrt{2}\right)^3} = \dfrac{1}{4}$.

21. $y = x^3 + 1 \Rightarrow y' = 3x^2 \Rightarrow y'' = 6x$, so $\kappa = \dfrac{|y''|}{\left[1 + (y')^2\right]^{3/2}} = \dfrac{|6x|}{\left[1 + (3x^2)^2\right]^{3/2}} = \dfrac{6|x|}{(1 + 9x^4)^{3/2}}$.

23. $y = \sin 2x \Rightarrow y' = 2\cos 2x \Rightarrow y'' = -4\sin 2x$, so $\kappa = \dfrac{|y''|}{\left[1 + (y')^2\right]^{3/2}} = \dfrac{|-4\sin 2x|}{\left[1 + (2\cos 2x)^2\right]^{3/2}} = \dfrac{4\,|\sin 2x|}{(1 + 4\cos^2 2x)^{3/2}}.$

25. $y = e^{-x^2} \Rightarrow y' = -2xe^{-x^2} \Rightarrow y'' = 2\left(2x^2 - 1\right)e^{-x^2}$, so

$$\kappa = \frac{|y''|}{\left[1 + (y')^2\right]^{3/2}} = \frac{2\left|2x^2 - 1\right|e^{-x^2}}{\left[1 + 4x^2 e^{-2x^2}\right]^{3/2}} = \frac{2\left|2x^2 - 1\right|e^{2x^2}}{\left(e^{2x^2} + 4x^2\right)^{3/2}}.$$

27. Using the result of Exercise 25, we see that $\kappa = \dfrac{2\left|2x^2 - 1\right|e^{2x^2}}{\left(e^{2x^2} + 4x^2\right)^{3/2}}$, so $\kappa = 0 \Rightarrow \left|2x^2 - 1\right| = 0 \Rightarrow x = \pm\frac{\sqrt{2}}{2}$, and the

required points are $\left(-\frac{\sqrt{2}}{2}, e^{-1/2}\right)$ and $\left(\frac{\sqrt{2}}{2}, e^{-1/2}\right).$

29. $y = e^x \Rightarrow y' = e^x \Rightarrow y'' = e^x$, so $\kappa = \dfrac{|y''|}{\left[1 + (y')^2\right]^{3/2}} = \dfrac{e^x}{(1 + e^{2x})^{3/2}}$. To find the maximum value of κ, we find

$\kappa'(x) = -\dfrac{\left(2e^{2x} - 1\right)e^x}{(1 + e^{2x})^{5/2}} = 0 \Rightarrow x = -\frac{1}{2}\ln 2$, the only critical number of κ. Since $\kappa'(x) > 0$ for $x < -\frac{1}{2}\ln 2$ and

$\kappa'(x) < 0$ for $x > -\frac{1}{2}\ln 2$, we see that $x = -\frac{1}{2}\ln 2 = \ln\frac{\sqrt{2}}{2}$ gives a relative maximum value for κ which is in fact the

absolute maximum. The required point is thus $\left(\ln\frac{\sqrt{2}}{2}, \frac{\sqrt{2}}{2}\right).$

31. $xy = 1 \Rightarrow y = \dfrac{1}{x} \Rightarrow y' = -\dfrac{1}{x^2} \Rightarrow y'' = \dfrac{2}{x^3}$, so $\kappa = \dfrac{|y''|}{\left[1 + (y')^2\right]^{3/2}} = \dfrac{\left|\frac{2}{x^3}\right|}{\left(1 + \frac{1}{x^4}\right)^{3/2}} = \dfrac{2\left|x^3\right|}{(1 + x^4)^{3/2}}$. By symmetry, it

suffices to consider the case where $x > 0$: $\kappa(x) = \dfrac{2x^3}{(1 + x^4)^{3/2}} \Rightarrow \kappa'(x) = -\dfrac{6x^2\left(x^4 - 1\right)}{(1 + x^4)^{5/2}} = 0 \Rightarrow x = 1$ is the only

critical number of κ in $(0, \infty)$. The First Derivative Test shows that 1 gives the relative (and therefore absolute) maximum

value for κ, so by symmetry the required points are $(-1, -1)$ and $(1, 1)$.

33. The answer is b. Note that the curvature of the curve is smallest at $x = 0$ and largest at the endpoints $x = -4$ and $x = 4$.

35. The answer is d. Note that the curvature of the curve is 0 between $x = -1$ and $x = 0$, and between $x = 0$ and $x = 1$,

because the concavity of the curve changes in these intervals. Also, the curvature approaches 0 as $x \to \pm\infty$.

37. Using the result of Exercise 25, we have $\kappa(x) = \dfrac{2\left|2x^2 - 1\right|e^{2x^2}}{\left(e^{2x^2} + 4x^2\right)^{3/2}}.$

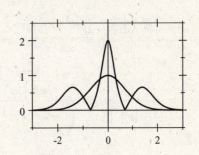

39. Let $\mathbf{r}(t) = f(t)\mathbf{i} + g(t)\mathbf{j} + 0\mathbf{k}$. Then $\mathbf{r}'(t) = f'(t)\mathbf{i} + g'(t)\mathbf{j}$ and $\mathbf{r}''(t) = f''(t)\mathbf{i} + g''(t)\mathbf{j}$,

so $\mathbf{r}'(t) \times \mathbf{r}''(t) = \begin{vmatrix} \mathbf{i} & \mathbf{j} & \mathbf{k} \\ f'(t) & g'(t) & 0 \\ f''(t) & g''(t) & 0 \end{vmatrix} = \left[f'(t)g''(t) - g'(t)f''(t)\right]\mathbf{k}$. Then

$$\kappa(t) = \frac{|\mathbf{r}'(t) \times \mathbf{r}''(t)|}{|\mathbf{r}'(t)|^3} = \frac{|f'(t)g''(t) - g'(t)f''(t)|}{\left\{[f'(t)]^2 + [g'(t)]^2\right\}^{3/2}}.$$

41. $x = t - \sin t$, $y = 1 - \cos t \Rightarrow x' = 1 - \cos t$, $y' = \sin t \Rightarrow x'' = \sin t$, $y'' = \cos t$. Thus,

$$\kappa(t) = \frac{|(1 - \cos t)\cos t - (\sin t)\sin t|}{\left[(1 - \cos t)^2 + (\sin t)^2\right]^{3/2}} = \frac{0}{(2 - 2\cos t)^{3/2}} = 0 \text{ provided that } t \neq 2n\pi, n \text{ an integer.}$$

43. a. $\dfrac{x^2}{9} + \dfrac{y^2}{4} = 1 \Rightarrow y^2 = \dfrac{4}{9}\left(9 - x^2\right) \Rightarrow y = \pm\dfrac{2}{3}\sqrt{9 - x^2}$. By symmetry, it suffices to look at $y = \dfrac{2}{3}\left(9 - x^2\right)^{1/2} \Rightarrow$

$$y' = \frac{-2x}{3\sqrt{9 - x^2}} \Rightarrow y'' = -\frac{2}{3}\left[\frac{\left(9 - x^2\right)^{1/2} - x\left(\frac{1}{2}\right)\left(9 - x^2\right)^{-1/2}(-2x)}{9 - x^2}\right] = -\frac{6}{\left(9 - x^2\right)^{3/2}}, \text{ so}$$

$$\kappa(x) = \frac{\left|-\dfrac{6}{\left(9 - x^2\right)^{3/2}}\right|}{\left[1 + \left(\dfrac{-2x}{3\sqrt{9 - x^2}}\right)^2\right]^{3/2}} = \frac{162}{\left(81 - 5x^2\right)^{3/2}}. \text{ The curvature is the same for } y = -\frac{2}{3}\sqrt{9 - x^2}.$$

b. The curvature at $(3, 0)$ is $\kappa(3) = \dfrac{162}{(81 - 5 \cdot 9)^{3/2}} = \dfrac{3}{4} \Rightarrow \rho = \dfrac{4}{3}$ and an

equation of the osculating circle is $\left(x - \dfrac{5}{3}\right)^2 + y^2 = \dfrac{16}{9}$. The curvature at

$(0, 2)$ is $\kappa(0) = \dfrac{162}{81^{3/2}} = \dfrac{2}{9} \Rightarrow \rho = \dfrac{9}{2}$ and an equation of the osculating

circle is $x^2 + \left(y + \dfrac{5}{2}\right)^2 = \dfrac{81}{4}$

c.

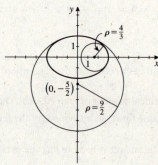

45. a. $\dfrac{dx}{dt} = \dfrac{d}{dt}\int_0^t \cos\frac{1}{2}\pi u^2\, du = \cos\frac{1}{2}\pi t^2$ and $\dfrac{dy}{dt} = \dfrac{d}{dt}\int_0^t \sin\frac{1}{2}\pi u^2\, du = \sin\frac{1}{2}\pi t^2$, so $\dfrac{dy}{dx} = \dfrac{\frac{dy}{dt}}{\frac{dx}{dt}} = \dfrac{\sin\frac{1}{2}\pi t^2}{\cos\frac{1}{2}\pi t^2} = \tan\frac{1}{2}\pi t^2$

$$\Rightarrow \frac{d^2y}{dx^2} = \frac{\frac{d}{dt}\left(\frac{dy}{dx}\right)}{\frac{dx}{dt}} = \frac{\left(\sec^2\frac{1}{2}\pi t^2\right)\pi t}{\cos\frac{1}{2}\pi t^2} = \frac{\pi t}{\cos^3\frac{1}{2}\pi t^2}.$$

b. $\kappa(t) = \dfrac{|y''|}{\left[1 + (y')^2\right]^{3/2}} = \dfrac{\left|\dfrac{\pi t}{\cos^3\frac{1}{2}\pi t^2}\right|}{\left(1 + \tan^2\frac{1}{2}\pi t^2\right)^{3/2}} = \dfrac{\left|\dfrac{\pi t}{\cos^3\frac{1}{2}\pi t^2}\right|}{\left|\sec^3\frac{1}{2}\pi t^2\right|} = \pi|t| = \pi t \text{ since } t \geq 0.$

47. $r = 1 + \sin\theta \Rightarrow r' = \cos\theta \Rightarrow r'' = -\sin\theta$, so

$$\kappa = \frac{\left|2\cos^2\theta - (1 + \sin\theta)(-\sin\theta) + (1 + \sin\theta)^2\right|}{\left[\cos^2\theta + (1 + \sin\theta)^2\right]^{3/2}} = \frac{\left|2\cos^2\theta + \sin\theta + \sin^2\theta + 1 + 2\sin\theta + \sin^2\theta\right|}{\left(\cos^2\theta + 1 + 2\sin\theta + \sin^2\theta\right)^{3/2}}$$

$$= \frac{|3 + 3\sin\theta|}{(2 + 2\sin\theta)^{3/2}} = \frac{3(1 + \sin\theta)}{2\sqrt{2}(1 + \sin\theta)^{3/2}} = \frac{3\sqrt{2}}{4\sqrt{1 + \sin\theta}}$$

49. Let $\mathbf{r}(t) = a\cos t\,\mathbf{i} + a\sin t\,\mathbf{j} + bt\,\mathbf{k}$. Then $\mathbf{r}'(t) = -a\sin t\,\mathbf{i} + a\cos t\,\mathbf{j} + b\,\mathbf{k}$ and

$$\mathbf{r}''(t) = -a\cos t\,\mathbf{i} - a\sin t\,\mathbf{j}, \text{ so } \mathbf{r}'(t) \times \mathbf{r}''(t) = \begin{vmatrix} \mathbf{i} & \mathbf{j} & \mathbf{k} \\ -a\sin t & a\cos t & b \\ -a\cos t & -a\sin t & 0 \end{vmatrix} = ab\sin t\,\mathbf{i} - ab\cos t\,\mathbf{j} + a^2\mathbf{k}.$$

Thus, $\left|\mathbf{r}'(t) \times \mathbf{r}''(t)\right|^2 = (ab\sin t)^2 + (-ab\cos t)^2 + \left(a^2\right)^2 = a^2\left(a^2 + b^2\right)$ and

$\left|\mathbf{r}'(t)\right|^2 = (-a\sin t)^2 + (a\cos t)^2 + b^2 = a^2 + b^2$. Finally, $\kappa = \dfrac{\left|\mathbf{r}'(t) \times \mathbf{r}''(t)\right|}{\left|\mathbf{r}'(t)\right|^3} = \dfrac{a\left(a^2 + b^2\right)^{1/2}}{\left(a^2 + b^2\right)^{3/2}} = \dfrac{a}{a^2 + b^2}$.

51. $\mathbf{r}(t) = t\cos t\,\mathbf{i} + t\sin t\,\mathbf{j} + t\,\mathbf{k} \Rightarrow \mathbf{r}'(t) = (\cos t - t\sin t)\,\mathbf{i} + (\sin t + t\cos t)\,\mathbf{j} + \mathbf{k} \Rightarrow$

$$\left|\mathbf{r}'(t)\right|^2 = (\cos t - t\sin t)^2 + (\sin t + t\cos t)^2 + 1$$

$$= \cos^2 t - 2t\cos t\sin t + t^2\sin^2 t + \sin^2 t + 2t\cos t\sin t + t^2\cos^2 t + 1 = 2\left(1 + t^2\right)$$

so

$$L = \int_a^b \left|\mathbf{r}'(t)\right|dt = \sqrt{2}\int_0^{2\pi}\sqrt{1 + t^2}\,dt = \sqrt{2}\left[\tfrac{1}{2}t\sqrt{1 + t^2} + \tfrac{1}{2}\ln\left(t + \sqrt{1 + t^2}\right)\right]_0^{2\pi}$$

$$= \sqrt{2}\left[\pi\sqrt{1 + 4\pi^2} + \tfrac{1}{2}\ln\left(2\pi + \sqrt{1 + 4\pi^2}\right)\right]$$

53. True. If C is smooth, then both $\frac{dx}{dt}$ and $\frac{dy}{dt}$ are continuous on I and furthermore $\frac{dx}{dt} \neq 0$ and $\frac{dy}{dt} \neq 0$ on I. Thus,

$\frac{dy}{dx} = \frac{dy/dt}{dx/dt}$ exists on I.

55. True. Since f has an inflection point at a, $f''(a) = 0$, so $\kappa(a) = \dfrac{\left|y''(a)\right|}{\left\{1 + \left[y'(a)\right]^2\right\}^{3/2}} = 0$.

57. True. The curve is a semicircle and hence has constant curvature.

12.4   Concept Questions   ET 11.4

1. a. See page 1011 (1007 in ET).   **b.** See page 1011 (1007 in ET).

12.4   Velocity and Acceleration   ET 11.4

1. $\mathbf{r}(t) = t\mathbf{i} + \left(4 - t^2\right)\mathbf{j} \Rightarrow \mathbf{v}(t) = \mathbf{r}'(t) = \mathbf{i} - 2t\mathbf{j} \Rightarrow$

$\mathbf{a}(t) = \mathbf{v}'(t) = -2\mathbf{j}$, so $\mathbf{r}(1) = \mathbf{i} + 3\mathbf{j}$, $\mathbf{v}(1) = \mathbf{i} - 2\mathbf{j}$,

$\mathbf{a}(1) = -2\mathbf{j}$, and $|\mathbf{v}(1)| = \sqrt{1 + (-2)^2} = \sqrt{5}$.

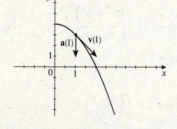

3. $\mathbf{r}(t) = \cos t\,\mathbf{i} + 3\sin t\,\mathbf{j} \Rightarrow \mathbf{v}(t) = -\sin t\,\mathbf{i} + 3\cos t\,\mathbf{j} \Rightarrow$

$\mathbf{a}(t) = -\cos t\,\mathbf{i} - 3\sin t\,\mathbf{j}$, so $\mathbf{r}\left(\frac{\pi}{4}\right) = \frac{\sqrt{2}}{2}\mathbf{i} + \frac{3\sqrt{2}}{2}\mathbf{j}$,

$\mathbf{v}\left(\frac{\pi}{4}\right) = -\frac{\sqrt{2}}{2}\mathbf{i} + \frac{3\sqrt{2}}{2}\mathbf{j}$, $\mathbf{a}\left(\frac{\pi}{4}\right) = -\frac{\sqrt{2}}{2}\mathbf{i} - \frac{3\sqrt{2}}{2}\mathbf{j}$, and

$\left|\mathbf{v}\left(\frac{\pi}{4}\right)\right| = \sqrt{\frac{1}{2} + \frac{9}{2}} = \sqrt{5}$.

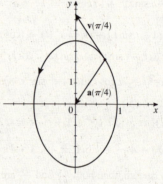

5. $\mathbf{r}(t) = \cos t\,\mathbf{i} + \sin t\,\mathbf{j} + t\mathbf{k} \Rightarrow$

$\mathbf{v}(t) = -\sin t\,\mathbf{i} + \cos t\,\mathbf{j} + \mathbf{k} \Rightarrow \mathbf{a}(t) = -\cos t\,\mathbf{i} - \sin t\,\mathbf{j}$,

so $\mathbf{r}\left(\frac{\pi}{2}\right) = \mathbf{j} + \frac{\pi}{2}\mathbf{k}$, $\mathbf{v}\left(\frac{\pi}{2}\right) = -\mathbf{i} + \mathbf{k}$, $\mathbf{a}\left(\frac{\pi}{2}\right) = -\mathbf{j}$, and

$\left|\mathbf{v}\left(\frac{\pi}{2}\right)\right| = \sqrt{1 + 1} = \sqrt{2}$.

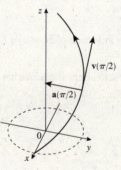

7. $\mathbf{r}(t) = t\mathbf{i} + t^2\mathbf{j} + \left(t^2 - 4\right)\mathbf{k} \Rightarrow \mathbf{v}(t) = \mathbf{i} + 2t\mathbf{j} + 2t\mathbf{k} \Rightarrow \mathbf{a}(t) = 2\mathbf{j} + 2\mathbf{k}$ and $|\mathbf{v}(t)| = \sqrt{1 + 4t^2 + 4t^2} = \sqrt{8t^2 + 1}$.

9. $\mathbf{r}(t) = t\mathbf{i} + t^2\mathbf{j} + \frac{1}{t}\mathbf{k} \Rightarrow \mathbf{v}(t) = \mathbf{i} + 2t\mathbf{j} - \frac{1}{t^2}\mathbf{k} \Rightarrow \mathbf{a}(t) = 2\mathbf{j} + \frac{2}{t^3}\mathbf{k}$ and $|\mathbf{v}(t)| = \sqrt{1 + 4t^2 + \frac{1}{t^4}} = \frac{\sqrt{4t^6 + t^4 + 1}}{t^2}$.

11. $\mathbf{r}(t) = e^t\langle\cos t, \sin t, 1\rangle \Rightarrow \mathbf{v}(t) = e^t\langle\cos t - \sin t, \cos t + \sin t, 1\rangle \Rightarrow \mathbf{a}(t) = e^t\langle -2\sin t, 2\cos t, 1\rangle$ and

$|\mathbf{v}(t)| = e^t\sqrt{(\cos t - \sin t)^2 + (\cos t + \sin t)^2 + 1} = \sqrt{3}\,e^t$.

13. $\mathbf{v}(t) = \int \mathbf{a}(t)\,dt = \int(-32\mathbf{k})\,dt = -32t\mathbf{k} + \mathbf{C}_1$ and $\mathbf{v}(0) = \mathbf{C}_1 = \mathbf{i} + 2\mathbf{j}$, so $\mathbf{v}(t) = \mathbf{i} + 2\mathbf{j} - 32t\mathbf{k}$.

$\mathbf{r}(t) = \int \mathbf{v}(t)\,dt = \int(\mathbf{i} + 2\mathbf{j} - 32t\mathbf{k})\,dt = t\mathbf{i} + 2t\mathbf{j} - 16t^2\mathbf{k} + \mathbf{C}_2$ and $\mathbf{r}(0) = \mathbf{C}_2 = 128\mathbf{k}$, so

$\mathbf{r}(t) = t\mathbf{i} + 2t\mathbf{j} + \left(128 - 16t^2\right)\mathbf{k}$.

15. $\mathbf{v}(t) = \int \mathbf{a}(t)\, dt = \int [\mathbf{i} - t\mathbf{j} + (1+t)\mathbf{k}]\, dt = t\mathbf{i} - \frac{1}{2}t^2\mathbf{j} + \left(t + \frac{1}{2}t^2\right)\mathbf{k} + \mathbf{C}_1$

and $\mathbf{v}(0) = \mathbf{C}_1 = \mathbf{i} + \mathbf{k}$, so $\mathbf{v}(t) = (t+1)\mathbf{i} - \frac{1}{2}t^2\mathbf{j} + \left(\frac{1}{2}t^2 + t + 1\right)\mathbf{k}$.

$\mathbf{r}(t) = \int \mathbf{v}(t)\, dt = \int \left[(t+1)\mathbf{i} - \frac{1}{2}t^2\mathbf{j} + \left(\frac{1}{2}t^2 + t + 1\right)\mathbf{k}\right] dt = \left(\frac{1}{2}t^2 + t\right)\mathbf{i} - \frac{1}{6}t^3\mathbf{j} + \left(\frac{1}{6}t^3 + \frac{1}{2}t^2 + t\right)\mathbf{k} + \mathbf{C}_2$ and

$\mathbf{r}(0) = \mathbf{C}_2 = \mathbf{j} + \mathbf{k}$, so $\mathbf{r}(t) = \left(\frac{1}{2}t^2 + t\right)\mathbf{i} + \left(1 - \frac{1}{6}t^3\right)\mathbf{j} + \left(\frac{1}{6}t^3 + \frac{1}{2}t^2 + t + 1\right)\mathbf{k}$.

17. $\mathbf{v}(t) = \int \mathbf{a}(t)\, dt = \int (-\cos t\, \mathbf{i} - \sin t\, \mathbf{j} + \mathbf{k})\, dt = -\sin t\, \mathbf{i} + \cos t\, \mathbf{j} + t\mathbf{k} + \mathbf{C}_1$ and

$\mathbf{v}(0) = \mathbf{j} + \mathbf{C}_1 = 2\mathbf{k} \Rightarrow \mathbf{C}_1 = -\mathbf{j} + 2\mathbf{k}$, so $\mathbf{v}(t) = -\sin t\, \mathbf{i} + (\cos t - 1)\mathbf{j} + (t+2)\mathbf{k}$.

$\mathbf{r}(t) = \int \mathbf{v}(t)\, dt = \int [-\sin t\, \mathbf{i} + (\cos t - 1)\mathbf{j} + (t+2)\mathbf{k}]\, dt = \cos t\, \mathbf{i} + (\sin t - t)\mathbf{j} + \left(\frac{1}{2}t^2 + 2t\right)\mathbf{k} + \mathbf{C}_2$ and

$\mathbf{r}(0) = \mathbf{i} + \mathbf{C}_2 = \mathbf{i} \Rightarrow \mathbf{C}_2 = \mathbf{0}$, so $\mathbf{r}(t) = \cos t\, \mathbf{i} + (\sin t - t)\mathbf{j} + \left(\frac{1}{2}t^2 + 2t\right)\mathbf{k}$.

19. Let $\mathbf{v}(t)$ denote the velocity of the particle. Then by assumption, $|\mathbf{v}(t)| = c \Leftrightarrow |\mathbf{v}|^2 = c^2$, where c is a constant. Since

$|\mathbf{v}|^2 = \mathbf{v} \cdot \mathbf{v}$, we have $\mathbf{v} \cdot \mathbf{v} = c^2$, so $\dfrac{d}{dt}(\mathbf{v} \cdot \mathbf{v}) = \mathbf{v} \cdot \mathbf{v}' + \mathbf{v}' \cdot \mathbf{v} = \dfrac{d}{dt}\left(c^2\right) = 0 \Leftrightarrow 2\mathbf{v} \cdot \mathbf{v}' = 2\mathbf{v} \cdot \mathbf{a} = 0 \Rightarrow \mathbf{v} \cdot \mathbf{a} = 0$. Thus, the

velocity and acceleration vectors are orthogonal.

21. Let $\mathbf{r}(t) = x(t)\mathbf{i} + y(t)\mathbf{j} + z(t)\mathbf{k}$. Then $\mathbf{v}(t) = \mathbf{r}'(t) = x'(t)\mathbf{i} + y'(t)\mathbf{j} + z'(t)\mathbf{k}$. Since

$\mathbf{v}(t)$ is always orthogonal to $\mathbf{r}(t)$, we have $\mathbf{v}(t) \cdot \mathbf{r}(t) = 0 \Leftrightarrow xx' + yy' + zz' = 0$. Now

$\dfrac{d}{dt}\left\{[x(t)]^2 + [y(t)]^2 + [z(t)]^2\right\} = 2x(t)x'(t) + 2y(t)y'(t) + 2z(t)z'(t)$ and so $xx' + yy' + zz' = 0$ implies that

$\dfrac{d}{dt}\left\{[x(t)]^2 + [y(t)]^2 + [z(t)]^2\right\} = 0 \Leftrightarrow x^2 + y^2 + z^2 = c$, a constant. Since this equation describes a sphere with radius

$\sqrt{c}$ centered at the origin, the trajectory of the particle lies on such a sphere.

23. a. $\mathbf{r}(t) = (v_0 \cos \alpha)\, t\mathbf{i} + \left[(v_0 \sin \alpha)\, t - \frac{1}{2}gt^2\right]\mathbf{j} = (1500 \cos 30°)\, t\mathbf{i} + \left[(1500 \sin 30°)\, t - 16t^2\right]\mathbf{j} = 750\sqrt{3}t\mathbf{i} +$

$\left(750t - 16t^2\right)\mathbf{j}$. The projectile strikes the ground when $750t - 16t^2 = 0 \Leftrightarrow t = 0$ or $t = \frac{375}{8}$. The range of the

projectile is $750\sqrt{3} \cdot \frac{375}{8} \approx 60{,}892$ ft.

b. Here $y(t) = 750t - 16t^2 \Rightarrow y'(t) = 750 - 32t = 0 \Rightarrow t = \frac{375}{16}$, so the maximum height attained by the projectile is

$y\left(\frac{375}{16}\right) = 750\left(\frac{375}{16}\right) - 16\left(\frac{375}{16}\right)^2 \approx 8789$ ft.

c. $\mathbf{r}'(t) = 750\sqrt{3}\mathbf{i} + (750 - 32t)\mathbf{j}$, so $\mathbf{v}\left(\frac{375}{8}\right) = 750\sqrt{3}\mathbf{i} + \left[750 - 32\left(\frac{375}{8}\right)\right]\mathbf{j} = 750\sqrt{3}\mathbf{i} - 750\mathbf{j}$. Therefore, the speed

of the projectile at impact is $\left|\mathbf{v}\left(\frac{375}{8}\right)\right| = \sqrt{\left(750\sqrt{3}\right)^2 + (-750)^2} = 1500$ ft/s.

25. a. $\mathbf{r}(t) = (v_0 \cos \alpha)\, t\mathbf{i} + \left[h + (v_0 \sin \alpha)\, t - \frac{1}{2}gt^2\right]\mathbf{j} = (1500 \cos 30°)\, t\mathbf{i} + \left[200 + (1500 \sin 30°)\, t - 16t^2\right]\mathbf{j} = 750\sqrt{3}t\mathbf{i} +$

$\left(200 + 750t - 16t^2\right)\mathbf{j} = x(t)\mathbf{i} + y(t)\mathbf{j}$. The projectile strikes the ground when $200 + 750t - 16t^2 = 0 \Leftrightarrow$

$8t^2 - 375t - 100 = 0$. Solving, we find $t = \dfrac{375 \pm \sqrt{(-375)^2 - 4(8)(-100)}}{16} \approx 47.1$ s. Therefore, the range is approximately

$x(47.1) = 750\sqrt{3}\,(47.1) \approx 61{,}185$ ft.

b. $y(t) = 200 + 750t - 16t^2 \Rightarrow y'(t) = 750 - 32t = 0 \Rightarrow t = \frac{375}{16}$, so the maximum height attained by the projectile is

$y\left(\frac{375}{16}\right) = 200 + 750\left(\frac{375}{16}\right) - 16\left(\frac{375}{16}\right)^2 \approx 8989$ ft.

c. $\mathbf{v}(t) = \mathbf{r}'(t) = 750\sqrt{3}\mathbf{i} + (750 - 32t)\mathbf{j}$. $\mathbf{v}(47.1) \approx 750\sqrt{3}\mathbf{i} + [750 - 32\,(47.1)]\mathbf{j} \approx 750\sqrt{3}\mathbf{i} - 757.2\mathbf{j}$, so the speed of

the projectile at impact is $|\mathbf{v}(47.1)| \approx \sqrt{\left(750\sqrt{3}\right)^2 + (-757.2)^2} \approx 1504$ ft/s.

27. $\mathbf{r}(t) = (500\cos\alpha)\,t\mathbf{i} + \left[(500\sin\alpha)\,t - 16t^2\right]\mathbf{j} = x(t)\mathbf{i} + y(t)\mathbf{j}$. We want $x(t) = 1200$ and $y(t) = 0$; that is,

$(500\cos\alpha)\,t = 1200$ and $(500\sin\alpha)\,t - 16t^2 = 0$. The second equation gives $t = \dfrac{125\sin\alpha}{4}$ (we reject $t = 0$). Thus,

$(500\cos\alpha)(125\sin\alpha) = 4800 \Leftrightarrow 2\sin\alpha\cos\alpha = \dfrac{4800}{250\cdot125} \Leftrightarrow \sin 2\alpha = 0.1536$. Thus, $\alpha \approx 4.4°$. The angle of elevation

is approximately $4.4°$.

29. a. $\mathbf{r}(t) = a\cos\omega t\,\mathbf{i} + a\sin\omega t\,\mathbf{j} \Rightarrow \mathbf{v}(t) = \mathbf{r}'(t) = -a\omega\sin\omega t\,\mathbf{i} + a\omega\cos\omega t\,\mathbf{j}$. Since

   $\mathbf{v}(t)\cdot\mathbf{r}(t) = -a^2\omega\cos\omega t\sin\omega t + a^2\omega\sin\omega t\cos\omega t = 0$, we see that $\mathbf{v}$ and $\mathbf{r}$ are orthogonal.

   b. $\mathbf{a}(t) = \mathbf{v}'(t) = -a\omega^2\cos\omega t\,\mathbf{i} - a\omega^2\sin\omega t\,\mathbf{j} = -\omega^2\mathbf{r}(t)$. Since $\mathbf{r}(t)$ points away from the origin, $\mathbf{a}(t)$ points toward
   the origin.

   c. The speed is $|\mathbf{v}(t)| = \sqrt{a^2\omega^2\sin^2\omega t + a^2\omega^2\cos^2\omega t} = a\omega$ and

   $|\mathbf{a}(t)| = \left|-\omega^2\mathbf{r}(t)\right| = \omega^2\,|\mathbf{r}(t)| = \omega^2\sqrt{a^2\cos^2\omega t + a^2\sin^2\omega t} = a\omega^2$.

31. $\mathbf{r}(t) = (v_0\cos 60°)\,t\mathbf{i} + \left[(v_0\sin 60°)\,t - 16t^2\right]\mathbf{j} = \frac{1}{2}v_0 t\mathbf{i} + \left(\frac{\sqrt{3}}{2}v_0 t - 16t^2\right)\mathbf{j} = x(t)\mathbf{i} + y(t)\mathbf{j}$. Setting $y(t) = 0$ gives

$t\left(\frac{\sqrt{3}}{2}v_0 - 16t\right) = 0 \Rightarrow t = 0$ or $t = \frac{\sqrt{3}}{32}v_0$, so $x(t) = \frac{1}{2}v_0\left(\frac{\sqrt{3}}{32}v_0\right) = \frac{\sqrt{3}}{64}v_0^2 \Rightarrow v_0 = \left(\frac{64\sqrt{3}}{3}x\right)^{1/2}$. We require that

$150 \le x \le 180$. Then $\sqrt{\frac{64\sqrt{3}}{3}\cdot150} \le v_0 \le \sqrt{\frac{64\sqrt{3}}{3}\cdot180} \Leftrightarrow 74.4$ ft/s $\le v_0 \le 81.6$ ft/s.

33. Let us assume without loss of generality that the model train is at the origin at $t = 0$ and is moving to the right with $v_0 > 0$.
Then the position of the ball bearing at time t is $\mathbf{r}(t) = (v\cos\alpha)\,t\mathbf{i} + \left[(v\sin\alpha)\,t - 16t^2\right]\mathbf{j}$ where $\cos\alpha = \dfrac{v_0}{v}$, $\sin\alpha = \dfrac{v_1}{v}$,

and $v = \sqrt{v_0^2 + v_1^2}$. The ball bearing returns to the train when $(v\sin\alpha - 16t)\,t = 0 \Rightarrow t = \dfrac{v\sin\alpha}{16} = \dfrac{vv_1}{16v} = \dfrac{v_1}{16}$. The

horizontal distance traveled by the ball bearing is $x\left(\dfrac{v_1}{16}\right) = (v\cos\alpha)\,t = v\left(\dfrac{v_0}{v}\right)\left(\dfrac{v_1}{16}\right) = \dfrac{v_0 v_1}{16}$, but the train will have

moved to the right by $v_0\left(\dfrac{v_1}{16}\right) = \dfrac{v_0 v_1}{16}$ ft. This means that the ball bearing will end up at the same location on the train
from which it was released.

35. $\mathbf{r}(t) = \left\langle t, t^2, t^3\right\rangle \Rightarrow \mathbf{r}'(t) = \left\langle 1, 2t, 3t^2\right\rangle$. At $t = 1$, $\mathbf{r}(1) = \langle 1, 1, 1\rangle$ and $\mathbf{r}'(1) = \langle 1, 2, 3\rangle$, so the trajectory of the particle is
$\mathbf{r}(t) = \langle 1 + t, 1 + 2t, 1 + 3t\rangle$. Its position vector at $t = 2$ is $\mathbf{r}(2) = \langle 3, 5, 7\rangle$; in other words, the particle is at the point
$(3, 5, 7)$.

37. True. $\mathbf{r}'(t) = \dfrac{\mathbf{r}'(t)}{|\mathbf{r}'(t)|}\cdot|\mathbf{r}'(t)| = |\mathbf{r}'(t)|\dfrac{\mathbf{r}'(t)}{|\mathbf{r}'(t)|} = |\mathbf{r}'(t)|\,\mathbf{T}(t)$ since $\mathbf{r}'(t) \ne \mathbf{0}$.

12.5 Concept Questions   ET 11.5

1. a. See page 1018 (1014 in ET).   **b.** See page 1018 (1014 in ET).

12.5 Tangential and Normal Components of Acceleration ET 11.5

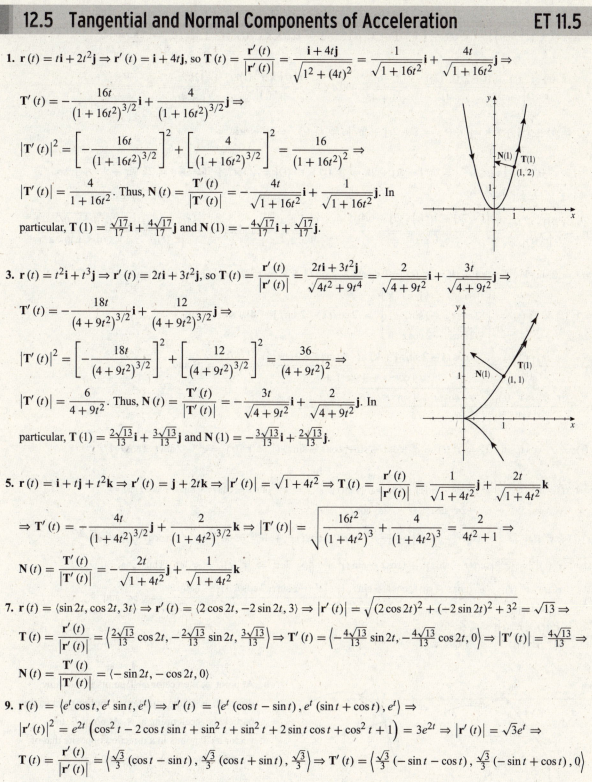

1. $\mathbf{r}(t) = t\mathbf{i} + 2t^2\mathbf{j} \Rightarrow \mathbf{r}'(t) = \mathbf{i} + 4t\mathbf{j}$, so $\mathbf{T}(t) = \dfrac{\mathbf{r}'(t)}{|\mathbf{r}'(t)|} = \dfrac{\mathbf{i} + 4t\mathbf{j}}{\sqrt{1^2 + (4t)^2}} = \dfrac{1}{\sqrt{1 + 16t^2}}\mathbf{i} + \dfrac{4t}{\sqrt{1 + 16t^2}}\mathbf{j} \Rightarrow$

$\mathbf{T}'(t) = -\dfrac{16t}{(1 + 16t^2)^{3/2}}\mathbf{i} + \dfrac{4}{(1 + 16t^2)^{3/2}}\mathbf{j} \Rightarrow$

$|\mathbf{T}'(t)|^2 = \left[-\dfrac{16t}{(1 + 16t^2)^{3/2}}\right]^2 + \left[\dfrac{4}{(1 + 16t^2)^{3/2}}\right]^2 = \dfrac{16}{(1 + 16t^2)^2} \Rightarrow$

$|\mathbf{T}'(t)| = \dfrac{4}{1 + 16t^2}$. Thus, $\mathbf{N}(t) = \dfrac{\mathbf{T}'(t)}{|\mathbf{T}'(t)|} = -\dfrac{4t}{\sqrt{1 + 16t^2}}\mathbf{i} + \dfrac{1}{\sqrt{1 + 16t^2}}\mathbf{j}$. In

particular, $\mathbf{T}(1) = \dfrac{\sqrt{17}}{17}\mathbf{i} + \dfrac{4\sqrt{17}}{17}\mathbf{j}$ and $\mathbf{N}(1) = -\dfrac{4\sqrt{17}}{17}\mathbf{i} + \dfrac{\sqrt{17}}{17}\mathbf{j}$.

3. $\mathbf{r}(t) = t^2\mathbf{i} + t^3\mathbf{j} \Rightarrow \mathbf{r}'(t) = 2t\mathbf{i} + 3t^2\mathbf{j}$, so $\mathbf{T}(t) = \dfrac{\mathbf{r}'(t)}{|\mathbf{r}'(t)|} = \dfrac{2t\mathbf{i} + 3t^2\mathbf{j}}{\sqrt{4t^2 + 9t^4}} = \dfrac{2}{\sqrt{4 + 9t^2}}\mathbf{i} + \dfrac{3t}{\sqrt{4 + 9t^2}}\mathbf{j} \Rightarrow$

$\mathbf{T}'(t) = -\dfrac{18t}{(4 + 9t^2)^{3/2}}\mathbf{i} + \dfrac{12}{(4 + 9t^2)^{3/2}}\mathbf{j} \Rightarrow$

$|\mathbf{T}'(t)|^2 = \left[-\dfrac{18t}{(4 + 9t^2)^{3/2}}\right]^2 + \left[\dfrac{12}{(4 + 9t^2)^{3/2}}\right]^2 = \dfrac{36}{(4 + 9t^2)^2} \Rightarrow$

$|\mathbf{T}'(t)| = \dfrac{6}{4 + 9t^2}$. Thus, $\mathbf{N}(t) = \dfrac{\mathbf{T}'(t)}{|\mathbf{T}'(t)|} = -\dfrac{3t}{\sqrt{4 + 9t^2}}\mathbf{i} + \dfrac{2}{\sqrt{4 + 9t^2}}\mathbf{j}$. In

particular, $\mathbf{T}(1) = \dfrac{2\sqrt{13}}{13}\mathbf{i} + \dfrac{3\sqrt{13}}{13}\mathbf{j}$ and $\mathbf{N}(1) = -\dfrac{3\sqrt{13}}{13}\mathbf{i} + \dfrac{2\sqrt{13}}{13}\mathbf{j}$.

5. $\mathbf{r}(t) = \mathbf{i} + t\mathbf{j} + t^2\mathbf{k} \Rightarrow \mathbf{r}'(t) = \mathbf{j} + 2t\mathbf{k} \Rightarrow |\mathbf{r}'(t)| = \sqrt{1 + 4t^2} \Rightarrow \mathbf{T}(t) = \dfrac{\mathbf{r}'(t)}{|\mathbf{r}'(t)|} = \dfrac{1}{\sqrt{1 + 4t^2}}\mathbf{j} + \dfrac{2t}{\sqrt{1 + 4t^2}}\mathbf{k}$

$\Rightarrow \mathbf{T}'(t) = -\dfrac{4t}{(1 + 4t^2)^{3/2}}\mathbf{j} + \dfrac{2}{(1 + 4t^2)^{3/2}}\mathbf{k} \Rightarrow |\mathbf{T}'(t)| = \sqrt{\dfrac{16t^2}{(1 + 4t^2)^3} + \dfrac{4}{(1 + 4t^2)^3}} = \dfrac{2}{4t^2 + 1} \Rightarrow$

$\mathbf{N}(t) = \dfrac{\mathbf{T}'(t)}{|\mathbf{T}'(t)|} = -\dfrac{2t}{\sqrt{1 + 4t^2}}\mathbf{j} + \dfrac{1}{\sqrt{1 + 4t^2}}\mathbf{k}$

7. $\mathbf{r}(t) = \langle \sin 2t, \cos 2t, 3t \rangle \Rightarrow \mathbf{r}'(t) = \langle 2\cos 2t, -2\sin 2t, 3 \rangle \Rightarrow |\mathbf{r}'(t)| = \sqrt{(2\cos 2t)^2 + (-2\sin 2t)^2 + 3^2} = \sqrt{13} \Rightarrow$

$\mathbf{T}(t) = \dfrac{\mathbf{r}'(t)}{|\mathbf{r}'(t)|} = \left\langle \dfrac{2\sqrt{13}}{13}\cos 2t, -\dfrac{2\sqrt{13}}{13}\sin 2t, \dfrac{3\sqrt{13}}{13} \right\rangle \Rightarrow \mathbf{T}'(t) = \left\langle -\dfrac{4\sqrt{13}}{13}\sin 2t, -\dfrac{4\sqrt{13}}{13}\cos 2t, 0 \right\rangle \Rightarrow |\mathbf{T}'(t)| = \dfrac{4\sqrt{13}}{13} \Rightarrow$

$\mathbf{N}(t) = \dfrac{\mathbf{T}'(t)}{|\mathbf{T}'(t)|} = \langle -\sin 2t, -\cos 2t, 0 \rangle$

9. $\mathbf{r}(t) = \langle e^t\cos t, e^t\sin t, e^t \rangle \Rightarrow \mathbf{r}'(t) = \langle e^t(\cos t - \sin t), e^t(\sin t + \cos t), e^t \rangle \Rightarrow$

$|\mathbf{r}'(t)|^2 = e^{2t}\left(\cos^2 t - 2\cos t\sin t + \sin^2 t + \sin^2 t + 2\sin t\cos t + \cos^2 t + 1\right) = 3e^{2t} \Rightarrow |\mathbf{r}'(t)| = \sqrt{3}\,e^t \Rightarrow$

$\mathbf{T}(t) = \dfrac{\mathbf{r}'(t)}{|\mathbf{r}'(t)|} = \left\langle \dfrac{\sqrt{3}}{3}(\cos t - \sin t), \dfrac{\sqrt{3}}{3}(\cos t + \sin t), \dfrac{\sqrt{3}}{3} \right\rangle \Rightarrow \mathbf{T}'(t) = \left\langle \dfrac{\sqrt{3}}{3}(-\sin t - \cos t), \dfrac{\sqrt{3}}{3}(-\sin t + \cos t), 0 \right\rangle$

$\Rightarrow |\mathbf{T}'(t)| = \dfrac{\sqrt{6}}{3} \Rightarrow \mathbf{N}(t) = \dfrac{\mathbf{T}'(t)}{|\mathbf{T}'(t)|} = \left\langle -\dfrac{\sqrt{2}}{2}(\sin t + \cos t), \dfrac{\sqrt{2}}{2}(\cos t - \sin t), 0 \right\rangle$

11. $\mathbf{r}(t) = t\mathbf{i} + \left(t^2 + 4\right)\mathbf{j} \Rightarrow \mathbf{r}'(t) = \mathbf{i} + 2t\mathbf{j} \Rightarrow \mathbf{r}''(t) = 2\mathbf{j} \Rightarrow \mathbf{r}'(t) \times \mathbf{r}''(t) = \begin{vmatrix} \mathbf{i} & \mathbf{j} & \mathbf{k} \\ 1 & 2t & 0 \\ 0 & 2 & 0 \end{vmatrix} = 2\mathbf{k}$, so

$$a_{\mathbf{T}} = \frac{\mathbf{r}'(t) \cdot \mathbf{r}''(t)}{|\mathbf{r}'(t)|} = \frac{(\mathbf{i} + 2t\mathbf{j}) \cdot (2\mathbf{j})}{\sqrt{1 + 4t^2}} = \frac{4t}{\sqrt{1 + 4t^2}} \text{ and } a_{\mathbf{N}} = \frac{|\mathbf{r}'(t) \times \mathbf{r}''(t)|}{|\mathbf{r}'(t)|} = \frac{|2\mathbf{k}|}{\sqrt{1 + 4t^2}} = \frac{2}{\sqrt{1 + 4t^2}}.$$

13. $\mathbf{r}(t) = t\mathbf{i} + t^2\mathbf{j} + t^3\mathbf{k} \Rightarrow \mathbf{r}'(t) = \mathbf{i} + 2t\mathbf{j} + 3t^2\mathbf{k} \Rightarrow \mathbf{r}''(t) = 2\mathbf{j} + 6t\mathbf{k} \Rightarrow$

$$\mathbf{r}'(t) \times \mathbf{r}''(t) = \begin{vmatrix} \mathbf{i} & \mathbf{j} & \mathbf{k} \\ 1 & 2t & 3t^2 \\ 0 & 2 & 6t \end{vmatrix} = 6t^2\mathbf{i} - 6t\mathbf{j} + 2\mathbf{k} \Rightarrow |\mathbf{r}'(t) \times \mathbf{r}''(t)| = \sqrt{36t^4 + 36t^2 + 4} = 2\sqrt{9t^4 + 9t^2 + 1}, \text{ so}$$

$$a_{\mathbf{T}} = \frac{\mathbf{r}'(t) \cdot \mathbf{r}''(t)}{|\mathbf{r}'(t)|} = \frac{\left(\mathbf{i} + 2t\mathbf{j} + 3t^2\mathbf{k}\right) \cdot (2\mathbf{j} + 6t\mathbf{k})}{\sqrt{9t^4 + 4t^2 + 1}} = \frac{18t^3 + 4t}{\sqrt{9t^4 + 4t^2 + 1}} \text{ and } a_{\mathbf{N}} = \frac{|\mathbf{r}'(t) \times \mathbf{r}''(t)|}{|\mathbf{r}'(t)|} = \frac{2\sqrt{9t^4 + 9t^2 + 1}}{\sqrt{9t^4 + 4t^2 + 1}}.$$

15. $\mathbf{r}(t) = 2\sin t\,\mathbf{i} + 2\cos t\,\mathbf{j} + t\mathbf{k} \Rightarrow \mathbf{r}'(t) = 2\cos t\,\mathbf{i} - 2\sin t\,\mathbf{j} + \mathbf{k} \Rightarrow \mathbf{r}''(t) = -2\sin t\,\mathbf{i} - 2\cos t\,\mathbf{j}$

$$\Rightarrow \mathbf{r}'(t) \times \mathbf{r}''(t) = \begin{vmatrix} \mathbf{i} & \mathbf{j} & \mathbf{k} \\ 2\cos t & -2\sin t & 1 \\ -2\sin t & -2\cos t & 0 \end{vmatrix} = 2\cos t\,\mathbf{i} - 2\sin t\,\mathbf{j} - 4\mathbf{k}, \text{ so}$$

$$a_{\mathbf{T}} = \frac{\mathbf{r}'(t) \cdot \mathbf{r}''(t)}{|\mathbf{r}'(t)|} = \frac{(2\cos t\,\mathbf{i} - 2\sin t\,\mathbf{j} + \mathbf{k}) \cdot (-2\sin t\,\mathbf{i} - 2\cos t\,\mathbf{j})}{\sqrt{4\cos^2 t + 4\sin^2 t + 1}} = \frac{0}{\sqrt{5}} = 0 \text{ and}$$

$$a_{\mathbf{N}} = \frac{|\mathbf{r}'(t) \times \mathbf{r}''(t)|}{|\mathbf{r}'(t)|} = \frac{\sqrt{4\cos^2 t + 4\sin^2 t + 16}}{\sqrt{5}} = \frac{2\sqrt{5}}{\sqrt{5}} = 2.$$

17. $\mathbf{r}(t) = e^t \langle \cos t, \sin t, 1 \rangle \Rightarrow \mathbf{r}'(t) = e^t \langle \cos t - \sin t, \cos t + \sin t, 1 \rangle \Rightarrow \mathbf{r}''(t) = e^t \langle -2\sin t, 2\cos t, 1 \rangle$

$$\Rightarrow \mathbf{r}'(t) \times \mathbf{r}''(t) = \begin{vmatrix} \mathbf{i} & \mathbf{j} & \mathbf{k} \\ e^t(\cos t - \sin t) & e^t(\cos t + \sin t) & e^t \\ -2e^t \sin t & 2e^t \cos t & e^t \end{vmatrix} = e^{2t}\langle \sin t - \cos t, -\sin t - \cos t, 2 \rangle \Rightarrow$$

$$|\mathbf{r}'(t) \times \mathbf{r}''(t)|^2 = \left[e^{2t}(\sin t - \cos t)\right]^2 + \left[-e^{2t}(\sin t + \cos t)\right]^2 + 4e^{4t} = 6e^{4t} \Rightarrow |\mathbf{r}'(t) \times \mathbf{r}''(t)| = \sqrt{6}e^{2t}.$$

Also, $|\mathbf{r}'(t)|^2 = e^{2t}\left[(\cos t - \sin t)^2 + (\cos t + \sin t)^2 + 1\right] = 3e^{2t} \Rightarrow |\mathbf{r}'(t)| = \sqrt{3}e^t.$ Thus,

$$a_{\mathbf{T}} = \frac{\mathbf{r}'(t) \cdot \mathbf{r}''(t)}{|\mathbf{r}'(t)|} = \frac{e^t \langle \cos t - \sin t, \cos t + \sin t, 1 \rangle \cdot \left[e^t \langle -2\sin t, 2\cos t, 1 \rangle\right]}{\sqrt{3}e^t} = \frac{3e^{2t}}{\sqrt{3}e^t} = \sqrt{3}e^t \text{ and}$$

$$a_{\mathbf{N}} = \frac{|\mathbf{r}'(t) \times \mathbf{r}''(t)|}{|\mathbf{r}'(t)|} = \frac{\sqrt{6}e^{2t}}{\sqrt{3}e^t} = \sqrt{2}e^t.$$

19. a.

b. At point A, the vector $a_{\mathbf{T}}\mathbf{T}$ points in the same direction as $\mathbf{T}$ (the direction of motion) and so the particle is accelerating at A. At point B, the vector $a_{\mathbf{T}}\mathbf{T}$ points in a direction opposite that of $\mathbf{T}$, and so the particle is decelerating at B.

21. a. We are given that at some time t_0, $\mathbf{r}'(t_0) = \mathbf{v}(t_0) = 2\mathbf{i} + 3\mathbf{j} - 6\mathbf{k}$, and $\mathbf{r}''(t_0) = \mathbf{a}(t_0) = -6\mathbf{i} - 4\mathbf{j} + 3\mathbf{k}$.

Thus, $a_{\mathbf{T}}(t_0) = \dfrac{\mathbf{r}'(t_0) \cdot \mathbf{r}''(t_0)}{|\mathbf{r}'(t_0)|} = \dfrac{(2\mathbf{i} + 3\mathbf{j} - 6\mathbf{k}) \cdot (-6\mathbf{i} - 4\mathbf{j} + 3\mathbf{k})}{\sqrt{2^2 + 3^2 + (-6)^2}} = -\dfrac{42}{7} = -6$,

$$\mathbf{r}'(t_0) \times \mathbf{r}''(t_0) = \begin{vmatrix} \mathbf{i} & \mathbf{j} & \mathbf{k} \\ 2 & 3 & -6 \\ -6 & -4 & 3 \end{vmatrix} = -15\mathbf{i} + 30\mathbf{j} + 10\mathbf{k}, \text{ and}$$

$a_{\mathbf{N}}(t_0) = \dfrac{|\mathbf{r}'(t_0) \times \mathbf{r}''(t_0)|}{|\mathbf{r}'(t_0)|} = \dfrac{\sqrt{(-15)^2 + 30^2 + 10^2}}{7} = \dfrac{35}{7} = 5$.

b. Since $a_{\mathbf{T}} < 0$, the particle is decelerating.

23. a. Here $x = \cos t^2$ and $y = \sin t^2$, so $x^2 + y^2 = \cos^2 t^2 + \sin^2 t^2 = 1$, an equation of a circle centered at the origin.

b. $\mathbf{v}(t) = \left\langle -2t \sin t^2, 2t \cos t^2 \right\rangle$ and $\mathbf{a}(t) = \left\langle -2 \sin t^2 - 4t^2 \cos t^2, 2 \cos t^2 - 4t^2 \sin t^2 \right\rangle$, so

$\mathbf{a} \cdot \mathbf{r} + \mathbf{v} \cdot \mathbf{v} = \left(-2 \sin t^2 - 4t^2 \cos t^2 \right) \cos t^2 + \left(2 \cos t^2 - 4t^2 \sin t^2 \right) \sin t^2 + \left(-2t \sin t^2 \right)^2 + \left(2t \cos t^2 \right)^2$

$\quad = -2 \sin t^2 \cos t^2 - 4t^2 \cos^2 t^2 + 2 \cos t^2 \sin t^2 - 4t^2 \sin^2 t^2 + 4t^2 \sin^2 t^2 + 4t^2 \cos^2 t^2 = 0$

Since $\mathbf{v} \cdot \mathbf{v} = |\mathbf{v}|^2 \geq 0$, we see that $\mathbf{a} \cdot \mathbf{r} \leq 0$.

25. Since the velocity is constant, we have $|\mathbf{v}|^2 = \mathbf{v} \cdot \mathbf{v} = c$, a positive constant. Then $\frac{d}{dt}(\mathbf{v} \cdot \mathbf{v}) = \frac{d}{dt}(c) \Rightarrow \mathbf{v} \cdot \mathbf{v}' + \mathbf{v}' \cdot \mathbf{v} = 0$
$\Rightarrow 2\mathbf{v} \cdot \mathbf{v}' = 0 \Leftrightarrow \mathbf{v} \cdot \mathbf{a} = 0$; that is, $\mathbf{a}$ is orthogonal to $\mathbf{v}$. But $\mathbf{v}$ is tangent to C, and so $\mathbf{a}$ is orthogonal to C.

27. Since the plane curve is the graph of a function f, it has representation $\mathbf{r}(t) = \langle t, f(t) \rangle$. Thus, $\mathbf{r}'(t) = \left\langle 1, f'(t) \right\rangle$ and

$\mathbf{r}''(t) = \left\langle 0, f''(t) \right\rangle$, so $\mathbf{r}'(t) \times \mathbf{r}''(t) = \begin{vmatrix} \mathbf{i} & \mathbf{j} & \mathbf{k} \\ 1 & f'(t) & 0 \\ 0 & f''(t) & 0 \end{vmatrix} = f''(t)\mathbf{k}$ and $a_{\mathbf{N}}(t) = \dfrac{|\mathbf{r}'(t) \times \mathbf{r}''(t)|}{|\mathbf{r}'(t)|} = \dfrac{|f''(t)|}{\sqrt{1 + [f'(t)]^2}}$. If the

point $(x_0, y_0) = (t_0, f(t_0))$ is an inflection point, then $f''(t_0) = 0$ and so $a_{\mathbf{N}}(t_0) = 0$.

29. $\mathbf{r}(t) = \left\langle e^t, e^{-t}, \sqrt{2}t \right\rangle \Rightarrow \mathbf{r}'(t) = \left\langle e^t, -e^{-t}, \sqrt{2} \right\rangle$, so $|\mathbf{r}'(t)| = \sqrt{e^{2t} + e^{-2t} + 2} = \dfrac{e^{2t} + 1}{e^t}$.

$\mathbf{T}(t) = \dfrac{\mathbf{r}'(t)}{|\mathbf{r}'(t)|} = \dfrac{e^t}{e^{2t} + 1} \left\langle e^t, -e^{-t}, \sqrt{2} \right\rangle = \left\langle \dfrac{e^{2t}}{e^{2t} + 1}, -\dfrac{1}{e^{2t} + 1}, \dfrac{\sqrt{2}e^t}{e^{2t} + 1} \right\rangle \Rightarrow$

$\mathbf{T}'(t) = \left\langle \dfrac{2e^{2t}}{(e^{2t} + 1)^2}, \dfrac{2e^{2t}}{(e^{2t} + 1)^2}, \dfrac{\sqrt{2}e^t (1 - e^{2t})}{(e^{2t} + 1)^2} \right\rangle$, so $\mathbf{T}(0) = \left\langle \frac{1}{2}, -\frac{1}{2}, \frac{\sqrt{2}}{2} \right\rangle$ and $\mathbf{T}'(0) = \left\langle \frac{1}{2}, \frac{1}{2}, 0 \right\rangle$. Since

$\mathbf{a} = 2\mathbf{T}(0) = \left\langle 1, -1, \sqrt{2} \right\rangle$ is parallel to $\mathbf{T}(0)$ and $\mathbf{b} = 2\mathbf{T}'(0) = \langle 1, 1, 0 \rangle$ is parallel to $\mathbf{N}(0)$, the normal to the osculating

plane is $\mathbf{n} = \mathbf{a} \times \mathbf{b} = \begin{vmatrix} \mathbf{i} & \mathbf{j} & \mathbf{k} \\ 1 & -1 & \sqrt{2} \\ 1 & 1 & 0 \end{vmatrix} = -\sqrt{2}\mathbf{i} + \sqrt{2}\mathbf{j} + 2\mathbf{k}$. Also the point $(1, 1, 0)$ is on the plane, so an equation of the

plane is $-\sqrt{2}(x - 1) + \sqrt{2}(y - 1) + 2z = 0 \Rightarrow -\sqrt{2}x + \sqrt{2}y + 2z = 0 \Leftrightarrow x - y - \sqrt{2}z = 0$.

31. $\mathbf{r}(t) = 2\cosh t\,\mathbf{i} + 2\sinh t\,\mathbf{j} + 2t\mathbf{k} \Rightarrow \mathbf{r}'(t) = 2\sinh t\,\mathbf{i} + 2\cosh t\,\mathbf{j} + 2\mathbf{k} \Rightarrow$

$|\mathbf{r}'(t)| = \sqrt{4\sinh^2 t + 4\cosh^2 t + 4} = 2\sqrt{2}\sqrt{1 + \sinh^2 t} = 2\sqrt{2}\cosh t$, so

$\mathbf{T}(t) = \dfrac{\mathbf{r}'(t)}{|\mathbf{r}'(t)|} = \dfrac{\sqrt{2}}{2}\tanh t\,\mathbf{i} + \dfrac{\sqrt{2}}{2}\mathbf{j} + \dfrac{\sqrt{2}}{2}\operatorname{sech} t\,\mathbf{k}$, $\mathbf{T}'(t) = \dfrac{\sqrt{2}}{2}\left(\operatorname{sech}^2 t\,\mathbf{i} - \operatorname{sech} t\tanh t\,\mathbf{k}\right)$,

$|\mathbf{T}'(t)| = \dfrac{\sqrt{2}}{2}\sqrt{\operatorname{sech}^4 t + \operatorname{sech}^2 t\tanh^2 t} = \dfrac{\sqrt{2}}{2}\operatorname{sech} t$, $\mathbf{N}(t) = \dfrac{\mathbf{T}'(t)}{|\mathbf{T}'(t)|} = \operatorname{sech} t\,\mathbf{i} - \tanh t\,\mathbf{k}$, and

$\mathbf{B} = \mathbf{T} \times \mathbf{N} = \dfrac{\sqrt{2}}{2}\begin{vmatrix} \mathbf{i} & \mathbf{j} & \mathbf{k} \\ \tanh t & 1 & \operatorname{sech} t \\ \operatorname{sech} t & 0 & -\tanh t \end{vmatrix} = \dfrac{\sqrt{2}}{2}\left(-\tanh t\,\mathbf{i} + \mathbf{j} - \operatorname{sech} t\,\mathbf{k}\right)$.

33. $\mathbf{r}(t) = \left\langle t, 2t^2, t^3 \right\rangle \Rightarrow \mathbf{r}'(t) = \left\langle 1, 4t, 3t^2 \right\rangle \Rightarrow \mathbf{r}''(t) = \langle 0, 4, 6t \rangle$, so $\mathbf{r}'(t) \times \mathbf{r}''(t) = \begin{vmatrix} \mathbf{i} & \mathbf{j} & \mathbf{k} \\ 1 & 4t & 3t^2 \\ 0 & 4 & 6t \end{vmatrix} = \left\langle 12t^2, -6t, 4 \right\rangle$

and $|\mathbf{r}'(t) \times \mathbf{r}''(t)| = \sqrt{(12t^2)^2 + (-6t)^2 + 4^2} = 2\sqrt{36t^4 + 9t^2 + 4}$. Thus,

$\mathbf{B}(t) = \dfrac{\mathbf{r}'(t) \times \mathbf{r}''(t)}{|\mathbf{r}'(t) \times \mathbf{r}''(t)|} = \dfrac{1}{\sqrt{36t^4 + 9t^2 + 4}}\left(6t^2\mathbf{i} - 3t\mathbf{j} + 2\mathbf{k}\right)$.

35. $\mathbf{r}(t) = \cos t\,\mathbf{i} + \sin t\,\mathbf{j} + t\mathbf{k} \Rightarrow \mathbf{r}'(t) = -\sin t\,\mathbf{i} + \cos t\,\mathbf{j} + \mathbf{k} \Rightarrow \mathbf{r}''(t) = -\cos t\,\mathbf{i} - \sin t\,\mathbf{j} \Rightarrow$

$\mathbf{r}'''(t) = \sin t\,\mathbf{i} - \cos t\,\mathbf{j}$. So $\mathbf{r}'(t) \times \mathbf{r}''(t) = \begin{vmatrix} \mathbf{i} & \mathbf{j} & \mathbf{k} \\ -\sin t & \cos t & 1 \\ -\cos t & -\sin t & 0 \end{vmatrix} = \sin t\,\mathbf{i} - \cos t\,\mathbf{j} + \mathbf{k}$ and

$\tau(t) = \dfrac{(\mathbf{r}' \times \mathbf{r}'') \cdot \mathbf{r}'''}{|\mathbf{r}' \times \mathbf{r}''|^2} = \dfrac{(\sin t\,\mathbf{i} - \cos t\,\mathbf{j} + \mathbf{k}) \cdot (\sin t\,\mathbf{i} - \cos t\,\mathbf{j})}{\sin^2 t + \cos^2 t + 1} = \dfrac{1}{2}$.

37. a. The area swept out by $\mathbf{r}(t)$ in the time interval $[t_0, t_1]$ is given by $A = \int_{\theta_0}^{\theta_1} r^2\,d\theta$ where $\theta(t_0) = \theta_0$ and $\theta(t_1) = \theta_1$.

By the Fundamental Theorem of Calculus, Part 1, we have $\dfrac{dA}{dt} = \left[\dfrac{d}{d\theta}\int_{\theta_0}^{\theta_1}\dfrac{1}{2}r^2\,d\theta\right]\dfrac{d\theta}{dt} = \dfrac{1}{2}r^2\dfrac{d\theta}{dt}$.

b. Since $\mathbf{r}$ lies in a plane, we can write $\mathbf{r} = r\cos\theta\,\mathbf{i} + r\sin\theta\,\mathbf{j}$. Let $\mathbf{c} = c\mathbf{k}$, where $c > 0$. Then

$\mathbf{c} = \mathbf{r} \times \mathbf{r}' = \begin{vmatrix} \mathbf{i} & \mathbf{j} & \mathbf{k} \\ r\cos\theta & r\sin\theta & 0 \\ r'\cos\theta - r\sin\theta\frac{d\theta}{dt} & r'\sin\theta + r\cos\theta\frac{d\theta}{dt} & 0 \end{vmatrix}$

$= \left(rr'\cos\theta\sin\theta + r^2\cos^2\theta\frac{d\theta}{dt} - rr'\cos\theta\sin\theta + r^2\sin^2\theta\frac{d\theta}{dt}\right)\mathbf{k} = r^2\frac{d\theta}{dt}\mathbf{k}$

Since $\mathbf{c} = c\mathbf{k}$, we have $c = r^2\dfrac{d\theta}{dt}$.

c. Using the results of parts a and b, we have $\dfrac{dA}{dt} = \dfrac{1}{2}r^2\dfrac{d\theta}{dt} = \dfrac{1}{2}c$.

39. The length of the major axis is

$2a = 2\left(a^3\right)^{1/3} = 2\sqrt[3]{\dfrac{GmT^2}{4\pi^2}} = 2\sqrt[3]{\dfrac{\left(6.67 \times 10^{-11}\right)\left(1.99 \times 10^{30}\right)(365.26 \cdot 24 \cdot 3600)^2}{4\pi^2}} \approx 2.99 \times 10^{11}$ m.

41. a. The position function of the projectile is $\mathbf{r}(t) = (v_0 \cos \alpha)\, t\mathbf{i} + \left[h + (v_0 \sin \alpha)\, t - \tfrac{1}{2}gt^2\right]\mathbf{j}$,

where g is the constant of acceleration due to gravity. (See Equation 2 on page 1020/1016
in ET.) Then $\mathbf{r}'(t) = (v_0 \cos \alpha)\mathbf{i} + (v_0 \sin \alpha - gt)\mathbf{j}$ and $\mathbf{r}''(t) = -g\mathbf{j}$. Therefore,

$$a_{\mathbf{T}} = \frac{\mathbf{r}'(t) \cdot \mathbf{r}''(t)}{|\mathbf{r}'(t)|} = \frac{[(v_0 \cos \alpha)\mathbf{i} + (v_0 \sin \alpha - gt)\mathbf{j}] \cdot (-g\mathbf{j})}{\sqrt{(v_0 \cos \alpha)^2 + (v_0 \sin \alpha - gt)^2}} = \frac{g(gt - v_0 \sin \alpha)}{\sqrt{(v_0 \cos \alpha)^2 + (v_0 \sin \alpha - gt)^2}}.$$

Next, $\mathbf{r}'(t) \times \mathbf{r}''(t) = \begin{vmatrix} \mathbf{i} & \mathbf{j} & \mathbf{k} \\ v_0 \cos \alpha & v_0 \sin \alpha - gt & 0 \\ 0 & -g & 0 \end{vmatrix} = -gv_0 \cos \alpha\, \mathbf{k}$, so

$$a_{\mathbf{N}} = \frac{|\mathbf{r}'(t) \times \mathbf{r}''(t)|}{|\mathbf{r}'(t)|} = \frac{gv_0 \cos \alpha}{\sqrt{(v_0 \cos \alpha)^2 + (v_0 \sin \alpha - gt)^2}}.$$

b. The projectile is at its maximum height when $v_0 \sin \alpha - gt = 0 \Leftrightarrow t = \dfrac{v_0 \sin \alpha}{g}$. At this point,

$$a_{\mathbf{T}} = a_{\mathbf{T}}(t)\big|_{t=(v_0 \sin \alpha)/g} = 0 \text{ and } a_{\mathbf{N}} = a_{\mathbf{N}}(t)\big|_{t=(v_0 \sin \alpha)/g} = \frac{gv_0 \cos \alpha}{\sqrt{(v_0 \cos \alpha)^2}} = g.$$

43. $\rho = \dfrac{1}{\kappa}$ where $\kappa = \dfrac{|\mathbf{r}' \times \mathbf{r}''|}{|\mathbf{r}'|^3} \Rightarrow \rho = \dfrac{|\mathbf{r}'|^3}{|\mathbf{r}' \times \mathbf{r}''|} = \dfrac{|\mathbf{r}'|^2}{\dfrac{|\mathbf{r}' \times \mathbf{r}''|}{|\mathbf{r}'|}}$. Let $\mathbf{r}(t) = \langle x(t), y(t), z(t) \rangle$. Then

$$\mathbf{r}' = \langle x', y', z' \rangle \text{ and } \mathbf{r}'' = \langle x'', y'', z'' \rangle. \quad v = |\mathbf{r}'| = \sqrt{(x')^2 + (y')^2 + (z')^2} \Rightarrow v' = \frac{x'x'' + y'y'' + z'z''}{\sqrt{(x')^2 + (y')^2 + (z')^2}},$$

$$\mathbf{r}' \times \mathbf{r}'' = \langle y'z'' - y''z', x''z' - x'z'', x'y'' - x''y' \rangle \Rightarrow$$

$$|\mathbf{r}' \times \mathbf{r}''|^2 = (y'z'' - y''z')^2 + (x''z' - x'z'')^2 + (x'y'' - x''y')^2$$

$$= (y')^2 (z'')^2 - 2y'y''z'z'' + (y'')^2 (z')^2 + (x'')^2 (z')^2 - 2x'x''z'z'' + (x')^2 (z'')^2$$

$$+ (x')^2 (y'')^2 - 2x'x''y'y'' + (x'')^2 (y')^2$$

$$= (x'')^2 \left[(x')^2 + (y')^2 + (z')^2\right] + (y'')^2 \left[(x')^2 + (y')^2 + (z')^2\right] + (z'')^2 \left[(x')^2 + (y')^2 + (z')^2\right]$$

$$- \left[(x')^2 (x'')^2 + (y')^2 (y'')^2 + (z')^2 (z'')^2 + 2x'x''y'y'' + 2x'x''z'z'' + 2y'y''z'z''\right]$$

Thus,

$$\frac{|\mathbf{r}' \times \mathbf{r}''|}{|\mathbf{r}'|} = \sqrt{(x'')^2 + (y'')^2 + (z'')^2 - \frac{(x'x'')^2 + (y'y'')^2 + (z'z'')^2 + 2x'x''y'y'' + 2x'x''z'z'' + 2y'y''z'z''}{(x')^2 + (y')^2 + (z')^2}}$$

$$= \sqrt{(x'')^2 + (y'')^2 + (z'')^2 - \left[\frac{x'x'' + y'y'' + z'z''}{\sqrt{(x')^2 + (y')^2 + (z')^2}}\right]^2} = \sqrt{(x'')^2 + (y'')^2 + (z'')^2 - (v')^2}$$

so $\rho = \dfrac{(x')^2 + (y')^2 + (z')^2}{\sqrt{(x'')^2 + (y'')^2 + (z'')^2 - (v')^2}}$.

45. True. $\mathbf{T} = \dfrac{\mathbf{r}'}{|\mathbf{r}'|} = \dfrac{\mathbf{r}'}{c} \Rightarrow \mathbf{T}' = \dfrac{1}{c}\mathbf{r}''$, so $\mathbf{N} = \dfrac{\mathbf{T}'}{|\mathbf{T}'|} = \dfrac{\frac{1}{c}\mathbf{r}''}{\left|\frac{1}{c}\mathbf{r}''\right|} = \dfrac{\mathbf{r}''}{|\mathbf{r}''|}$.

47. False. Take $\mathbf{r}(t) = \left\langle 1, t^3 \right\rangle$. Then $\mathbf{r}'(t) = \left\langle 0, 3t^2 \right\rangle$ and $\mathbf{r}''(t) = \langle 0, 6t \rangle$.

$s(t) = \int_0^t |\mathbf{r}'(u)| \, du = \int_0^t \sqrt{(3u^2)^2} \, du = 3 \int_0^t u^2 \, du = t^3$, so $t = s^{1/3}$ and $\mathbf{r}(t(s)) = \langle 1, s \rangle$. Then $\mathbf{r}'(s) = \langle 0, 1 \rangle$ and $\mathbf{r}''(s) = \langle 0, 0 \rangle$, so $\mathbf{r}''(t)$ is not a scalar multiple of $\mathbf{r}''(s)$.

Chapter 12   Review   ET 11

Concept Review

1. a. $\langle f(t), g(t), h(t) \rangle$; real-valued; t; parameter
    b. parameter interval; real numbers

3. a. $\lim\limits_{t \to a} f(t); \lim\limits_{t \to a} g(t); \lim\limits_{t \to a} h(t)$
    b. $\mathbf{r}(a)$; continuous

5. a. $\mathbf{u}'(t) \cdot \mathbf{v}(t) + \mathbf{u}(t) \cdot \mathbf{v}'(t);$
    b. $\mathbf{u}'(f(t)) f'(t)$
$\mathbf{u}'(t) \times \mathbf{v}(t) + \mathbf{u}(t) \times \mathbf{v}'(t)$

7. $\int_a^b \sqrt{[f'(t)]^2 + [g'(t)]^2 + [h'(t)]^2} \, dt$

9. a. $\left| \dfrac{d\mathbf{T}}{ds} \right|$
  b. $\dfrac{|\mathbf{T}'(t)|}{|\mathbf{r}'(t)|}$
  c. $\dfrac{|\mathbf{r}'(t) \times \mathbf{r}''(t)|}{|\mathbf{r}'(t)|^3}$
  d. $\dfrac{|y''|}{\left[1 + (y')^2\right]^{3/2}}$

    e. radius of curvature; radius; tangent line; circle of curvature

11. a. $\dfrac{\mathbf{r}'(t)}{|\mathbf{r}'(t)|}; \dfrac{\mathbf{T}'(t)}{|\mathbf{T}'(t)|}$
    b. $\mathbf{T}; \mathbf{N}; v'; \kappa v^2$; tangential; normal

    c. $\dfrac{\mathbf{r}'(t) \cdot \mathbf{r}''(t)}{|\mathbf{r}'(t)|}; \dfrac{|\mathbf{r}'(t) \times \mathbf{r}''(t)|}{|\mathbf{r}'(t)|}$

Review Exercises

1. $x = 2 + 3t$ and $y = 2t - 1$, so

$y = 2 \left(\dfrac{x - 2}{3} \right) - 1 = \frac{2}{3}x - \frac{7}{3}$. As t increases, both x and

y increase and so the orientation is upward.

3. $x = \cos t - 1, y = \sin t + 2$, and $z = 2$, so

$1 = \cos^2 t + \sin^2 t = (x + 1)^2 + (y - 2)^2$. This shows

that the curve lies on the circle with radius 1 centered at

$(-1, 2, 2)$ on the plane $z = 2$.

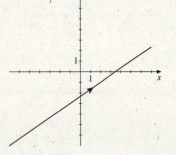

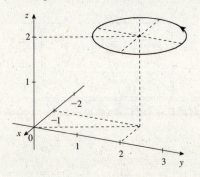

5. We require that $5 - t > 0, t \neq 0$, and $1 + t > 0$ simultaneously. This gives the domain of $\mathbf{r}$ as $(-1, 0) \cup (0, 5)$.

7. $\sqrt{t+1}$ is continuous on $[-1, \infty)$, $\dfrac{e^t}{\sqrt{2-t}}$ is continuous on $(-\infty, 2)$, and $\dfrac{t^2}{(t-1)^2}$ is continuous on $(-\infty, 1)$ and $(1, \infty)$, so $\mathbf{r}(t)$ is continuous on $[-1, 1)$ and $(1, 2)$.

9. $\mathbf{r}(t) = \sqrt{t}\,\mathbf{i} + t^2\mathbf{j} + \dfrac{1}{t+1}\mathbf{k} \Rightarrow \mathbf{r}'(t) = \dfrac{1}{2\sqrt{t}}\mathbf{i} + 2t\mathbf{j} - \dfrac{1}{(t+1)^2}\mathbf{k} \Rightarrow \mathbf{r}''(t) = -\dfrac{1}{4t^{3/2}}\mathbf{i} + 2\mathbf{j} + \dfrac{2}{(t+1)^3}\mathbf{k}$

11. $\mathbf{r}(t) = \left(t^2+1\right)\mathbf{i} + 2t\mathbf{j} + \ln t\,\mathbf{k} \Rightarrow \mathbf{r}'(t) = 2t\mathbf{i} + 2\mathbf{j} + \dfrac{1}{t}\mathbf{k} \Rightarrow \mathbf{r}''(t) = 2\mathbf{i} - \dfrac{1}{t^2}\mathbf{k}$

13. Let $\mathbf{r}(t) = \left(t^2+1\right)\mathbf{i} + (2t-3)\mathbf{j} + \left(t^3+1\right)\mathbf{k}$. Then $\mathbf{r}'(t) = 2t\mathbf{i} + 2\mathbf{j} + 3t^2\mathbf{k} \Rightarrow \mathbf{r}'(0) = 2\mathbf{j}$. Also, $\mathbf{r}(0) = \mathbf{i} - 3\mathbf{j} + \mathbf{k}$, so parametric equations of the required tangent line are $x = 1$, $y = -3 + 2t$, $z = 1$.

15. $\int \left(\sqrt{t}\,\mathbf{i} + e^{-2t}\mathbf{j} + \dfrac{1}{t+1}\mathbf{k}\right)dt = \frac{2}{3}t^{3/2}\mathbf{i} - \frac{1}{2}e^{-2t}\mathbf{j} + \ln|t+1|\,\mathbf{k} + \mathbf{C}$

17. $\mathbf{r}(t) = \int \mathbf{r}'(t)\,dt = \int \left(2\sqrt{t}\,\mathbf{i} + 3\cos 2\pi t\,\mathbf{j} - e^{-t}\mathbf{k}\right)dt = \frac{4}{3}t^{3/2}\mathbf{i} + \frac{3}{2\pi}\sin 2\pi t\,\mathbf{j} + e^{-t}\mathbf{k} + \mathbf{C}$. Next, $\mathbf{r}(0) = \mathbf{i} + 2\mathbf{j} \Rightarrow$
$0\mathbf{i} + 0\mathbf{j} + \mathbf{k} + \mathbf{C} = \mathbf{i} + 2\mathbf{j} \Rightarrow \mathbf{C} = \mathbf{i} + 2\mathbf{j} - \mathbf{k}$, so $\mathbf{r}(t) = \left(\frac{4}{3}t^{3/2} + 1\right)\mathbf{i} + \left(\frac{3}{2\pi}\sin 2\pi t + 2\right)\mathbf{j} + \left(e^{-t} - 1\right)\mathbf{k}$.

19. $\mathbf{r}(t) = t\mathbf{i} + t^2\mathbf{j} + t^3\mathbf{k} \Rightarrow \mathbf{r}'(t) = \mathbf{i} + 2t\mathbf{j} + 3t^2\mathbf{k} \Rightarrow |\mathbf{r}'(t)| = \sqrt{1 + 4t^2 + 9t^4}$

$\Rightarrow \mathbf{T}(t) = \dfrac{\mathbf{r}'(t)}{|\mathbf{r}'(t)|} = \dfrac{1}{\sqrt{1+4t^2+9t^4}}\left(\mathbf{i} + 2t\mathbf{j} + 3t^2\mathbf{k}\right) \Rightarrow$

$\mathbf{T}'(t) = \dfrac{1}{\left(1+4t^2+9t^4\right)^{3/2}}\left[-2t\left(2+9t^2\right)\mathbf{i} + \left(2-18t^4\right)\mathbf{j} + 6t\left(1+2t^2\right)\mathbf{k}\right]$. $\mathbf{r}'(1) = \mathbf{i} + 2\mathbf{j} + 3\mathbf{k}$

and $\mathbf{T}'(1) = \frac{1}{98}\left(-11\sqrt{14}\mathbf{i} - 8\sqrt{14}\mathbf{j} + 9\sqrt{14}\mathbf{k}\right)$, so $|\mathbf{r}'(1)| = \sqrt{14}$ and $|\mathbf{T}'(1)| = \frac{\sqrt{19}}{7}$. Therefore,

$\mathbf{T}(1) = \dfrac{\mathbf{r}'(1)}{|\mathbf{r}'(1)|} = \frac{\sqrt{14}}{14}(\mathbf{i} + 2\mathbf{j} + 3\mathbf{k})$ and $\mathbf{N}(1) = \dfrac{\mathbf{T}'(1)}{|\mathbf{T}'(1)|} = \frac{\sqrt{266}}{266}(-11\mathbf{i} - 8\mathbf{j} + 9\mathbf{k})$.

21. $\mathbf{r}(t) = 2\sin 2t\,\mathbf{i} + 2\cos 2t\,\mathbf{j} + 3t\mathbf{k} \Rightarrow \mathbf{r}'(t) = 4\cos 2t\,\mathbf{i} - 4\sin 2t\,\mathbf{j} + 3\mathbf{k}$, so
$L = \int_a^b |\mathbf{r}'(t)|\,dt = \int_0^2 \sqrt{(4\cos 2t)^2 + (-4\sin 2t)^2 + 3^2}\,dt = \int_0^2 5\,dt = 10$.

23. $\mathbf{r}(t) = t\mathbf{i} + t^2\mathbf{j} + t^3\mathbf{k} \Rightarrow \mathbf{r}'(t) = \mathbf{i} + 2t\mathbf{j} + 3t^2\mathbf{k} \Rightarrow \mathbf{r}''(t) = 2\mathbf{j} + 6t\mathbf{k}$, so $|\mathbf{r}'(t)| = \sqrt{1 + 4t^2 + 9t^4}$,

$\mathbf{r}'(t) \times \mathbf{r}''(t) = \begin{vmatrix} \mathbf{i} & \mathbf{j} & \mathbf{k} \\ 1 & 2t & 3t^2 \\ 0 & 2 & 6t \end{vmatrix} = 6t^2\mathbf{i} - 6t\mathbf{j} + 2\mathbf{k}$, and $|\mathbf{r}'(t) \times \mathbf{r}''(t)| = 2\sqrt{1 + 9t^2 + 9t^4}$. Thus,

$\kappa = \dfrac{|\mathbf{r}'(t) \times \mathbf{r}''(t)|}{|\mathbf{r}'(t)|^3} = \dfrac{2\sqrt{1 + 9t^2 + 9t^4}}{\left(1 + 4t^2 + 9t^4\right)^{3/2}}$.

25. $y = x - \frac{1}{4}x^2 \Rightarrow y' = 1 - \frac{1}{2}x \Rightarrow y'' = -\frac{1}{2} \Rightarrow \kappa = \dfrac{|y''|}{\left[1 + (y')^2\right]^{3/2}} = \dfrac{\frac{1}{2}}{\left[1 + \left(1 - \frac{1}{2}x\right)^2\right]^{3/2}} = \dfrac{4}{(x^2 - 4x + 8)^{3/2}}$. Next,

$\kappa' = 4\left(-\frac{3}{2}\right)\left(x^2 - 4x + 8\right)^{-5/2}(2x - 4) = -\dfrac{12(x-2)}{(x^2 - 4x + 8)^{5/2}}$. Setting $\kappa' = 0$ gives $x = 2$ and so 2 is a critical number of κ. The First Derivative Test implies that κ attains a maximum at $(2, 1)$.

27. $\mathbf{r}(t) = 2t\mathbf{i} + e^{-2t}\mathbf{j} + \cos t\,\mathbf{k} \Rightarrow \mathbf{v}(t) = \mathbf{r}'(t) = 2\mathbf{i} - 2e^{-2t}\mathbf{j} - \sin t\,\mathbf{k} \Rightarrow \mathbf{a}(t) = \mathbf{v}'(t) = \mathbf{r}''(t) = 4e^{-2t}\mathbf{j} - \cos t\,\mathbf{k}$ and
$|\mathbf{v}(t)| = \sqrt{4 + 4e^{-4t} + \sin^2 t}$.

29. $\mathbf{v}(t) = \int \mathbf{a}(t)\,dt = \int \left(t\mathbf{i} + \frac{1}{3}t^2\mathbf{j} + 3\mathbf{k}\right)dt = \frac{1}{2}t^2\mathbf{i} + \frac{1}{9}t^3\mathbf{j} + 3t\mathbf{k} + \mathbf{C}_1$ and

$\mathbf{v}(0) = 2\mathbf{i} + 3\mathbf{j} + \mathbf{k} \Rightarrow \mathbf{C}_1 = 2\mathbf{i} + 3\mathbf{j} + \mathbf{k}$, so $\mathbf{v}(t) = \left(\frac{1}{2}t^2 + 2\right)\mathbf{i} + \left(\frac{1}{9}t^3 + 3\right)\mathbf{j} + (3t+1)\mathbf{k}$.

$\mathbf{r}(t) = \int \mathbf{v}(t)\,dt = \int \left[\left(\frac{1}{2}t^2 + 2\right)\mathbf{i} + \left(\frac{1}{9}t^3 + 3\right)\mathbf{j} + (3t+1)\mathbf{k}\right]dt = \left(\frac{1}{6}t^3 + 2t\right)\mathbf{i} + \left(\frac{1}{36}t^4 + 3t\right)\mathbf{j} + \left(\frac{3}{2}t^2 + t\right)\mathbf{k} + \mathbf{C}_2$

and $\mathbf{r}(0) = \mathbf{0} \Rightarrow \mathbf{C}_2 = \mathbf{0}$, so $\mathbf{r}(t) = \left(\frac{1}{6}t^3 + 2t\right)\mathbf{i} + \left(\frac{1}{36}t^4 + 3t\right)\mathbf{j} + \left(\frac{3}{2}t^2 + t\right)\mathbf{k}$.

31. $\mathbf{r}(t) = \mathbf{i} + t\mathbf{j} + t^2\mathbf{k} \Rightarrow \mathbf{r}'(t) = \mathbf{j} + 2t\mathbf{k} \Rightarrow \mathbf{r}''(t) = 2\mathbf{k} \Rightarrow \mathbf{r}'(t) \times \mathbf{r}''(t) = \begin{vmatrix} \mathbf{i} & \mathbf{j} & \mathbf{k} \\ 0 & 1 & 2t \\ 0 & 0 & 2 \end{vmatrix} = 2\mathbf{i}$, so

$a_\mathbf{T} = \dfrac{\mathbf{r}'(t) \cdot \mathbf{r}''(t)}{|\mathbf{r}'(t)|} = \dfrac{4t}{\sqrt{1+4t^2}}$ and $a_\mathbf{N} = \dfrac{|\mathbf{r}'(t) \times \mathbf{r}''(t)|}{|\mathbf{r}'(t)|} = \dfrac{2}{\sqrt{1+4t^2}}$.

33. $\mathbf{r}(t) = \cos t\,\mathbf{i} + \sin 2t\,\mathbf{j} \Rightarrow \mathbf{r}'(t) = -\sin t\,\mathbf{i} + 2\cos 2t\,\mathbf{j} \Rightarrow \mathbf{r}''(t) = -\cos t\,\mathbf{i} - 4\sin 2t\,\mathbf{j} \Rightarrow$

$\mathbf{r}'(t) \times \mathbf{r}''(t) = \begin{vmatrix} \mathbf{i} & \mathbf{j} & \mathbf{k} \\ -\sin t & 2\cos 2t & 0 \\ -\cos t & -4\sin 2t & 0 \end{vmatrix} = (4\sin t \sin 2t + 2\cos t \cos 2t)\mathbf{k} = 2(2\sin t \sin 2t + \cos t \cos 2t)\mathbf{k}$, so

$a_\mathbf{T} = \dfrac{\mathbf{r}'(t) \cdot \mathbf{r}''(t)}{|\mathbf{r}'(t)|} = \dfrac{\sin t \cos t - 8\sin 2t \cos 2t}{\sqrt{\sin^2 t + 4\cos^2 2t}}$ and $a_\mathbf{N} = \dfrac{|\mathbf{r}'(t) \times \mathbf{r}''(t)|}{|\mathbf{r}'(t)|} = \dfrac{2|2\sin t \sin 2t + \cos t \cos 2t|}{\sqrt{\sin^2 t + 4\cos^2 2t}}$.

35. a. We use Formula 2 from Section 12.4 (11.4 in ET) with $v_0 = 40$, $\alpha = 45°$, and $h = 7$, obtaining

$\mathbf{r}(t) = (40\cos 45°)\mathbf{i} + \left[7 + (40\sin 45°)t - \frac{1}{2}(32)t^2\right]\mathbf{j} = 20\sqrt{2}t\mathbf{i} + \left(7 + 20\sqrt{2}t - 16t^2\right)\mathbf{j}$.

b. The shot hits the ground when $7 + 20\sqrt{2}t - 16t^2 = 0 \Rightarrow t = \dfrac{-20\sqrt{2} \pm \sqrt{\left(20\sqrt{2}\right)^2 - 4(-16)(7)}}{2(-16)} \approx -0.22009$ or

1.98785. We reject the negative root, so her put is approximately $x(1.98785) \approx 20\sqrt{2}(1.98785) \approx 56.2$ ft.

Challenge Problems

1. a. Parametric equations for C are $x = a_1 t^2 + b_1 t + c$, $y = a_2 t^2 + b_2 t + c$, $z = a_3 t^2 + b_3 t + c$.

Put $t = 0$, $t = -1$, and $t = 1$ in succession to obtain the three points $P_0(c, c, c)$,

$P_1(a_1 - b_1 + c, a_2 - b_2 + c, a_3 - b_3 + c)$, and $P_2(a_1 + b_1 + c, a_2 + b_2 + c, a_3 + b_3 + c)$

respectively. Consider the plane containing these three points. The normal to this plane is

$\mathbf{n} = \overrightarrow{P_0 P_1} \times \overrightarrow{P_0 P_2} = \begin{vmatrix} \mathbf{i} & \mathbf{j} & \mathbf{k} \\ a_1 - b_1 & a_2 - b_2 & a_3 - b_3 \\ a_1 + b_1 & a_2 + b_2 & a_3 + b_3 \end{vmatrix} = 2(a_2 b_3 - a_3 b_2)\mathbf{i} - 2(a_1 b_3 - b_1 a_3)\mathbf{j} + 2(a_1 b_2 - a_2 b_1)\mathbf{k}$, so

an equation of the plane is

$$(a_2 b_3 - a_3 b_2)(x - c) - (a_1 b_3 - b_1 a_3)(y - c) + (a_1 b_2 - a_2 b_1)(z - c) = 0 \qquad (1)$$

To show that C lies in this plane, we substitute the expressions for x, y and z into (1). The left-hand side is

$(a_2 b_3 - a_3 b_2)\left(a_1 t^2 + b_1 t\right) - (a_1 b_3 - a_3 b_1)\left(a_2 t^2 + b_2 t\right) + (a_1 b_2 - a_2 b_1)\left(a_3 t^2 + b_3 t\right)$ and, upon simplification,

this is seen to be equal to 0. This shows that C lies on the plane.

b. Equation (1) is seen to be equivalent to $\begin{vmatrix} x - c & y - c & z - c \\ a_1 & a_2 & a_3 \\ b_1 & b_2 & b_3 \end{vmatrix} = 0$ by expanding the latter.

3. a. The position of the bullet is $\mathbf{r}_1(t) = (v_0 \cos \theta) \, t\mathbf{i} + \left[(v_0 \sin \theta) \, t - \frac{1}{2}gt^2\right]\mathbf{j}$, but

$$\cos \theta = \frac{d}{\sqrt{d^2 + h^2}} \text{ and } \sin \theta = \frac{h}{\sqrt{d^2 + h^2}}, \text{ so}$$

$\mathbf{r}_1(t) = \dfrac{v_0 d}{\sqrt{d^2 + h^2}}t\mathbf{i} + \left[\dfrac{v_0 h}{\sqrt{d^2 + h^2}}t - \dfrac{g}{2}t^2\right]\mathbf{j}$. The position of the target is

$\mathbf{r}_2(t) = d\mathbf{i} + \left(h - \frac{1}{2}gt^2\right)\mathbf{j}$. In order to hit the target, the bullet must reach the line

$x = d$ before the target hits the ground. But the target strikes the ground when $\mathbf{r}_2(t) = d\mathbf{i}$, that is, when t satisfies

$h - \frac{1}{2}gt^2 = 0 \Rightarrow t = \sqrt{2h/g}$ (we reject the negative root). Thus, the target is on or above the ground for $t \leq \sqrt{2h/g}$.

Now the bullet reaches $x = d$ when $\dfrac{v_0 d}{\sqrt{d^2 + h^2}}t = d \Leftrightarrow t = \dfrac{\sqrt{d^2 + h^2}}{v_0}$. Since $t \leq \sqrt{2h/g}$, we have $\dfrac{\sqrt{d^2 + h^2}}{v_0} \leq \sqrt{\dfrac{2h}{g}}$

$\Rightarrow v_0 \geq \sqrt{\dfrac{g\left(d^2 + h^2\right)}{2h}}$ (1). This inequality gives the initial speed the bullet must have in order to hit the target.

Suppose condition (1) is satisfied. Then the position of the bullet at $t = \dfrac{\sqrt{d^2 + h^2}}{v_0}$ is

$$\mathbf{r}_1\left(\frac{\sqrt{d^2 + h^2}}{v_0}\right) = \frac{v_0 d}{\sqrt{d^2 + h^2}}\left(\frac{\sqrt{d^2 + h^2}}{v_0}\right)\mathbf{i} + \left[\frac{v_0 h}{\sqrt{d^2 + h^2}}\left(\frac{\sqrt{d^2 + h^2}}{v_0}\right) - \frac{g}{2}\left(\frac{\sqrt{d^2 + h^2}}{v_0}\right)^2\right]\mathbf{j}$$

$$= d\mathbf{i} + \left[h - \frac{g\left(d^2 + h^2\right)}{2v_0^2}\right]\mathbf{j}$$

The position of the target is

$$\mathbf{r}_2\left(\frac{\sqrt{d^2 + h^2}}{v_0}\right) = d\mathbf{i} + \left[h - \frac{g}{2}\left(\frac{\sqrt{d^2 + h^2}}{v_0}\right)^2\right]\mathbf{j} = d\mathbf{i} + \left[h - \frac{g\left(d^2 + h^2\right)}{2v_0^2}\right]\mathbf{j} = \mathbf{r}_1\left(\frac{\sqrt{d^2 + h^2}}{v_0}\right), \text{ showing that}$$

the bullet and the target are at the same point at $t = \dfrac{\sqrt{d^2 + h^2}}{v_0}$. Thus, the bullet hits the target at that point in time.

b. The distance the target has fallen when it is hit is $\frac{1}{2}gt^2 = \frac{1}{2}g\left(\dfrac{\sqrt{d^2 + h^2}}{v_0}\right)^2 = \dfrac{g\left(d^2 + h^2\right)}{2v_0^2}$.

5. a. We use the result of Exercise 4 with $\mathbf{v}_0 = v_0 \cos \alpha \, \mathbf{i} + v_0 \sin \alpha \, \mathbf{j}$:

$$\mathbf{r}(t) = \frac{m}{k}\left[1 - e^{-(m/k)t}\right](v_0 \cos \alpha \, \mathbf{i} + v_0 \sin \alpha \, \mathbf{j}) + \left\{\frac{m^2 g}{k^2}\left[1 - e^{-(k/m)t}\right] - \frac{mg}{k}t\right\}\mathbf{j}$$

$$= \frac{mv_0 \cos \alpha}{k}\left[1 - e^{-(k/m)t}\right]\mathbf{i} + \left\{\left(\frac{mv_0 \sin \alpha}{k} + \frac{m^2 g}{k^2}\right)\left[1 - e^{-(k/m)t}\right] - \frac{mg}{k}t\right\}\mathbf{j}$$

Thus, the required parametric equations are

$$x(t) = \frac{mv_0 \cos \alpha}{k}\left[1 - e^{-(k/m)t}\right], \, y(t) = \left(\frac{m^2 g}{k^2} + \frac{mv_0 \sin \alpha}{k}\right)\left[1 - e^{-(k/m)t} - \frac{mg}{k}t\right].$$

b. We have $\dfrac{kx}{mv_0 \cos\alpha} = 1 - e^{-(k/m)t} \Rightarrow e^{-(k/m)t} = 1 - \dfrac{kx}{mv_0 \cos\alpha} = \dfrac{mv_0 \cos\alpha - kx}{mv_0 \cos\alpha} \Leftrightarrow$

$t = -\dfrac{m}{k} \ln\left(\dfrac{mv_0 \cos\alpha - kx}{mv_0 \cos\alpha}\right) = \dfrac{m}{k} \ln\left(\dfrac{mv_0 \cos\alpha}{mv_0 \cos\alpha - kx}\right)$. Therefore,

$$y = \left(\dfrac{m^2 g}{k^2} + \dfrac{mv_0 \sin\alpha}{k}\right)\dfrac{kx}{mv_0 \cos\alpha} - \dfrac{mg}{k} \cdot \dfrac{m}{k} \ln\left(\dfrac{mv_0 \cos\alpha}{mv_0 \cos\alpha - kx}\right)$$

$$= \left(\dfrac{mg}{kv_0 \cos\alpha} + \tan\alpha\right)x + \dfrac{m^2 g}{k^2} \ln\left(\dfrac{mv_0 \cos\alpha - kx}{mv_0 \cos\alpha}\right)$$

$$= \left(\dfrac{mg}{kv_0 \cos\alpha} + \tan\alpha\right)x + \dfrac{m^2 g}{k^2} \ln\left(1 - \dfrac{kx}{mv_0 \cos\alpha}\right)$$

c. Here $m = \dfrac{1600}{32} = 50$ slugs, $v_0 = 1200\left(\dfrac{88}{60}\right) = 1760$ ft/s, and $\alpha = 30°$.

Thus,

$$y = \left[\dfrac{1600}{k(1760)\cos 30°} + \tan 30°\right]x$$

$$+ \dfrac{50^2 \cdot 32}{k^2} \ln\left(1 - \dfrac{kx}{50(1760)\left(\frac{\sqrt{3}}{2}\right)}\right)$$

$$= \left(\dfrac{20\sqrt{3}}{33k} + \dfrac{\sqrt{3}}{3}\right)x + \dfrac{80{,}000}{k^2} \ln\left(1 - \dfrac{\sqrt{3}kx}{132{,}000}\right)$$

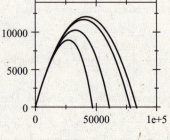

d. We use the Taylor series expansion from Table 9.1 in Section 9.8 (8.8 in ET): $\ln(1 + x) = x - \dfrac{x^2}{2} + \dfrac{x^3}{3} - \dfrac{x^4}{4} + \cdots$,

$-1 < x \le 1$. Thus,

$$y = \left(\dfrac{mg}{kv_0 \cos\alpha} + \tan\alpha\right)x + \dfrac{m^2 g}{k^2} \ln\left(1 - \dfrac{kx}{mv_0 \cos\alpha}\right)$$

$$= (\tan\alpha)x + \dfrac{mg}{kv_0 \cos\alpha}x$$

$$+ \dfrac{m^2 g}{k^2}\left[-\dfrac{kx}{mv_0 \cos\alpha} - \dfrac{1}{2}\left(-\dfrac{kx}{mv_0 \cos\alpha}\right)^2 + \dfrac{1}{3}\left(-\dfrac{kx}{mv_0 \cos\alpha}\right)^3 - \dfrac{1}{4}\left(-\dfrac{kx}{mv_0 \cos\alpha}\right)^4 + \cdots\right]$$

$$= (\tan\alpha)x + \dfrac{mg}{kv_0 \cos\alpha}x - \dfrac{mg}{kv_0 \cos\alpha}x - \dfrac{1}{2}\dfrac{m^2 g}{k^2}\left(\dfrac{kx}{mv_0 \cos\alpha}\right)^2$$

$$+ \dfrac{1}{3}\dfrac{m^2 g}{k^2}\left(-\dfrac{kx}{mv_0 \cos\alpha}\right)^3 - \dfrac{1}{4}\dfrac{m^2 g}{k^2}\left(\dfrac{kx}{mv_0 \cos\alpha}\right)^4 + \cdots$$

$$= (\tan\alpha)x - \dfrac{1}{2}\dfrac{g}{(v_0 \cos\alpha)^2}x^2 - \dfrac{1}{3}\dfrac{kg}{m(v_0 \cos\alpha)^3}x^3 - \cdots$$

So if k is very small, then $y \approx (\tan\alpha)x - \dfrac{1}{2}\dfrac{g}{(v_0 \cos\alpha)^2}x^2$, a quadratic in x. In this case, the trajectory of the projectile

closely resembles a parabola.

7. a. $\mathbf{r} = r\cos\theta\,\mathbf{i} + r\sin\theta\,\mathbf{j}$, so $\mathbf{u}_r = \dfrac{\mathbf{r}}{|\mathbf{r}|} = \dfrac{r\cos\theta\,\mathbf{i} + r\sin\theta\,\mathbf{j}}{r} = \cos\theta\,\mathbf{i} + \sin\theta\,\mathbf{j}$ and

$\mathbf{u}_\theta = \dfrac{d\mathbf{r}/d\theta}{|d\mathbf{r}/d\theta|} = \dfrac{-r\sin\theta\,\mathbf{i} + r\cos\theta\,\mathbf{j}}{r} = -\sin\theta\,\mathbf{i} + \cos\theta\,\mathbf{j}$.

b. $\mathbf{r} = r\mathbf{u}_r$, so $\mathbf{v}(t) = \dfrac{d\mathbf{r}}{dt} = \dfrac{dr}{dt}\mathbf{u}_r + r\dfrac{d\mathbf{u}_r}{dt} = \dfrac{dr}{dt}\mathbf{u}_r + r\left(-\sin\theta\dfrac{d\theta}{dt}\mathbf{i} + \cos\theta\dfrac{d\theta}{dt}\mathbf{j}\right) = \dfrac{dr}{dt}\mathbf{u}_r + r\dfrac{d\theta}{dt}\mathbf{u}_\theta$ (since

$\mathbf{u}_\theta = -\sin\theta\,\mathbf{i} + \cos\theta\,\mathbf{j}$).

$$\mathbf{a}(t) = \dfrac{d\mathbf{v}}{dt} = \dfrac{d^2r}{dt^2}\mathbf{u}_r + \dfrac{dr}{dt}\dfrac{d\mathbf{u}_r}{dt} + \dfrac{dr}{dt}\dfrac{d\theta}{dt}\mathbf{u}_\theta + r\dfrac{d^2\theta}{dt^2}\mathbf{u}_\theta + r\dfrac{d\theta}{dt}\dfrac{d\mathbf{u}_\theta}{dt}$$

$$= \dfrac{d^2r}{dt^2}\mathbf{u}_r + \dfrac{dr}{dt}\dfrac{d\mathbf{u}_r}{d\theta}\dfrac{d\theta}{dt} + \dfrac{dr}{dt}\dfrac{d\theta}{dt}\mathbf{u}_\theta + r\dfrac{d^2\theta}{dt^2}\mathbf{u}_\theta + r\dfrac{d\theta}{dt}\dfrac{d\mathbf{u}_\theta}{d\theta}\dfrac{d\theta}{dt}$$

$$= \dfrac{d^2r}{dt^2}\mathbf{u}_r + \dfrac{dr}{dt}\dfrac{d\theta}{dt}\mathbf{u}_\theta + \dfrac{dr}{dt}\dfrac{d\theta}{dt}\mathbf{u}_\theta + r\dfrac{d^2\theta}{dt^2}\mathbf{u}_\theta - r\left(\dfrac{d\theta}{dt}\right)^2\mathbf{u}_r$$

$$= \dfrac{d^2r}{dt^2}\mathbf{u}_r + 2\dfrac{dr}{dt}\dfrac{d\theta}{dt}\mathbf{u}_\theta + r\dfrac{d^2\theta}{dt^2}\mathbf{u}_\theta - r\left(\dfrac{d\theta}{dt}\right)^2\mathbf{u}_r = \left[\dfrac{d^2r}{dt^2} - r\left(\dfrac{d\theta}{dt}\right)^2\right]\mathbf{u}_r + \left[r\dfrac{d^2\theta}{dt^2} + 2\dfrac{dr}{dt}\dfrac{d\theta}{dt}\right]\mathbf{u}_\theta$$

9. a. Here $\dfrac{dr}{dt} = 2 \Rightarrow r = 2t + C$, where C is a constant. But $r = 0$ at $t = 0$, and so $C = 0$. Therefore, $r = 2t$. Also,

$\dfrac{d\theta}{dt} = \dfrac{30\cdot 2\pi}{60} = \pi$ radians per second. Using the result of Exercise 7, we have $\mathbf{v} = \dfrac{dr}{dt}\mathbf{u}_r + r\dfrac{d\theta}{dt}\mathbf{u}_\theta = 2\mathbf{u}_r + \pi r\mathbf{u}_\theta$

and $\mathbf{a} = \left[\dfrac{d^2r}{dt^2} - r\left(\dfrac{d\theta}{dt}\right)^2\right]\mathbf{u}_r + \left[r\dfrac{d^2\theta}{dt^2} + 2\dfrac{dr}{dt}\dfrac{d\theta}{dt}\right]\mathbf{u}_\theta = -2t\left(\pi^2\right)\mathbf{u}_r + 2\,(2)\,(\pi)\,\mathbf{u}_\theta = -2\pi^2 t\mathbf{u}_r + 4\pi\mathbf{u}_\theta$.

At $t = 3$, $r = 6$, so $\mathbf{v} = 2\mathbf{u}_r + 6\pi\mathbf{u}_\theta$ and $\mathbf{a} = -6\pi^2\mathbf{u}_r + 4\pi\mathbf{u}_\theta$. Therefore, the speed of

the ant at $t = 3$ is $|\mathbf{v}| = \sqrt{4 + 36\pi^2} \approx 18.96$ cm/s. The magnitude of its acceleration is

$|\mathbf{a}| = \sqrt{36\pi^4 + 16\pi^2} = 2\pi\sqrt{9\pi^2 + 4} \approx 60.54$ cm/s^2.

b. The Coriolis acceleration at $t = 3$ is $2\dfrac{dr}{d\theta}\dfrac{d\theta}{dt}\mathbf{u}_\theta = 2\,(2)\,(\pi)\,\mathbf{u}_\theta = 4\pi\mathbf{u}_\theta$, so its magnitude is 4π cm/s^2.

11. a. By the Chain Rule, $\dfrac{d\mathbf{T}}{dt} = \dfrac{d\mathbf{T}}{ds}\dfrac{ds}{dt} \Rightarrow \dfrac{d\mathbf{T}}{ds} = \dfrac{d\mathbf{T}/dt}{ds/dt}$. Multiplying numerator and denominator by $\left|\dfrac{d\mathbf{T}}{dt}\right|$ gives

$$\dfrac{d\mathbf{T}}{ds} = \dfrac{\dfrac{d\mathbf{T}}{dt}\left|\dfrac{d\mathbf{T}}{dt}\right|}{\dfrac{ds}{dt}\left|\dfrac{d\mathbf{T}}{dt}\right|} = \dfrac{\dfrac{d\mathbf{T}}{dt}}{\left|\dfrac{d\mathbf{T}}{dt}\right|} \cdot \dfrac{\left|\dfrac{d\mathbf{T}}{dt}\right|}{\dfrac{ds}{dt}} = \dfrac{\left|\dfrac{d\mathbf{T}}{dt}\right|}{\dfrac{ds}{dt}}\mathbf{N} = \left|\dfrac{d\mathbf{T}}{ds}\right|\mathbf{N} = \kappa\mathbf{N}.$$

b. $\mathbf{T} \times \mathbf{B} = \mathbf{T} \times (\mathbf{T} \times \mathbf{N}) = (\mathbf{T}\cdot\mathbf{N})\mathbf{T} - (\mathbf{T}\cdot\mathbf{T})\mathbf{N} = -\mathbf{N}$ since $\mathbf{T}\cdot\mathbf{N} = 0$ and $\mathbf{T}\cdot\mathbf{T} = |\mathbf{T}|^2 = 1$. Thus,

$\mathbf{N} = -\mathbf{T} \times \mathbf{B} = \mathbf{B} \times \mathbf{T}$. Next, $\mathbf{N} \times \mathbf{B} = \mathbf{N} \times (\mathbf{T} \times \mathbf{N}) = (\mathbf{N}\cdot\mathbf{N})\mathbf{T} - (\mathbf{N}\cdot\mathbf{T})\mathbf{N} = \mathbf{T}$ since $\mathbf{N}\cdot\mathbf{N} = |\mathbf{N}|^2 = 1$ and $\mathbf{N}\cdot\mathbf{T} = 0$.

c. From part b, we have $\mathbf{N} = \mathbf{B} \times \mathbf{T}$, so

$$\dfrac{d\mathbf{N}}{ds} = \dfrac{d}{ds}(\mathbf{B} \times \mathbf{T}) = \dfrac{d\mathbf{B}}{ds} \times \mathbf{T} + \mathbf{B} \times \dfrac{d\mathbf{T}}{ds} \qquad \text{(Theorem 12.2.2/11.2.2 in ET)}$$

$$= (-\tau\mathbf{N}) \times \mathbf{T} + \mathbf{B} \times (\kappa\mathbf{N}) \qquad \text{(Exercise 12.5.34/11.5.34 in ET and part a)}$$

$$= -\tau\,(\mathbf{N} \times \mathbf{T}) + \kappa\,(\mathbf{B} \times \mathbf{N}) \qquad \text{(Theorem 12.2.2/11.2.2 in ET)}$$

From part b, we have $\mathbf{N} \times \mathbf{B} = \mathbf{T} \Rightarrow \mathbf{B} \times \mathbf{N} = -\mathbf{T}$. Also, $\mathbf{N} \times \mathbf{T} = -\mathbf{T} \times \mathbf{N} = -\mathbf{B}$, so

$$\dfrac{d\mathbf{N}}{ds} = -\tau\,(-\mathbf{B}) + \kappa\,(-\mathbf{T}) = -\kappa\mathbf{T} + \tau\mathbf{B}.$$

13 Functions of Several Variables ET 12

13.1 Concept Questions ET 12.1

1. See page 1034 (1030 in ET). Answers will vary.   3. See page 1039 (1035 in ET). Answers will vary.

13.1 Functions of Two or More Variables ET 12.1

1. **a.** $f(1, 2) = 1^2 + 3(1)(2) - 2(1) + 3 = 8$

   b. $f(2, 1) = 2^2 + 3(2)(1) - 2(2) + 3 = 9$

   c. $f(2h, 3k) = (2h)^2 + 3(2h)(3k) - 2(2h) + 3 = 4h^2 + 18hk - 4h + 3$

   d. $f(x + h, y) = (x + h)^2 + 3(x + h)y - 2(x + h) + 3 = x^2 + 2xh + h^2 + 3xy + 3hy - 2x - 2h + 3$

   e. $f(x, y + k) = x^2 + 3x(y + k) - 2x + 3 = x^2 + 3xy + 3xk - 2x + 3$

3. **a.** $f(1, 2, 3) = \sqrt{1^2 + 2(2)^2 + 3(3^2)} = 6$

   b. $f(0, 2, -1) = \sqrt{0^2 + 2(2)^2 + 3(-1)^2} = \sqrt{11}$

   c. $f(t, -t, t) = \sqrt{t^2 + 2(-t)^2 + 3t^2} = \sqrt{6}\,|t|$

   d. $f(u, u - 1, u + 1) = \sqrt{u^2 + 2(u - 1)^2 + 3(u + 1)^2} = \sqrt{6u^2 + 2u + 5}$

   e. $f(-x, x, -2x) = \sqrt{(-x)^2 + 2x^2 + 3(-2x)^2} = \sqrt{15}\,|x|$

5. Since $f(x, y) = x + 3y - 1$ is defined for all pairs (x, y) of real numbers, the domain of f is
   $\{(x, y) \mid -\infty < x < \infty, -\infty < y < \infty\}$. The range of f is $\{z \mid -\infty < z < \infty\}$.

7. Since $u \neq v$, the domain of f is $\{(u, v) \mid u \neq v\}$. The range of f is $\{z \mid -\infty < z < \infty\}$.

9. We require that the radicand $4 - x^2 - y^2 \geq 0 \Leftrightarrow x^2 + y^2 \leq 4$, so the domain of g is $\left\{(x, y) \mid x^2 + y^2 \leq 4\right\}$. The range of
   g is $\{z \mid 0 \leq z \leq 2\}$.

11. We require that the radicand $9 - x^2 - y^2 - z^2 \geq 0 \Leftrightarrow x^2 + y^2 + z^2 \leq 9$, so the domain of f is
    $\left\{(x, y, z) \mid x^2 + y^2 + z^2 \leq 9\right\}$. The range of f is $\{z \mid 0 \leq z \leq 3\}$.

13. Since $\tan x$ is not defined for $x = \frac{\pi}{2} + n\pi$, n any integer, the domain of h is $\left\{(u, v, w) \mid u \neq \frac{\pi}{2} + n\pi, n \text{ an integer}\right\}$. The
    range of h is $\{z \mid -\infty < z < \infty\}$.

103

15. The domain of f is
$$\{(x, y) \mid x \geq 0 \text{ and } y \geq 0\}$$

17. The domain of f is
$$\{(u, v) \mid u \neq v, u \neq -v\}$$

19. The domain of f is
$$\{(x, y) \mid x > 0, y > 0\}$$

21. The domain of f is
$$\left\{(x, y, z) \mid x^2 + y^2 + z^2 \leq 9\right\}$$

23.

25.

27.

29.

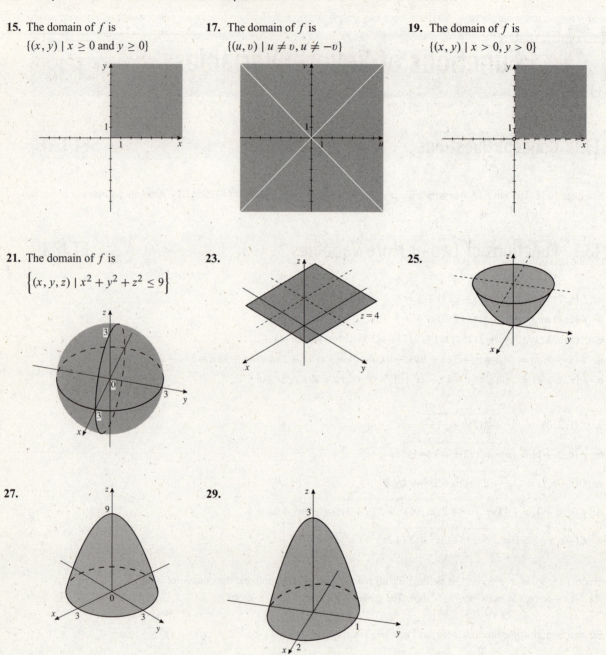

31. a. The altitude at A is 200 ft. It is 400 ft at B.

b. If you move north from A, you will be ascending. You will also be ascending if you move east from B, because you will be moving toward a level curve with a larger value.

c. It is steeper at C where the level curves are closer together than at A.

33. c.   **35. a.**   **37. d.**

39.

41.

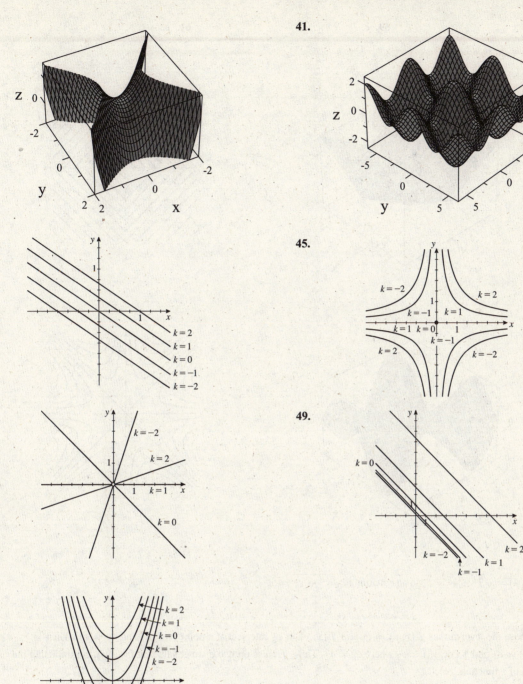

43.

45.

47.

49.

51.

53. The level surfaces $2x + 4y - 3z + 1 = k$ are a family of parallel planes with normal vector $\langle 2, 4, -3 \rangle$.

55. The level surfaces $x^2 + y^2 - z^2 = k$ consist of a cone with vertex the origin and axis the z-axis (if $k = 0$), a family of hyperboloids of one sheet with axis the z-axis (if $k > 0$), and a family of hyperboloids of two sheets with axis the z-axis (if $k < 0$).

57. a.   **59. c.**   **61. e.**

63. a.   **b.**

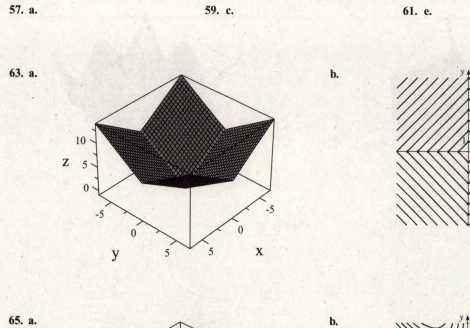

65. a.   **b.**

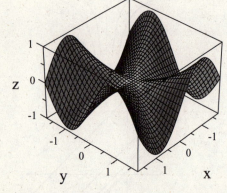

67. $k = f(3, 4) = \sqrt{3^2 + 4^2} = 5$, so an equation is $\sqrt{x^2 + y^2} = 5$.

69. No. Suppose the level curves $f(x, y) = c_1$ and $f(x, y) = c_2$ intersect at a point (x_0, y_0) and $c_1 \neq c_2$. Then $f(x_0, y_0) = c_1$ and $f(x_0, y_0) = c_2$ where $c_1 \neq c_2$. Thus, f takes on two distinct values at (x_0, y_0), contradicting the definition of a function.

71. a.

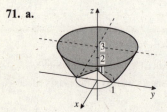

b. For $k = 0$, we have $x^2 + y^2 - 1 = 0 \Leftrightarrow x^2 + y^2 = 1$. For $k = \frac{1}{2}$, we have $1 - \sqrt{x^2 + y^2} = \frac{1}{2} \Leftrightarrow x^2 + y^2 = \frac{1}{4}$ or $x^2 + y^2 - 1 = \frac{1}{2} \Leftrightarrow x^2 + y^2 = \frac{3}{2}$. For $k = 1$, we have $1 - \sqrt{x^2 + y^2} = 1 \Leftrightarrow (x, y) = (0, 0)$ or $x^2 + y^2 - 1 = 1 \Leftrightarrow x^2 + y^2 = 2$. For $k = 3$, we have $x^2 + y^2 - 1 = 3 \Leftrightarrow x^2 + y^2 = 4$.

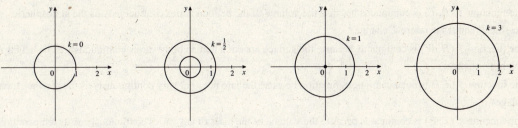

73. $R = f(4, 0.1) = \dfrac{k(4)}{0.1^4} = 40{,}000k$ dynes

75. a. $f(px, py) = a(px)^b (py)^{1-b} = ap^b x^b p^{1-b} y^{1-b} = pax^b y^{1-b} = pf(x, y)$

    b. $f\left(\left(1 + \frac{r}{100}\right)x, \left(1 + \frac{r}{100}\right)y\right) = \left(1 + \frac{r}{100}\right)f(x, y)$, showing that the output is also increased by $r\%$.

77. After the 60th payment, their principal repayment will be $B = 280{,}000 \dfrac{\left(1 + \frac{0.06}{12}\right)^{60} - 1}{\left(1 + \frac{0.06}{12}\right)^{12(30)} - 1} \approx 19{,}447.80$, so they will owe

$\$260{,}552.20$. After the 240th payment, their principal repayment will be $B = 280{,}000 \dfrac{\left(1 + \frac{0.06}{12}\right)^{240} - 1}{\left(1 + \frac{0.06}{12}\right)^{12(30)} - 1} \approx 128{,}789.96$,

so they will owe $\$151{,}210.04$.

79. $F = f(M, 600, 10) = \dfrac{\pi^2 (600)^2 M(10)}{900} \approx 39{,}478.42M$ dynes or $\dfrac{39{,}478.42M}{980M} \approx 40.28$ times the earth's gravity.

81. For yacht A, we have $f(20.95, 277.3, 17.56) = \dfrac{20.95 + 1.25\left(277.3^{1/2}\right) - 9.80\left(17.56^{1/3}\right)}{0.388} \approx 41.993$.

Since this is less than 42, yacht A satisfies the formula. For yacht B, we have

$f(21.87, 311.78, 22.48) = \dfrac{21.87 + 1.25\left(311.78^{1/2}\right) - 9.80\left(22.48^{1/3}\right)}{0.388} \approx 41.967$. Since this is less than 42, yacht B

satisfies the formula as well.

83. The level surfaces of F are described by $F = \dfrac{Gm_1 m_2}{x^2 + y^2 + z^2} = k \Leftrightarrow x^2 + y^2 + z^2 = \dfrac{Gm_1 m_2}{k}$, $k > 0$. Thus, the level

surfaces of F are a family of concentric spheres centered at the origin. A spherical level surface gives the set of points at which the force exerted by m_1 (located at the origin) on m_2 (located on the surface) has the same magnitude F.

85. $F = \left(\dfrac{1100 - 100}{1100 + 50}\right)(500) \approx 435$ Hz

87. False. Let $f(x, y) = x^2 + y^2$. Then $f(1, 1) = 2 = f(-1, -1)$, and evidently $P_1(1, 1) \neq P_2(-1, -1)$.

89. False. Let $f(x, y) = x^2 + y^2$. Then $f(x, y) = k$ is not defined for $k < 0$.

13.2   Concept Questions

1. a. See page 1049 (1045 in ET).

b. See page 1049 (1045 in ET).

3. a. The function $f(P, T)$ is continuous because the volume of the balloon varies continuously as the atmospheric pressure P and temperature T change.

b. The function $f(H, W)$ is continuous because the surface area of a human body varies continuously as its height H and weight W change.

c. The function $f(d, t)$ is not a continuous function because the fare does not vary continuously as the distance traveled changes.

d. The function $f(T, P)$ is continuous because the volume of the mass of gas varies continuously as its temperature T and pressure P change.

13.2   Limits and Continuity

1. Along $y = 0$, $\displaystyle\lim_{(x,y)\to(0,0)} \frac{x^2 - y^2}{2x^2 + y^2} = \lim_{x\to 0} \frac{x^2}{2x^2} = \lim_{x\to 0} \frac{1}{2} = \frac{1}{2}$. Along $x = 0$,

$\displaystyle\lim_{(x,y)\to(0,0)} \frac{x^2 - y^2}{2x^2 + y^2} = \lim_{y\to 0} \frac{-y^2}{y^2} = \lim_{y\to 0} (-1) = -1$. Because these two limits are not equal, the given limit does not exist.

3. Along $y = 0$, $\displaystyle\lim_{(x,y)\to(0,0)} \frac{3xy}{3x^2 + y^2} = \lim_{x\to 0} \frac{0}{3x^2} = \lim_{x\to 0} = 0$. Along $y = x$,

$\displaystyle\lim_{(x,y)\to(0,0)} \frac{3xy}{3x^2 + y^2} = \lim_{x\to 0} \frac{3x^2}{3x^2 + x^2} = \lim_{x\to 0} \frac{3}{4} = \frac{3}{4}$. Because these two limits are not equal, the given limit does not exist.

5. Along $y = 0$, $\displaystyle\lim_{(x,y)\to(0,0)} \frac{2xy}{\sqrt{x^4 + y^4}} = \lim_{x\to 0} \frac{0}{\sqrt{x^4}} = \lim_{x\to 0} 0 = 0$. Along $y = x$,

$\displaystyle\lim_{(x,y)\to(0,0)} \frac{2xy}{\sqrt{x^4 + y^4}} = \lim_{x\to 0} \frac{2x^2}{\sqrt{x^4 + x^4}} = \lim_{x\to 0} \sqrt{2} = \sqrt{2}$. Because these two limits are not equal, the given limit does not exist.

7. Along $y = 0$, $\displaystyle\lim_{(x,y)\to(1,0)} \frac{2xy - 2y}{x^2 + y^2 - 2x + 1} = \lim_{x\to 1} \frac{0}{x^2 - 2x + 1} = \lim_{x\to 1} 0 = 0$.

Along $y = x - 1$,

$\displaystyle\lim_{(x,y)\to(1,0)} \frac{2xy - 2y}{x^2 + y^2 - 2x + 1} = \lim_{(x,y)\to(1,0)} \frac{2y(x-1)}{(x-1)^2 + y^2} = \lim_{y\to 0} \frac{2y^2}{y^2 + y^2}$

$\qquad\qquad = \lim_{y\to 0} 1 = 1$

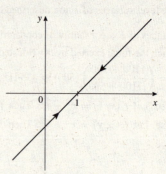

Because these two limits are not equal, the given limit does not exist.

9. Along the x-axis, $\displaystyle\lim_{(x,y,z)\to(0,0,0)} \frac{xy + yz + xz}{x^2 + y^2 + z^2} = \lim_{x\to 0} \frac{0}{x^2} = 0$. Let C denote the curve with parametric equations

$x = t$, $y = t$, $z = t$. Then along C, $\displaystyle\lim_{(x,y,z)\to(0,0,0)} \frac{xy + yz + xz}{x^2 + y^2 + z^2} = \lim_{t\to 0} \frac{3t^2}{3t^2} = \lim_{t\to 0} 1 = 1$. Because these two limits are not equal, the given limit does not exist.

11. Along the z-axis, $\displaystyle\lim_{(x,y,z)\to(0,0,0)} \frac{xz^2+2y^2}{x^2+2y^2+z^4} = \lim_{z\to0}\frac{0}{z^4} = \lim_{z\to0}0 = 0$. Let $C: x=t^2, y=t^2, z=t$. Then along C,

$$\lim_{(x,y,z)\to(0,0,0)} \frac{xz^2+2y^2}{x^2+2y^2+z^4} = \lim_{t\to0}\frac{t^4+2t^4}{t^4+2t^4+t^4} = \lim_{t\to0}\frac{3}{4} = \frac{3}{4}.$$ Because these two limits are not equal, the given limit

does not exist.

13. $\displaystyle\lim_{(x,y)\to(1,2)}\left(x^2+2y^2\right) = 1^2 + 2\left(2^2\right) = 9$

15. $\displaystyle\lim_{(x,y)\to(1,2)}\frac{2x^2-3y^3+4}{3-xy} = \frac{2\,(1)^2-3\,(2)^3+4}{3-(1)\,(2)} = -18$

17. $\displaystyle\lim_{(x,y)\to(1,-2)}\frac{3xy}{2x^2-y^2} = \frac{3\,(1)\,(-2)}{2\,(1)-(-2)^2} = 3$

19. $\displaystyle\lim_{(x,y)\to(0^+,0^+)}\frac{e^{\sqrt{x+y}}}{x+y-1} = \frac{e^0}{-1} = -1$

21. $\displaystyle\lim_{(x,y)\to(1,1)} = \frac{\tan^{-1}(x/y)}{\cos^{-1}(x-2y)} = \frac{\tan^{-1}1}{\cos^{-1}(-1)} = \frac{\pi/4}{\pi} = \frac{1}{4}$

23. $\displaystyle\lim_{(x,y)\to(2,1)}\ln\left(x^2-3y\right) = \ln(4-3) = \ln1 = 0$

25. $\displaystyle\lim_{(x,y,z)\to(1,2,3)}\frac{xy+yz+xz}{xyz-3} = \frac{2+6+3}{6-3} = \frac{11}{3}$

27. $\displaystyle\lim_{(x,y)\to(0,0)}\frac{x^3+y^3}{x^2+y^2} = \lim_{r\to0^+}\frac{r^3\left(\cos^3\theta+\sin^3\theta\right)}{r^2} = 0$

29. $\displaystyle\lim_{(x,y)\to(0,0)}\left(x^2+y^2\right)\ln\left(x^2+y^2\right) = \lim_{r\to0^+}r^2\ln r^2 = \lim_{u\to0}u\ln u = \lim_{u\to0}\frac{\ln u}{1/u} = \lim_{u\to0}\frac{\frac{d}{du}(\ln u)}{\frac{d}{du}(1/u)}$ (using l'Hôpital's Rule)

$$= \lim_{u\to0}(-u) = 0$$

31. The function f is a rational function and is continuous at all points where its denominator is nonzero. Thus, f is continuous on $\{(x,y)\mid 2x+3y\neq1\}$.

33. The function g is continuous for all (x,y) such that $x+y\geq0$ and $x-y\geq0$; that is, on $\{(x,y)\mid x\geq0,\,|y|\leq x\}$.

35. We require that $x\geq0$ and $y\neq0$. Thus, F is continuous on $\{(x,y)\mid x\geq0,\,y\neq0\}$.

37. The function f is rational, and thus continuous at points at which its denominator is nonzero. Thus, f is continuous on $\left\{(x,y)\mid x^2+y^2+z^2\neq4\right\}$.

39. The logarithmic function $f(u)=\ln u$ is continuous for $u>0$, so h is continuous on $\{(x,y,z)\mid yz>1\}$.

41. a. Since xy is continuous for all values of x and y and the sine function is continuous everywhere, we see that f is continuous in the plane except possibly at $xy=0$ where the denominator of $f(x,y)$ is equal to zero. To see what happens when $xy=0$, let (a,b) be any point in the plane such that $ab=0$. Then if we put $t=xy$, then $t\to0$ as $xy\to0$. Thus

$$\lim_{(x,y)\to(a,b)}\frac{\sin xy}{xy} = \lim_{t\to0}\frac{\sin t}{t} = 1$$

so $\displaystyle\lim_{(x,y)\to(0,0)}f(x,y) = f(a,b)$ for all (a,b), and f is continuous on the plane.

b.

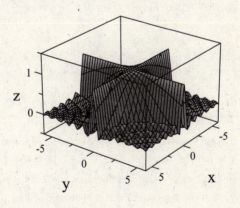

43. $h(x, y) = g(f(x, y)) = f(x, y)\cos f(x, y) + \sin f(x, y) = \left(x^2 - xy + y^2\right)\cos\left(x^2 - xy + y^2\right) + \sin\left(x^2 - xy + y^2\right).$

The polynomial function $f(x, y) = x^2 - xy + y^2$ is continuous everywhere and the function $g(t) = t\cos t + \sin t$ is continuous for all values of t, so $h = g \circ f$ is continuous on the entire xy-plane.

45. $h(x, y) = g(f(x, y)) = \dfrac{f(x, y) + 2}{f(x, y) - 1} = \dfrac{2x - y + 2}{2x - y - 1}.$ The polynomial function $f(x, y) = 2x - y$ is continuous

everywhere, and the rational function $g(t) = \dfrac{t + 2}{t - 1}$ is discontinuous at $t = 1$, so the function $h = g \circ f$ is discontinuous

where $2x - y = 1$; that is, h is continuous on $\{(x, y) \mid 2x - y \neq 1\}.$

47. $h(x, y) = g(f(x, y)) = \cos f(x, y) = \cos(x\tan y).$ The function $f(x, y) = x\tan y$ is discontinuous at

$y = \frac{\pi}{2} + n\pi$, n an integer. The function $g(t) = \cos t$ is continuous everywhere. Thus, $h = g \circ f$ is continuous on

$\left\{(x, y) \mid y \neq \frac{\pi}{2} + n\pi, n \text{ an integer}\right\}.$

49. Let $\varepsilon > 0$ be given and consider $|f(x, y) - c| = |c - c| = 0 < \varepsilon$. This is true for all (x, y). Therefore, we can pick $\boldsymbol{\delta}$ to be

any positive number, and then $0 < \sqrt{(x - a)^2 + (y - b)^2} < \boldsymbol{\delta} \Rightarrow |f(x, y) - c| < \varepsilon$. Because ε is arbitrary, the proof is complete.

51. Let $\varepsilon > 0$ be given and consider $|f(x, y) - 0| = \left|\dfrac{3xy^3}{x^2 + y^2}\right| = 3|xy|\dfrac{y^2}{x^2 + y^2} < 3|xy|$ for $(x, y) \neq (0, 0)$. Using the

fact that $(|x| - |y|)^2 \geq 0$, we can write $3|xy| \leq 3\left(x^2 + y^2\right) = 3\left(\sqrt{x^2 + y^2}\right)^2$. Thus, if we pick $\boldsymbol{\delta} = \sqrt{\frac{1}{3}\varepsilon}$, then

$\sqrt{x^2 + y^2} < \boldsymbol{\delta}$ implies that $|f(x, y) - 0| < 3\left(\frac{1}{3}\varepsilon\right) = \varepsilon$. Since ε is arbitrary, the proof is complete.

53. True. If $\lim\limits_{(x,y)\to(a,b)} f(x, y) = L_1$ along a path C_1 and $\lim\limits_{(x,y)\to(a,b)} f(x, y) = L_2$ along a different path C_2, where

$L_1 \neq L_2$, then in any neighborhood of (a, b) there are values $f(x, y)$ that are close to L_1 and values that are close to L_2,

and so $\lim\limits_{(x,y)\to(a,b)} f(x, y)$ cannot exist.

55. True. The function g is continuous at (a, b); so is the function h. Therefore, the product of g and h defined by

$f(x, y) = g(x)h(y)$ is also continuous at (a, b).

57. True. This follows from the definition of continuity.

13.3   Concept Questions   ET 12.3

1. a. See page 1060 (1056 in ET).   **b.** See page 1060 (1056 in ET).

3. $F(x, y, z) = 0$, so $F_x\dfrac{\partial x}{\partial z} + F_z = 0 \Rightarrow \dfrac{\partial x}{\partial z} = -\dfrac{F_z}{F_x}$. Answers will vary.

13.3   Partial Derivatives   ET 12.3

1. a. $f_x(x, y) = \frac{\partial}{\partial x}\left(x^2 + 2y^2\right) = 2x \Rightarrow f_x(2, 1) = 2(2) = 4$ and $f_y(x, y) = \frac{\partial}{\partial y}\left(x^2 + 2y^2\right) = 4y \Rightarrow$

$f_y(2, 1) = 4(1) = 4.$

b. $f_x(2, 1) = 4$ says that the slope of the tangent line to the curve of intersection of the surface $z = x^2 + 2y^2$ and the plane

$y = 1$ at the point $(2, 1, 6)$ is 4. $f_y(2, 1) = 4$ says that the slope of the tangent line to the curve of intersection of the

surface $z = x^2 + 2y^2$ and the plane $x = 2$ at the point $(2, 1, 6)$ is 4.

c. $f_x(2, 1) = 4$ says that the rate of change of $f(x, y)$ with respect to x with y fixed at 1 is 4 units per unit change in x.

$f_y(2, 1) = 4$ says that the rate of change of $f(x, y)$ with respect to y with x fixed at 2 is 4 units per unit change in y.

3. At the point P, $\frac{\partial f}{\partial x} < 0$ and $\frac{\partial f}{\partial y} < 0$. At the point Q, both $\frac{\partial f}{\partial x}$ and $\frac{\partial f}{\partial y}$ are 0. At the point R, $\frac{\partial f}{\partial x} < 0$ and $\frac{\partial f}{\partial y} > 0$.

5. $f_x(3, 2) \approx \dfrac{f(3.7, 2) - f(3, 2)}{3.7 - 3} = \dfrac{95 - 100}{0.7} \approx -7.1$, so the temperature at $(3, 2)$ is dropping at the rate of approximately

$7.1°\text{F/in.}$ in the positive x-direction.

$f_y(3, 2) \approx \dfrac{f(3, 3.3) - f(3, 2)}{3.3 - 2} = \dfrac{95 - 100}{1.3} \approx -3.8$, so the temperature at $(3, 2)$ is dropping at the rate of approximately

$3.8°\text{F/in.}$ in the positive y-direction.

7. $f_x(x, y) = \frac{\partial}{\partial x}\left(2x^2 - 3xy + y^2\right) = 4x - 3y$ and $f_y(x, y) = \frac{\partial}{\partial y}\left(2x^2 - 3xy + y^2\right) = -3x + 2y$.

9. $\dfrac{\partial z}{\partial x} = \dfrac{\partial}{\partial x}\left(x\sqrt{y}\right) = \sqrt{y}$ and $\dfrac{\partial z}{\partial y} = \dfrac{\partial}{\partial y}\left(xy^{1/2}\right) = \tfrac{1}{2}xy^{-1/2} = \dfrac{x}{2\sqrt{y}}$.

11. $g_r(r, s) = \frac{\partial}{\partial r}\left(\sqrt{r} + s^2\right) = \tfrac{1}{2}r^{-1/2} = \dfrac{1}{2\sqrt{r}}$ and $g_s(r, s) = \frac{\partial}{\partial s}\left(\sqrt{r} + s^2\right) = 2s$.

13. $f_x(x, y) = \frac{\partial}{\partial x}\left(xe^{y/x}\right) = e^{y/x} + xe^{y/x}\left(-\dfrac{y}{x^2}\right) = e^{y/x}\left(1 - \dfrac{y}{x}\right)$ and $f_y(x, y) = \frac{\partial}{\partial y}\left(xe^{y/x}\right) = xe^{y/x}\left(\dfrac{1}{x}\right) = e^{y/x}$.

15. $\dfrac{\partial z}{\partial x} = \dfrac{\partial}{\partial x}\left[\tan^{-1}\left(x^2 + y^2\right)\right] = \dfrac{\frac{\partial}{\partial x}\left(x^2 + y^2\right)}{1 + \left(x^2 + y^2\right)^2} = \dfrac{2x}{1 + \left(x^2 + y^2\right)^2}$ and

$\dfrac{\partial z}{\partial y} = \dfrac{\partial}{\partial y}\left[\tan^{-1}\left(x^2 + y^2\right)\right] = \dfrac{2y}{1 + \left(x^2 + y^2\right)^2}$.

17. $g_u(u, v) = \dfrac{\partial}{\partial u}\left(\dfrac{uv}{u^2 + v^3}\right) = \dfrac{\left(u^2 + v^3\right)(v) - (uv)(2u)}{\left(u^2 + v^3\right)^2} = \dfrac{v\left(v^3 - u^2\right)}{\left(u^2 + v^3\right)^2}$ and

$g_v(u, v) = \dfrac{\partial}{\partial v}\left(\dfrac{uv}{u^2 + v^3}\right) = \dfrac{\left(u^2 + v^3\right)(u) - (uv)\left(3v^2\right)}{\left(u^2 + v^3\right)^2} = \dfrac{u\left(u^2 - 2v^3\right)}{\left(u^2 + v^3\right)^2}$.

19. $g_x(x, y) = \dfrac{\partial}{\partial x}\left(x^2\cosh\dfrac{x}{y}\right) = 2x\cosh\dfrac{x}{y} + \left(x^2\sinh\dfrac{x}{y}\right)\left(\dfrac{1}{y}\right) = 2x\cosh\dfrac{x}{y} + \dfrac{x^2}{y}\sinh\dfrac{x}{y}$ and

$g_y(x, y) = \dfrac{\partial}{\partial y}\left(x^2\cosh\dfrac{x}{y}\right) = \left(x^2\sinh\dfrac{x}{y}\right)\left(-\dfrac{x}{y^2}\right) = -\dfrac{x^3}{y^2}\sinh\dfrac{x}{y}$.

21. $f_x(x, y) = \frac{\partial}{\partial x}y^x = y^x\ln y$ and $f_y(x, y) = \frac{\partial}{\partial y}y^x = xy^{x-1}$.

23. $f_x(x, y) = \frac{\partial}{\partial x}\int_x^y te^{-t}\,dt = -\frac{\partial}{\partial x}\int_y^x te^{-t}\,dt = -xe^{-x}$ and $f_y(x, y) = \frac{\partial}{\partial y}\int_x^y te^{-t}\,dt = ye^{-y}$.

25. $g_x(x, y, z) = \frac{\partial}{\partial x}\left(x^{1/2}y^{1/2}z^{1/2}\right) = \tfrac{1}{2}x^{-1/2}y^{1/2}z^{1/2} = \dfrac{\sqrt{xyz}}{2x}$. By symmetry, $g_y(x, y, z) = \dfrac{\sqrt{xyz}}{2y}$ and

$g_z(x, y, z) = \dfrac{\sqrt{xyz}}{2z}$.

27. $\dfrac{\partial u}{\partial x} = \dfrac{\partial}{\partial x}\left(xe^{y/z} - z^2\right) = e^{y/z}$, $\dfrac{\partial u}{\partial y} = \dfrac{\partial}{\partial y}\left(xe^{y/z} - z^2\right) = xe^{y/z}\left(\dfrac{1}{z}\right) = \dfrac{x}{z}e^{y/z}$, and

$\dfrac{\partial u}{\partial z} = \dfrac{\partial}{\partial z}\left(xe^{y/z} - z^2\right) = xe^{y/z}\left(-\dfrac{y}{z^2}\right) - 2z = -\dfrac{xy}{z^2}e^{y/z} - 2z$.

29. $f_r(r, s, t) = \dfrac{\partial}{\partial r}rs\ln st = s\ln st$, $f_s(r, s, t) = \dfrac{\partial}{\partial s}rs\ln st = r\ln st + rs\left(\dfrac{1}{s}\right) = r\ln st + r$, and

$f_t(r, s, t) = \dfrac{\partial}{\partial t}rs\ln st = rs\dfrac{\partial}{\partial t}(\ln s + \ln t) = \dfrac{rs}{t}$.

31. $\frac{\partial}{\partial x}\left(xe^y + ye^{-x} + e^z\right) = \frac{\partial}{\partial x}(10) \Rightarrow e^y - ye^{-x} + e^z\frac{\partial z}{\partial x} = 0 \Rightarrow \frac{\partial z}{\partial x} = \frac{ye^{-x} - e^y}{e^z}$ and $\frac{\partial}{\partial y}\left(xe^y + ye^{-x} + e^z\right) = \frac{\partial}{\partial y}(10)$

$\Rightarrow xe^y + e^{-x} + e^z\frac{\partial z}{\partial y} = 0 \Rightarrow \frac{\partial z}{\partial y} = -\frac{xe^y + e^{-x}}{e^z}$.

33. $\frac{\partial}{\partial x}\left[\ln\left(x^2 + z^2\right) + yz^3 + 2x^2\right] = \frac{\partial}{\partial x}(10) \Rightarrow \frac{2x + 2z\frac{\partial z}{\partial x}}{x^2 + z^2} + 3yz^2\frac{\partial z}{\partial x} + 4x = 0 \Rightarrow \frac{\partial z}{\partial x} = -\frac{2x\left(2x^2 + 2z^2 + 1\right)}{z\left(3yz^3 + 3x^2yz + 2\right)}$ and

$\frac{\partial}{\partial y}\left[\ln\left(x^2 + z^2\right) + yz^3 + 2x^2\right] = \frac{\partial}{\partial y}(10) \Rightarrow \frac{2z\frac{\partial z}{\partial y}}{x^2 + z^2} + z^3 + 3yz^2\frac{\partial z}{\partial y} = 0 \Rightarrow \frac{\partial z}{\partial y} = -\frac{z^2\left(x^2 + z^2\right)}{3yz^3 + 3x^2yz + 2}$.

35. $g_x(x, y) = \frac{\partial}{\partial x}\left(x^3y^2 + xy^3 - 2x + 3y + 1\right) = 3x^2y^2 + y^3 - 2$,

$g_y(x, y) = \frac{\partial}{\partial y}\left(x^3y^2 + xy^3 - 2x + 3y + 1\right) = 2x^3y + 3xy^2 + 3, g_{xx}(x, y) = \frac{\partial}{\partial x}\left(3x^2y^2 + y^3 - 2\right) = 6xy^2$,

$g_{yy} = \frac{\partial}{\partial y}\left(2x^3y + 3xy^2 + 3\right) = 2x^3 + 6xy, g_{xy} = \frac{\partial}{\partial y}\left(3x^2y^2 + y^3 - 2\right) = 6x^2y + 3y^2$, and

$g_{yx} = \frac{\partial}{\partial x}\left(2x^3y + 3xy^2 + 3\right) = 6x^2y + 3y^2$.

37. $\frac{\partial w}{\partial u} = \frac{\partial}{\partial u}\left[\cos(2u - v) + \sin(2u + v)\right] = -2\sin(2u - v) + 2\cos(2u + v)$,

$\frac{\partial w}{\partial v} = \frac{\partial}{\partial v}\left[\cos(2u - v) + \sin(2u + v)\right] = \sin(2u - v) + \cos(2u + v), \frac{\partial^2 w}{\partial u^2} = -4\cos(2u - v) - 4\sin(2u + v)$,

$\frac{\partial^2 w}{\partial v^2} = -\cos(2u - v) - \sin(2u + v)$, and $\frac{\partial^2 w}{\partial u\,\partial v} = \frac{\partial^2 w}{\partial v\,\partial u} = 2\cos(2u - v) - 2\sin(2u + v)$.

39. $h_x(x, y) = \frac{\partial}{\partial x}\left(\tan^{-1}\frac{y}{x}\right) = \frac{-y/x^2}{1 + (y/x)^2} = -\frac{y}{x^2 + y^2}, h_y(x, y) = \frac{\partial}{\partial y}\left(\tan^{-1}\frac{y}{x}\right) = \frac{1/x}{1 + (y/x)^2} = \frac{x}{x^2 + y^2}$,

$h_{xx}(x, y) = \frac{\partial}{\partial x}\left[-y\left(x^2 + y^2\right)^{-1}\right] = -y(-1)\left(x^2 + y^2\right)^{-2}(2x) = \frac{2xy}{\left(x^2 + y^2\right)^2}$,

$h_{yy}(x, y) = \frac{\partial}{\partial y}\left(\frac{x}{x^2 + y^2}\right) = -\frac{2xy}{\left(x^2 + y^2\right)^2}$, and

$h_{yx}(x, y) = h_{xy}(x, y) = \frac{\partial}{\partial y}\left[-y\left(x^2 + y^2\right)^{-1}\right] = -\left(x^2 + y^2\right)^{-1} - y(-1)\left(x^2 + y^2\right)^{-2}(2y) = \frac{y^2 - x^2}{\left(x^2 + y^2\right)^2}$.

41. $f_x(x, y, z) = \frac{\partial}{\partial x}\left(x^4 - 2x^2y^2 + xy^3 + 2y^4\right) = 4x^3 - 4xy^2 + y^3 \Rightarrow f_{xy}(x, y, z) = \frac{\partial}{\partial y}\left(4x^3 - 4xy^2 + y^3\right) = -8xy + 3y^2$

$\Rightarrow f_{xyx}(x, y, z) = \frac{\partial}{\partial x}\left(-8xy + 3y^2\right) = -8y$

43. $\frac{\partial z}{\partial x} = \frac{\partial}{\partial x}(x\cos y + y\sin x) = \cos y + y\cos x \Rightarrow \frac{\partial^2 z}{\partial y\,\partial x} = \frac{\partial}{\partial y}(\cos y + y\cos x) = -\sin y + \cos x \Rightarrow$

$\frac{\partial^3 z}{\partial x\,\partial y\,\partial x} = \frac{\partial}{\partial x}(-\sin y + \cos x) = -\sin x$

45. $h_z(x, y, z) = \frac{\partial}{\partial z}\left[e^x\cos(y + 2z)\right] = -2e^x\sin(y + 2z) \Rightarrow h_{zz}(x, y, z) = -4e^x\cos(y + 2z) \Rightarrow$

$h_{zzy}(x, y, z) = 4e^x\sin(y + 2z)$

47. $f_x(x, y) = \frac{\partial}{\partial x}\left(x\sin^2 y + y^2\cos x\right) = \sin^2 y - y^2\sin x \Rightarrow f_{xy}(x, y) = \frac{\partial}{\partial y}\left(\sin^2 y - y^2\sin x\right) = 2\sin y\cos y - 2y\sin x$

and $f_y(x, y) = \frac{\partial}{\partial y}\left(x\sin^2 y + y^2\cos x\right) = 2x\sin y\cos y + 2y\cos x \Rightarrow$

$f_{yx}(x, y) = \frac{\partial}{\partial x}(2x\sin y\cos y + 2y\cos x) = 2\sin y\cos y - 2y\sin x$, so $f_{xy} = f_{yx}$.

49. $f_x(x, y) = \dfrac{\partial}{\partial x}\left[\tan^{-1}\left(x^2 + y^3\right)\right] = \dfrac{2x}{1 + \left(x^2 + y^3\right)^2} \Rightarrow$

$f_{xy}(x, y) = \dfrac{\partial}{\partial y}\left\{2x\left[1 + \left(x^2 + y^3\right)^2\right]^{-1}\right\} = 2x(-1)\left[1 + \left(x^2 + y^3\right)^2\right]^{-2}(2)\left(x^2 + y^3\right)\left(3y^2\right) = -\dfrac{12xy^2\left(x^2 + y^3\right)}{\left[1 + \left(x^2 + y^3\right)^2\right]^2}$

and $f_y(x, y) = \dfrac{\partial}{\partial y}\left[\tan^{-1}\left(x^2 + y^3\right)\right] = \dfrac{3y^2}{1 + \left(x^2 + y^3\right)^2} \Rightarrow$

$f_{yx}(x, y) = \dfrac{\partial}{\partial x}\left\{3y^2\left[1 + \left(x^2 + y^3\right)^2\right]^{-1}\right\} = 3y^2(-1)\left[1 + \left(x^2 + y^3\right)^2\right]^{-2}(2)\left(x^2 + y^3\right)(2x) = -\dfrac{12xy^2\left(x^2 + y^3\right)}{\left[1 + \left(x^2 + y^3\right)^2\right]^2}$,

so $f_{xy} = f_{yx}$.

51. $f_x(x, y, z) = \dfrac{\partial}{\partial x}\left(9 - x^2 - 2y^2 - z^2\right)^{1/2} = -x\left(9 - x^2 - 2y^2 - z^2\right)^{-1/2} \Rightarrow$

$f_{xy}(x, y, z) = -2xy\left(9 - x^2 - 2y^2 - z^2\right)^{-3/2} \Rightarrow f_{xyz}(x, y, z) = -\dfrac{6xyz}{\left(9 - x^2 - 2y^2 - z^2\right)^{5/2}}$,

$f_y(x, y, z) = \dfrac{\partial}{\partial y}\left(9 - x^2 - 2y^2 - z^2\right)^{1/2} = -2y\left(9 - x^2 - 2y^2 - z^2\right)^{-1/2} \Rightarrow$

$f_{yx}(x, y, z) = -2xy\left(9 - x^2 - 2y^2 - z^2\right)^{-3/2} \Rightarrow f_{yxz}(x, y, z) = -\dfrac{6xyz}{\left(9 - x^2 - 2y^2 - z^2\right)^{5/2}}$,

and $f_z(x, y, z) = \dfrac{\partial}{\partial z}\left(9 - x^2 - 2y^2 - z^2\right)^{1/2} = -z\left(9 - x^2 - 2y^2 - z^2\right)^{-1/2} \Rightarrow$

$f_{zy}(x, y, z) = -2yz\left(9 - x^2 - 2y^2 - z^2\right)^{-3/2} \Rightarrow f_{zyx}(x, y, z) = -\dfrac{6xyz}{\left(9 - x^2 - 2y^2 - z^2\right)^{5/2}}$, so $f_{xyz} = f_{yxz} = f_{zyx}$.

53. $f_x(x, y, z) = \dfrac{\partial}{\partial x}\left(e^{-x}\cos yz\right) = -e^{-x}\cos yz \Rightarrow f_{xy}(x, y, z) = ze^{-x}\sin yz \Rightarrow f_{xyz}(x, y, z) = e^{-x}(\sin yz + yz\cos yz)$,

$f_y(x, y, z) = -ze^{-x}\sin yz \Rightarrow f_{yx}(x, y, z) = ze^{-x}\sin yz \Rightarrow f_{yxz}(x, y, z) = e^{-x}(\sin yz + yz\cos yz)$, and

$f_z(x, y, z) = -ye^{-x}\sin yz \Rightarrow f_{zy}(x, y, z) = -e^{-x}(\sin yz + yz\cos yz) \Rightarrow f_{zyx}(x, y, z) = e^{-x}(\sin yz + yz\cos yz)$, so

$f_{xyz} = f_{yxz} = f_{zyx}$.

55. $u_t = \dfrac{\partial}{\partial t}\left(e^{-t}\sin\dfrac{x}{c}\right) = -e^{-t}\sin\dfrac{x}{c}$ and $u_x = \dfrac{1}{c}e^{-t}\cos\dfrac{x}{c} \Rightarrow u_{xx} = -\dfrac{1}{c^2}e^{-t}\sin\dfrac{x}{c}$, so $u_t = c^2 u_{xx}$ as was to be shown.

57. $u_t = \dfrac{\partial}{\partial t}\left[\cos\left(x - ct\right) + 2\sin\left(x + ct\right)\right] = c\sin\left(x - ct\right) + 2c\cos\left(x + ct\right) \Rightarrow$

$u_{tt} = -c^2\cos\left(x - ct\right) - 2c^2\sin\left(x + ct\right) = -c^2\left[\cos\left(x - ct\right) + 2\sin\left(x + ct\right)\right]$ and $u_x = -\sin\left(x - ct\right) + 2\cos\left(x + ct\right)$

$\Rightarrow u_{xx} = -\cos\left(x - ct\right) - 2\sin\left(x + ct\right)$, so $u_{tt} = c^2 u_{xx}$ as was to be shown.

59. $u_x = \dfrac{\partial}{\partial x}\left(3x^2y - y^3\right) = 6xy \Rightarrow u_{xx} = \dfrac{\partial}{\partial x}\left(6xy\right) = 6y$ and $u_y = \dfrac{\partial}{\partial y}\left(3x^2y - y^3\right) = 3x^2 - 3y^2 \Rightarrow$

$u_{yy} = \dfrac{\partial}{\partial y}\left(3x^2 - 3y^2\right) = -6y$, so $u_{xx} + u_{yy} = 6y - 6y = 0$.

61. $u = \ln\sqrt{x^2 + y^2} = \ln\left(x^2 + y^2\right)^{1/2} = \dfrac{1}{2}\ln\left(x^2 + y^2\right)$, so $u_x = \dfrac{1}{2}\dfrac{\partial}{\partial x}\ln\left(x^2 + y^2\right) = \dfrac{x}{x^2 + y^2} \Rightarrow$

$u_{xx} = \dfrac{\left(x^2 + y^2\right) - x\left(2x\right)}{\left(x^2 + y^2\right)^2} = \dfrac{y^2 - x^2}{\left(x^2 + y^2\right)^2}$. By symmetry, $u_y = \dfrac{y}{x^2 + y^2}$ and $u_{yy} = \dfrac{x^2 - y^2}{\left(x^2 + y^2\right)^2}$, so

$u_{xx} + u_{yy} = \dfrac{y^2 - x^2}{\left(x^2 + y^2\right)^2} + \dfrac{x^2 - y^2}{\left(x^2 + y^2\right)^2} = 0$.

63. $u_x = \dfrac{\partial}{\partial x}\left(\tan^{-1}\dfrac{y}{x}\right) = \dfrac{-y/x^2}{1+(y/x)^2} = -\dfrac{y}{x^2+y^2} \Rightarrow u_{xx} = \dfrac{2xy}{\left(x^2+y^2\right)^2}$ and $u_y = \dfrac{1/x}{1+(y/x)^2} = \dfrac{x}{x^2+y^2} \Rightarrow$

$u_{yy} = -\dfrac{2xy}{\left(x^2+y^2\right)^2}$, so $u_{xx}+u_{yy} = \dfrac{2xy}{\left(x^2+y^2\right)^2} - \dfrac{2xy}{\left(x^2+y^2\right)^2} = 0$ as was to be shown.

65. $u_x = \dfrac{\partial}{\partial x}\left(x^2+3xy+2y^2-3z^2+4xyz\right) = 2x+3y+4yz \Rightarrow u_{xx} = 2, u_y = 3x+4y+4xz \Rightarrow u_{yy} = 4$, and

$u_z = -6z+4xy \Rightarrow u_{zz} = -6$, so $u_{xx}+u_{yy}+u_{zz} = 2+4-6 = 0$ as was to be shown.

67. $\dfrac{\partial z}{\partial x} = \dfrac{\partial}{\partial x}\left[\left(x^2+y^2\right)^{1/2}\tan^{-1}\dfrac{y}{x}\right] = x\left(x^2+y^2\right)^{-1/2}\tan^{-1}\dfrac{y}{x} + \left(x^2+y^2\right)^{1/2}\cdot\dfrac{-y/x^2}{1+(y/x)^2} = \dfrac{x\tan^{-1}(y/x)-y}{\sqrt{x^2+y^2}}$

and $\dfrac{\partial z}{\partial y} = y\left(x^2+y^2\right)^{-1/2}\tan^{-1}\dfrac{y}{x} + \left(x^2+y^2\right)^{1/2}\cdot\dfrac{1/x}{1+(y/x)^2} = \dfrac{y\tan^{-1}(y/x)+x}{\sqrt{x^2+y^2}}$, so

$x\dfrac{\partial z}{\partial x}+y\dfrac{\partial z}{\partial y} = \dfrac{x^2\tan^{-1}(y/x)-xy}{\sqrt{x^2+y^2}} + \dfrac{y^2\tan^{-1}(y/x)+xy}{\sqrt{x^2+y^2}} = \sqrt{x^2+y^2}\,\tan^{-1}\dfrac{y}{x} = z$, as was to be shown.

69. If $k = 8.314$, then $V = \dfrac{8.314T}{P}$, so $\left.\dfrac{\partial V}{\partial T}\right|_{T=300,P=125} = \left.\dfrac{8.314}{P}\right|_{P=125} = 0.066512$, which tells us that with pressure held

constant at 125 Pa, the volume increases at a rate of 0.066512 liters per $1\,^{\circ}$K increase in temperature when the temperature

is 300 $^{\circ}$K.

$\left.\dfrac{\partial V}{\partial P}\right|_{T=300,P=125} = \left.-\dfrac{8.314T}{P^2}\right|_{T=300,P=125} = -0.1596288$, which tells us that with temperature held constant at

300 $^{\circ}$K, the volume decreases at a rate of 0.1596288 liters per 1 Pa increase in pressure when the pressure is 125 Pa.

71. $\dfrac{1}{R} = \dfrac{1}{R_1}+\dfrac{1}{R_2}+\dfrac{1}{R_3} \Rightarrow \dfrac{\partial}{\partial R_1}\left(R^{-1}\right) = \dfrac{\partial}{\partial R_1}\left(\dfrac{1}{R_1}+\dfrac{1}{R_2}+\dfrac{1}{R_3}\right) \Rightarrow -R^{-2}\left(\dfrac{\partial R}{\partial R_1}\right) = -\dfrac{1}{R_1^2} \Rightarrow \dfrac{\partial R}{\partial R_1} = \dfrac{R^2}{R_1^2} = \left(\dfrac{R}{R_1}\right)^2$,

which says that with R_2 and R_3 held constant, the rate of change of the total resistance with respect to R_1 is equal to the

square of the ratio of R to R_1.

73. $\dfrac{\partial P}{\partial x} = \dfrac{\partial}{\partial x}\left(-0.02x^2-15y^2+xy+39x+25y-15{,}000\right) = -0.04x+y+39$, so

$\dfrac{\partial P}{\partial x}(5000,200) = -0.04(5000)+200+39 = \$39$ per \$1000. This says that with a fixed floor space of 200,000 ft^2, the

profit increases at a rate of \$39 per \$1000 of inventory when the inventory level is \$5,000,000.

$\dfrac{\partial P}{\partial y}(5000,200) = (-30y+x+25)|_{x=5000,y=200} = -30(200)+5000+25 = -\$975$ per 1000 ft^2. This says that with

a fixed inventory of \$5,000,000, the profit decreases at a rate of \$975 per 1000 ft^2 increase of floor space when the floor

space is 200,000 ft^2.

75. The required rate of change is

$\left.\dfrac{\partial V}{\partial x}(1,2,3) = kQ\dfrac{\partial}{\partial x}\left(x^2+y^2+z^2\right)^{-1/2}\right|_{(1,2,3)} = \left.-\dfrac{kQx}{\left(x^2+y^2+z^2\right)^{3/2}}\right|_{(1,2,3)} = -\dfrac{\sqrt{14}\,kQ}{196}$ volts per meter.

77. $\dfrac{\partial N}{\partial x} = \dfrac{\partial}{\partial x} \dfrac{120\left(1000 + 0.03x^2y\right)^{1/2}}{(5 + 0.2y)^2} = \dfrac{120\left(\frac{1}{2}\right)\left(1000 + 0.03x^2y\right)^{-1/2}(0.06xy)}{(5 + 0.2y)^2}$, so

$\dfrac{\partial N}{\partial x}(100, 20) = \dfrac{3.6xy}{(5 + 0.2y)^2\sqrt{1000 + 0.03x^2y}}\Bigg|_{(100,20)} \approx 1.06.$ This means that with the level of reinvestment held

constant at 20 cents per dollar deposited, the number of suspicious fires will grow at the rate of approximately 1 fire per

increase of 1 person per census tract when the number of people per census tract is 100. Next,

$\dfrac{\partial N}{\partial y}(100, 20) = 120\,\dfrac{\partial}{\partial y}\left[\left(1000 + 0.03x^2y\right)^{1/2}(5 + 0.2y)^{-2}\right]\Bigg|_{(100,20)}$

$= 120\left[\tfrac{1}{2}\left(1000 + 0.03x^2y\right)^{-1/2}(0.03x^2)(5 + 0.2y)^{-2} + \left(1000 + 0.03x^2y\right)^{1/2}(-2)(5 + 0.2y)^{-3}(0.2)\right]\Bigg|_{(100,20)}$

$= \dfrac{9x^2 - 1.08x^2y - 48{,}000}{(5 + 0.2y)^3\sqrt{1000 + 0.03x^2y}}\Bigg|_{(100,20)} \approx -2.85$

which tells us that if the number of people per census tract is constant at 100 per tract, the number of suspicious fires

decreases at a rate of approximately 2.9 per increase of 1 cent per dollar deposited for reinvestment when the level of

reinvestment is 20 cents per dollar deposited.

79. a. $T = f(32, 20) = 35.74 + 0.6125(32) - 35.75\left(20^{0.16}\right) + 0.4275(32)\left(20^{0.16}\right) \approx 19.70°\text{F}$

b. $\dfrac{\partial T}{\partial s} = -35.75\left(0.16S^{-0.84}\right) + 0.4275t\left(0.16S^{-0.84}\right) = 0.16(-35.75 + 0.4275t)\,s^{-0.84} \Rightarrow$

$\dfrac{\partial T}{\partial s}\Bigg|_{(32,20)} = 0.16\,[-35.75 + 0.4275(32)]\,20^{-0.84} \approx -0.285;$ that is, the wind chill will drop by 0.3° for each 1 mi/h

increase in wind speed.

81. a. $g'(a) = \lim\limits_{h \to 0} \dfrac{g(a + h) - g(a)}{h} = \lim\limits_{h \to 0} \dfrac{f(a + h, b) - f(a, b)}{h} = f_x(a, b)$

b. $h'(b) = \lim\limits_{h \to 0} \dfrac{h(b + h) - h(b)}{h} = \lim\limits_{h \to 0} \dfrac{f(a, b + h) - f(a, b)}{h} = f_y(a, b)$

83. Put $g(x) = f(x, 1) = \ln\left(e^x + \cos\sqrt{x^2 + 1}\right)$. Then, using a calculator or computer, we find $g'(2) \approx 0.9872$, so

$f_x(2, 1) \approx 0.9872$.

85. $\dfrac{\partial P}{\partial x} = \dfrac{\partial}{\partial x}kx^\alpha y^{1-\alpha} = k\alpha x^{\alpha-1}y^{1-\alpha} = k\alpha\left(\dfrac{y}{x}\right)^{1-\alpha}$ and $\dfrac{\partial P}{\partial y} = k(1-\alpha)x^\alpha y^{-\alpha} = k(1-\alpha)\left(\dfrac{x}{y}\right)^\alpha$. Therefore,

$x\dfrac{\partial P}{\partial x} + y\dfrac{\partial P}{\partial y} = \dfrac{k\alpha xy^{1-\alpha}}{x^{1-\alpha}} + \dfrac{k(1-\alpha)yx^\alpha}{y^\alpha} = k\alpha x^\alpha y^{1-\alpha} + k(1-\alpha)x^\alpha y^{1-\alpha} = kx^\alpha y^{1-\alpha} = P$, as was to be

shown.

87. Here $f(x, y) = x^2 + \tfrac{1}{4}y^2$ and $P(x_0, y_0, z_0) = P(1, 2, 2)$, so $f_x(1, 2) = 2x|_{(1,2)} = 2$ and $f_y(1, 2) = \tfrac{1}{2}y\big|_{(1,2)} = 1$, so

an equation of the tangent plane is $z - 2 = 2(x - 1) + (y - 2) \Rightarrow 2x + y - z = 2$.

89. $\dfrac{\partial T}{\partial x} = \dfrac{\partial}{\partial x}\left(\dfrac{2}{\pi}\tan^{-1}\dfrac{\sin x}{\sinh y}\right) = \dfrac{2}{\pi}\dfrac{\frac{\cos x}{\sinh y}}{1+\left(\frac{\sin x}{\sinh y}\right)^2}$, so $\dfrac{\partial T}{\partial x}\left(\dfrac{\pi}{2},1\right) = 0$. This says that the rate of change of temperature in

the x-direction, with y held constant at 1, is $0°$ per unit change in x when $x = \dfrac{\pi}{2}$.

$\dfrac{\partial T}{\partial y} = \dfrac{\partial}{\partial y}\left(\dfrac{2}{\pi}\tan^{-1}\dfrac{\sin x}{\sinh y}\right) = \dfrac{2}{\pi}\dfrac{(\sin x)(-\operatorname{csch} y\coth y)}{1+\left(\frac{\sin x}{\sinh y}\right)^2}$, so $\dfrac{\partial T}{\partial y}\left(\dfrac{\pi}{2},1\right) = -\dfrac{2}{\pi}\dfrac{(\operatorname{csch} 1)(\coth 1)}{1+\operatorname{csch}^2 1} = -\dfrac{4e}{\pi\left(e^2+1\right)}$

after simplification. This says that the rate of change of temperature in the y-direction, with x held constant at $\dfrac{\pi}{2}$, is

$\left(-\dfrac{4e}{\pi\left(e^2+1\right)}\right)^{\circ}$ per unit change in y when $y = 1$.

91. $f_{xy} = \dfrac{\partial}{\partial y}\left[e^{2x}\left(2\cos xy - y\sin xy\right)\right] = e^{2x}\left(-2x\sin xy - \sin xy - xy\cos xy\right) = -e^{2x}\left(2x\sin xy + \sin xy + xy\cos xy\right)$

and $f_{yx} = \dfrac{\partial}{\partial x}\left[-ye^{2x}\sin xy\right] = -\left[2ye^{2x}\sin xy + ye^{2x}\left(y\cos xy\right)\right] = -ye^{2x}\left(2\sin xy + y\cos xy\right)$. Evidently,

$f_{yx} \neq f_{xy}$ for all (x,y). But if f has continuous second-order partial derivatives, then f_{xy} must be equal to f_{yx}, according

to Theorem 1. We conclude, therefore, that there is no function satisfying the stated condition.

93. True. This follows immediately from the definition if we put $h = x - a$.

95. True. This is the condition for downward concavity.

13.4   Concept Questions   ET 12.4

1. See page 1075 (1071 in ET).

3. a. See pages 1079 and 1080 (1075 and 1076 in ET).   **b.** See page 1080 (1076 in ET).   **c.** See page 1080 (1076 in ET).

13.4   Differentials   ET 12.4

1. a. $\Delta z = \left[2\left(2.01\right)^2 + 3\left(-0.98\right)^2\right] - \left[2\left(2\right)^2 + 3\left(-1\right)^2\right] = -0.0386$

   b. $\dfrac{\partial z}{\partial x} = \dfrac{\partial}{\partial x}\left(2x^2 + 3y^2\right) = 4x$ and $\dfrac{\partial z}{\partial y} = 6y$, so with $x = 2$, $y = -1$, $dx = 0.01$, and $dy = 0.02$, we have

      $dz = 4\left(2\right)\left(0.01\right) + 6\left(-1\right)\left(0.02\right) = -0.04$.

   c. dz is a good approximation of Δz.

3. $dz = \dfrac{\partial f}{\partial x}\Delta x + \dfrac{\partial f}{\partial y}\Delta y = \dfrac{\partial}{\partial x}\left(3x^2y^3\right)dx + \dfrac{\partial}{\partial y}\left(3x^2y^3\right)dy = 6xy^3\,dx + 9x^2y^2\,dy$

5. $dz = \dfrac{\partial}{\partial x}\left(\dfrac{x+y}{x-y}\right)dx + \dfrac{\partial}{\partial y}\left(\dfrac{x+y}{x-y}\right)dy = -\dfrac{2y}{(x-y)^2}\,dx + \dfrac{2x}{(x-y)^2}\,dy$

7. $\dfrac{\partial z}{\partial x} = \dfrac{\partial}{\partial x}\left(2x^2y + 3y^3\right)^3 = 3\left(2x^2y + 3y^3\right)^2\left(4xy\right) = 12xy^3\left(2x^2 + 3y^2\right)^2$ and

   $\dfrac{\partial z}{\partial y} = \dfrac{\partial}{\partial y}\left(2x^2y + 3y^3\right)^3 = 3\left(2x^2y + 3y^3\right)^2\left(2x^2 + 9y^2\right) = 3y^2\left(2x^2 + 9y^2\right)\left(2x^2 + 3y^2\right)^2$, so

   $dz = 12xy^3\left(2x^2 + 3y^2\right)^2\,dx + 3y^2\left(2x^2 + 9y^2\right)\left(2x^2 + 3y^2\right)^2\,dy$.

9. $dw = \dfrac{\partial}{\partial x}\left(ye^{x^2-y^2}\right)dx + \dfrac{\partial}{\partial y}\left(ye^{x^2-y^2}\right)dy = 2xye^{x^2-y^2}\,dx + \left(1 - 2y^2\right)e^{x^2-y^2}\,dy$

11. $\dfrac{\partial w}{\partial x} = \dfrac{\partial}{\partial x}x^2\ln\left(x^2 + y^2\right) = 2x\ln\left(x^2 + y^2\right) + \dfrac{2x^3}{x^2 + y^2}$ and $\dfrac{\partial w}{\partial y} = \dfrac{2x^2y}{x^2 + y^2}$, so

   $dw = \left[2x\ln\left(x^2 + y^2\right) + \dfrac{2x^3}{x^2 + y^2}\right]dx + \dfrac{2x^2y}{x^2 + y^2}\,dy$.

13. $dz = \frac{\partial}{\partial x}\left(e^{2x}\cos 3y\right)dx + \frac{\partial}{\partial y}\left(e^{2x}\cos 3y\right)dy = 2e^{2x}\cos 3y\,dx - 3e^{2x}\sin 3y\,dy$

15. $dw = \frac{\partial}{\partial x}\left(x^2 + xy + z^2\right)dx + \frac{\partial}{\partial y}\left(x^2 + xy + z^2\right)dy + \frac{\partial}{\partial z}\left(x^2 + xy + z^2\right)dz = (2x + y)\,dx + x\,dy + 2z\,dz$

17. $dw = \frac{\partial}{\partial x}\left(x^2 e^{-yz}\right)dx + \frac{\partial}{\partial y}\left(x^2 e^{-yz}\right)dy + \frac{\partial}{\partial z}\left(x^2 e^{-yz}\right)dz = 2xe^{-yz}\,dx - x^2 z e^{-yz}\,dy - x^2 y e^{-yz}\,dz$

19. $\frac{\partial w}{\partial x} = \frac{\partial}{\partial x}\left(x^2 e^y + y\ln z\right) = 2xe^y$, $\frac{\partial w}{\partial y} = x^2 e^y + \ln z$, and $\frac{\partial w}{\partial z} = \frac{y}{z}$, so $dw = 2xe^y\,dx + \left(x^2 e^y + \ln z\right)dy + \frac{y}{z}\,dz$.

21. $\frac{\partial f}{\partial x} = \frac{\partial}{\partial x}\left(x^4 - 3x^2 y^2 + y^3 - 2y + 4\right) = 4x^3 - 6xy^2$ and $\frac{\partial f}{\partial y} = \frac{\partial}{\partial y}\left(x^4 - 3x^2 y^2 + y^3 - 2y + 4\right) = -6x^2 y + 3y^2 - 2$,

so $df = \frac{\partial f}{\partial x}\,dx + \frac{\partial f}{\partial y}\,dy = \left(4x^3 - 6xy^2\right)dx + \left(-6x^2 y + 3y^2 - 2\right)dy$. Then with $x = 2$, $y = 2$, $dx = \Delta x = -0.02$, and

$dy = \Delta y = 0.01$, we find $\Delta f \approx df = \left[4\left(2^3\right) - 6\,(2)\left(2^2\right)\right](-0.02) + \left[-6\left(2^2\right)(2) + 3\left(2^2\right) - 2\right](0.01) = -0.06$.

23. $\frac{\partial f}{\partial x} = \frac{\partial}{\partial x}\left[\ln(2x - y) + e^{2xz}\right] = \frac{2}{2x - y} + 2ze^{2xz}$, $\frac{\partial f}{\partial y} = -\frac{1}{2x - y}$, and $\frac{\partial f}{\partial z} = 2xe^{2xz}$, so

$df = \left(\frac{2}{2x - y} + 2ze^{2xz}\right)dx - \frac{1}{2x - y}\,dy + 2xe^{2xz}\,dz$. Then with $x = 2$, $y = 3$, $z = 0$, $dx = 0.01$, $dy = -0.03$, and

$dz = 0.04$, we find $\Delta f \approx df = 2\,(0.01) - (-0.03) + 4\,(0.04) = 0.21$.

25. Let x, y, and z denote the dimensions of the box. Then its volume is $V = xyz$, so

$dV = \frac{\partial}{\partial x}(xyz)\,dx + \frac{\partial}{\partial y}(xyz)\,dy + \frac{\partial}{\partial z}(xyz)\,dz = yz\,dx + xz\,dy + xy\,dz$. Then with $x = 30$, $y = 40$, $z = 60$, and

$dx = dy = dz = \pm 0.2$, we find

$|\Delta V| \approx |dV| = |(40)(60)(\pm 0.2) + (30)(60)(\pm 0.2) + (30)(40)(\pm 0.2)|$

$\leq |(40)(60)(\pm 0.2)| + |(30)(60)(\pm 0.2)| + |(30)(40)(\pm 0.2)| = 1080\text{ in}^3$

so the maximum error is 1080 in^3.

27. The area of the land is $A = f(a, b, \theta) = \frac{1}{2}ab\sin\theta$, so

$dA = \frac{\partial}{\partial a}\left(\frac{1}{2}ab\sin\theta\right)da + \frac{\partial}{\partial b}\left(\frac{1}{2}ab\sin\theta\right)db + \frac{\partial}{\partial \theta}\left(\frac{1}{2}ab\sin\theta\right)d\theta$

$= \frac{1}{2}b\sin\theta\,da + \frac{1}{2}a\sin\theta\,db + \frac{1}{2}ab\cos\theta\,d\theta$

If $a = 100$, $b = 80$, $\theta = \frac{\pi}{3}$, $da = db = \pm 0.3$, and $d\theta = \pm\frac{\pi}{180}$, then

$|\Delta A| \approx |dA| = \left|\frac{1}{2}(80)\left(\frac{\sqrt{3}}{2}\right)(\pm 0.3) + \frac{1}{2}(100)\left(\frac{\sqrt{3}}{2}\right)(\pm 0.3) + \frac{1}{2}(100)(80)\left(\frac{1}{2}\right)\left(\pm\frac{\pi}{180}\right)\right|$

$\leq \left|\frac{1}{2}(80)\left(\frac{\sqrt{3}}{2}\right)(\pm 0.3)\right| + \left|\frac{1}{2}(100)\left(\frac{\sqrt{3}}{2}\right)(\pm 0.3)\right| + \left|\frac{1}{2}(100)(80)\left(\frac{1}{2}\right)\left(\pm\frac{\pi}{180}\right)\right| = 6\sqrt{3} + 7.5\sqrt{3} + \frac{100\pi}{9} \approx 58.3$

so the maximum error is 58.3 ft^2.

29. $dP = \frac{\partial}{\partial V}\left(\frac{8.314T}{V}\right)dV + \frac{\partial}{\partial T}\left(\frac{8.314T}{V}\right)dT = -\frac{8.314T}{V^2}\,dV + \frac{8.314}{V}\,dT$. With $V = 20$, $T = 300$, $dV = 0.2$,

and $dT = -5$, we find $\Delta P \approx dP = -\frac{8.314\,(300)}{20^2}\,(0.2) + \frac{8.314}{20}\,(-5) = -3.3256$, so the pressure decreases by

approximately 3.3256 Pa.

31. $\dfrac{\partial S}{\partial W} = \dfrac{\partial}{\partial W}\left(0.1091\,W^{0.425}H^{0.725}\right) = 0.0463675\left(\dfrac{H^{0.725}}{W^{0.575}}\right)$ and $\dfrac{\partial S}{\partial H} = 0.0790975\left(\dfrac{W^{0.425}}{H^{0.275}}\right)$, so

$dS = \dfrac{\partial S}{\partial W}\,dW + \dfrac{\partial S}{\partial H}\,dH = 0.0463675\left(\dfrac{H^{0.725}}{W^{0.575}}\right)dW + 0.0790975\left(\dfrac{W^{0.425}}{H^{0.275}}\right)dH \Rightarrow$

$\left|\dfrac{\Delta S}{S}\right| \approx \left|\dfrac{dS}{S}\right| = \left|0.425\left(\dfrac{dW}{W}\right) + 0.725\left(\dfrac{dH}{H}\right)\right| \leq 0.425\left|\dfrac{dW}{W}\right| + 0.725\left|\dfrac{dH}{H}\right| = 0.425\,(0.03) + 0.725\,(0.02)$

$\qquad = 0.02725$

so the maximum error is approximately 2.73%.

33. $dF = \dfrac{\partial}{\partial R}\left(\dfrac{\pi P R^4}{8kL}\right)dR + \dfrac{\partial}{\partial L}\left(\dfrac{\pi P R^4}{8kL}\right)dL = \dfrac{\pi P R^3}{2kL}\,dR - \dfrac{\pi P R^4}{8kL^2}\,dL$, so $\dfrac{\Delta F}{F} \approx \dfrac{dF}{F} = 4\dfrac{dR}{R} - \dfrac{dL}{L}$. With

$\left|\dfrac{dL}{L}\right| \leq 0.01$ and $\left|\dfrac{dR}{R}\right| \leq 0.02$, we find $\left|\dfrac{\Delta F}{F}\right| \approx \left|\dfrac{dF}{F}\right| = \left|4\dfrac{dR}{R} - \dfrac{dL}{L}\right| \leq 4\left|\dfrac{dR}{R}\right| + \left|\dfrac{dL}{L}\right| = 4\,(0.02) + (0.01) = 0.09$,

and so the relative error is 9%.

35. $\dfrac{\partial T}{\partial L} = \dfrac{\partial}{\partial L}\left(2\pi L^{1/2}g^{-1/2}\right) = \dfrac{\pi}{L^{1/2}g^{1/2}}$ and $\dfrac{\partial T}{\partial g} = -\dfrac{\pi L^{1/2}}{g^{3/2}}$, so with

$\dfrac{dL}{L} = \dfrac{4.05 - 4}{4.05} = \dfrac{0.05}{4.05}$ and $\dfrac{dg}{g} = \dfrac{32.2 - 32}{32.2} = \dfrac{0.2}{32.2}$, we have

$\dfrac{\Delta T}{T} \approx \dfrac{dT}{T} = \dfrac{1}{2\pi L^{1/2}g^{-1/2}}\left(\dfrac{\pi}{L^{1/2}g^{1/2}}\,dL - \dfrac{\pi L^{1/2}}{g^{3/2}}\,dg\right) = \dfrac{1}{2}\dfrac{dL}{L} - \dfrac{1}{2}\dfrac{dg}{g} = \dfrac{1}{2}\left(\dfrac{0.05}{4.05}\right) - \dfrac{1}{2}\left(\dfrac{0.2}{32.2}\right) \approx 0.0031$. The

error in the measurement of T is approximately 0.3%.

37. Using the results of Exercise 13.3.71 (12.3.71 in ET), we have $\dfrac{\partial R}{\partial R_1} = \left(\dfrac{R}{R_1}\right)^2$, $\dfrac{\partial R}{\partial R_2} = \left(\dfrac{R}{R_2}\right)^2$, and $\dfrac{\partial R}{\partial R_3} = \left(\dfrac{R}{R_3}\right)^2$, so

$|\Delta R| \approx |dR| = \left|\dfrac{\partial R}{\partial R_1}\,dR_1 + \dfrac{\partial R}{\partial R_2}\,dR_2 + \dfrac{\partial R}{\partial R_3}\,dR_3\right| \leq \left|\dfrac{\partial R}{\partial R_1}\right||dR_1| + \left|\dfrac{\partial R}{\partial R_2}\right||dR_2| + \left|\dfrac{\partial R}{\partial R_3}\right||dR_3|$

$\qquad = \left(\dfrac{R}{R_1}\right)^2|dR_1| + \left(\dfrac{R}{R_2}\right)^2|dR_2| + \left(\dfrac{R}{R_3}\right)^2|dR_3|$

Now $R_1 = 20$, $R_2 = 30$, $R_3 = 50$, $|dR_1| \leq 0.5$, $|dR_2| \leq 0.5$, and $|dR_3| \leq 0.5$, so $\dfrac{1}{R} = \dfrac{1}{20} + \dfrac{1}{30} + \dfrac{1}{50} = \dfrac{31}{300} \Rightarrow$

$R = \dfrac{300}{31}$, and $|\Delta R| \approx |dR| \leq \left(\dfrac{300}{31}\right)^2\left(\dfrac{0.5}{20^2} + \dfrac{0.5}{30^2} + \dfrac{0.5}{50^2}\right) \approx 0.1878$, so the maximum error in the calculated value of R is about 0.19 Ω.

39. $\dfrac{\partial H}{\partial W} = \dfrac{\partial}{\partial W}\left(\dfrac{WL^2}{8a}\right) = \dfrac{L^2}{8a}$, $\dfrac{\partial H}{\partial L} = \dfrac{WL}{4a}$, and $\dfrac{\partial H}{\partial a} = -\dfrac{WL^2}{8a^2}$, so with $\left|\dfrac{dW}{W}\right| = 0.01$, $\left|\dfrac{dL}{L}\right| = 0.02$, and $\left|\dfrac{da}{a}\right| = 0.02$, we

have $\left|\dfrac{\Delta H}{H}\right| \approx \left|\dfrac{dH}{H}\right| = \dfrac{8a}{WL^2}\left|\dfrac{L^2}{8a}\,dW + \dfrac{WL}{4a}\,dL - \dfrac{WL^2}{8a^2}\,da\right| \leq \left|\dfrac{dW}{W}\right| + 2\left|\dfrac{dL}{L}\right| + \left|\dfrac{da}{a}\right| = 0.01 + 2\,(0.02) + 0.02 = 0.07$,

so the maximum error in calculating H is no greater than 7%.

41. We write $z = f(x, y) = x^2 - y^2$ and let (x, y) be any point on the plane. Then

$\Delta z = f(x + \Delta x, y + \Delta y) - f(x, y) = (x + \Delta x)^2 - (y + \Delta y)^2 - \left(x^2 - y^2\right) = 2x\,\Delta x - 2y\,\Delta y + (\Delta x)^2 - (\Delta y)^2$.

Since $f_x = 2x$ and $f_y = -2y$, we can write $\Delta z = f_x(x, y)\,\Delta x + f_y(x, y)\,\Delta y + \varepsilon_1\,\Delta x + \varepsilon_2\,\Delta y$, where $\varepsilon_1 = \Delta x$ and

$\varepsilon_2 = \Delta y$. Since $\varepsilon_1 \to 0$ and $\varepsilon_2 \to 0$ as $(\Delta x, \Delta y) \to (0, 0)$, it follows that f is differentiable at (x, y). But (x, y) is any

point on the plane, and so f is differentiable everywhere.

43. $f_x(0,0) = \lim\limits_{\Delta x \to 0} \dfrac{f(0 + \Delta x, 0) - f(0,0)}{\Delta x} = \lim\limits_{\Delta x \to 0} \dfrac{0 - 0}{\Delta x} = 0$ and

$f_y(0,0) = \lim\limits_{\Delta y \to 0} \dfrac{f(0, 0 + \Delta y) - f(0,0)}{\Delta y} = \lim\limits_{\Delta y \to 0} \dfrac{0}{\Delta y} = 0$, so both $f_x(0,0)$ and $f_y(0,0)$ exist. Next, recall from

Example 13.2.2 that $\lim\limits_{(x,y) \to (0,0)} f(x,y) = \lim\limits_{(x,y) \to (0,0)} \dfrac{xy}{x^2 + y^2}$ does not exist. This shows that f is not continuous at

$(0,0)$, and so by Theorem 3, f is not differentiable at $(0,0)$.

45. True. If f is differentiable at (a,b), then f is continuous at (a,b) and the statement is just an affirmation of this fact.

47. False. Since $\lim\limits_{(x,y) \to (0,0)} f(x,y) = 0 \neq 1 = f(0,0)$, we see that f is not continuous at $(0,0)$ and by Theorem 3, f is not

differentiable at $(0,0)$.

13.5   Concept Questions   ET 12.5

1. See page 1085 (1081 in ET).   **3.** See page 1089 (1085 in ET).

13.5   The Chain Rule   ET 12.5

1. $\dfrac{dw}{dt} = \dfrac{\partial w}{\partial x}\dfrac{dx}{dt} + \dfrac{\partial w}{\partial y}\dfrac{dy}{dt} = (2x)\,(2t) + (-2y)\left(3t^2 + 1\right) = 4xt - 6yt^2 - 2y$

3. $\dfrac{dw}{dt} = \dfrac{\partial w}{\partial r}\dfrac{dr}{dt} + \dfrac{\partial w}{\partial s}\dfrac{ds}{dt} = (\cos s + s\cos r)\left(-2e^{-2t}\right) + (-r\sin s + \sin r)\left(3t^2 - 2\right)$

$= -2\,(\cos s + s\cos r)\,e^{-2t} + (\sin r - r\sin s)\left(3t^2 - 2\right)$

5. $\dfrac{dw}{dt} = \dfrac{\partial w}{\partial x}\dfrac{dx}{dt} + \dfrac{\partial w}{\partial y}\dfrac{dy}{dt} + \dfrac{\partial w}{\partial z}\dfrac{dz}{dt} = \left(6x^2 y^2 z\right)(1) + \left(4x^3 yz\right)(-\sin t) + \left(2x^3 y^2\right)(\sin t + t\cos t)$

$= 2x^2 y\left[3yz - 2xz\sin t + xy\,(\sin t + t\cos t)\right]$

7. $\dfrac{dw}{dt} = \dfrac{\partial w}{\partial x}\dfrac{dx}{dt} + \dfrac{\partial w}{\partial y}\dfrac{dy}{dt} + \dfrac{\partial w}{\partial z}\dfrac{dz}{dt} = \left[\dfrac{z}{1+(xz)^2}\right](1) + \left(-\dfrac{z}{y^2}\right)(2t) + \left[\dfrac{x}{1+(xz)^2} + \dfrac{1}{y}\right]\cosh t$

$= \dfrac{-2tz + y^2 z - 2tx^2 z^3 + \cosh t\left(y + xy^2 + x^2 yz^2\right)}{y^2\left(1 + x^2 z^2\right)}$

9. $\dfrac{\partial w}{\partial u} = \dfrac{\partial w}{\partial x}\dfrac{\partial x}{\partial u} + \dfrac{\partial w}{\partial y}\dfrac{\partial y}{\partial u} = \left(3x^2\right)(2u) + \left(3y^2\right)(2v) = 6\left(x^2 u + y^2 v\right)$ and

$\dfrac{\partial w}{\partial v} = \dfrac{\partial w}{\partial x}\dfrac{\partial x}{\partial v} + \dfrac{\partial w}{\partial y}\dfrac{\partial y}{\partial v} = \left(3x^2\right)(2v) + \left(3y^2\right)(2u) = 6\left(x^2 v + y^2 u\right)$

11. $\dfrac{\partial w}{\partial u} = \dfrac{\partial w}{\partial x}\dfrac{\partial x}{\partial u} + \dfrac{\partial w}{\partial y}\dfrac{\partial y}{\partial u} = \left(e^x \cos y\right)\left(\dfrac{2u}{u^2 + v^2}\right) + \left(-e^x \sin y\right)\left(\dfrac{1}{2}\dfrac{\sqrt{uv}}{u}\right) = e^x\left(\dfrac{2u\cos y}{u^2 + v^2} - \dfrac{\sqrt{uv}\sin y}{2u}\right)$ and

$\dfrac{\partial w}{\partial v} = \dfrac{\partial w}{\partial x}\dfrac{\partial x}{\partial v} + \dfrac{\partial w}{\partial y}\dfrac{\partial y}{\partial v} = \left(e^x \cos y\right)\left(\dfrac{2v}{u^2 + v^2}\right) + \left(-e^x \sin y\right)\left(\dfrac{1}{2}\dfrac{\sqrt{uv}}{v}\right) = e^x\left(\dfrac{2v\cos y}{u^2 + v^2} - \dfrac{\sqrt{uv}\sin y}{2v}\right)$.

13. $\dfrac{\partial w}{\partial u} = \dfrac{\partial w}{\partial x}\dfrac{\partial x}{\partial u} + \dfrac{\partial w}{\partial y}\dfrac{\partial y}{\partial u} + \dfrac{\partial w}{\partial z}\dfrac{\partial z}{\partial u} = \left(\tan^{-1} yz\right)\left(\dfrac{1}{2\sqrt{u}}\right) + \dfrac{xz}{1+(yz)^2}(0) + \dfrac{xy}{1+(yz)^2}(-v\sin u)$

$= \dfrac{\left(\tan^{-1} yz\right)\sqrt{u}}{2u} - \dfrac{xyv\sin u}{1 + y^2 z^2}$ and

$\dfrac{\partial w}{\partial v} = \dfrac{\partial w}{\partial x}\dfrac{\partial x}{\partial v} + \dfrac{\partial w}{\partial y}\dfrac{\partial y}{\partial v} + \dfrac{\partial w}{\partial z}\dfrac{\partial z}{\partial v} = \left(\tan^{-1} yz\right)(0) + \dfrac{xz}{1+y^2 z^2}\left(-2e^{-2v}\right) + \dfrac{xy}{1+y^2 z^2}(\cos u) = \dfrac{x\left(y\cos u - 2ze^{-2v}\right)}{1 + y^2 z^2}$.

15. $\dfrac{dw}{dt} = \dfrac{\partial w}{\partial r}\dfrac{dr}{dt} + \dfrac{\partial w}{\partial s}\dfrac{ds}{dt} + \dfrac{\partial w}{\partial u}\dfrac{du}{dt} + \dfrac{\partial w}{\partial v}\dfrac{dv}{dt}$ **17.** $\dfrac{\partial w}{\partial t} = \dfrac{\partial w}{\partial x}\dfrac{\partial x}{\partial t} + \dfrac{\partial w}{\partial y}\dfrac{\partial y}{\partial t} + \dfrac{\partial w}{\partial z}\dfrac{\partial z}{\partial t}$

19. $\dfrac{dw}{dt} = \dfrac{\partial w}{\partial x}\dfrac{dx}{dt} + \dfrac{\partial w}{\partial y}\dfrac{dy}{dt} + \dfrac{\partial w}{\partial z}\dfrac{dz}{dt} = (2x + y)(2) + (x + 2y)\left(e^t\right) + 3z^2\,(-2\sin 2t) = 4x + 2y + xe^t + 2ye^t - 6z^2 \sin 2t$

21. $\dfrac{du}{dt} = \dfrac{\partial u}{\partial x}\dfrac{dx}{dt} + \dfrac{\partial u}{\partial y}\dfrac{dy}{dt} = \dfrac{y^2 - x^2}{\left(x^2 + y^2\right)^2} \cdot 2\sec 2t \tan 2t - \dfrac{2xy}{\left(x^2 + y^2\right)^2} \cdot \sec^2 t.$ If $t = 0$, then $x = 1$ and $y = 0$, so

$\dfrac{du}{dt}\bigg|_{t=0} = \dfrac{-1}{1} \cdot 0 - 0 = 0.$

23. $\dfrac{\partial u}{\partial s} = \dfrac{\partial u}{\partial x}\dfrac{\partial x}{\partial s} + \dfrac{\partial u}{\partial y}\dfrac{\partial y}{\partial s} + \dfrac{\partial u}{\partial z}\dfrac{\partial z}{\partial s} = (\csc yz)\, r - xz \csc yz \cot yz\, (2st) - xy \csc yz \cot yz \left(\dfrac{1}{t^2}\right)$

$\quad = \dfrac{\csc yz \left(rt^2 - 2xzst^3 \cot yz - xy \cot yz\right)}{t^2}$ and

$\dfrac{\partial u}{\partial t} = \dfrac{\partial u}{\partial x}\dfrac{\partial x}{\partial t} + \dfrac{\partial u}{\partial y}\dfrac{\partial y}{\partial t} + \dfrac{\partial u}{\partial z}\dfrac{\partial z}{\partial t} = (\csc yz)\,(0) - (xz \csc yz \cot yz)\, s^2 - xy \csc yz \cot yz \left(-\dfrac{2s}{t^3}\right)$

$\quad = \dfrac{sx \csc yz \cot yz \left(2y - st^3 z\right)}{t^3}$

25. $\dfrac{\partial w}{\partial r} = \dfrac{\partial w}{\partial x}\dfrac{\partial x}{\partial r} + \dfrac{\partial w}{\partial y}\dfrac{\partial y}{\partial r} + \dfrac{\partial w}{\partial z}\dfrac{\partial z}{\partial r} = \dfrac{2xy}{z^2}e^{st} + \dfrac{x^2}{z^2}ste^{rt} + \left(-\dfrac{2x^2 y}{z^3}\right)ste^{rst}$ and

$\dfrac{\partial w}{\partial t} = \dfrac{\partial w}{\partial x}\dfrac{\partial x}{\partial t} + \dfrac{\partial w}{\partial y}\dfrac{\partial y}{\partial t} + \dfrac{\partial w}{\partial z}\dfrac{\partial z}{\partial t} = \dfrac{2xy}{z^2}rse^{st} + \dfrac{x^2}{z^2}rse^{rt} + \left(-\dfrac{2x^2 y}{z^3}\right)rse^{rst}.$ If $r = 1$, $s = 2$, and $t = 0$, then $x = 1$,

$y = 2$, and $z = 1$, so $\dfrac{\partial w}{\partial r} = 4 + 0 + (-4)\,(0) = 4$ and $\dfrac{\partial w}{\partial t} = 4\,(2) + 1\,(2) - 4\,(2) = 2.$

27. Differentiating each equation in the system with respect to x, we obtain $\dfrac{\partial}{\partial x}(x) = \dfrac{\partial}{\partial x}\left(u^2 + v^2\right) = 2u\dfrac{\partial u}{\partial x} + 2v\dfrac{\partial v}{\partial x}$

and $\dfrac{\partial}{\partial x}(y) = \dfrac{\partial}{\partial x}\left(u^2 - v^2\right) = 2u\dfrac{\partial u}{\partial x} + 2v\dfrac{\partial v}{\partial x}.$ Since y is an independent variable, $\dfrac{\partial}{\partial x}(y) = 0$, so we have

$2u\dfrac{\partial u}{\partial x} + 2v\dfrac{\partial v}{\partial x} = 1$ and $2u\dfrac{\partial u}{\partial x} - 2v\dfrac{\partial v}{\partial x} = 0.$ Using Cramer's Rule, we obtain $\dfrac{\partial u}{\partial x} = \dfrac{\begin{vmatrix} 1 & 2v \\ 0 & -2v \end{vmatrix}}{\begin{vmatrix} 2u & 2v \\ 2u & -2v \end{vmatrix}} = \dfrac{-2v}{-8uv} = \dfrac{1}{4u}$ and

$\dfrac{\partial v}{\partial x} = \dfrac{\begin{vmatrix} 2u & 1 \\ 2u & 0 \end{vmatrix}}{\begin{vmatrix} 2u & 2v \\ 2u & -2v \end{vmatrix}} = \dfrac{-2u}{-8uv} = \dfrac{1}{4v}.$ Similarly, we obtain $\dfrac{\partial u}{\partial y} = \dfrac{1}{4u}$ and $\dfrac{\partial v}{\partial y} = -\dfrac{1}{4v}.$

29. Here $F(x, y) = x^3 - 2xy + y^3 - 4 = 0$, so $\dfrac{dy}{dx} = -\dfrac{F_x}{F_y} = -\dfrac{3x^2 - 2y}{-2x + 3y^2} = \dfrac{3x^2 - 2y}{2x - 3y^2}.$

31. Here $F(x, y) = 2x^2 + 3x^{1/2}y^{1/2} - 2y - 4 = 0$, so $\dfrac{dy}{dx} = -\dfrac{F_x}{F_y} = -\dfrac{4x + \frac{3}{2}x^{-1/2}y^{1/2}}{\frac{3}{2}x^{1/2}y^{-1/2} - 2} = \dfrac{8x\sqrt{xy} + 3y}{4\sqrt{xy} - 3x}.$

33. Here $F(x, y, z) = x^2 + xy - x^2 z + yz^2 = 0$, so $\dfrac{\partial z}{\partial x} = -\dfrac{F_x}{F_z} = -\dfrac{2x + y - 2xz}{-x^2 + 2yz} = \dfrac{2x + y - 2xz}{x^2 - 2yz}$ and

$\dfrac{\partial z}{\partial y} = -\dfrac{F_y}{F_z} = -\dfrac{x + z^2}{-x^2 + 2yz} = \dfrac{x + z^2}{x^2 - 2yz}.$

35. Here $F(x, y, z) = xe^y + ye^{xz} + x^2e^{x/y} - 10 = 0$, so

$$\frac{\partial z}{\partial x} = -\frac{F_x}{F_z} = -\frac{e^y + yze^{xz} + \left(2x + \frac{x^2}{y}\right)e^{x/y}}{xye^{xz}} = -\frac{ye^y + y^2ze^{xz} + x(2y + x)e^{x/y}}{xy^2e^{xz}}$$ and

$$\frac{\partial z}{\partial y} = -\frac{F_y}{F_z} = -\frac{xe^y + e^{xz} + x^2\left(-\frac{x}{y^2}\right)e^{x/y}}{xye^{xz}} = \frac{x^3e^{x/y} - xy^2e^y - y^2e^{xz}}{xy^3e^{xz}}.$$

37. We write $F(x, y) = x^3 + y^3 - 3axy = 0$, so $\dfrac{dy}{dx} = -\dfrac{F_x}{F_y} = -\dfrac{3x^2 - 3ay}{3y^2 - 3ax} = -\dfrac{x^2 - ay}{y^2 - ax}.$

39. Denote the radius of the base and the height of the cone by r and h,

respectively. Then the area of its lateral surface is $S(r, h) = \pi r\sqrt{r^2 + h^2}$,

so $\dfrac{dS}{dt} = \dfrac{\partial S}{\partial r}\dfrac{dr}{dt} + \dfrac{\partial S}{\partial h}\dfrac{dh}{dt} = \dfrac{\pi\left(2r^2 + h^2\right)}{\sqrt{r^2 + h^2}}\dfrac{dr}{dt} + \dfrac{\pi rh}{\sqrt{r^2 + h^2}}\dfrac{dh}{dt}$. We are

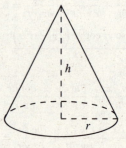

given that $r = 10$, $h = 18$, $\frac{dr}{dt} = -0.2$, and $\frac{dh}{dt} = 0.1$. Therefore,

$$\frac{dS}{dt} = \frac{\pi\left(2 \cdot 10^2 + 18^2\right)}{\sqrt{10^2 + 18^2}}(-0.2) + \frac{\pi \cdot 10 \cdot 18}{\sqrt{10^2 + 18^2}}(0.1) \approx -13.2, \text{ and so}$$

the lateral surface area is decreasing at a rate of about 13.2 in^2/min.

41. The distance between the two cars is $D = f(x, y) = \sqrt{x^2 + y^2}$, so

$$\frac{dD}{dt} = \frac{\partial D}{\partial x}\frac{dx}{dt} + \frac{\partial D}{\partial y}\frac{dy}{dt} = \frac{x}{\sqrt{x^2 + y^2}}\frac{dx}{dt} + \frac{y}{\sqrt{x^2 + y^2}}\frac{dy}{dt}. \text{ With}$$

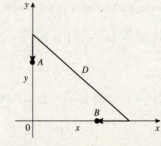

$x = 0.3$, $y = 0.4$, $\dfrac{dx}{dt} = -30$, and $\dfrac{dy}{dt} = -45$, we find

$$\frac{dD}{dt} = \frac{0.3(-30)}{\sqrt{(0.3)^2 + (0.4)^2}} + \frac{0.4(-45)}{\sqrt{(0.3)^2 + (0.4)^2}} = -54, \text{ so the distance}$$

between the two cars is decreasing at a rate of 54 mi/h.

43. We differentiate both sides of the equation $\dfrac{1}{R} = \dfrac{1}{R_1} + \dfrac{1}{R_2} + \cdots + \dfrac{1}{R_n}$ with respect to R_k. Thus,

$$\frac{\partial}{\partial R_k}\left(R^{-1}\right) = \frac{\partial}{\partial R_k}\left(R_1^{-1}\right) + \frac{\partial}{\partial R_k}\left(R_2^{-1}\right) + \cdots + \frac{\partial}{\partial R_k}\left(R_k^{-1}\right) + \cdots + \frac{\partial}{\partial R_k}\left(R_n^{-1}\right) \Leftrightarrow \left(-R^{-2}\right)\frac{\partial R}{\partial R_k} = -R_k^{-2}, \text{ and so}$$

$$\frac{\partial R}{\partial R_k} = \left(\frac{R}{R_k}\right)^2.$$

45. Parametric equations giving the position of the insect are $x = 2t$, $y = t^2$,

$z = t^3$. The rate of change in temperature experienced by the insect is

$$\frac{dT}{dt} = \frac{\partial T}{\partial x}\frac{dx}{dt} + \frac{\partial T}{\partial y}\frac{dy}{dt} + \frac{\partial T}{\partial z}\frac{dz}{dt} = \frac{-120x}{\left(1 + x^2 + y^2 + z^2\right)^2}(2) + \frac{-120y}{\left(1 + x^2 + y^2 + z^2\right)^2}(2t) + \frac{-120z}{\left(1 + x^2 + y^2 + z^2\right)^2}\left(3t^2\right).$$

When $t = 2$, $x = 4$, $y = 4$, and $z = 8$, this is equal to

$$\frac{dT}{dt}\bigg|_{t=2} = \frac{-120(4)}{(1 + 16 + 16 + 64)^2}(2) + \frac{-120(4)}{(1 + 16 + 16 + 64)^2}(4) + \frac{-120(8)}{(1 + 16 + 16 + 64)^2}(12) \approx -1.53. \text{ The insect}$$

experiences a temperature drop of approximately 1.53°F/s.

47. $\dfrac{\partial u}{\partial r} = \dfrac{\partial u}{\partial x}\dfrac{\partial x}{\partial r} + \dfrac{\partial u}{\partial y}\dfrac{\partial y}{\partial r} = \dfrac{\partial u}{\partial x}e^r\cos\theta + \dfrac{\partial u}{\partial y}e^r\sin\theta = e^r\left(\dfrac{\partial u}{\partial x}\cos\theta + \dfrac{\partial u}{\partial y}\sin\theta\right)$ and

$\dfrac{\partial u}{\partial \theta} = \dfrac{\partial u}{\partial x}\dfrac{\partial x}{\partial \theta} + \dfrac{\partial u}{\partial y}\dfrac{\partial y}{\partial \theta} = \dfrac{\partial u}{\partial x}(-e^r\sin\theta) + \dfrac{\partial u}{\partial y}e^r\cos\theta = e^r\left(-\dfrac{\partial u}{\partial x}\sin\theta + \dfrac{\partial u}{\partial y}\cos\theta\right)$, so

$$e^{-2r}\left[\left(\dfrac{\partial u}{\partial r}\right)^2 + \left(\dfrac{\partial u}{\partial \theta}\right)^2\right] = e^{-2r}\left[e^{2r}\left(\dfrac{\partial u}{\partial x}\cos\theta + \dfrac{\partial u}{\partial y}\sin\theta\right)^2 + e^{2r}\left(\dfrac{\partial u}{\partial y}\cos\theta - \dfrac{\partial u}{\partial x}\sin\theta\right)^2\right]$$

$$= \left(\dfrac{\partial u}{\partial x}\right)^2\left(\cos^2\theta + \sin^2\theta\right) + \left(\dfrac{\partial u}{\partial y}\right)^2\left(\cos^2\theta + \sin^2\theta\right) = \left(\dfrac{\partial u}{\partial x}\right)^2 + \left(\dfrac{\partial u}{\partial y}\right)^2$$

49. $\dfrac{\partial z}{\partial u} = \dfrac{\partial z}{\partial x}\dfrac{\partial x}{\partial u} + \dfrac{\partial z}{\partial y}\dfrac{\partial y}{\partial u} = \dfrac{\partial z}{\partial x}(1) + \dfrac{\partial z}{\partial y}(-1) = \dfrac{\partial z}{\partial x} - \dfrac{\partial z}{\partial y}$ and $\dfrac{\partial z}{\partial v} = \dfrac{\partial z}{\partial x}\dfrac{\partial x}{\partial v} + \dfrac{\partial z}{\partial y}\dfrac{\partial y}{\partial v} = \dfrac{\partial z}{\partial x}(-1) + \dfrac{\partial z}{\partial y}(1) = -\dfrac{\partial z}{\partial x} + \dfrac{\partial z}{\partial y}$,

so $\dfrac{\partial z}{\partial u} + \dfrac{\partial z}{\partial v} = \dfrac{\partial z}{\partial x} - \dfrac{\partial z}{\partial y} - \dfrac{\partial z}{\partial x} + \dfrac{\partial z}{\partial y} = 0$.

51. Let $u = x + at$ and $v = x - at$. Then $z = f(x + at) + g(x - at) = f(u) + g(v)$,

so $\dfrac{\partial z}{\partial t} = \dfrac{\partial z}{\partial u}\dfrac{\partial u}{\partial t} + \dfrac{\partial z}{\partial v}\dfrac{\partial v}{\partial t} = f'(u)(a) + g'(v)(-a) = a[f'(u) - g'(v)] \Rightarrow$

$\dfrac{\partial^2 z}{\partial t^2} = a\dfrac{\partial}{\partial t}[f'(u) - g'(v)] = a[f''(u)(a) - g''(v)(-a)] = a^2[f''(u) + g''(v)]$

and $\dfrac{\partial z}{\partial x} = \dfrac{\partial z}{\partial u}\dfrac{\partial u}{\partial x} + \dfrac{\partial z}{\partial v}\dfrac{\partial v}{\partial x} = f'(u)(1) + g'(v)(1) = f'(u) + g'(v) \Rightarrow$

$\dfrac{\partial^2 z}{\partial x^2} = \dfrac{\partial}{\partial x}[f'(u) + g'(v)] = f''(u)(1) + g''(v)(1) = f''(u) + g''(v)$. Therefore,

$\dfrac{\partial^2 z}{\partial t^2} = a^2[f''(u) + g''(v)] = a^2\dfrac{\partial^2 z}{\partial x^2}$.

53. $\dfrac{\partial z}{\partial x} = \dfrac{\partial z}{\partial u}\dfrac{\partial u}{\partial x} + \dfrac{\partial z}{\partial v}\dfrac{\partial v}{\partial x} \Rightarrow$

$\dfrac{\partial^2 z}{\partial x^2} = \dfrac{\partial}{\partial x}\left(\dfrac{\partial z}{\partial u}\dfrac{\partial u}{\partial x} + \dfrac{\partial z}{\partial v}\dfrac{\partial v}{\partial x}\right) = \dfrac{\partial}{\partial x}\left(\dfrac{\partial z}{\partial u}\dfrac{\partial u}{\partial x}\right) + \dfrac{\partial}{\partial x}\left(\dfrac{\partial z}{\partial v}\dfrac{\partial v}{\partial x}\right)$

$= \dfrac{\partial}{\partial x}\left(\dfrac{\partial z}{\partial u}\right)\cdot\dfrac{\partial u}{\partial x} + \dfrac{\partial z}{\partial u}\cdot\dfrac{\partial}{\partial x}\left(\dfrac{\partial u}{\partial x}\right) + \dfrac{\partial}{\partial x}\left(\dfrac{\partial z}{\partial v}\right)\cdot\dfrac{\partial v}{\partial x} + \dfrac{\partial z}{\partial v}\cdot\dfrac{\partial}{\partial x}\left(\dfrac{\partial v}{\partial x}\right)$

$= \left(\dfrac{\partial^2 z}{\partial u^2}\cdot\dfrac{\partial u}{\partial x} + \dfrac{\partial^2 z}{\partial v\,\partial u}\dfrac{\partial v}{\partial x}\right)\dfrac{\partial u}{\partial x} + \dfrac{\partial z}{\partial u}\dfrac{\partial^2 u}{\partial x^2} + \left(\dfrac{\partial^2 z}{\partial u\,\partial v}\cdot\dfrac{\partial u}{\partial x} + \dfrac{\partial^2 z}{\partial v^2}\cdot\dfrac{\partial v}{\partial x}\right)\dfrac{\partial v}{\partial x} + \dfrac{\partial z}{\partial v}\dfrac{\partial^2 v}{\partial x^2}$

$= \dfrac{\partial^2 z}{\partial u^2}\left(\dfrac{\partial u}{\partial x}\right)^2 + \dfrac{\partial^2 z}{\partial v\,\partial u}\dfrac{\partial v}{\partial x}\dfrac{\partial u}{\partial x} + \dfrac{\partial z}{\partial u}\dfrac{\partial^2 u}{\partial x^2} + \dfrac{\partial^2 z}{\partial u\,\partial v}\dfrac{\partial u}{\partial x}\dfrac{\partial v}{\partial x} + \dfrac{\partial^2 z}{\partial v^2}\left(\dfrac{\partial v}{\partial x}\right)^2 + \dfrac{\partial z}{\partial v}\dfrac{\partial^2 v}{\partial x^2}$

$= \dfrac{\partial^2 z}{\partial u^2}\left(\dfrac{\partial u}{\partial x}\right)^2 + \left(\dfrac{\partial^2 z}{\partial v\,\partial u} + \dfrac{\partial^2 z}{\partial u\,\partial v}\right)\dfrac{\partial u}{\partial x}\dfrac{\partial v}{\partial x} + \dfrac{\partial^2 z}{\partial v^2}\left(\dfrac{\partial v}{\partial x}\right)^2 + \dfrac{\partial z}{\partial u}\dfrac{\partial^2 u}{\partial x^2} + \dfrac{\partial z}{\partial v}\dfrac{\partial^2 v}{\partial x^2}$

55. Differentiating both sides of the equation $f(tx, ty) = t^n f(x, y)$ with respect to t, we obtain

$\dfrac{\partial}{\partial t}[f(tx, ty)] = \dfrac{\partial}{\partial t}[t^n f(x, y)] = nt^{n-1} f(x, y)$. To calculate the expression on the left, let $u = tx$ and $v = ty$. Then

$\dfrac{\partial}{\partial t}f(u, v) = \dfrac{\partial f}{\partial u}\dfrac{\partial u}{\partial t} + \dfrac{\partial f}{\partial v}\dfrac{\partial v}{\partial t} = \dfrac{\partial f}{\partial u}\cdot x + \dfrac{\partial f}{\partial v}\cdot y$. Since the equation holds for all t, we may put $t = 1$, obtaining

$x\dfrac{\partial f}{\partial x} + y\dfrac{\partial f}{\partial y} = nf(x, y)$.

57. $f(tx, ty) = \dfrac{(tx)(ty)^2}{\sqrt{(tx)^2 + (ty)^2}} = \dfrac{t^3 x y^2}{\sqrt{t^2(x^2 + y^2)}} = t^2 \left(\dfrac{x y^2}{\sqrt{x^2 + y^2}}\right) = t^2 f(x, y)$, so f is homogeneous

of degree 2. Next, $\dfrac{\partial f}{\partial x} = \dfrac{\partial}{\partial x}\left(\dfrac{x y^2}{\sqrt{x^2 + y^2}}\right) = \dfrac{y^4}{(x^2 + y^2)^{3/2}}$ and $\dfrac{\partial f}{\partial y} = \dfrac{2x^3 y + x y^3}{(x^2 + y^2)^{3/2}}$, so

$x \dfrac{\partial f}{\partial x} + y \dfrac{\partial f}{\partial y} = \dfrac{x y^4}{(x^2 + y^2)^{3/2}} + \dfrac{2x^3 y^2 + x y^4}{(x^2 + y^2)^{3/2}} = \dfrac{2x y^2}{\sqrt{x^2 + y^2}} = 2f(x, y)$.

59. $f(tx, ty) = e^{(tx)/(ty)} = e^{x/y} = t^0 f(x, y)$, so f is homogeneous of degree 0. Next, $\dfrac{\partial f}{\partial x} = \dfrac{\partial}{\partial x} e^{x/y} = \dfrac{1}{y} e^{x/y}$ and

$\dfrac{\partial f}{\partial y} = -\dfrac{x}{y^2} e^{x/y}$, and so $x \dfrac{\partial f}{\partial x} + y \dfrac{\partial f}{\partial y} = \dfrac{x}{y} e^{x/y} - \dfrac{x}{y} e^{x/y} = 0 = 0 \cdot f(x, y)$.

61. $\dfrac{\partial u}{\partial x} = \dfrac{\partial}{\partial x}\left(\ln\sqrt{x^2 + y^2}\right) = \dfrac{x}{x^2 + y^2}$ and $\dfrac{\partial v}{\partial y} = \dfrac{\partial}{\partial y} \tan^{-1}\dfrac{y}{x} = \dfrac{1/x}{1 + (y/x)^2} = \dfrac{x}{x^2 + y^2}$, so $\dfrac{\partial u}{\partial x} = \dfrac{\partial v}{\partial y}$.

$\dfrac{\partial u}{\partial y} = \dfrac{\partial}{\partial y}\left(\ln\sqrt{x^2 + y^2}\right) = \dfrac{y}{x^2 + y^2}$, $\dfrac{\partial v}{\partial x} = \dfrac{\partial}{\partial x} \tan^{-1}\dfrac{y}{x} = \dfrac{-y/x^2}{1 + (y/x)^2} = -\dfrac{y}{x^2 + y^2}$, so $\dfrac{\partial u}{\partial y} = -\dfrac{\partial v}{\partial x}$.

63. a. $\dfrac{dy}{dx} = -\dfrac{f_x}{f_y}$, so

$\dfrac{d^2 y}{dx^2} = -\dfrac{d}{dx}\left(\dfrac{f_x}{f_y}\right) = -\dfrac{f_y \frac{d}{dx}(f_x) - f_x \frac{d}{dx}(f_y)}{f_y^2} = -\dfrac{f_y\left(f_{xx} + f_{xy}\frac{dy}{dx}\right) - f_x\left(f_{yx} + f_{yy}\frac{dy}{dx}\right)}{f_y^2}$

$= -\dfrac{f_y f_{xx} + f_y f_{xy}(-f_x/f_y) - f_x f_{yx} - f_x f_{yy}(-f_x/f_y)}{f_y^2} = -\dfrac{f_y^2 f_{xx} - f_x f_y f_{xy} - f_y f_x f_{yx} + f_x^2 f_{yy}}{f_y^3}$

$= -\dfrac{f_x^2 f_{yy} - 2 f_x f_y f_{xy} + f_y^2 f_{xx}}{f_y^3}$ since $f_{xy} = f_{yx}$.

b. Here $f(x, y) = x^3 + y^3 - 3xy = 0$. We find $f_x = 3x^2 - 3y$, $f_y = 3y^2 - 3x$, $f_{xx} = 6x$, $f_{xy} = -3$, and $f_{yy} = 6y$, so

$\dfrac{d^2 y}{dx^2} = -\dfrac{\left(3x^2 - 3y\right)^2(6y) - 2\left(3x^2 - 3y\right)\left(3y^2 - 3x\right)(-3) + \left(3y^2 - 3x\right)^2(6x)}{\left(3y^2 - 3x\right)^3}$

$= \dfrac{-54\left[\left(x^2 - y\right)^2 y + \left(x^2 - y\right)\left(y^2 - x\right) + \left(y^2 - x\right)^2 x\right]}{27\left(y^2 - x\right)^3} = \dfrac{2xy\left(x^3 + y^3 - 3xy + 1\right)}{\left(x - y^2\right)^3} = \dfrac{2xy}{\left(x - y^2\right)^3}$

since $x^3 + y^3 - 3xy = 0$. Its domain is $\left\{(x, y) \mid x \neq y^2\right\}$.

65. True. $\dfrac{d}{dy} F(x, y) = \dfrac{d}{dy}(0) \Rightarrow F_x \dfrac{dx}{dy} + F_y = 0 \Rightarrow \dfrac{dx}{dy} = -\dfrac{F_y}{F_x} \ (F_x \neq 0)$.

13.6 Concept Questions   ET 12.6

1. a. See page 1097 (1093 in ET). We need to use a unit vector to indicate the direction. Otherwise, the directional derivative would not be unique.

b. See page 1103 (1099 in ET).

3. a. If $\nabla f(x, y) = \mathbf{0}$, then $D_{\mathbf{u}} f(x, y) = 0$ for every $\mathbf{u}$.

b. See page 1101 (1097 in ET).

13.6   Directional Derivatives and Gradient Vectors   ET 12.6

1. Here $\mathbf{u} = \cos\frac{\pi}{6}\mathbf{i} + \sin\frac{\pi}{6}\mathbf{j} = \frac{\sqrt{3}}{2}\mathbf{i} + \frac{1}{2}\mathbf{j}$, $D_{\mathbf{u}}f(x,y) = f_x(x,y)u_1 + f_y(x,y)u_2 = \left(3x^2 - 4x\right)\left(\frac{\sqrt{3}}{2}\right) + \left(3y^2\right)\left(\frac{1}{2}\right)$,

and $D_{\mathbf{u}}f(1,2) = \left[3(1)^2 - 4(1)\right]\left(\frac{\sqrt{3}}{2}\right) + \left[3(2)^2\right]\left(\frac{1}{2}\right) = \frac{12-\sqrt{3}}{2}$.

3. Here $\mathbf{u} = \cos\frac{\pi}{2}\mathbf{i} + \sin\frac{\pi}{2}\mathbf{j} = \mathbf{j}$, so $D_{\mathbf{u}}f(3,0) = \frac{\partial f}{\partial y}(3,0) = (x+1)e^y\big|_{(3,0)} = 4$.

5. $f_x(x,y) = \frac{\partial}{\partial x}(2x + 3xy - 3y + 4) = 2 + 3y$ and $f_y(x,y) = 3x - 3$, so

$\nabla f(2,1) = \left[(2 + 3y)\mathbf{i} + 3(x-1)\mathbf{j}\right]_{(2,1)} = 5\mathbf{i} + 3\mathbf{j}$.

7. $f_x(x,y) = \frac{\partial}{\partial x}(x\sin y + y\cos x) = \sin y - y\sin x$ and $f_y(x,y) = x\cos y + \cos x$, so

$\nabla f\left(\frac{\pi}{4}, \frac{\pi}{2}\right) = \left[(\sin y - y\sin x)\mathbf{i} + (x\cos y + \cos x)\mathbf{j}\right]_{(\pi/4,\pi/2)} = \left[1 - \frac{\pi}{2}\left(\frac{\sqrt{2}}{2}\right)\right]\mathbf{i} + \frac{\sqrt{2}}{2}\mathbf{j} = \frac{4 - \sqrt{2}\pi}{4}\mathbf{i} + \frac{\sqrt{2}}{2}\mathbf{j}$.

9. $f_x(x,y,z) = \frac{\partial}{\partial x}\left(xe^{yz}\right) = e^{yz}$, $f_y(x,y,z) = xze^{yz}$, and $f_z(x,y,z) = xye^{yz}$, so

$\nabla f(1,0,2) = \left(e^{yz}\mathbf{i} + xze^{yz}\mathbf{j} + xye^{yz}\mathbf{k}\right)\big|_{(1,0,2)} = \mathbf{i} + 2\mathbf{j}$.

11. Here $\mathbf{u} = \frac{\mathbf{v}}{|\mathbf{v}|} = \frac{\mathbf{i} - 2\mathbf{j}}{\sqrt{1 + (-2)^2}} = \frac{\sqrt{5}}{5}\mathbf{i} - \frac{2\sqrt{5}}{5}\mathbf{j}$, $f_x(x,y) = 3x^2 - 2xy^2 + y$, and $f_y(x,y) = -2x^2y + x + 2y$, so

$$D_{\mathbf{u}}f(1,-1) = f_x(1,-1)u_1 + f_y(1,-1)u_2$$
$$= \left[3(1)^2 - 2(1)(-1)^2 + (-1)\right]\left(\frac{\sqrt{5}}{5}\right) + \left[-2(1)^2(-1) + 1 + 2(-1)\right]\left(-\frac{2\sqrt{5}}{5}\right) = -\frac{2\sqrt{5}}{5}$$

13. Here $\mathbf{u} = \mathbf{v} = -\mathbf{i}$, $f_x(x,y) = -\frac{y}{x^2}$, and $f_y(x,y) = \frac{1}{x}$, so $D_{\mathbf{u}}f(3,1) = f_x(3,1)(-1) + f_y(3,1)(0) = -\frac{1}{3^2}(-1) = \frac{1}{9}$.

15. Here $\mathbf{u} = \frac{\mathbf{v}}{|\mathbf{v}|} = \frac{-\mathbf{i} + 3\mathbf{j}}{\sqrt{(-1)^2 + 3^2}} = -\frac{\sqrt{10}}{10}\mathbf{i} + \frac{3\sqrt{10}}{10}\mathbf{j}$, $f_x(x,y) = -\frac{2y}{(x-y)^2}$, and $f_y(x,y) = \frac{2x}{(x-y)^2}$, so

$D_{\mathbf{u}}f(2,1) = f_x(2,1)\left(-\frac{\sqrt{10}}{10}\right) + f_y(2,1)\left(\frac{3\sqrt{10}}{10}\right) = -2\left(-\frac{\sqrt{10}}{10}\right) + 4\left(\frac{3\sqrt{10}}{10}\right) = \frac{7\sqrt{10}}{5}$.

17. Here $\mathbf{u} = \frac{\mathbf{v}}{|\mathbf{v}|} = \frac{-2\mathbf{i} + 3\mathbf{j}}{\sqrt{(-2)^2 + 3^2}} = -\frac{2\sqrt{13}}{13}\mathbf{i} + \frac{3\sqrt{13}}{13}\mathbf{j}$, $f_x(x,y) = \sin^2 y$, and $f_y(x,y) = 2x\sin y\cos y = x\sin 2y$, so

$D_{\mathbf{u}}f\left(-1, \frac{\pi}{4}\right) = f_x\left(-1, \frac{\pi}{4}\right)\left(-\frac{2\sqrt{13}}{13}\right) + f_y\left(-1, \frac{\pi}{4}\right)\left(\frac{3\sqrt{13}}{13}\right) = \frac{1}{2}\left(-\frac{2\sqrt{13}}{13}\right) + (-1)\left(\frac{3\sqrt{13}}{13}\right) = -\frac{4\sqrt{13}}{13}$.

19. Here $\mathbf{u} = \frac{\mathbf{v}}{|\mathbf{v}|} = \frac{\mathbf{i} + \mathbf{j} + \mathbf{k}}{\sqrt{1^2 + 1^2 + 1^2}} = \frac{\sqrt{3}}{3}(\mathbf{i} + \mathbf{j} + \mathbf{k})$, $f_x(x,y,z) = 2xy^3z^4$,

$f_y(x,y,z) = 3x^2y^2z^4$, and $f_z(x,y,z) = 4x^2y^3z^3$, so

$D_{\mathbf{u}}f(3,-2,1) = 2(3)(-2)^3(1)^4\left(\frac{\sqrt{3}}{3}\right) + 3(3)^2(-2)^2(1)^4\left(\frac{\sqrt{3}}{3}\right) + 4(3)^2(-2)^3(1)^3\left(\frac{\sqrt{3}}{3}\right) = -76\sqrt{3}$.

21. Here $\mathbf{u} = \frac{\mathbf{v}}{|\mathbf{v}|} = \frac{2\mathbf{i} - 4\mathbf{j} + 4\mathbf{k}}{\sqrt{2^2 + (-4)^2 + 4^2}} = \frac{1}{3}\mathbf{i} - \frac{2}{3}\mathbf{j} + \frac{2}{3}\mathbf{k}$, $f_x(x,y,z) = \frac{yz}{2\sqrt{xyz}}$, $f_y(x,y,z) = \frac{xz}{2\sqrt{xyz}}$, and

$f_z(x,y,z) = \frac{xy}{2\sqrt{xyz}}$, so $D_{\mathbf{u}}f(4,2,2) = \frac{1}{2}\left(\frac{1}{3}\right) + 1\left(-\frac{2}{3}\right) + 1\left(\frac{2}{3}\right) = \frac{1}{6}$.

23. Here $\mathbf{u} = \frac{\mathbf{v}}{|\mathbf{v}|} = \frac{\mathbf{i} - 2\mathbf{j} + 3\mathbf{k}}{\sqrt{1^2 + (-2)^2 + 3^2}} = \frac{\sqrt{14}}{14}(\mathbf{i} - 2\mathbf{j} + 3\mathbf{k})$, $f_x(x,y,z) = 2xe^{yz}$, $f_y(x,y,z) = x^2ze^{yz}$, and

$f_z(x,y,z) = x^2ye^{yz}$, so $D_{\mathbf{u}}f(2,3,0) = 4\left(\frac{\sqrt{14}}{14}\right) + 0\left(-\frac{2\sqrt{14}}{14}\right) + 12\left(\frac{3\sqrt{14}}{14}\right) = \frac{20\sqrt{14}}{7}$.

25. Here $\mathbf{u} = \dfrac{\mathbf{v}}{|\mathbf{v}|} = \dfrac{\mathbf{i} - \mathbf{j} + \mathbf{k}}{\sqrt{1^2 + (-1)^2 + 1^2}} = \dfrac{\sqrt{3}}{3}(\mathbf{i} - \mathbf{j} + \mathbf{k})$, $f_x(x, y, z) = 2xy \cos 2z$, $f_y(x, y, z) = x^2 \cos 2z$,

and $f_z(x, y, z) = -2x^2 y \sin 2z$, so $D_{\mathbf{u}}f\left(-1, 2, \frac{\pi}{4}\right) = f_x\left(-1, 2, \frac{\pi}{4}\right)\left(\frac{\sqrt{3}}{3}\right) + f_y\left(-1, 2, \frac{\pi}{4}\right)\left(-\frac{\sqrt{3}}{3}\right) +$

$f_z\left(-1, 2, \frac{\pi}{4}\right)\left(\frac{\sqrt{3}}{3}\right) = 0\left(\frac{\sqrt{3}}{3}\right) + 0\left(-\frac{\sqrt{3}}{3}\right) - 4\left(\frac{\sqrt{3}}{3}\right) = -\frac{4\sqrt{3}}{3}$.

27. Here $\mathbf{u} = \dfrac{\mathbf{v}}{|\mathbf{v}|} = \dfrac{\mathbf{i} + 2\mathbf{j} - \mathbf{k}}{\sqrt{1 + 4 + 1}} = \dfrac{\sqrt{6}}{6}\mathbf{i} + \dfrac{\sqrt{6}}{3}\mathbf{j} - \dfrac{\sqrt{6}}{6}\mathbf{k}$, $f_x(x, y, z) = \tan^{-1}\dfrac{y}{z}$, $f_y(x, y, z) = \dfrac{xz}{y^2 + z^2}$,

and $f_z(x, y, z) = -\dfrac{xy}{y^2 + z^2}$, so $D_{\mathbf{u}}f(3, -2, 2) = f_x(3, -2, 2)\left(\frac{\sqrt{6}}{6}\right) + f_y(3, -2, 2)\left(\frac{\sqrt{6}}{3}\right) +$

$f_z(3, -2, 2)\left(-\frac{\sqrt{6}}{6}\right) = -\frac{\pi}{4}\left(\frac{\sqrt{6}}{6}\right) + \frac{3}{4}\left(\frac{\sqrt{6}}{3}\right) + \frac{3}{4}\left(-\frac{\sqrt{6}}{6}\right) = \frac{\sqrt{6}}{24}(3 - \pi)$.

29. Here $\mathbf{u} = \dfrac{\overrightarrow{PQ}}{|\overrightarrow{PQ}|} = \dfrac{\mathbf{i} + 3\mathbf{j}}{\sqrt{1^2 + 3^2}} = \dfrac{\sqrt{10}}{10}\mathbf{i} + \dfrac{3\sqrt{10}}{10}\mathbf{j}$, $f_x(x, y) = 3x^2$, and $f_y(x, y) = 3y^2$, so

$D_{\mathbf{u}}f(1, 2) = f_x(1, 2)\left(\frac{\sqrt{10}}{10}\right) + f_y(1, 2)\left(\frac{3\sqrt{10}}{10}\right) = 3\left(\frac{\sqrt{10}}{10}\right) + 12\left(\frac{3\sqrt{10}}{10}\right) = \frac{39\sqrt{10}}{10}$.

31. Here $\mathbf{u} = \dfrac{\overrightarrow{PQ}}{|\overrightarrow{PQ}|} = \dfrac{2\mathbf{i} + \frac{\pi}{4}\mathbf{j} - \frac{\pi}{6}\mathbf{k}}{\sqrt{2^2 + \left(\frac{\pi}{4}\right)^2 + \left(\frac{\pi}{6}\right)^2}} = \dfrac{12}{\sqrt{576 + 13\pi^2}}\left(2\mathbf{i} + \frac{\pi}{4}\mathbf{j} - \frac{\pi}{6}\mathbf{k}\right)$, $f_x(x, y, z) = \sin(2y + 3z)$,

$f_y(x, y, z) = 2x \cos(2y + 3z)$, and $f_z(x, y, z) = 3x \cos(2y + 3z)$, so

$D_{\mathbf{u}}f\left(1, \frac{\pi}{4}, -\frac{\pi}{12}\right) = f_x\left(1, \frac{\pi}{4}, -\frac{\pi}{12}\right) \cdot \dfrac{24}{\sqrt{576 + 13\pi^2}} + f_y\left(1, \frac{\pi}{4}, -\frac{\pi}{12}\right) \cdot \dfrac{3\pi}{\sqrt{576 + 13\pi^2}}$

$+ f_z\left(1, \frac{\pi}{4}, -\frac{\pi}{12}\right) \cdot \left(-\dfrac{2\pi}{\sqrt{576 + 13\pi^2}}\right)$

$= \dfrac{\sqrt{2}}{2}\left(\dfrac{24}{\sqrt{576 + 13\pi^2}}\right) + \sqrt{2}\left(\dfrac{3\pi}{\sqrt{576 + 13\pi^2}}\right) + \dfrac{3\sqrt{2}}{2}\left(-\dfrac{2\pi}{\sqrt{576 + 13\pi^2}}\right) = \dfrac{12\sqrt{2}}{\sqrt{13\pi^2 + 576}}$

33. $\nabla f(x, y) = \dfrac{\partial}{\partial x}\sqrt{2x + 3y^2}\,\mathbf{i} + \dfrac{\partial}{\partial y}\sqrt{2x + 3y^2}\,\mathbf{j} = \dfrac{1}{\sqrt{2x + 3y^2}}\mathbf{i} + \dfrac{3y}{\sqrt{2x + 3y^2}}\mathbf{j}$, so $\nabla f(3, 2) = \dfrac{\sqrt{2}}{6}\mathbf{i} + \sqrt{2}\mathbf{j}$ and a

vector in the desired direction is $\mathbf{v} = 3\sqrt{2}\nabla f(3, 2) = \mathbf{i} + 6\mathbf{j}$. The maximum rate of increase of f at $P(3, 2)$ is

$|\nabla f(3, 2)| = \sqrt{\left(\frac{\sqrt{2}}{6}\right)^2 + \left(\sqrt{2}\right)^2} = \dfrac{\sqrt{74}}{6}$.

35. $f_x(x, y, z) = \dfrac{\partial}{\partial x}\left(x^3 + 2xz + 2yz^2 + z^3\right) = 3x^2 + 2z$, $f_y(x, y, z) = 2z^2$, and $f_z(x, y, z) = 2x + 4yz + 3z^2$,

so $\nabla f(-1, 3, 2) = f_x(-1, 3, 2)\mathbf{i} + f_y(-1, 3, 2)\mathbf{j} + f_z(-1, 3, 2)\mathbf{k} = 7\mathbf{i} + 8\mathbf{j} + 34\mathbf{k}$ and a vector in

the desired direction is $\mathbf{v} = \nabla f(-1, 3, 2) = 7\mathbf{i} + 8\mathbf{j} + 34\mathbf{k}$. The maximum rate of increase of f at P is

$|\nabla f(-1, 3, 2)| = \sqrt{7^2 + 8^2 + 34^2} = 3\sqrt{141}$.

37. $\nabla f(x, y) = f_x(x, y)\mathbf{i} + f_y(x, y)\mathbf{j} = \dfrac{2}{1 + (2x + y)^2}\mathbf{i} + \dfrac{1}{1 + (2x + y)^2}\mathbf{j}$, so $\nabla f(0, 0) = 2\mathbf{i} + \mathbf{j}$ and a vector in the

desired direction is $\mathbf{v} = -\nabla f(0, 0) = -2\mathbf{i} - \mathbf{j}$. The maximum rate of decrease of f at P is $|-\nabla f(0, 0)| = \sqrt{4 + 1} = \sqrt{5}$.

39. $\nabla f(x, y, z) = \dfrac{1}{y}\mathbf{i} + \left(-\dfrac{x}{y^2} + \dfrac{1}{z}\right)\mathbf{j} - \dfrac{y}{z^2}\mathbf{k} \Rightarrow \nabla f(1, -1, 2) = -\mathbf{i} - \frac{1}{2}\mathbf{j} + \frac{1}{4}\mathbf{k}$ and a vector in the

desired direction is $\mathbf{v} = -4\nabla f(1, -1, 2) = 4\mathbf{i} + 2\mathbf{j} - \mathbf{k}$. The maximum rate of decrease of f at P is

$|-\nabla f(1, -1, 2)| = \sqrt{(-1)^2 + \left(-\frac{1}{2}\right)^2 + \left(\frac{1}{4}\right)^2} = \dfrac{\sqrt{21}}{4}$.

41. The direction is that of $\nabla h (1, 1)$, but

$$\nabla h (x, y) = f_x (x, y) \mathbf{i} + f_y (x, y) \mathbf{j} = 20 (-8x + 2y + 28) \mathbf{i} + 20 (-6y + 2x - 18) \mathbf{j}, \text{ and so}$$

$\nabla h (1, 1) = 440 \mathbf{i} - 440 \mathbf{j}$. Therefore, the desired direction is $\mathbf{v} = \frac{1}{440} \nabla h (1, 1) = \mathbf{i} - \mathbf{j}$. The steepest slope is

$|\nabla h (1, 1)| = |440 (\mathbf{i} - \mathbf{j})| = 440 |\mathbf{i} - \mathbf{j}| = 440 \sqrt{2}$.

43. a. You will be ascending if you start from A and proceed in a southwesterly direction, because you will be moving from a point on the level curve with value 200 toward one with value 300. If you start from B, you will be descending.

    b. You will be moving along the tangent line to the level curve at C, and so you will be neither ascending nor descending.

    c. You should move in an easterly direction. This is perpendicular to the curve at D.

    d. You would start from the east. The level curves are spaced further apart and so it takes a greater (horizontal) distance to reach the same altitude.

45. a. $T (x, y) = \dfrac{200}{\pi} \tan^{-1} \dfrac{2y}{1 - x^2 - y^2} \Rightarrow$

$$\nabla T \left(\frac{\sqrt{7}}{4}, \frac{1}{4} \right) = \frac{200}{\pi} \left[\frac{4xy}{\left(1 - x^2 - y^2\right)^2 + 4y^2} \mathbf{i} + \frac{2 \left(y^2 - x^2 + 1\right)}{\left(1 - x^2 - y^2\right)^2 + 4y^2} \mathbf{j} \right]_{\left(\frac{\sqrt{7}}{4}, \frac{1}{4}\right)} = \frac{100\sqrt{7}}{\pi} \mathbf{i} + \frac{500}{\pi} \mathbf{j}.$$

This vector gives the direction in which T is increasing most rapidly

at $\left(\frac{\sqrt{7}}{4}, \frac{1}{4} \right)$.

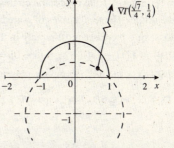

    b. $T \left(\frac{\sqrt{7}}{4}, \frac{1}{4} \right) = \frac{200}{\pi} \tan^{-1} 1 = 50$, so the level curve of T passing

through $\left(\frac{\sqrt{7}}{4}, \frac{1}{4} \right)$ is $\dfrac{200}{\pi} \tan^{-1} \dfrac{2y}{1 - x^2 - y^2} = 50 \Leftrightarrow$

$\dfrac{2y}{1 - x^2 - y^2} = \tan \dfrac{\pi}{4} = 1 \Leftrightarrow x^2 + (y + 1)^2 = 2$.

47. The isotherms of T are the level surfaces of T; that is, they are concentric spheres with equations of the form $x^2 + y^2 + z^2 = c$, where c is a positive constant. So the temperature function is $T (x, y, z) = k \left(x^2 + y^2 + z^2\right)$, where k is a constant. $\nabla T (x, y, z) = k (2x\mathbf{i} + 2y\mathbf{j} + 2z\mathbf{k}) = 2k (x\mathbf{i} + y\mathbf{j} + z\mathbf{k})$ is a vector that points away from the center of the spheres if $k > 0$ and toward the center if $k < 0$.

49. a. Let the path of the insect be described by the position function $\mathbf{r} (t) = x (t) \mathbf{i} + y (t) \mathbf{j}$, $\mathbf{r} (0) = \mathbf{0}$. Next, $\nabla T (x, y) = -12x\mathbf{i} - 4y\mathbf{j}$. Since the insect moves in the direction of maximum decrease in temperature, we have $\mathbf{v} (t) = -k \nabla T (x, y)$, where k is a scalar function of t. That is,

$\dfrac{dx}{dt} \mathbf{i} + \dfrac{dy}{dt} \mathbf{j} = 12kx\mathbf{i} + 4ky\mathbf{j} \Leftrightarrow \left\{ \dfrac{dx}{dt} = 12kx, \dfrac{dy}{dt} = 4ky \right\}$. Thus,

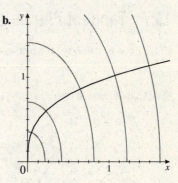

$\dfrac{dy}{dx} = \dfrac{dy/dt}{dx/dt} = \dfrac{4ky}{12kx} = \dfrac{1}{3} \dfrac{y}{x}$. This is a first-order separable differential

equation. We find $\displaystyle\int \dfrac{dy}{y} = \dfrac{1}{3} \int \dfrac{dx}{x} \Rightarrow$

$\ln |y| = \frac{1}{3} \ln |x| + |\ln c_1| \Leftrightarrow 3 \ln |y| = \ln |x| + \ln |c_2|$ (where $c_2 = c_1^3$) $\Leftrightarrow \ln \left|y^3\right| = \ln |cx| \Leftrightarrow y^3 = cx$. Using the initial condition $y (1) = 1$, we find that $c = 1$, so the required path is $x = y^3$.

51. $f (P_0) = 20$ and $f (P_1) = 25$. Also, $d (P_0, P_1) \approx 0.8$, so $D_{\mathbf{u}} f (P_0) \approx \dfrac{f (P_1) - f (P_0)}{d (P_0, P_1)} \approx \dfrac{25 - 20}{0.8} = 6.25$.

53. The vector from the origin to $(-3, 4)$ is $\mathbf{v} = -3\mathbf{i} + 4\mathbf{j}$, and a unit vector in the same direction is $\mathbf{u} = -\frac{3}{5}\mathbf{i} + \frac{4}{5}\mathbf{j}$. We are

given that $D_{\mathbf{u}}f(0, 0) = \nabla f(0, 0) \cdot \mathbf{u} = 5 \Leftrightarrow [f_x(0, 0)\mathbf{i} + f_y(0, 0)\mathbf{j}] \cdot \left(-\frac{3}{5}\mathbf{i} + \frac{4}{5}\mathbf{j}\right) = 5$ and, by Theorem 3, the direction

of $\nabla f(0, 0)$ is the same as that of $\mathbf{u}$. Thus, $f_x(0, 0)\mathbf{i} + f_y(0, 0)\mathbf{j} = c\left(-\frac{3}{5}\mathbf{i} + \frac{4}{5}\mathbf{j}\right)$ for $c > 0$. We have the system

$$\left.\begin{array}{r} -\frac{3}{5}f_x(0, 0) + \frac{4}{5}f_y(0, 0) = 5 \\ f_x(0, 0) = -\frac{3}{5}c \\ f_y(0, 0) = \frac{4}{5}c \end{array}\right\}$$ whose solution is $c = 5$, $f_x(0, 0) = -3$, $f_y(0, 0) = 4$. Thus, $\nabla f(0, 0) = -3\mathbf{i} + 4\mathbf{j}$.

55. a.

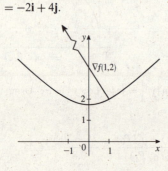

b. We need to show that the gradients of f and g are orthogonal at each point (x, y) in the plane where the curves intersect. $\nabla f(x, y) = 2x\mathbf{i} - 2y\mathbf{j}$ and $\nabla g(x, y) = y\mathbf{i} + x\mathbf{j}$, so since $\nabla f(x, y) \cdot \nabla g(x, y) = (2x\mathbf{i} - 2y\mathbf{j}) \cdot (y\mathbf{i} + x\mathbf{j}) = 2xy - 2yx = 0$ for all (x, y), the result follows.

57. The level curves of f are concentric circles centered at the origin, and so f increases most rapidly in any direction away from the origin. The level curves of g are hyperbolas centered at the origin with transverse axes lying on the x-axis or the y-axis, so g increases most rapidly in the x-direction (positive or negative). Theorem 3 is not applicable because $\nabla f(0, 0) = (2x\mathbf{i} + 2y\mathbf{j})|_{(0,0)} = 0\mathbf{i} + 0\mathbf{j} = \mathbf{0}$ and $\nabla g(0, 0) = (2x\mathbf{i} - 2y\mathbf{j})|_{(0,0)} = 0\mathbf{i} - 0\mathbf{j} = \mathbf{0}$.

59. True. $D_{\mathbf{u}}f(x, y) = \nabla f(x, y) \cdot \mathbf{u}$ and so it depends only on $\mathbf{u}$ and $\nabla f(x, y) = f_x(x, y)\mathbf{i} + f_y(x, y)\mathbf{j}$.

61. True. $D_{\mathbf{u}}f(x, y) \leq |D_{\mathbf{u}}f(x, y)| = |\nabla f(x, y) \cdot \mathbf{u}| = |\nabla f(x, y)| = \sqrt{f_x^2(x, y) + f_y^2(x, y)}$.

13.7 Concept Questions   ET 12.7

1. a. See page 1109 (1105 in ET).   **b.** See page 1111 (1107 in ET).

13.7 Tangent Planes and Normal Lines   ET 12.7

1. $f(1, 2) = 4 - 1 = 3$, so an equation of the required level curve is $y^2 - x^2 = 3$.

$\nabla f(1, 2) = f_x(1, 2)\mathbf{i} + f_y(1, 2)\mathbf{j} = (-2x\mathbf{i} + 2y\mathbf{j})|_{(1,2)}$

$\quad = -2\mathbf{i} + 4\mathbf{j}$.

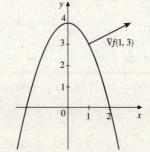

3. $f(1, 3) = \left(x^2 + y\right)\Big|_{(1,3)} = 1^2 + 3 = 4$, so an equation of the required level curve is $x^2 + y = 4$.

$\nabla f(1, 3) = (2x\mathbf{i} + \mathbf{j})|_{(1,3)} = 2\mathbf{i} + \mathbf{j}$.

5. $F(x, y) = \dfrac{x^2}{9} + \dfrac{y^2}{16} \Rightarrow \nabla F\left(\dfrac{3\sqrt{3}}{2}, 2\right) = \left(\dfrac{2}{9}x\mathbf{i} + \dfrac{1}{8}y\mathbf{j}\right)\Big|_{\left(3\sqrt{3}/2,2\right)} = \dfrac{\sqrt{3}}{3}\mathbf{i} + \dfrac{1}{4}\mathbf{j}$ is normal to the level curve

$F(x, y) = \dfrac{x^2}{9} + \dfrac{y^2}{16} = 1$ at $\left(\dfrac{3\sqrt{3}}{2}, 2\right)$. From this, we see that the slope of the required normal line is $m = \dfrac{3}{4\sqrt{3}} = \dfrac{\sqrt{3}}{4}$. So

an equation of the required normal line is $y - 2 = \dfrac{\sqrt{3}}{4}\left(x - \dfrac{3\sqrt{3}}{2}\right) \Leftrightarrow y = \dfrac{\sqrt{3}}{4}x + \dfrac{7}{8}$. The slope of the required tangent line

is $m = -\dfrac{4}{\sqrt{3}} = -\dfrac{4\sqrt{3}}{3}$, and so an equation of the tangent line is $y - 2 = -\dfrac{4\sqrt{3}}{3}\left(x - \dfrac{3\sqrt{3}}{2}\right) \Leftrightarrow y = -\dfrac{4\sqrt{3}}{3}x + 8$.

7. $F(x, y) = x^4 + 2x^2y^2 + y^4 - 9x^2 + 9y^2 \Rightarrow$

$\nabla F\left(\sqrt{5}, -1\right) = \left[\left(4x^3 + 4xy^2 - 18x\right)\mathbf{i} + \left(4x^2y + 4y^3 + 18y\right)\right]\mathbf{j}\Big|_{\left(\sqrt{5},-1\right)} = 6\sqrt{5}\mathbf{i} - 42\mathbf{j}$ is normal to the level

curve $F(x, y) = x^4 + 2x^2y^2 + y^4 - 9x^2 + 9y^2 = 0$ at $\left(\sqrt{5}, -1\right)$, so the slope of the required normal line is

$m = -\dfrac{42}{6\sqrt{5}} = -\dfrac{7\sqrt{5}}{5}$ and an equation of the normal line is $y - (-1) = -\dfrac{7\sqrt{5}}{5}\left(x - \sqrt{5}\right) \Leftrightarrow y = -\dfrac{7\sqrt{5}}{5}x + 6$. The slope

of the required tangent line is $m = -\dfrac{1}{-\frac{7\sqrt{5}}{5}} = \dfrac{\sqrt{5}}{7}$, and an equation of the tangent line is $y - (-1) = \dfrac{\sqrt{5}}{7}\left(x - \sqrt{5}\right) \Leftrightarrow$

$y = \dfrac{\sqrt{5}}{7}x - \dfrac{12}{7}$.

9. $F(1, 2, 2) = \left(x^2 + y^2 + z^2\right)\Big|_{(1,2,2)} = 1 + 4 + 4 = 9$, so

an equation of the required level surface is

$x^2 + y^2 + z^2 = 9$.

$\nabla F(1, 2, 2) = (2x\mathbf{i} + 2y\mathbf{j} + 2z\mathbf{k})|_{(1,2,2)} = 2\mathbf{i} + 4\mathbf{j} + 4\mathbf{k}$.

11. $F(0, 2, 4) = \left(x^2 + y^2\right)\Big|_{(0,2,4)} = 0 + 4 = 4$, so an

equation of the required level surface is $x^2 + y^2 = 4$.

$\nabla F(0, 2, 4) = (2x\mathbf{i} + 2y\mathbf{j})|_{(0,2,4)} = 4\mathbf{j}$.

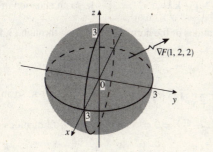

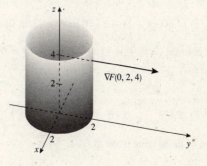

13. $F(1, 3, 2) = \left(-x^2 + y^2 - z^2\right)\Big|_{(1,3,2)} = -1 + 9 - 4 = 4$,

so an equation of the required level surface is

$-x^2 + y^2 - z^2 = 4 \Leftrightarrow \dfrac{y^2}{2^2} - \dfrac{x^2}{2^2} - \dfrac{z^2}{2^2} = 1$.

$\nabla F(1, 3, 2) = (-2x\mathbf{i} + 2y\mathbf{j} - 2z\mathbf{k})|_{(1,3,2)}$

$= -2\mathbf{i} + 6\mathbf{j} - 4\mathbf{k}$.

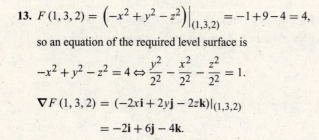

15. $F(x, y, z) = x^2 + 4y^2 + 9z^2 - 17 \Rightarrow \nabla F(2, 1, 1) = (2x\mathbf{i} + 8y\mathbf{j} + 18z\mathbf{k})|_{(2,1,1)} = 4\mathbf{i} + 8\mathbf{j} + 18\mathbf{k}$, so an equation of the

tangent plane at $(2, 1, 1)$ is $4(x - 2) + 8(y - 1) + 18(z - 1) = 0 \Leftrightarrow 2x + 4y + 9z = 17$. Equations of the normal line

passing through $(2, 1, 1)$ are $\dfrac{x - 2}{4} = \dfrac{y - 1}{8} = \dfrac{z - 1}{18} \Leftrightarrow \dfrac{x - 2}{2} = \dfrac{y - 1}{4} = \dfrac{z - 1}{9}$.

17. $F(x, y, z) = x^2 - 2y^2 - 4z^2 - 4 = 0 \Rightarrow \nabla F(4, -2, -1) = (2x\mathbf{i} - 4y\mathbf{j} - 8z\mathbf{k})|_{(4,-2,-1)} = 8\mathbf{i} + 8\mathbf{j} + 8\mathbf{k} = 8(\mathbf{i} + \mathbf{j} + \mathbf{k})$,
so an equation of the tangent plane at $(4, -2, -1)$ is $(x - 4) + (y + 2) + (z + 1) = 0 \Leftrightarrow x + y + z = 1$. Equations of the
normal line passing through $(4, -2, -1)$ are $\dfrac{x - 4}{1} = \dfrac{y + 2}{1} = \dfrac{z + 1}{1} \Leftrightarrow x - 4 = y + 2 = z + 1$.

19. $F(x, y, z) = xy + yz + xz - 11 = 0 \Rightarrow \nabla F(1, 2, 3) = \left[(y + z)\mathbf{i} + (x + z)\mathbf{j} + (x + y)\mathbf{k} \right]_{(1,2,3)} = 5\mathbf{i} + 4\mathbf{j} + 3\mathbf{k}$, so an
equation of the tangent plane at $(1, 2, 3)$ is $5(x - 1) + 4(y - 2) + 3(z - 3) = 0 \Leftrightarrow 5x + 4y + 3z = 22$. Equations of the
normal line passing through $(1, 2, 3)$ are $\dfrac{x - 1}{5} = \dfrac{y - 2}{4} = \dfrac{z - 3}{3}$.

21. $F(x, y, z) = 9x^2 + 4y^2 - z = 0 \Rightarrow \nabla F(-1, 2, 25) = (18x\mathbf{i} + 8y\mathbf{j} - \mathbf{k})|_{(-1,2,25)} = -18\mathbf{i} + 16\mathbf{j} - \mathbf{k}$, so an equation of
the tangent plane at $(-1, 2, 25)$ is $-18(x + 1) + 16(y - 2) - (z - 25) = 0 \Leftrightarrow 18x - 16y + z = -25$. Equations of the
normal line passing through $(-1, 2, 25)$ are $\dfrac{x + 1}{-18} = \dfrac{y - 2}{16} = \dfrac{z - 25}{-1}$.

23. $F(x, y, z) = xz^2 + yx^2 + y^2 - 2x + 3y + 6 = 0 \Rightarrow$
$\nabla F(-2, 1, 3) = \left[\left(z^2 + 2xy - 2 \right)\mathbf{i} + \left(x^2 + 2y + 3 \right)\mathbf{j} + 2xz\mathbf{k} \right]_{(-2,1,3)} = 3\mathbf{i} + 9\mathbf{j} - 12\mathbf{k} = 3(\mathbf{i} + 3\mathbf{j} - 4\mathbf{k})$, so an
equation of the tangent plane at $(-2, 1, 3)$ is $(x + 2) + 3(y - 1) - 4(z - 3) = 0 \Leftrightarrow x + 3y - 4z = -11$. Equations of the
normal line passing through $(-2, 1, 3)$ are $\dfrac{x + 2}{1} = \dfrac{y - 1}{3} = \dfrac{z - 3}{-4} \Leftrightarrow x + 2 = \dfrac{y - 1}{3} = \dfrac{z - 3}{-4}$.

25. $F(x, y, z) = xe^y - z = 0 \Rightarrow \nabla F(2, 0, 2) = \left(e^y\mathbf{i} + xe^y\mathbf{j} - \mathbf{k} \right)|_{(2,0,2)} = \mathbf{i} + 2\mathbf{j} - \mathbf{k}$, so an equation of the tangent plane at
$(2, 0, 2)$ is $(x - 2) + 2y - (z - 2) = 0 \Leftrightarrow x + 2y - z = 0$. Equations of the normal line passing through $(2, 0, 2)$ are
$x - 2 = \dfrac{y}{2} = \dfrac{z - 2}{-1}$.

27. $F(x, y, z) = \ln(xy + 1) - z = 0 \Rightarrow \nabla F(3, 0, 0) = \left(\dfrac{y}{xy + 1}\mathbf{i} + \dfrac{x}{xy + 1}\mathbf{j} - \mathbf{k} \right)\Big|_{(3,0,0)} = 3\mathbf{j} - \mathbf{k}$, so an equation of the
tangent plane at $(3, 0, 0)$ is $3(y - 0) - (z - 0) = 0 \Leftrightarrow 3y - z = 0$. Equations of the normal line passing through $(3, 0, 0)$
are $x = 3, \dfrac{y}{3} = \dfrac{z}{-1}$.

29. $F(x, y, z) = \tan^{-1}\dfrac{y}{x} - z = 0 \Rightarrow \nabla F\left(1, 1, \dfrac{\pi}{4}\right) = \left(-\dfrac{y}{x^2 + y^2}\mathbf{i} + \dfrac{x}{x^2 + y^2}\mathbf{j} - \mathbf{k} \right)\Big|_{(1,1,\pi/4)} = -\dfrac{1}{2}\mathbf{i} + \dfrac{1}{2}\mathbf{j} - \mathbf{k}$, so an
equation of the tangent plane at $\left(1, 1, \dfrac{\pi}{4}\right)$ is $-\dfrac{1}{2}(x - 1) + \dfrac{1}{2}(y - 1) - \left(z - \dfrac{\pi}{4}\right) = 0 \Leftrightarrow x - y + 2z = \dfrac{\pi}{2}$. Equations of the
normal line passing through $\left(1, 1, \dfrac{\pi}{4}\right)$ are $\dfrac{x - 1}{-\frac{1}{2}} = \dfrac{y - 1}{\frac{1}{2}} = \dfrac{z - \frac{\pi}{4}}{-1} \Leftrightarrow \dfrac{x - 1}{-1} = \dfrac{y - 1}{-1} = \dfrac{z - \frac{\pi}{4}}{2}$.

31. $F(x, y, z) = \sin xy + 3z - 3 = 0 \Rightarrow \nabla F(0, 3, 1) = (y \cos xy\,\mathbf{i} + x \cos xy\,\mathbf{j} + 3\mathbf{k})|_{(0,3,1)} = 3\mathbf{i} + 3\mathbf{k} = 3(\mathbf{i} + \mathbf{k})$, so an
equation of the tangent plane at $(0, 3, 1)$ is $(x - 0) + (z - 1) = 0 \Leftrightarrow x + z = 1$. Equations of the normal line passing
through $(0, 3, 1)$ are $\dfrac{x}{1} = \dfrac{z - 1}{1}, y = 3$.

33. $F(x, y, z) = \dfrac{x^2}{a^2} + \dfrac{y^2}{b^2} + \dfrac{z^2}{c^2} - 1 \Rightarrow \nabla F(x_0, y_0, z_0) = \left(\dfrac{2x}{a^2}\mathbf{i} + \dfrac{2y}{b^2}\mathbf{j} + \dfrac{2z}{c^2}\mathbf{k} \right)\Big|_{(x_0,y_0,z_0)} = \dfrac{2x_0}{a^2}\mathbf{i} + \dfrac{2y_0}{b^2}\mathbf{j} + \dfrac{2z_0}{c^2}\mathbf{k}$,
so an equation of the tangent plane at (x_0, y_0, z_0) is $\dfrac{2x_0}{a^2}(x - x_0) + \dfrac{2y_0}{b^2}(y - y_0) + \dfrac{2z_0}{c^2}(z - z_0) = 0 \Leftrightarrow$
$\dfrac{x_0 x}{a^2} + \dfrac{y_0 y}{b^2} + \dfrac{z_0 z}{c^2} - \left(\dfrac{x_0^2}{a^2} + \dfrac{y_0^2}{b^2} + \dfrac{z_0^2}{c^2} \right) = 0$. But (x_0, y_0, z_0) lies on the ellipsoid, so the expression in parentheses is
equal to 1 and we have $\dfrac{xx_0}{a^2} + \dfrac{yy_0}{b^2} + \dfrac{zz_0}{c^2} = 1$, as was to be shown.

35. $F(x, y, z) = \dfrac{x^2}{a^2} - \dfrac{y^2}{b^2} - \dfrac{z^2}{c^2} - 1 \Rightarrow \nabla F(x_0, y_0, z_0) = \left. \left(\dfrac{2x}{a^2}\mathbf{i} - \dfrac{2y}{b^2}\mathbf{j} - \dfrac{2z}{c^2}\mathbf{k} \right) \right|_{(x_0, y_0, z_0)} = \dfrac{2x_0}{a^2}\mathbf{i} - \dfrac{2y_0}{b^2}\mathbf{j} - \dfrac{2z_0}{c^2}\mathbf{k},$

so an equation of the tangent plane at (x_0, y_0, z_0) is $\dfrac{2x_0}{a^2}(x - x_0) - \dfrac{2y_0}{b^2}(y - y_0) - \dfrac{2z_0}{c^2}(z - z_0) = 0 \Leftrightarrow$

$\dfrac{x_0 x}{a^2} - \dfrac{y_0 y}{b^2} - \dfrac{z_0 z}{c^2} - \left(\dfrac{x_0^2}{a^2} - \dfrac{y_0^2}{b^2} - \dfrac{z_0^2}{c^2} \right) = 0.$ But (x_0, y_0, z_0) lies on the hyperboloid, so the expression in parentheses is

equal to 1 and we have $\dfrac{x x_0}{a^2} - \dfrac{y y_0}{b^2} - \dfrac{z z_0}{c^2} = 1.$

37. We write $F(x, y, z) = x^2 + y^2 + z^2 - 14$. Then the normal to the tangent plane to the sphere at the point (x_0, y_0, z_0) is
$\nabla F(x_0, y_0, z_0) = 2x_0\mathbf{i} + 2y_0\mathbf{j} + 2z_0\mathbf{k}$. Since the tangent plane is parallel to the plane $x + 2y + 3z = 12$ whose normal is
$\mathbf{i} + 2\mathbf{j} + 3\mathbf{k}$, we see that $x_0\mathbf{i} + y_0\mathbf{j} + z_0\mathbf{k} = c(\mathbf{i} + 2\mathbf{j} + 3\mathbf{k})$, where c is a constant. This equation implies that $x_0 = c$,
$y_0 = 2c$, and $z_0 = 3c$, and substituting these values into the equation of the sphere gives $c^2 + (2c)^2 + (3c)^2 = 14 \Leftrightarrow$
$14c^2 = 14 \Leftrightarrow c = \pm 1$. Thus, the required points are $(-1, -2, -3)$ and $(1, 2, 3)$.

39. We write $F(x, y, z) = 2x^2 - y^2 + z^2 - 1$. A vector normal to the hyperboloid at the point (x_0, y_0, z_0) is
$\nabla F(x_0, y_0, z_0) = 4x_0\mathbf{i} - 2y_0\mathbf{j} + 2z_0\mathbf{k}$. A vector in the direction of the line passing through $(-1, 1, 2)$ and $(3, 3, 3)$ is
$\mathbf{v} = (3 + 1)\mathbf{i} + (3 - 1)\mathbf{j} + (3 - 2)\mathbf{k} = 4\mathbf{i} + 2\mathbf{j} + \mathbf{k}$. Since $\nabla F(x_0, y_0, z_0)$ is parallel to $\mathbf{v}$, there exists a constant c such
that $\nabla F = c\mathbf{v} \Rightarrow 4x_0\mathbf{i} - 2y_0\mathbf{j} + 2z_0\mathbf{k} = 4c\mathbf{i} + 2c\mathbf{j} + c\mathbf{k} \Rightarrow x_0 = c$, $y_0 = -c$, and $z_0 = \frac{1}{2}c$. Thus, $F(x_0, y_0, z_0) = 0$
gives $2c^2 - (-c)^2 + \left(\frac{1}{2}c \right)^2 - 1 = 0 \Leftrightarrow \frac{5}{4}c^2 = 1 \Leftrightarrow c = \pm \frac{2\sqrt{5}}{5}$, so the required points are $\left(-\frac{2\sqrt{5}}{5}, \frac{2\sqrt{5}}{5}, -\frac{\sqrt{5}}{5} \right)$ and
$\left(\frac{2\sqrt{5}}{5}, -\frac{2\sqrt{5}}{5}, \frac{\sqrt{5}}{5} \right)$.

41. It suffices to show that the normals to each of the two tangent planes at $(1, 2, 1)$ are parallel to each other. Writing
$F(x, y, z) = 2x^2 + y^2 - z - 5$ and $G(x, y, z) = x^2 + y^2 + z^2 - 6x - 8y - z + 17$, we see that the normals to the
planes $F(x, y, z) = 0$ and $G(x, y, z) = 0$ at $(1, 2, 1)$ are $\nabla F(1, 2, 1) = (4x\mathbf{i} + 2y\mathbf{j} - \mathbf{k})|_{(1,2,1)} = 4\mathbf{i} + 4\mathbf{j} - \mathbf{k}$ and
$\nabla G(1, 2, 1) = \left[(2x - 6)\mathbf{i} + (2y - 8)\mathbf{j} + (2z - 1)\mathbf{k} \right]_{(1,2,1)} = -4\mathbf{i} - 4\mathbf{j} + \mathbf{k}$. Since $\nabla F(1, 2, 1) = -\nabla G(1, 2, 1)$, the
proof is complete.

43. We write $F(x, y) = \dfrac{x^2}{a^2} + \dfrac{y^2}{b^2} - 1 = 0$. Then

$\nabla F(x_0, y_0) = \dfrac{2x_0}{a^2}\mathbf{i} + \dfrac{2y_0}{b^2}\mathbf{j}$ is normal to the ellipse at (x_0, y_0). Its slope

is $\dfrac{2y_0/b^2}{2x_0/a^2} = \dfrac{a^2 y_0}{b^2 x_0}$, so the slope of the tangent line is $-\dfrac{b^2 x_0}{a^2 y_0}$. Thus, an

equation of the tangent line at that point is $y - y_0 = -\dfrac{b^2 x_0}{a^2 y_0}(x - x_0) \Leftrightarrow$

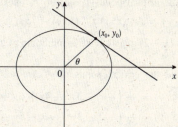

$\dfrac{x x_0}{a^2} + \dfrac{y y_0}{b^2} = \dfrac{x_0^2}{a^2} + \dfrac{y_0^2}{b^2} = 1.$ But $x_0 = a \cos \theta$ and $y_0 = b \sin \theta$ for

$0 \le \theta < 2\pi$, so we have $\dfrac{ax \cos \theta}{a^2} + \dfrac{by \sin \theta}{b^2} = 1 \Leftrightarrow (b \cos \theta) x + (a \sin \theta) y = ab, 0 \le \theta < 2\pi.$

45. $\nabla F(x_0, y_0, z_0)$ is normal to $F(x, y, z) = 0$ at P and $\nabla G(x_0, y_0, z_0)$ is normal to $G(x, y, z) = 0$ at P, so $\nabla F(x_0, y_0, z_0) \times \nabla G(x_0, y_0, z_0)$ is parallel to the tangent line to C at P. Write $F(x, y, z) = x^2 + y^2 + z^2 - 2$ and $G(x, y, z) = x^2 + y^2 - z$. Then $\nabla F\left(-\frac{\sqrt{2}}{2}, \frac{\sqrt{2}}{2}, 1\right) = \left.(2x\mathbf{i} + 2y\mathbf{j} + 2z\mathbf{k})\right|_{\left(-\sqrt{2}/2, \sqrt{2}/2, 1\right)} = -\sqrt{2}\mathbf{i} + \sqrt{2}\mathbf{j} + 2\mathbf{k}$

and $\nabla G\left(-\frac{\sqrt{2}}{2}, \frac{\sqrt{2}}{2}, 1\right) = \left.(2x\mathbf{i} + 2y\mathbf{j} - \mathbf{k})\right|_{\left(-\sqrt{2}/2, \sqrt{2}/2, 1\right)} = -\sqrt{2}\mathbf{i} + \sqrt{2}\mathbf{j} - \mathbf{k}$, so

$$\nabla F\left(-\tfrac{\sqrt{2}}{2}, \tfrac{\sqrt{2}}{2}, 1\right) \times \nabla G\left(-\tfrac{\sqrt{2}}{2}, \tfrac{\sqrt{2}}{2}, 1\right) = \begin{vmatrix} \mathbf{i} & \mathbf{j} & \mathbf{k} \\ -\sqrt{2} & \sqrt{2} & 2 \\ -\sqrt{2} & \sqrt{2} & -1 \end{vmatrix} = -3\sqrt{2}\mathbf{i} - 3\sqrt{2}\mathbf{j} = -3\sqrt{2}\,(\mathbf{i} + \mathbf{j}).$$ We may

take the vector parallel to C at $\left(-\frac{\sqrt{2}}{2}, \frac{\sqrt{2}}{2}, 1\right)$ to be $\mathbf{v} = \mathbf{i} + \mathbf{j}$. Then parametric equations of the tangent line are $x = -\frac{\sqrt{2}}{2} + t, y = \frac{\sqrt{2}}{2} + t, z = 1$.

47. True. A vector parallel to the given line is $\mathbf{v} = 4\mathbf{i} + 6\mathbf{j} - 2\mathbf{k}$. A normal to the plane $2x + 3y - z = 4$ is $\mathbf{n} = 2\mathbf{i} + 3\mathbf{j} - \mathbf{k}$. Since $\mathbf{v} = 2\mathbf{n}$, we see that the two vectors are parallel, showing that the given line is perpendicular to the given plane.

49. True. The equations of the normal line are $\dfrac{x - x_0}{F_x(x_0, y_0, z_0)} = \dfrac{y - y_0}{F_y(x_0, y_0, z_0)} = \dfrac{z - z_0}{F_z(x_0, y_0, z_0)} = t \Leftrightarrow$ $x = x_0 + tF_x(x_0, y_0, z_0), y = y_0 + tF_y(x_0, y_0, z_0), z = z_0 + tF_z(x_0, y_0, z_0)$, which can be written as $\mathbf{r}(t) = \langle x_0, y_0, z_0 \rangle + t\nabla F(x_0, y_0, z_0)$.

13.8 Concept Questions   ET 12.8

1. a. See page 1116 (1112 in ET).   **b.** See page 1117 (1113 in ET).

3. a. See page 1122 (1118 in ET).   **b.** See page 1122 (1118 in ET).

13.8 Extrema of Functions of Two Variables   ET 12.8

1. To find the critical points of f, we solve the system $\left.\begin{array}{l} f_x(x, y) = \frac{\partial}{\partial x}\left(x^2 + y^2 - 2x + 4y\right) = 2x - 2 = 0 \\ f_y(x, y) = \frac{\partial}{\partial y}\left(x^2 + y^2 - 2x + 4y\right) = 2y + 4 = 0 \end{array}\right\}$ obtaining

the sole critical point $(1, -2)$ of f. Next, we use the SDT on the critical point:
$D(x, y) = f_{xx}(x, y)\, f_{yy}(x, y) - f_{xy}^2(x, y) = 2 \cdot 2 - 0^2 = 4$. Since $D(1, -2) = 4 > 0$ and $f_{xx}(1, -2) = 2 > 0$, the point $(1, -2)$ gives a relative minimum of f with value $f(1, -2) = 1^2 + (-2)^2 - 2(1) + 4(-2) = -5$.

3. To find the critical points of f, we solve the system

$\left.\begin{array}{l} f_x(x, y) = \frac{\partial}{\partial x}\left(-x^2 - 3y^2 + 4x - 6y + 8\right) = -2x + 4 = 0 \\ f_y(x, y) = \frac{\partial}{\partial y}\left(-x^2 - 3y^2 + 4x - 6y + 8\right) = -6y - 6 = 0 \end{array}\right\}$ obtaining the sole critical point $(2, -1)$ of f.

Next, we use the SDT on the critical point: $D(x, y) = f_{xx}(x, y)\, f_{yy}(x, y) - f_{xy}^2(x, y) = (-2)(-6) - 0^2 = 12$. Since $D(2, -1) = 12 > 0$ and $f_{xx}(2, -1) = -2 < 0$, the point $(2, -1)$ gives a relative maximum of f with value $f(2, -1) = -2^2 - 3(-1)^2 + 4(2) - 6(-1) + 8 = 15$.

5. To find the critical points of f, we solve the system $\left.\begin{array}{l} f_x(x,y) = \frac{\partial}{\partial x}\left(x^2 + 3xy + 3y^2\right) = 2x + 3y = 0 \\ f_y(x,y) = \frac{\partial}{\partial y}\left(x^2 + 3xy + 3y^2\right) = 3x + 6y = 0 \end{array}\right\}$ obtaining

the sole critical point $(0,0)$ of f. Next, we use the SDT on the critical point:

$D(x,y) = f_{xx}(x,y)\, f_{yy}(x,y) - f_{xy}^2(x,y) = 2 \cdot 6 - 3^2 = 3$. Since $D(0,0) = 3 > 0$ and $f_{xx}(0,0) = 2 > 0$, the point $(0,0)$ gives a relative minimum of f with value $f(0,0) = 0$.

7. $\left.\begin{array}{l} f_x(x,y) = \frac{\partial}{\partial x}\left(2x^2 + y^2 - 2xy - 8x - 2y + 2\right) = 4x - 2y - 8 = 0 \\ f_y(x,y) = \frac{\partial}{\partial y}\left(2x^2 + y^2 - 2xy - 8x - 2y + 2\right) = 2y - 2x - 2 = 0 \end{array}\right\} \Rightarrow x = 5,\ y = 6$, so $(5,6)$ is

the sole critical point of f. Next, $D(x,y) = f_{xx}(x,y)\, f_{yy}(x,y) - f_{xy}^2(x,y) = 4 \cdot 2 - (-2)^2 = 4$.

Since $D(5,6) = 4 > 0$ and $f_{xx}(5,6) = 4 > 0$, the point $(5,6)$ gives a relative minimum of f with value

$f(5,6) = 2(5)^2 + 6^2 - 2(5)(6) - 8(5) - 2(6) + 2 = -24$.

9. $\left.\begin{array}{l} f_x(x,y) = \frac{\partial}{\partial x}\left(x^2 + 2y^2 + x^2 y + 3\right) = 2x + 2xy = 0 \\ f_y(x,y) = \frac{\partial}{\partial y}\left(x^2 + 2y^2 + x^2 y + 3\right) = 4y + x^2 = 0 \end{array}\right\}$ The first equation gives $x = 0$ or

$y = -1$. Substituting $x = 0$ into the second equation gives $y = 0$; substituting $y = -1$ into the

second equation gives $x = \pm 2$. Thus, f has critical points $(0,0)$, $(-2,-1)$, and $(2,-1)$. Next,

$D(x,y) = f_{xx}(x,y)\, f_{yy}(x,y) - f_{xy}^2(x,y) = (2 + 2y)(4) - (2x)^2 = -4x^2 + 8y + 8$.

At $(0,0)$: $D(0,0) = 8 > 0$ and $f_{xx}(0,0) = 2 > 0$, so $(0,0)$ gives a relative minimum of f with value $f(0,0) = 3$.

At $(-2,-1)$: $D(-2,-1) = -4(-2)^2 + 8(-1) + 8 = -16 < 0$, so $(-2,-1,5)$ is a saddle point of f.

At $(2,-1)$: $D(2,-1) = -4(2)^2 + 8(-1) + 8 = -16 < 0$, so $(2,-1,5)$ is also a saddle point of f.

11. $\left.\begin{array}{l} f_x(x,y) = \frac{\partial}{\partial x}\left(x^2 + 5y^2 + x^2 y + 2y^3\right) = 2x + 2xy = 2x(y + 1) = 0 \\ f_y(x,y) = \frac{\partial}{\partial y}\left(x^2 + 5y^2 + x^2 y + 2y^3\right) = 10y + x^2 + 6y^2 = 0 \end{array}\right\}$ The first equation gives $x = 0$ or $y = -1$.

Substituting $x = 0$ into the second equation gives $10y + 6y^2 = 2y(3y + 5) = 0 \Rightarrow y = 0$ or $-\frac{5}{3}$; substituting $y = -1$ into

the second equation gives $x^2 - 4 = 0 \Rightarrow x = \pm 2$. Thus, f has critical points $(0,0)$, $\left(0, -\frac{5}{3}\right)$, $(-2,-1)$, and $(2,-1)$. Next,

$D(x,y) = f_{xx}(x,y)\, f_{yy}(x,y) - f_{xy}^2(x,y) = 2(y + 1)(10 + 12y) - (2x)^2 = 4(y + 1)(5 + 6y) - 4x^2$.

At $(0,0)$: $D(0,0) = 4(1)(5) - 0 = 20 > 0$ and $f_{xx}(0,0) = 2 > 0$, so f has a relative minimum at $(0,0)$ with value

$f(0,0) = 0$.

At $\left(0, -\frac{5}{3}\right)$: $D\left(0, -\frac{5}{3}\right) = 4\left(-\frac{5}{3} + 1\right)\left[5 + 6\left(-\frac{5}{3}\right)\right] - 4(0) = \frac{40}{3} > 0$ and $f_{xx}\left(0, -\frac{5}{3}\right) = 2\left(-\frac{5}{3} + 1\right) = -\frac{4}{3} < 0$, so

f has a relative maximum at $\left(0, -\frac{5}{3}\right)$ with value $f\left(0, -\frac{5}{3}\right) = \frac{125}{27}$.

At $(-2,-1)$: $D(-2,-1) = 0 - 4(-2)^2 = -16 < 0$, and so $(-2,-1,3)$ is a saddle point of f.

At $(2,-1)$: $D(2,-1) = 0 - 4(2)^2 = -16 < 0$, and so $(2,-1,3)$ is also a saddle point of f.

13. $\left.\begin{array}{l} f_x(x,y) = \frac{\partial}{\partial x}\left(x^2 - 6x - x\sqrt{y} + y\right) = 2x - 6 - \sqrt{y} = 0 \\ f_y(x,y) = \frac{\partial}{\partial y}\left(x^2 - 6x - xy^{1/2} + y\right) = -\frac{x}{2\sqrt{y}} + 1 = 0 \end{array}\right\}$ From the first equation, we see that

$\sqrt{y} = 2x - 6$. Substituting this into the second equation gives $-x + 2(2x - 6) = 0 \Rightarrow x = 4$. Substituting

this into the first equation gives $8 - 6 = \sqrt{y} \Rightarrow y = 4$, so the sole critical point of f is $(4,4)$. Next,

$D(x,y) = f_{xx}(x,y)\, f_{yy}(x,y) - f_{xy}^2(x,y) = 2\left(\frac{x}{4y^{3/2}}\right) - \left(-\frac{1}{2\sqrt{y}}\right)^2 = \frac{x}{2y^{3/2}} - \frac{1}{4y}$. Since

$D(4,4) = \frac{4}{2(8)} - \frac{1}{4(4)} = \frac{3}{16} > 0$ and $f_{xx}(4,4) = 2 > 0$, the point $(4,4)$ gives a relative minimum of f with value

$f(4,4) = -12$.

15. $f_x(x, y) = \dfrac{\partial}{\partial x}\left(xy - \dfrac{2}{x} - \dfrac{4}{y}\right) = y + \dfrac{2}{x^2} = \dfrac{x^2 y + 2}{x^2} = 0$

$\left.\begin{array}{l} f_y(x, y) = \dfrac{\partial}{\partial y}\left(xy - \dfrac{2}{x} - \dfrac{4}{y}\right) = \dfrac{xy^2 + 4}{y^2} = 0 \end{array}\right\} \Rightarrow \left.\begin{array}{l} x^2 y + 2 = 0 \\ xy^2 + 4 = 0 \end{array}\right\}$ From the first equation,

$y = -\dfrac{2}{x^2}$, and substituting this into the second equation yields $x\left(-\dfrac{2}{x^2}\right)^2 + 4 = 0 \Leftrightarrow 4\left(1 + x^3\right) = 0 \Rightarrow x = -1$.

Substituting into the first equation gives $y = -2$, so $(-1, -2)$ is the only critical point of f. Next, $f_{xx}(x, y) = -\dfrac{4}{x^3}$,

$f_{xy}(x, y) = 1$, and $f_{yy}(x, y) = -\dfrac{8}{y^3}$, so $D(x, y) = f_{xx}(x, y)\,f_{yy}(x, y) - f_{xy}^2(x, y) = \dfrac{32}{x^3 y^3} - 1$. Since

$D(-1, -2) = 3 > 0$ and $f_{xx}(-1, -2) = 4 > 0$, the point $(-1, -2)$ gives a relative minimum of f with value

$f(-1, -2) = (-1)(-2) - \dfrac{2}{-1} - \dfrac{4}{-2} = 6$.

17. $\left.\begin{array}{l} f_x(x, y) = \dfrac{\partial}{\partial x}\left(e^{-x^2 - y^2}\right) = -2xe^{-x^2 - y^2} \\ f_y(x, y) = \dfrac{\partial}{\partial y}\left(e^{-x^2 - y^2}\right) = -2ye^{-x^2 - y^2} \end{array}\right\} \Rightarrow f_{xx}(x, y) = 2\left(2x^2 - 1\right)e^{-x^2 - y^2}$, $f_{xy}(x, y) = 4xye^{-x^2 - y^2}$, and

$f_{yy}(x, y) = 2\left(2y^2 - 1\right)e^{-x^2 - y^2}$. Setting $\left.\begin{array}{l} f_x(x, y) = 0 \\ f_y(x, y) = 0 \end{array}\right\} \Rightarrow x = 0$ and $y = 0$, so $(0, 0)$ is the sole critical point of f.

Next, $D(0, 0) = f_{xx}(0, 0)\,f_{yy}(0, 0) - f_{xy}^2(0, 0) = (-2)(-2) - 0^2 = 4 > 0$ and $f_{xx}(0, 0) = -2 < 0$, we see that f has

a relative maximum at $(0, 0)$ with value $f(0, 0) = 1$.

19. $\left.\begin{array}{l} f_x(x, y) = \dfrac{\partial}{\partial x}(x \sin y) = \sin y = 0 \\ f_y(x, y) = \dfrac{\partial}{\partial y}(x \sin y) = x \cos y = 0 \end{array}\right\} \Rightarrow (0, 0), (0, \pi),$ and $(0, 2\pi)$ are critical points of f.

Next, $D(x, y) = f_{xx}(x, y)\,f_{yy}(x, y) - f_{xy}^2(x, y) = 0(-x \sin y) - (\cos y)^2 = -\cos^2 y$. Since

$D(0, n\pi) = -\cos^2 n\pi = -1 < 0$ for $n = 0, 1,$ or 2, we see that $(0, 0, 0), (0, \pi, 0),$ and $(0, 2\pi, 0)$ are saddle points of f.

21. $\left.\begin{array}{l} f_x(x, y) = \dfrac{\partial}{\partial x}\left(e^{-x} \cos y\right) = -e^{-x} \cos y = 0 \\ f_y(x, y) = \dfrac{\partial}{\partial y}\left(e^{-x} \cos y\right) = -e^{-x} \sin y = 0 \end{array}\right\}$ has no solution, so f has no critical point and thus no relative extremum

or saddle point.

23. a. It appears that f has a saddle point at $(0, 0, 0)$ and relative minima at $\left(\frac{3}{2}, -\frac{3}{2}\right)$ and $\left(\frac{3}{2}, \frac{3}{2}\right)$.

b. Set $f_x = 3x^2 - 3y^2 = 0$ (1) and $f_y = -6xy + 4y^3 = 0$ (2). From (1), we find $y = \pm x$. Substituting $y = -x$

into (2) gives $6x^2 - 4x^3 = 2x^2(3 - 2x) = 0 \Leftrightarrow x = 0$ or $x = \frac{3}{2}$. Substituting $y = x$ into (2) gives

$-6x^2 + 4x^3 = -2x^2(3 - 2x) = 0 \Leftrightarrow x = 0$ or $x = \frac{3}{2}$. The corresponding values of y are 0 (for $x = 0$) and $\pm\frac{3}{2}$ (for

$x = \frac{3}{2}$). Thus, f has critical points $(0, 0)$, $\left(\frac{3}{2}, -\frac{3}{2}\right)$, and $\left(\frac{3}{2}, \frac{3}{2}\right)$. Next, $f_{xx} = 6x$, $f_{xy} = -6y$, and $f_{yy} = -6x + 12y^2$,

so $D(x, y) = f_{xx}f_{yy} - f_{xy}^2 = 72xy^2 - 36x^2 - 36y^2$. We use the SDT:

At $(0, 0)$, $D(0, 0) = 0$, so the test does not apply. But observe that for $x = 0$, $f(0, y) = y^4$, and so the values of f are

positive in the y-direction in the neighborhood of $(0, 0)$. On the other hand, for $y = 0$, $f(x, 0) = x^3$, and we see that

$f(x, y) < 0$ for $x < 0$ and $f(x, y) > 0$ for $x > 0$ near $(0, 0)$. This shows that $(0, 0, 0)$ is a saddle point of f.

At $\left(\frac{3}{2}, -\frac{3}{2}\right)$, $D\left(\frac{3}{2}, -\frac{3}{2}\right) = 81 > 0$, and since $f_{xx}\left(\frac{3}{2}, -\frac{3}{2}\right) = 6x|_{x=3/2} = 9 > 0$, we see that $\left(\frac{3}{2}, -\frac{3}{2}\right)$ gives a relative

minimum value of $f\left(\frac{3}{2}, -\frac{3}{2}\right) = -\frac{27}{16}$.

At $\left(\frac{3}{2}, \frac{3}{2}\right)$, $D\left(\frac{3}{2}, \frac{3}{2}\right) = 81 > 0$, and since $f_{xx}\left(\frac{3}{2}, \frac{3}{2}\right) = 9 > 0$, we see that $\left(\frac{3}{2}, \frac{3}{2}\right)$ gives a relative minimum value of

$f\left(\frac{3}{2}, \frac{3}{2}\right) = -\frac{27}{16}$.

25.

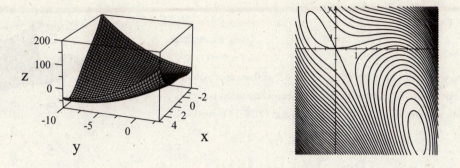

We see that f has a saddle point at $(0, 0)$ and relative minima at $(-1, 2)$ and $(4, -8)$. We set
$f_x = 2x^3 - 6x^2 + 4y = 0$ (1) and $f_y = 4x + 2y = 0$ (2). From (2), we have $y = -2x$. Substituting into (1) gives
$2x^3 - 6x^2 - 8x = 2x\left(x^2 - 3x - 4\right) = 2x\left(x - 4\right)\left(x + 1\right) = 0 \Rightarrow x = -1, 0,$ or 4. The corresponding values of y are
$y = 2, 0,$ and -8, so f has critical points $(-1, 2)$, $(0, 0)$, and $(4, -8)$. $f_{xx} = 6x^2 - 12x = 6x\left(x - 2\right)$, $f_{xy} = 4$, and
$f_{yy} = 2$, so $D\left(x, y\right) = f_{xx}f_{yy} - f_{xy}^2 = 12x\left(x - 2\right) - 16 = 4\left(3x^2 - 6x - 4\right)$.
At $(-1, 2)$, $D(-1, 2) = 4\left(3 + 6 - 4\right) = 20 > 0$, and since $f_{xx}\left(-1, 2\right) = 18 > 0$, we see that $(-1, 2)$ gives a relative
minimum with value $f\left(-1, 2\right) = -\frac{3}{2}$.
At $(0, 0)$, $D\left(0, 0\right) = -16 < 0$, so $(0, 0, 0)$ is a saddle point.
At $(4, -8)$, $D\left(4, -8\right) = 80 > 0$, and since $f_{xx}\left(4, -8\right) = 48 > 0$, we see that $(4, -8)$ gives a relative minimum with value
$f\left(4, -8\right) = -64$.

27.

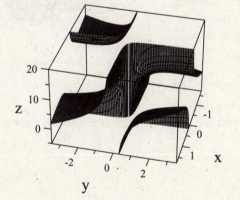
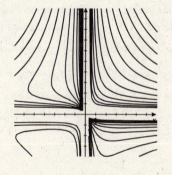

We see that f has a relative minimum at $(-1, -2)$. We set
$$\begin{cases} f_x = y + \dfrac{2}{x^2} = 0 \\ f_y = x + \dfrac{4}{y^2} = 0 \end{cases}$$
The first equation gives
$y = -\dfrac{2}{x^2}$, and substituting into the second gives $x + \dfrac{4}{\left(-2/x^2\right)^2} = 0 \Leftrightarrow x + x^4 = x\left(1 + x^3\right) = 0 \Leftrightarrow x = -1$ (since
$x \neq 0$). The corresponding value of y is -2, and so $(-1, -2)$ is the only critical point of f. We find $f_{xx} = -4/x^3$,
$f_{xy} = 1$, and $f_{yy} = -8/y^3$, so $D\left(-1, -2\right) = f_{xx}\left(-1, -2\right)f_{yy}\left(-1, -2\right) - f_{xy}^2\left(-1, -2\right) = 4 \cdot 1 - 1 = 3 > 0$, and since
$f_{xx}\left(-1, -2\right) = 4 > 0$, we see that f has a relative minimum at $(-1, -2)$ with value $f\left(-1, -2\right) = 14$.

29. $f(x, y) = 2x^4 - 8x^2 + y^2 + 4x - 2y - 5 \Rightarrow$

$$\begin{cases} f_x = 8x^3 - 16x + 4 = 4\left(2x^3 - 4x + 1\right) = 0 \\ f_y = 2y - 2 = 0 \end{cases}$$ Using a calculator

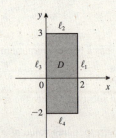

or computer, we see that the roots of $2x^3 - 4x + 1 = 0$ are approximately
-1.526, 0.259, and 1.267. The solution of the second equation is $y = 1$,
so f has critical points $(-1.526, 1)$, $(0.259, 1)$, and $(1.267, 1)$. Next,

$f_{xx} = 24x^2 - 16 = 8\left(3x^2 - 2\right)$, $f_{xy} = 0$, and $f_{yy} = 2$, so

$D(x, y) = f_{xx}f_{yy} - f_{xy}^2 = 8\left(3x^2 - 2\right)(2) = 16\left(3x^2 - 2\right)$.

$D(-1.526, 1) = 16\left[3(-1.526)^2 - 2\right] \approx 79.776 > 0$ and since $f_{xx}(-1.526, 1) \approx 39.888 > 0$, we see that $(-1.526, 1)$
gives a relative minimum with value $f(-1.526, 1) \approx -19.888$.

$D(0.259, 1) = 16\left[3(0.259)^2 - 2\right] \approx -28.78 < 0$, and so $(0.259, 1, -5.492)$ is a saddle point.

$D(1.267, 1) \approx 45.054 > 0$, and since $f_{xx}(1.267, 1) \approx 22.527 > 0$, we see that $(1.267, 1)$ gives a relative minimum with
value $f(1.267, 1) \approx -8.620$.

31. $f(x, y) = -x^4 - y^4 + 2x^2 y + x^2 + y - 2 \Rightarrow$

$$\begin{cases} f_x = -4x^3 + 4xy + 2x = -2x\left(2x^2 - 2y - 1\right) = 0 & (1) \\ f_y = -4y^3 + 2x^2 + 1 = 0 & (2) \end{cases}$$

From (1), we see that $x = 0$ or $y = \frac{1}{2}\left(2x^2 - 1\right)$. If $x = 0$, then (2) gives

$y \approx 0.630$. If $2x^2 = 2y + 1$, then (2) gives $-4y^3 + 2y + 2 = 0$. Solving
this equation, we find $y = 1$. The corresponding values of x are
approximately ± 1.225, so f has critical points $(-1.225, 1)$, $(0, 0.630)$,
and $(1.225, 1)$. Next, $f_{xx} = -12x^2 + 4y + 2$, $f_{xy} = 4x$, and

$f_{yy} = -12y^2$, so $D(x, y) = f_{xx}f_{yy} - f_{xy}^2 = 12y^2\left(12x^2 - 4y - 2\right) - 16x^2$.

$D(0, 0.630) \approx -21.528 < 0$, and so $(0, 0.630, -1.528)$ is a saddle point.

$D(\pm 1.225, 1) = 120.08 > 0$ and $f_{xx}(\pm 1.225, 1) \approx -12.008 < 0$, and so $(\pm 1.225, 1)$ give relative maxima of f with
value $f(\pm 1.225, 1) \approx 0.250$.

33. Since $f_x(x, y) = \frac{\partial}{\partial x}(2x + 3y - 6) = 2$ and

$f_y(x, y) = \frac{\partial}{\partial y}(2x + 3y - 6) = 3$ are never equal to 0, f has no critical

point on D.

On ℓ_1, $x = 2$ and $y = y$, so $g(y) = f(2, y) = 3y - 2$ for $-2 \leq y \leq 3$.
We see that g has an absolute minimum value of -8 at $(2, -2)$ and an
absolute maximum value of 7 at $(2, 3)$.

On ℓ_2, $x = x$ and $y = 3$, so $h(x) = f(x, 3) = 2x + 3$ for $0 \leq x \leq 2$. We see that h has an absolute minimum value of 3 at
$(0, 3)$ and an absolute maximum value of 7 at $(2, 3)$.

On ℓ_3, $x = 0$ and $y = y$, so $s(y) = f(0, y) = 3y - 6$ for $-2 \leq y \leq 3$. We see that s has an absolute minimum value of -12 at $(0, -2)$ and an absolute maximum value of 3 at $(0, 3)$.

On ℓ_4, $x = x$ and $y = -2$, so $t(x) = f(x, -2) = 2x - 12$ for $0 \leq x \leq 2$. We see that t has an absolute minimum value of -12 at $(0, -2)$ and an absolute maximum value of -8 at $(2, -2)$.

The extreme values of f on each boundary of D are summarized below.

	ℓ_1		ℓ_2		ℓ_3		ℓ_4	
(x, y)	$(2, -2)$	$(2, 3)$	$(0, 3)$	$(2, 3)$	$(0, -2)$	$(0, 3)$	$(0, -2)$	$(2, -2)$
Extreme value	-8	7	3	7	-12	3	-12	-8

We see that f has an absolute minimum value of $f(0, -2) = -12$ and an absolute maximum value of $f(2, 3) = 7$ on D.

35. $f_x(x, y) = 3$ and $f_y(x, y) = 4$, so f has no critical point.

On ℓ_1, $x = x$ and $y = 0$, so $g(y) = f(x, 0) = 3x - 12$ for $0 \leq x \leq 3$.

We see that f has an absolute minimum value of -12 and an absolute

maximum value of -3 on ℓ_1.

On ℓ_2, $x = 3$ and $y = y$, so $h(y) = f(3, y) = 4y - 3$ for $0 \leq y \leq 4$. We

see that f has an absolute minimum value of -3 and an absolute

maximum value of 13 on ℓ_2.

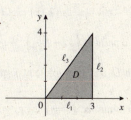

On ℓ_3, $y = \frac{4}{3}x$, so $s(x) = f\left(x, \frac{4}{3}x\right) = 3x + 4\left(\frac{4}{3}x\right) - 12 = \frac{25}{3}x - 12$ for $0 \leq x \leq 3$. We see that f has an absolute

minimum value of -12 and an absolute maximum value of 13 on ℓ_3.

From these calculations, we see that f has an absolute minimum value of -12 and an absolute maximum value of 13 on D.

37. $\begin{aligned} f_x(x, y) &= \frac{\partial}{\partial x}\left(xy - x^2\right) = y - 2x = 0 \\ f_y(x, y) &= \frac{\partial}{\partial y}\left(xy - x^2\right) = x = 0 \end{aligned}\right\} \Rightarrow x = 0, y = 0$, so f has

no critical point in the interior of D.

On C_1, $y = x^2$, so $g(x) = f\left(x, x^2\right) = x^3 - x^2$ for $-2 \leq x \leq 2$.

$g'(x) = 3x^2 - 2x = x(3x - 2) = 0 \Rightarrow x = 0$ or $x = \frac{2}{3}$, so 0 and $\frac{2}{3}$ are

critical numbers of g on $(-2, 2)$.

From the table, we see that f has an absolute minimum value of -12 and

an absolute maximum value of 4 on C_1.

On C_2, $x = x$ and $y = 4$, so $h(x) = f(x, 4) = 4x - x^2$ for $-2 \leq x \leq 2$.

$h'(x) = 4 - 2x = 0 \Rightarrow x = 2$, an endpoint. We find $h(-2) = -12$ and

$g(2) = 4$, so f has an absolute minimum value of -12 and an absolute

maximum value of 4 on C_2.

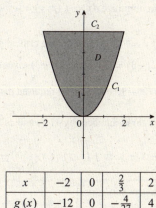

x	-2	0	$\frac{2}{3}$	2
$g(x)$	-12	0	$-\frac{4}{27}$	4

We conclude that f has an absolute minimum value of -12 and an absolute maximum value of 4 on D.

39.
$$\left. \begin{array}{l} f_x\,(x,y) = \frac{\partial}{\partial x}\left(x^2 + 4y^2 + 3x - 1\right) = 2x + 3 = 0 \\ f_y\,(x,y) = \frac{\partial}{\partial y}\left(x^2 + 4y^2 + 3x - 1\right) = 8y = 0 \end{array} \right\} \Rightarrow x = -\tfrac{3}{2} \text{ and }$$

$y = 0$, so f has the critical point $\left(-\tfrac{3}{2}, 0\right)$ in D with $f\left(-\tfrac{3}{2}, 0\right) = -\tfrac{13}{4}$.

Next we consider the boundary of D. On C_1, $y = \sqrt{4 - x^2}$, so

$$g\,(x) = f\left(x, \sqrt{4 - x^2}\right) = x^2 + 4\left(\sqrt{4 - x^2}\right)^2 + 3x - 1 = -3x^2 + 3x + 15$$

for $-2 \le x \le 2$. $g'\,(x) = -6x + 3 = 0 \Rightarrow x = \tfrac{1}{2}$, so f has a critical point

at $\left(\tfrac{1}{2}, \tfrac{\sqrt{15}}{2}\right)$. From the table, we see that f has an absolute minimum

value of -3 and an absolute maximum value of $\tfrac{63}{4}$ on C_1.

x	-2	$\tfrac{1}{2}$	2
$g\,(x)$	-3	$\tfrac{63}{4}$	9

On C_2, $y = -\sqrt{4 - x^2}$, so $h\,(x) = f\left(x, -\sqrt{4 - x^2}\right) = x^2 + 4\left(-\sqrt{4 - x^2}\right)^2 + 3x - 1$, which is the same as $g\,(x)$. So h

has a critical point at $\left(\tfrac{1}{2}, -\tfrac{\sqrt{15}}{2}\right)$. We conclude that f has an absolute minimum value of $-\tfrac{13}{4}$ attained at the critical point

$\left(-\tfrac{3}{2}, 0\right)$ on D, and an absolute maximum value of $\tfrac{63}{4}$ attained at the points $\left(\tfrac{1}{2}, \pm\tfrac{\sqrt{15}}{2}\right)$ on the boundary of D.

41. We want to minimize $d^2 = f\,(x,y) = x^2 + y^2 + z^2 = x^2 + y^2 + (4 - x - 2y)^2 = 2x^2 + 5y^2 + 4xy - 8x - 16y + 16$.

$$\left. \begin{array}{l} f_x\,(x,y) = \frac{\partial}{\partial x}\left(2x^2 + 5y^2 + 4xy - 8x - 16y + 16\right) = 4x + 4y - 8 = 0 \\ f_y\,(x,y) = \frac{\partial}{\partial y}\left(2x^2 + 5y^2 + 4xy - 8x - 16y + 16\right) = 10y + 4x - 16 = 0 \end{array} \right\} \Rightarrow x = \tfrac{2}{3},\, y = \tfrac{4}{3}, \text{ so } \left(\tfrac{2}{3}, \tfrac{4}{3}\right) \text{ is a critical}$$

point of f. $D\,(x,y) = f_{xx}\,(x,y)\, f_{yy}\,(x,y) - f_{xy}^2\,(x,y) = 4 \cdot 10 - 4^2 = 24 > 0$ and $f_{xx}\left(\tfrac{2}{3}, \tfrac{4}{3}\right) = 4 > 0$, so

$\left(\tfrac{2}{3}, \tfrac{4}{3}\right)$ gives the only relative minimum (and thus the absolute minimum) of f in the plane. The required distance is

$$d = \sqrt{f\left(\tfrac{2}{3}, \tfrac{4}{3}\right)} = \left[\left(\tfrac{2}{3}\right)^2 + \left(\tfrac{4}{3}\right)^2 + \left(4 - \tfrac{2}{3} - \tfrac{8}{3}\right)^2\right]^{1/2} = \tfrac{2\sqrt{6}}{3}.$$

43. We want to minimize $d^2 = f\,(x,y) = x^2 + y^2 + z^2 = x^2 + y^2 + xy - x + 4y + 21$. $\left. \begin{array}{l} f_x\,(x,y) = 2x + y - 1 = 0 \\ f_y\,(x,y) = 2y + x + 4 = 0 \end{array} \right\} \Rightarrow$

$x = 2$ and $y = -3$, so $(2, -3)$ is a critical point of f. $D\,(x,y) = f_{xx}\,(x,y)\, f_{yy}\,(x,y) - f_{xy}^2\,(x,y) = 2 \cdot 2 - 1^2 = 3 > 0$

and $f_{xx}\,(2, -3) = 2 > 0$, so $(2, -3)$ gives the only relative minimum (and thus the absolute minimum) of f. The required

points are $(2, -3, \pm 1)$, and the required distance is $d = \sqrt{f\,(2, -3)} = \sqrt{2^2 + (-3)^2 + 1^2} = \sqrt{14}$.

45. Denote the three numbers by x, y, and z. We want to maximize $P = xyz$. But $x + y + z = 500 \Leftrightarrow$

$z = 500 - x - y$, so $f\,(x,y) = xy\,(500 - x - y) = -x^2 y - xy^2 + 500xy$. $\left. \begin{array}{l} f_x\,(x,y) = -2xy - y^2 + 500y = 0 \\ f_y\,(x,y) = -x^2 - 2xy + 500x = 0 \end{array} \right\}$

$\Rightarrow \left. \begin{array}{l} 2x + y - 500 = 0 \\ x + 2y - 500 = 0 \end{array} \right\} \Rightarrow x = y = \tfrac{500}{3}$, so $\left(\tfrac{500}{3}, \tfrac{500}{3}\right)$ is a critical point of f.

$D\,(x,y) = f_{xx}\,(x,y)\, f_{yy}\,(x,y) - f_{xy}^2\,(x,y) = (-2y)\,(-2x) + (-2x - 2y + 500)^2 = 4xy + (2x + 2y - 500)^2 > 0$

and $f_{xx}\left(\tfrac{500}{3}, \tfrac{500}{3}\right) = -\tfrac{1000}{3} < 0$, so $\left(\tfrac{500}{3}, \tfrac{500}{3}\right)$ gives a relative maximum (and thus the absolute maximum) of f.

$z = 500 - x - y = 500 - \tfrac{500}{3} - \tfrac{500}{3} = \tfrac{500}{3}$, so all three numbers are $\tfrac{500}{3}$.

47. Denote the length, width, and height of the box (in feet) by x, y, and z, respectively. We want to maximize

$V = xyz$, but $2xy + 2yz + 2xz = 48 \Leftrightarrow z = \dfrac{24 - xy}{x + y}$, so $V = f(x, y) = xy\left(\dfrac{24 - xy}{x + y}\right) = \dfrac{24xy - x^2y^2}{x + y} \Rightarrow$

$f_x(x, y) = \dfrac{24y^2 - x^2y^2 - 2xy^3}{(x + y)^2}$ and $f_y(x, y) = \dfrac{24x^2 - x^2y^2 - 2x^3y}{(x + y)^2}$. Setting $f_x(x, y) = 0$ and $f_y(x, y) = 0$ and

keeping in mind that $x > 0$ and $y > 0$, we have $\left.\begin{array}{l} x^2 + 2xy = 24 \\ y^2 + 2xy = 24 \end{array}\right\} \Rightarrow x = y = 2\sqrt{2}$, so $\left(2\sqrt{2}, 2\sqrt{2}\right)$ is the only critical

point of f. Since a maximum must exist, we see that the dimensions of the rectangular box are $x = 2\sqrt{2}$, $y = 2\sqrt{2}$, and

$z = \dfrac{24 - \left(2\sqrt{2}\right)^2}{2\sqrt{2} + 2\sqrt{2}} = 2\sqrt{2}.$

49. The length, width, and height of the box in the first octant are x, y, and z, respectively. Thus, $V = xyz$, where V

is one eighth of the volume of the box. $z = \frac{2}{3}\sqrt{36 - 9x^2 - 4y^2}$, so $V = f(x, y) = \frac{2}{3}xy\sqrt{36 - 9x^2 - 4y^2} \Rightarrow$

$f_x(x, y) = \dfrac{2y\left(36 - 18x^2 - 4y^2\right)}{3\sqrt{36 - 9x^2 - 4y^2}}$ and $f_y(x, y) = \dfrac{2x\left(36 - 9x^2 - 8y^2\right)}{3\sqrt{36 - 9x^2 - 4y^2}}$. Setting $f_x(x, y) = 0$ and $f_y(x, y) = 0$

simultaneously leads to $\left.\begin{array}{l} 36 - 18x^2 - 4y^2 = 0 \\ 36 - 9x^2 - 8y^2 = 0 \end{array}\right\} \Rightarrow x = \frac{2\sqrt{3}}{3}$ and $y = \sqrt{3}$, so $\left(\frac{2\sqrt{3}}{3}, \sqrt{3}\right)$ is the only critical point of f.

Since a maximum must exist, we conclude that the dimensions of the rectangular box are $2x = \frac{4\sqrt{3}}{3}$, $2y = 2\sqrt{3}$, and

$2z = \frac{4}{3}\sqrt{36 - 9\left(\frac{2\sqrt{3}}{3}\right)^2 - 4\left(\sqrt{3}\right)^2} = \frac{8\sqrt{3}}{3}$; that is, $\frac{4\sqrt{3}}{3} \times 2\sqrt{3} \times \frac{8\sqrt{3}}{3}$.

51. $V = xyz$, but $z = 6 - 2x - 3y$, so

$V = f(x, y) = xy(6 - 2x - 3y) = 6xy - 2x^2y - 3xy^2 \Rightarrow$

$\left.\begin{array}{l} f_x(x, y) = 6y - 4xy - 3y^2 = 0 \\ f_y(x, y) = 6x - 2x^2 - 6xy = 0 \end{array}\right\} \Rightarrow \left.\begin{array}{l} 6 - 4x - 3y = 0 \\ 6 - 2x - 6y = 0 \end{array}\right\} \Rightarrow x = 1$

and $y = \frac{2}{3}$, so $\left(1, \frac{2}{3}\right)$ is a critical point of f. Since the maximum must

exist, we conclude that the volume of the required box is

$V = (1)\left(\frac{2}{3}\right)\left[6 - 2(1) - 3\left(\frac{2}{3}\right)\right] = \frac{4}{3}$ and the dimensions of the box are

$1 \times \frac{2}{3} \times 2$.

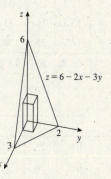

53. Denote the length, width, and height of the box (in feet) by x, y, and z, respectively. We may assume, without affecting

the conclusion, that the cost of the material for the sides of the box is $1/\text{ft}^2$. We want to minimize the cost function

$C = 3xy + 2xz + 2yz$, but $xyz = 12 \Leftrightarrow z = \dfrac{12}{xy}$, so $C = f(x, y) = 3xy + \dfrac{24}{y} + \dfrac{24}{x}$. $\left.\begin{array}{l} f_x(x, y) = 3y - \dfrac{24}{x^2} = 0 \\ f_x(x, y) = 3x - \dfrac{24}{y^2} = 0 \end{array}\right\} \Rightarrow$

$x = y = 2$, so $(2, 2)$ is a critical point of f. Since the minimum cost clearly exists, we see that the dimensions of the box

are $x = 2$, $y = 2$, and $z = \dfrac{12}{2 \cdot 2} = 3$; that is, $2' \times 2' \times 3'$.

55. The sums of the squares of the distances from the proposed site of the radio station to the three communities is

$$D = f(x, y) = [d(P, A)]^2 + [d(P, B)]^2 + [d(P, C)]^2$$

$$= \left[(x-2)^2 + (y-4)^2\right] + \left[(x-20)^2 + (y-8)^2\right] + \left[(x-4)^2 + (y-24)^2\right]$$

$$\left. \begin{array}{l} f_x(x, y) = 2(x-2) + 2(x-20) + 2(x-4) = 0 \\ f_y(x, y) = 2(y-4) + 2(y-8) + 2(y-24) = 0 \end{array} \right\} \Rightarrow x = \frac{26}{3} \text{ and } y = 12, \text{ so } \left(\frac{26}{3}, 12\right) \text{ is a critical point of } f. \text{ Since it}$$

is clear that D must attain a minimum, we see that the station should be located at $\left(\frac{26}{3}, 12\right)$.

57. $g(m, b) = \sum_{k=1}^{n} (y_k - mx_k - b)^2$, so we require that $\left. \begin{array}{l} g_m(m, b) = -2\sum_{k=1}^{n} x_k (y_k - mx_k - b) = 0 \\ g_b(m, b) = -2\sum_{k=1}^{n} (y_k - mx_k - b) = 0 \end{array} \right\} \Rightarrow$

$$\left. \begin{array}{l} \left(\sum_{k=1}^{n} x_k^2\right) m + \left(\sum_{k=1}^{n} x_k\right) b = \sum_{k=1}^{n} x_k y_k \\ \left(\sum_{k=1}^{n} x_k\right) m + \left(\sum_{k=1}^{n} 1\right) b = \sum_{k=1}^{n} y_k \end{array} \right\} \Rightarrow \left. \begin{array}{l} \left(\sum_{k=1}^{n} x_k^2\right) m + \left(\sum_{k=1}^{n} x_k\right) b = \sum_{k=1}^{n} x_k y_k \\ \left(\sum_{k=1}^{n} x_k\right) m + nb = \sum_{k=1}^{n} y_k \end{array} \right\} \text{ It is clear}$$

that a minimum value of g must exist, so the critical point (m_0, b_0) satisfying the system gives the least-squares line $y = m_0 x + b_0$ that we seek.

59.

t	y	t^2	ty
0	6.8	0	0
1	8.3	1	8.3
2	9.8	4	19.6
3	11.3	9	33.9
4	12.8	16	51.2
5	14.9	25	74.5
15	63.9	55	187.5

a. The normal equations are $\left. \begin{array}{l} 6b + 15m = 63.9 \\ 15b + 55m = 187.5 \end{array} \right\}$ The solutions

are $m \approx 1.59$ and $b \approx 6.69$. Therefore, the required equation is $y = 1.59t + 6.69$.

b. $y = 1.59(8) + 6.69 = 19.41$, or $\$19.41$ billion.

61.

x	y	x^2	xy
5	40	25	200
6	43.2	36	259.2
7	47.4	49	331.8
8	50.5	64	404
9	53.7	81	483.3
10	56.8	100	568
45	291.6	355	2246.3

a. The data are summarized in the table. The normal equations are

$$\left. \begin{array}{l} 6b + 45m = 291.6 \\ 45b + 355m = 2246.3 \end{array} \right\}$$

The solutions are $m \approx 3.39$ and $b \approx 23.19$. The required equation is $y = 3.39x + 23.19$.

b. $\$3.39$ billion/yr

c. $y(11) = 3.39(11) + 23.19 = 60.48$, representing about $\$60.5$ billion.

63. a. The domain of $f(x, y)$ is $\{(x, y) \mid -\infty < x < \infty \text{ and } -\infty < y < \infty\}$. We complete the squares in x and y, obtaining

$$f(x, y) = -3\left[x^2 - 2x + (-1)^2\right] - 4\left[y^2 + y + \left(\frac{1}{2}\right)^2\right] - 3 + 3 + 1 = -3(x-1)^2 - 4\left(y + \frac{1}{2}\right)^2 + 1. \text{ Since } x \text{ and } y$$

are arbitrary, f is unbounded below. This shows that f has no minimum value.

b. $\begin{cases} f_x(x,y) = -6x + 6 = -6(x-1) = 0 \\ f_y(x,y) = -8y - 4 = -4(2y+1) = 0 \end{cases} \Rightarrow x = 1,\ y = -\frac{1}{2}$, so $\left(1, -\frac{1}{2}\right)$ is the only critical point in D, and

$f\left(1, -\frac{1}{2}\right) = 1$. Next, observe that the boundary of D has parametric representation $x = \cos t$, $y = \sin t$ for

$0 \le t \le 2\pi$. We find

$$F(t) = f(\cos t, \sin t) = -3\cos^2 t + 6\cos t - 4\sin^2 t - 4\sin t - 3$$

$$= -3\left(\cos^2 t + \sin^2 t\right) + 6\cos t - \sin^2 t - 4\sin t - 3 = -\sin^2 t + 6\cos t - 4\sin t - 6$$

Using a calculator or computer, we see that f has a minimum value of approximately -13.5833 attained at $t \approx 2.4154$

$\Rightarrow x = \cos 2.4154 \approx -0.75$ and $y = \sin 2.4154 \approx 0.66$. Its maximum value of approximately 0.9560 is attained at

$t \approx 5.8082 \Rightarrow x = \cos 5.8082 \approx 0.89$ and $y = \sin 5.8082 \approx -0.46$. We conclude that the required maximum value of f

is 1 attained at the critical point $\left(1, -\frac{1}{2}\right)$, and its minimum value is approximately -13.5833 attained at the point

$(-0.75, 0.66)$ on the boundary.

65. False. $f(x, y) = 1 - |x| - |y|$ has an absolute maximum value of 1 attained at $(0, 0)$, but f_x and f_y do not exist at $(0, 0)$.

67. False. The function $f(x, y) = x^2 - y^2$ satsifies $\nabla f(0, 0) = 0$, but $(0, 0, 0)$ is a saddle point.

13.9 Concept Questions   ET 12.9

1. See page 1128 (1124 in ET). Answers will vary.

3. There are four (critical) points where the graph of f is tangent to a level curve of g. We see that the values of f attained at these points are 4, -6, 6, and -2, so the minimum value is -6 and the maximum value is 4.

13.9 Lagrange Multipliers   ET 12.9

1. $\nabla f = 3\mathbf{i} + 4\mathbf{j}$ and $\nabla g = 2x\mathbf{i} + 2y\mathbf{j}$, so $\nabla f = \lambda \nabla g \Rightarrow$

$3\mathbf{i} + 4\mathbf{j} = 2\lambda x\mathbf{i} + 2\lambda y\mathbf{j}$. Together with the constraint equation, we have

$$\left.\begin{array}{r} 3 = 2\lambda x \\ 4 = 2\lambda y \\ x^2 + y^2 = 1 \end{array}\right\}$$

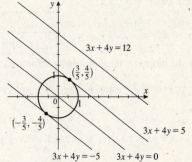

Solving the system, we find $\lambda = \frac{5}{2}$, $x = \frac{3}{5}$, and $y = \frac{4}{5}$ or $\lambda = -\frac{5}{2}$,

$x = -\frac{3}{5}$, and $y = -\frac{4}{5}$. The minimum value of f subject to $x^2 + y^2 = 1$ is

$f\left(-\frac{3}{5}, -\frac{4}{5}\right) = -5$ and the maximum value of f subject to $x^2 + y^2 = 1$

is $f\left(\frac{3}{5}, \frac{4}{5}\right) = 5$.

3. $\nabla f = 2x\mathbf{i} + 2y\mathbf{j}$ and $\nabla g = y\mathbf{i} + x\mathbf{j}$, so $\nabla f = \lambda\nabla g \Rightarrow$
$2x\mathbf{i} + 2y\mathbf{j} = \lambda y\mathbf{i} + \lambda x\mathbf{j}$. Together with the constraint equation, we have

$$\left.\begin{array}{r} 2x = \lambda y \\ 2y = \lambda x \\ xy = 1 \end{array}\right\}$$

Solving the system, we find $\lambda = 2$, $x = -1$, and $y = -1$ or $\lambda = 2$, $x = 1$,
and $y = 1$. The minimum value of f subject to $xy = 1$ is
$f(-1,-1) = f(1,1) = 2$.

5. $f(x,y) = xy \Rightarrow \nabla f(x,y) = y\mathbf{i} + x\mathbf{j}$ and $g(x,y) = 2x + 3y - 6 \Rightarrow \nabla g(x,y) = 2\mathbf{i} + 3\mathbf{j}$, so $\nabla f = \lambda\nabla g \Rightarrow$

$y\mathbf{i} + x\mathbf{j} = 2\lambda\mathbf{i} + 3\lambda\mathbf{j}$, and we solve $\left.\begin{array}{rr} y = 2\lambda & (1) \\ x = 3\lambda & (2) \\ 2x + 3y = 6 & (3) \end{array}\right\} \Rightarrow \lambda = \frac{1}{2}, x = \frac{3}{2}$, and $y = 1$. The maximum value of f is

$f\left(\frac{3}{2},1\right) = \frac{3}{2}$.

7. $f(x,y) = xy \Rightarrow \nabla f(x,y) = y\mathbf{i} + x\mathbf{j}$ and $g(x,y) = x^2 + 4y^2 - 1 \Rightarrow \nabla g(x,y) = 2x\mathbf{i} + 8y\mathbf{j}$, so $\nabla f = \lambda\nabla g \Rightarrow$

$y\mathbf{i} + x\mathbf{j} = 2\lambda x\mathbf{i} + 8\lambda y\mathbf{j}$, and we solve $\left.\begin{array}{rr} y = 2\lambda x & (1) \\ x = 8\lambda y & (2) \\ x^2 + 4y^2 = 1 & (3) \end{array}\right\}$ Substituting (1) into (2) gives $x = 8\lambda(2\lambda x) \Rightarrow$

$\lambda = \pm\frac{1}{4}$. [Note that $(x,y) \neq (0,0)$ since this violates (3).] If $\lambda = -\frac{1}{4}$, then (1) gives $y = -\frac{1}{2}x$ and so (3) gives $y = \pm\frac{\sqrt{2}}{4}$

and $x = \mp\frac{\sqrt{2}}{2}$; if $\lambda = \frac{1}{4}$, then (1) gives $y = \frac{1}{2}x$ and so (3) gives $\left(\frac{\sqrt{2}}{2},\frac{\sqrt{2}}{4}\right)$ and $\left(-\frac{\sqrt{2}}{2},-\frac{\sqrt{2}}{4}\right)$.

(x,y)	$\left(-\frac{\sqrt{2}}{2},\frac{\sqrt{2}}{4}\right)$	$\left(\frac{\sqrt{2}}{2},-\frac{\sqrt{2}}{4}\right)$	$\left(\frac{\sqrt{2}}{2},\frac{\sqrt{2}}{4}\right)$	$\left(-\frac{\sqrt{2}}{2},-\frac{\sqrt{2}}{4}\right)$
$f(x,y)$	$-\frac{1}{4}$	$-\frac{1}{4}$	$\frac{1}{4}$	$\frac{1}{4}$

From the table, we see that f has a minimum value of $-\frac{1}{4}$ and a maximum value of $\frac{1}{4}$.

9. $f(x,y) = x^2 + xy + y^2 \Rightarrow \nabla f(x,y) = (2x+y)\mathbf{i} + (x+2y)\mathbf{j}$ and $g(x,y) = x^2 + y^2 - 8 \Rightarrow \nabla g(x,y) = 2x\mathbf{i} + 2y\mathbf{j}$,

so $\nabla f = \lambda\nabla g \Rightarrow (2x+y)\mathbf{i} + (x+2y)\mathbf{j} = 2\lambda x\mathbf{i} + 2\lambda y\mathbf{j}$, and we solve $\left.\begin{array}{r} 2x + y = 2\lambda x \\ x + 2y = 2\lambda y \\ x^2 + y^2 = 8 \end{array}\right\} \Rightarrow \left.\begin{array}{r} x = 2y(\lambda - 1) \\ y = 2x(\lambda - 1) \end{array}\right\} \Rightarrow$

$y = \pm x$ (because $\lambda = 1 \Rightarrow x = y = 0$, violating the third equation). Substituting $y = \pm x$ into the third equation gives
$x = \pm 2$ and $y = \pm 2$.

(x,y)	$(-2,-2)$	$(-2,2)$	$(2,-2)$	$(2,2)$
$f(x,y)$	12	4	4	12

From the table, we see that f has a minimum value of 4 and a maximum value of 12.

11. $f(x,y,z) = x + 2y + z \Rightarrow \nabla f(x,y,z) = \mathbf{i} + 2\mathbf{j} + \mathbf{k}$ and $g(x,y,z) = x^2 + 4y^2 - z \Rightarrow \nabla g(x,y,z) = 2x\mathbf{i} + 8y\mathbf{j} - \mathbf{k}$, so

$\nabla f = \lambda\nabla g \Rightarrow \mathbf{i} + 2\mathbf{j} + \mathbf{k} = 2\lambda x\mathbf{i} + 8\lambda y\mathbf{j} - \lambda\mathbf{k}$, and we solve $\left.\begin{array}{r} 1 = 2\lambda x \\ 2 = 8\lambda y \\ 1 = -\lambda \\ x^2 + 4y^2 - z = 0 \end{array}\right\}$ We find $\lambda = -1$, $x = -\frac{1}{2}$,

$y = -\frac{1}{4}$, and $z = \frac{1}{2}$. We see that f has a minimum value of $f\left(-\frac{1}{2},-\frac{1}{4},\frac{1}{2}\right) = -\frac{1}{2}$.

13. $f(x, y, z) = x + 2y - 2z \Rightarrow \nabla f(x, y, z) = \mathbf{i} + 2\mathbf{j} - 2\mathbf{k}$ and $g(x, y, z) = x^2 + 2y^2 + 4z^2 - 1 \Rightarrow$
$\nabla g(x, y, z) = 2x\mathbf{i} + 4y\mathbf{j} + 8z\mathbf{k}$, so $\nabla f = \lambda \nabla g \Rightarrow \mathbf{i} + 2\mathbf{j} - 2\mathbf{k} = 2\lambda x\mathbf{i} + 4\lambda y\mathbf{j} + 8\lambda z\mathbf{k}$, and we solve

$$\left.\begin{array}{r} 1 = 2\lambda x \\ 2 = 4\lambda y \\ -2 = 8\lambda z \\ x^2 + 2y^2 + 4z^2 = 1 \end{array}\right\}$$ We see that $x = \dfrac{1}{2\lambda}$, $y = \dfrac{1}{2\lambda}$, and $z = -\dfrac{1}{4\lambda}$, which we substitute into the fourth equation:

$\left(\dfrac{1}{2\lambda}\right)^2 + 2\left(\dfrac{1}{2\lambda}\right)^2 + 4\left(-\dfrac{1}{4\lambda}\right)^2 = 1 \Rightarrow \lambda = \pm 1$. If $\lambda = -1$, then $x = -\frac{1}{2}$, $y = -\frac{1}{2}$, and $z = \frac{1}{4}$; if $\lambda = 1$, then $x = \frac{1}{2}$,

$y = \frac{1}{2}$, and $z = -\frac{1}{4}$. We see that the minimum value of f is $f\left(-\frac{1}{2}, -\frac{1}{2}, \frac{1}{4}\right) = -2$ and the maximum value of f is

$f\left(\frac{1}{2}, \frac{1}{2}, -\frac{1}{4}\right) = 2$.

15. $f(x, y, z) = xyz \Rightarrow \nabla f(x, y, z) = yz\mathbf{i} + xz\mathbf{j} + xy\mathbf{k}$ and $g(x, y, z) = x^2 + 2y^2 + \frac{1}{2}z^2 = 6 \Rightarrow$
$\nabla g(x, y, z) = 2x\mathbf{i} + 4y\mathbf{j} + z\mathbf{k}$, so $\nabla f = \lambda \nabla g \Rightarrow yz\mathbf{i} + xz\mathbf{j} + xy\mathbf{k} = 2\lambda x\mathbf{i} + 4\lambda y\mathbf{j} + \lambda z\mathbf{k}$, and we solve

$$\left.\begin{array}{r} yz = 2\lambda x \\ xz = 4\lambda y \\ xy = \lambda z \\ x^2 + 2y^2 + \frac{1}{2}z^2 = 6 \end{array}\right\}$$ Solving the first three equations for λ gives $\lambda = \dfrac{yz}{2x} = \dfrac{xz}{4y} = \dfrac{xy}{z}$, so $x^2 = 2y^2$ and

$z^2 = 4y^2$. Substituting these into the last equation in the system gives $2y^2 + 2y^2 + \frac{1}{2}\left(4y^2\right) = 6 \Rightarrow y^2 = 1 \Rightarrow y = \pm 1$.

Thus, $x = \pm\sqrt{2}$ and $z = \pm 2$. Evaluating $f(x, y, z)$ at $\left(\pm\sqrt{2}, -1, -2\right)$, $\left(\pm\sqrt{2}, -1, 2\right)$, $\left(\pm\sqrt{2}, 1, -2\right)$, and $\left(\pm\sqrt{2}, 1, 2\right)$,

we see that f has a minimum value of $-2\sqrt{2}$ and a maximum value of $2\sqrt{2}$.

17. $f(x, y, z) = 2x + y \Rightarrow \nabla f(x, y, z) = 2\mathbf{i} + \mathbf{j}$, $g(x, y, z) = x + y + z - 1 \Rightarrow \nabla g(x, y, z) = \mathbf{i} + \mathbf{j} + \mathbf{k}$, and
$h(x, y, z) = y^2 + z^2 - 9 \Rightarrow \nabla h(x, y, z) = 2y\mathbf{j} + 2z\mathbf{k}$, so $\nabla f = \lambda \nabla g + \mu \nabla h$ along with the constraints $g(x, y, z) = 0$

and $h(x, y, z) = 0$ give the system $\left.\begin{array}{rl} 2 = \lambda & (1) \\ 1 = \lambda + 2\mu y & (2) \\ 0 = \lambda + 2\mu z & (3) \\ x + y + z = 1 & (4) \\ y^2 + z^2 = 9 & (5) \end{array}\right\}$ From (1), (2), and (3), we obtain $\mu = -\dfrac{1}{2y} = -\dfrac{1}{z}$

$\Rightarrow z = 2y$. Substituting this into (5) gives $y^2 + (2y)^2 = 5y^2 = 9 \Leftrightarrow y = \pm\dfrac{3\sqrt{5}}{5} \Rightarrow z = \pm\dfrac{6\sqrt{5}}{5}$. Then (4) gives

$x = 1 - \left(\pm\dfrac{3\sqrt{5}}{5}\right) - \left(\pm\dfrac{6\sqrt{5}}{5}\right) = \dfrac{5 \mp 9\sqrt{5}}{5}$, so f has minimum value $f\left(\dfrac{5 - 9\sqrt{5}}{5}, \dfrac{3\sqrt{5}}{5}, \dfrac{6\sqrt{5}}{5}\right) = 2 - 3\sqrt{5}$ and

maximum value $f\left(\dfrac{5 + 9\sqrt{5}}{5}, -\dfrac{3\sqrt{5}}{5}, -\dfrac{6\sqrt{5}}{5}\right) = 2 + 3\sqrt{5}$.

19. $f(x, y, z) = yz + xz \Rightarrow \nabla f(x, y, z) = z\mathbf{i} + z\mathbf{j} + (x + y)\mathbf{k}$, $g(x, y, z) = xz - 1 \Rightarrow \nabla g(x, y, z) = z\mathbf{i} + x\mathbf{k}$,

and $h(x, y, z) = y^2 + z^2 - 1 \Rightarrow \nabla h(x, y, z) = 2y\mathbf{j} + 2z\mathbf{k}$, so $\nabla f = \lambda \nabla g + \mu \nabla h$ along with the constraints

$$
\begin{aligned}
z &= \lambda z & (1) \\
z &= 2\mu y & (2) \\
x + y &= \lambda x + 2\mu z & (3) \\
xz &= 1 & (4) \\
y^2 + z^2 &= 1 & (5)
\end{aligned}
$$

$g(x, y, z) = 0$ and $h(x, y, z) = 0$ give the system (with the brace after (5))   From (1) we have $\lambda = 1$ since

$z \neq 0$ by (4). From (2) we have $\mu = \dfrac{z}{2y}$. Substituting these into (3) gives $x + y = x + 2\left(\dfrac{z}{2y}\right)z \Rightarrow y^2 = z^2 \Rightarrow$

$z = \pm y$. Using (5), we obtain $y^2 + y^2 = 1 \Rightarrow y = \pm\frac{\sqrt{2}}{2} \Rightarrow z = \pm\frac{\sqrt{2}}{2}$, and by (4), $x = \pm\sqrt{2}$. Because x and z

must have the same sign by (4), we consider the points $\left(\sqrt{2}, -\frac{\sqrt{2}}{2}, \frac{\sqrt{2}}{2}\right)$, $\left(\sqrt{2}, \frac{\sqrt{2}}{2}, \frac{\sqrt{2}}{2}\right)$, $\left(-\sqrt{2}, -\frac{\sqrt{2}}{2}, -\frac{\sqrt{2}}{2}\right)$, and

$\left(-\sqrt{2}, \frac{\sqrt{2}}{2}, -\frac{\sqrt{2}}{2}\right)$. We see that f has minimum value $f\left(-\sqrt{2}, \frac{\sqrt{2}}{2}, -\frac{\sqrt{2}}{2}\right) = f\left(\sqrt{2}, -\frac{\sqrt{2}}{2}, \frac{\sqrt{2}}{2}\right) = \frac{1}{2}$ and maximum

value $f\left(-\sqrt{2}, -\frac{\sqrt{2}}{2}, -\frac{\sqrt{2}}{2}\right) = f\left(\sqrt{2}, \frac{\sqrt{2}}{2}, \frac{\sqrt{2}}{2}\right) = \frac{3}{2}$.

21. $\begin{aligned} f_x(x, y) &= \frac{\partial}{\partial x}\left(3x^2 + 2y^2 - 2x - 1\right) = 6x - 2 = 0 \\ f_y(x, y) &= \frac{\partial}{\partial y}\left(3x^2 + 2y^2 - 2x - 1\right) = 4y = 0 \end{aligned}$ $\Rightarrow x = \frac{1}{3}$ and $y = 0$, so f has the critical point $\left(\frac{1}{3}, 0\right)$ in the disk

$D = \left\{(x, y) \mid x^2 + y^2 \leq 9\right\}$. Next, we use the method of Lagrange to find the critical points of f on the boundary of D.

We write $g(x, y) = x^2 + y^2 - 9 = 0$. Then $\nabla f(x, y) = (6x - 2)\mathbf{i} + 4y\mathbf{j}$ and $\nabla g(x, y) = 2x\mathbf{i} + 2y\mathbf{j}$. The equation

$$
\begin{aligned}
6x - 2 &= 2\lambda x & (1) \\
4y &= 2\lambda y & (2) \\
x^2 + y^2 &= 9 & (3)
\end{aligned}
$$

$\nabla f = \lambda \nabla g$ and the constraint equation $g(x, y) = 0$ give the system   Equation (2) gives

$y = 0$ or $\lambda = 2$. If $y = 0$, then (3) gives $x = \pm 3$; if $\lambda = 2$, then (1) gives $x = 1$. Substituting this value of x into (3) gives

$y = \pm 2\sqrt{2}$.

(x, y)	$\left(\frac{1}{3}, 0\right)$	$(-3, 0)$	$(3, 0)$	$\left(1, -2\sqrt{2}\right)$	$\left(1, 2\sqrt{2}\right)$
$f(x, y)$	$-\frac{4}{3}$	32	20	16	16

From the table, we see that f has a minimum value of $f\left(\frac{1}{3}, 0\right) = -\frac{4}{3}$ and a maximum value of $f(-3, 0) = 32$.

23. The distance from the origin to a point (x, y, z) on the plane $x + 2y + z = 4$ is $D = \sqrt{x^2 + y^2 + z^2}$. To minimize

D, we just need to minimize $f(x, y, z) = D^2 = x^2 + y^2 + z^2$ subject to $g(x, y, z) = x + 2y + z - 4 = 0$.

$\nabla f(x, y, z) = 2x\mathbf{i} + 2y\mathbf{j} + 2z\mathbf{k}$ and $\nabla g(x, y, z) = \mathbf{i} + 2\mathbf{j} + \mathbf{k}$, so $\nabla f = \lambda \nabla g$ and $g(x, y, z) = 0$ give the system

$$
\begin{aligned}
2x &= \lambda \\
2y &= 2\lambda \\
2z &= \lambda \\
x + 2y + z &= 4
\end{aligned}
\Leftrightarrow
\begin{aligned}
x &= \tfrac{1}{2}\lambda \\
y &= \lambda \\
z &= \tfrac{1}{2}\lambda \\
x + 2y + z &= 4
\end{aligned}
\Rightarrow \tfrac{1}{2}\lambda + 2\lambda + \tfrac{1}{2}\lambda = 4, \text{ giving } \lambda = \tfrac{4}{3}. \text{ Therefore, the required point is}
$$

$\left(\frac{2}{3}, \frac{4}{3}, \frac{2}{3}\right)$.

25. The distance from a point (x, y, z) on the plane $x + 2y - z = 5$ to the point $(2, 3, -1)$

is $D = \sqrt{(x-2)^2 + (y-3)^2 + (z+1)^2}$. To minimize D, it suffices to minimize

$D^2 = f(x, y) = (x-2)^2 + (y-3)^2 + (z+1)^2$ subject to $g(x, y) = x + 2y - z - 5 = 0$.

$\nabla f(x, y) = 2(x-2)\mathbf{i} + 2(y-3)\mathbf{j} + 2(z+1)\mathbf{k}$ and $\nabla g(x, y) = \mathbf{i} + 2\mathbf{j} - \mathbf{k}$, so $\nabla f = \lambda \nabla g$ together with the constraint

equation gives
$$\left. \begin{array}{r} 2x - 4 = \lambda \\ 2y - 6 = 2\lambda \\ 2z + 2 = -\lambda \\ x + 2y - z = 5 \end{array} \right\}$$
Solving the first three equations for x, y, and z in terms of λ, we find $x = \dfrac{\lambda + 4}{2}$,

$y = \lambda + 3$, and $z = -\dfrac{\lambda + 2}{2}$. Substituting these into the last equation in the system gives $\dfrac{\lambda + 4}{2} + 2(\lambda + 3) + \dfrac{\lambda + 2}{2} = 5$

$\Leftrightarrow \lambda = -\dfrac{4}{3}$, so the required point is $\left(\dfrac{4}{3}, \dfrac{5}{3}, -\dfrac{1}{3} \right)$.

27. The distance from the origin to a point (x, y, z) on the surface $xy^2z - 4 = 0$ is $D = \sqrt{x^2 + y^2 + z^2}$. To

minimize D, it suffices to minimize $D^2 = f(x, y, z) = x^2 + y^2 + z^2$ subject to $g(x, y, z) = xy^2z - 4 = 0$.

$\nabla f(x, y, z) = 2x\mathbf{i} + 2y\mathbf{j} + 2z\mathbf{k}$ and $\nabla g(x, y, z) = y^2z\mathbf{i} + 2xyz\mathbf{j} + xy^2\mathbf{k}$, so $\nabla f = \lambda \nabla g$ together with $g(x, y, z) = 0$

leads to the system
$$\left. \begin{array}{rl} 2x = \lambda y^2 z & (1) \\ 2y = 2\lambda xyz & (2) \\ 2z = \lambda xy^2 & (3) \\ xy^2z = 4 & (4) \end{array} \right\}$$
From (2), we have $y(\lambda xz - 1) = 0$. Now $y \neq 0$ because of (4), so

$\lambda = \dfrac{1}{xz}$. Substituting this into (1) and (3) successively gives $2x^2 = y^2$ and $2z^2 = y^2$. We see that $z^2 = x^2 \Leftrightarrow z = \pm x$.

Now $z \neq -x$, because then (4) gives $-x^2y^2 = 4$, which is impossible. If $z = x$, then (4) gives $x(2x^2)x = 4 \Leftrightarrow$

$x = \pm 2^{1/4} \Rightarrow y = \pm 2^{3/4}$ and $z = \pm 2^{1/4}$.

(x, y, z)	$\left(-2^{1/4}, -2^{3/4}, -2^{1/4}\right)$	$\left(-2^{1/4}, 2^{3/4}, -2^{1/4}\right)$	$\left(2^{1/4}, -2^{3/4}, 2^{1/4}\right)$	$\left(2^{1/4}, 2^{3/4}, 2^{1/4}\right)$
$f(x, y, z)$	$4\sqrt{2}$	$4\sqrt{2}$	$4\sqrt{2}$	$4\sqrt{2}$

From the table, we see that there are four such points: $\left(-2^{1/4}, \pm 2^{3/4}, -2^{1/4}\right)$ and $\left(2^{1/4}, \pm 2^{3/4}, 2^{1/4}\right)$. The shortest

distance is $D = \sqrt{4\sqrt{2}} = 2\sqrt[4]{2}$.

29. Denote the length, width, and height of the base by x, y, and z respectively. Then we want to maximize

$V = f(x, y, z) = xyz$ subject to $2(xy + yz + xz) - 48 = 0 \Rightarrow g(x, y) = xy + yz + xz - 24 = 0$.

$\nabla f(x, y, z) = yz\mathbf{i} + xz\mathbf{j} + xy\mathbf{k}$ and $\nabla g(x, y, z) = (y+z)\mathbf{i} + (x+z)\mathbf{j} + (x+y)\mathbf{k}$, so $\nabla f = \lambda \nabla g$ together

with the constraint equation gives the system
$$\left. \begin{array}{rl} yz = \lambda(y+z) & (1) \\ xz = \lambda(x+z) & (2) \\ xy = \lambda(x+y) & (3) \\ xy + yz + xz = 24 & (4) \end{array} \right\}$$
Equations (1) and (2) give

$\lambda = \dfrac{yz}{y+z} = \dfrac{xz}{x+z} \Rightarrow xyz + yz^2 = xyz + xz^2 \Rightarrow z^2(y - x) = 0$. Since $z > 0$, we see that $y = x$. Equations (2) and

(3) give $\lambda = \dfrac{xz}{x+z} = \dfrac{xy}{x+y}$, and as above, $z = y$. Substituting into (4) gives $3x^2 = 24 \Rightarrow x = 2\sqrt{2}$. The required

dimensions are $2\sqrt{2}' \times 2\sqrt{2}' \times 2\sqrt{2}'$.

31. Denote the length, width, and height of the base by x, y, and z respectively. Then we want to minimize $S = f(x, y, z) = xy + 2yz + 2xz$ subject to $g(x, y, z) = xyz - 108 = 0$.
$\nabla f(x, y, z) = (y + 2z)\,\mathbf{i} + (x + 2z)\,\mathbf{j} + (2x + 2y)\,\mathbf{k}$ and $\nabla g(x, y, z) = yz\mathbf{i} + xz\mathbf{j} + xy\mathbf{k}$, so $\nabla f = \lambda \nabla g$ together with

the constraint equation gives the system
$$\left.\begin{array}{rl} y + 2z = \lambda yz & (1) \\ x + 2z = \lambda xz & (2) \\ 2x + 2y = \lambda xy & (3) \\ xyz = 108 & (4) \end{array}\right\}$$
Equations (1) and (2) give $\lambda = \dfrac{y + 2z}{yz} = \dfrac{x + 2z}{xz}$

$\Rightarrow xyz + 2xz^2 = xyz + 2yz^2 \Rightarrow 2z^2(x - y) = 0 \Leftrightarrow x = y$. Equations (2) and (3) give $\lambda = \dfrac{x + 2z}{xz} = \dfrac{2x + 2y}{xy} \Rightarrow$

$x^2y + 2xyz = 2x^2z + 2xyz \Rightarrow x^2(y - 2z) = 0 \Rightarrow z = \frac{1}{2}y$. Substituting these values into (4) gives $y(y)\left(\frac{1}{2}y\right) = 108 \Rightarrow$

$y^3 = 216 \Rightarrow y = 6$, so $x = 6$ and $z = 3$. Therefore, the required dimensions are $6'' \times 6'' \times 3''$.

33. Let (x, y, z) denote the corner point of the inscribed rectangular box in the first octant. Then the volume of the box is

$V = f(x, y, z) = 8xyz$, where x, y, and z satisfy $g(x, y, z) = \dfrac{x^2}{a^2} + \dfrac{y^2}{b^2} + \dfrac{z^2}{c^2} - 1 = 0$. We want to maximize f subject to

$g(x, y, z) = 0$, so we calculate $\nabla f(x, y, z) = 8yz\mathbf{i} + 8xz\mathbf{j} + 8xy\mathbf{k}$ and $\nabla g(x, y, z) = \dfrac{2x}{a^2}\mathbf{i} + \dfrac{2y}{b^2}\mathbf{j} + \dfrac{2z}{c^2}\mathbf{k}$. $\nabla f = \lambda \nabla g$

and the constraint lead to
$$\left.\begin{array}{rl} 8yz = 2\lambda x/a^2 & (1) \\ 8xz = 2\lambda y/b^2 & (2) \\ 8xy = 2\lambda z/c^2 & (3) \\ \dfrac{x^2}{a^2} + \dfrac{y^2}{b^2} + \dfrac{z^2}{c^2} = 1 & (4) \end{array}\right\}$$
From (1) and (2), we have $\lambda = \dfrac{4a^2yz}{x} = \dfrac{4b^2xz}{y} \Rightarrow$

$\left(a^2y^2 - b^2x^2\right)z = 0 \Rightarrow x^2 = \dfrac{a^2}{b^2}y^2$. From (2) and (3), we have $\lambda = \dfrac{4b^2xz}{y} = \dfrac{4c^2xy}{z} \Rightarrow \left(b^2z^2 - c^2y^2\right)x = 0 \Rightarrow$

$z^2 = \dfrac{c^2}{b^2}y^2$. Substituting these values into (4) gives $\dfrac{1}{a^2}\left(\dfrac{a^2}{b^2}y^2\right) + \dfrac{1}{b^2}y^2 + \dfrac{1}{c^2}\left(\dfrac{c^2}{b^2}y^2\right) = 1 \Rightarrow \dfrac{3}{b^2}y^2 = 1 \Rightarrow y = \dfrac{\sqrt{3}}{3}b$,

so $x = \dfrac{\sqrt{3}}{3}a$ and $z = \dfrac{\sqrt{3}}{3}c$. Therefore, the dimensions of the box are $\dfrac{2\sqrt{3}}{3}a \times \dfrac{2\sqrt{3}}{3}b \times \dfrac{2\sqrt{3}}{3}c$.

35. The volume of the inscribed box is $V = f(x, y, z) = xyz$, where x, y, and z satisfy $g(x, y, z) = \dfrac{x}{a} + \dfrac{y}{b} + \dfrac{z}{c} - 1 = 0$. We

want to maximize f subject to $g(x, y, z) = 0$. $\nabla f(x, y, z) = yz\mathbf{i} + xz\mathbf{j} + xy\mathbf{k}$ and $\nabla g(x, y, z) = \dfrac{1}{a}\mathbf{i} + \dfrac{1}{b}\mathbf{j} + \dfrac{1}{c}\mathbf{k}$.

$\nabla f = \lambda \nabla g$ together with the constraint equation gives
$$\left.\begin{array}{rl} yz = \lambda/a & (1) \\ xz = \lambda/b & (2) \\ xy = \lambda/c & (3) \\ \dfrac{x}{a} + \dfrac{y}{b} + \dfrac{z}{c} = 1 & (4) \end{array}\right\}$$
From (1) and (2), we have

$\lambda = ayz = bxz \Rightarrow z(ay - bx) = 0 \Rightarrow x = \dfrac{a}{b}y$. From (2) and (3), we find $\lambda = bxz = cxy \Rightarrow x(bz - cy) = 0 \Rightarrow$

$z = \dfrac{c}{b}y$. Substituting into (4), we find $\dfrac{1}{a}\left(\dfrac{a}{b}y\right) + \dfrac{1}{b}y + \dfrac{1}{c}\left(\dfrac{c}{b}y\right) = 1 \Rightarrow y = \dfrac{b}{3}$. Thus, $x = \dfrac{a}{3}$ and $z = \dfrac{c}{3}$, and the required

dimensions are $\dfrac{a}{3} \times \dfrac{b}{3} \times \dfrac{c}{3}$. The volume of the box is $\dfrac{abc}{27}$.

37. Let x, y, and z denote the length, width, and height of the box. We can assume without loss of generality that the cost of the material for constructing the sides and top is \$$1/\text{ft}^2$. Then the total cost is $C = f(x, y, z) = 3xy + 2xz + 2yz$. We want to minimize f subject to the constraint $g(x, y, z) = xyz - 16 = 0$. $\nabla f(x, y, z) = (3y + 2z)\mathbf{i} + (3x + 2z)\mathbf{j} + (2x + 2y)\mathbf{k}$

and $\nabla g(x, y, z) = yz\mathbf{i} + xz\mathbf{j} + xy\mathbf{k}$, so $\nabla f = \lambda \nabla g$ together with the constraint gives

$$\left. \begin{array}{rl} 3y + 2z = \lambda yz & (1) \\ 3x + 2z = \lambda xz & (2) \\ 2x + 3y = \lambda xy & (3) \\ xyz = 16 & (4) \end{array} \right\}$$

From (1) and (2), we find $\lambda = \dfrac{3y + 2z}{yz} = \dfrac{3x + 2z}{xz} \Rightarrow 3xyz + 2xz^2 = 3xyz + 2yz^2 \Rightarrow x = y$. From (2) and (3), we have $\lambda = \dfrac{3x + 2z}{xz} = \dfrac{2x + 2y}{xy} \Rightarrow 3x^2y + 2xyz = 2x^2z + 2xyz \Rightarrow z = \frac{3}{2}y$. Substituting into (4), we have $y(y)\left(\frac{3}{2}y\right) = 16 \Rightarrow y^3 = \frac{32}{3} \Rightarrow y = \frac{2}{3}\sqrt[3]{36}$. Thus, $x = \frac{2}{3}\sqrt[3]{36}$ and $z = \sqrt[3]{36}$. The dimensions of the box are $\frac{2}{3}\sqrt[3]{36}' \times \frac{2}{3}\sqrt[3]{36}' \times \sqrt[3]{36}'$.

39. $f(x, y) = 100x^{3/4}y^{1/4}$ and $g(x, y) = 200x + 300y - 60{,}000$. $\nabla f(x, y) = 75x^{-1/4}y^{1/4}\mathbf{i} + 25x^{3/4}y^{-3/4}\mathbf{j}$ and $\nabla g(x, y) = 200\mathbf{i} + 300\mathbf{j}$. The equation $\nabla f = \lambda \nabla g$ together with the constraint gives

$$\left. \begin{array}{rl} 75(y/x)^{1/4} = 200\lambda & (1) \\ 25(x/y)^{3/4} = 300\lambda & (2) \\ 200x + 300y = 60{,}000 & (3) \end{array} \right\}$$
Dividing (2) by (1) gives $\dfrac{25(x/y)^{3/4}}{75(x/y)^{-1/4}} = \dfrac{300}{200} \Rightarrow \dfrac{x}{y} = \dfrac{9}{2} \Rightarrow x = \frac{9}{2}y$. Substituting

this into (3) gives $200\left(\frac{9}{2}y\right) + 300y = 60{,}000 \Rightarrow y = 50$, and so $x = 225$. We need 225 units of labor and 50 units of capital.

41. a. $\nabla f(x, y) = \mathbf{i} - \mathbf{j}$ and $\nabla g(x, y) = \left(1 + 5x^4\right)\mathbf{i} - \mathbf{j}$, so $\nabla f = \lambda \nabla g$ together with the constraint constant lead to the

system $\left. \begin{array}{rl} 1 = \lambda\left(1 + 5x^4\right) & (1) \\ -1 = -\lambda & (2) \\ x + x^5 - y = 1 & (3) \end{array} \right\}$ From (2), we have $\lambda = 1$, which we substitute into (1) to get

$1 = 1 + 5x^4 \Rightarrow x = 0$. Substituting this into (3) gives $y = -1$, so f may have a relative maximum or a relative minimum at the point $(0, -1)$.

b.

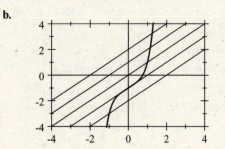

Observe that the point $(0, -1)$, at which the level curve $x - y = 1$ is tangent to the constraint curve $g(x, y) = 1$, is an inflection point of the graph of g. In a neighborhood of $(0, -1)$, $f(x, y)$ takes on values greater than 1 if (x, y) lies on the graph of g but with $x < 0$. On the other hand, $f(x, y)$ takes on values less than 1 if (x, y) lies on the graph of g but with $x > 0$. Thus, f does not have a relative maximum or a relative minimum at $(0, -1)$.

c. Solving the constraint equation $x + x^5 - y = 1$ for y and substituting this value of y into the expression for f gives $F(x) = f(x, y) = x - y = x - \left(x + x^5 - 1\right) = -x^5 + 1$. Since $F'(x) = -5x^4 \Rightarrow x = 0$, we see that 0 is a critical number of F. But F' does not change sign as we move across 0, so F has no relative extremum at 0.

43. We want to minimize $f(x, y) = x^2 + y^2$ subject to the system $\left.\begin{array}{r} x + 2y - 3z = 9 \\ 2x - 3y + z = 4 \end{array}\right\}$ Write the

constraint equations in the form $g(x, y, z) = x + 2y - 3z = 9$ and $h(x, y, z) = 2x - 3y + z = 4$.
Then the equation $\nabla f(x, y, z) = \lambda \nabla g(x, y, z) + \mu \nabla h(x, y, z)$ becomes
$2x\mathbf{i} + 2y\mathbf{j} = \lambda(\mathbf{i} + 2\mathbf{j} - 3\mathbf{k}) + \mu(2\mathbf{i} - 3\mathbf{j} + \mathbf{k}) = (\lambda + 2\mu)\mathbf{i} + (2\lambda - 3\mu)\mathbf{j} + (-3\lambda + \mu)\mathbf{k}$. Equating like components

and rewriting the constraint equations leads to

$$\left.\begin{array}{rl} 2x = & \lambda + 2\mu \quad (1) \\ 2y = & 2\lambda - 3\mu \quad (2) \\ 0 = & -3\lambda - 3\mu \quad (3) \\ x + 2y - 3z = & 9 \quad (4) \\ 2x - 3y + z = & 4 \quad (5) \end{array}\right\} \quad \text{Equation (3) gives } \mu = 3\lambda.$$

Substituting this into (1) and (2) gives $2x = 7\lambda$ and $2y = -7\lambda \Rightarrow y = -x$. Substituting this into (4) and (5) gives

$$\left.\begin{array}{r} x - 2x - 3z = 9 \\ 2x + 3x + z = 4 \end{array}\right\} \Rightarrow \left.\begin{array}{r} -x - 3z = 9 \\ 5x + z = 4 \end{array}\right\} \Rightarrow \left.\begin{array}{r} -5x - 15z = 45 \\ 5x + z = 4 \end{array}\right\} \Rightarrow -14z = 49 \Rightarrow z = -\tfrac{7}{2} \Rightarrow x = \tfrac{3}{2} \Rightarrow$$

$y = -\tfrac{3}{2}$. Thus, the required point is $\left(\tfrac{3}{2}, -\tfrac{3}{2}, -\tfrac{7}{2}\right)$.

45. a. The maximum distance is the distance between $(0, 0)$ and the point of intersection of $y = x$ and $x^3 + y^3 - 3axy = 0$.
We find $x^3 + x^3 - 3ax^2 = 0 \Rightarrow x^2(2x - 3a) = 0 \Rightarrow x = 0$ or $x = \tfrac{3}{2}a$, and so $y = \tfrac{3}{2}a$. The required distance is

$$D = \sqrt{\left(\tfrac{3}{2}a\right)^2 + \left(\tfrac{3}{2}a\right)^2} = \tfrac{3\sqrt{2}}{2}a.$$

b. We want to maximize $f(x, y) = x^2 + y^2$ subject to $g(x, y) = x^3 + y^3 - 3axy = 0$. $\nabla f(x, y) = 2x\mathbf{i} + 2y\mathbf{j}$
and $\nabla g(x, y) = 3\left(x^2 - ay\right)\mathbf{i} + 3\left(y^2 - ax\right)\mathbf{j}$, so $\nabla f = \lambda \nabla g$ together with the constraint equation gives

$$\left.\begin{array}{rl} 2x = & 3\lambda\left(x^2 - ay\right) \quad (1) \\ 2y = & 3\lambda\left(y^2 - ax\right) \quad (2) \\ x^3 + y^3 - 3axy = & 0 \quad (3) \end{array}\right\} \quad \text{From (1) and (2), we see that } \lambda = \frac{2x}{3\left(x^2 - ay\right)} = \frac{2y}{3\left(y^2 - ax\right)} \Rightarrow$$

$x^2(y + a) = y^2(x + a)$. By inspection, we see that $y = x$ satisfies the last equation. Substituting this into (3) gives the
result obtained earlier.

47. a. $\nabla f = \left\langle x^{p-1}, y^{q-1}\right\rangle$ and with $g(x, y) = xy$ we have $\nabla g = \langle y, x\rangle$. Thus, $\nabla f = \lambda \nabla g$ and the constraint

$$\left.\begin{array}{rl} x^{p-1} = & \lambda y \quad (1) \\ y^{q-1} = & \lambda x \quad (2) \\ xy = & c \quad (3) \end{array}\right\} \quad \text{From (1) and (2), we have } \lambda = \frac{x^{p-1}}{y} = \frac{y^{q-1}}{x} \Leftrightarrow$$

equation $g(x, y) = xy - c = 0$ give

$x^p = y^q \Leftrightarrow x = y^{q/p}$. Using (3), we have $xy = y^{q/p}y = y^{(q/p)+1} = y^{(q+p)/p} = c \Leftrightarrow y = c^{p/(p+q)}$ and

$x = \left[c^{p/(p+q)}\right]^{q/p} = c^{q/(p+q)}$. Therefore, the minimum value of f is $f\left(c^{q/(p+q)}, c^{p/(p+q)}\right) = \dfrac{\left[c^{q/(p+q)}\right]^p}{p} +$

$\dfrac{\left[c^{p/(p+q)}\right]^q}{q} = \left(\dfrac{1}{p} + \dfrac{1}{q}\right)c^{pq/(p+q)} = c^{pq/(p+q)} = c^{[(1/p)+(1/q)]^{-1}} = c.$

b. Using the result of part a, we have $f(x, y) = \dfrac{x^p}{p} + \dfrac{y^q}{q} \geq c = xy$.

49. Let d_1 denote the distance between P and O and d_2 the distance between Q and O. Then $d_1 = \dfrac{a}{\cos\theta_1}$ and

$d_2 = \dfrac{b}{\cos\theta_2}$, where θ_1 and θ_2 lie between 0 and $\dfrac{\pi}{2}$. The time it takes for a ray of light to travel from P to Q is

$t = \dfrac{d_1}{v_1} + \dfrac{d_2}{v_2} = \dfrac{a}{v_1\cos\theta_1} + \dfrac{b}{v_2\cos\theta_2}$. But $k = a\tan\theta_1 + b\tan\theta_2$, so the problem can be solved by minimizing

$t = f(\theta_1,\theta_2) = \dfrac{a}{v_1\cos\theta_1} + \dfrac{b}{v_2\cos\theta_2} = \dfrac{a}{v_1}\sec\theta_1 + \dfrac{b}{v_2}\sec\theta_2$ subject to $g(\theta_1,\theta_2) = a\tan\theta_1 + b\tan\theta_2 - k = 0$:

$\nabla f(\theta_1,\theta_2) = \dfrac{a}{v_1}\sec\theta_1\tan\theta_1\,\mathbf{i} + \dfrac{b}{v_2}\sec\theta_2\tan\theta_2\,\mathbf{j}$ and $\nabla g(\theta_1,\theta_2) = a\sec^2\theta_1\,\mathbf{i} + b\sec^2\theta_2\,\mathbf{j}$. Then $\nabla f = \lambda\nabla g$

together with the constraint equation gives
$$
\left.\begin{array}{c}
\dfrac{a}{v_1}\sec\theta_1\tan\theta_1 = \lambda a\sec^2\theta_1 \\[2mm]
\dfrac{b}{v_2}\sec\theta_2\tan\theta_2 = \lambda b\sec^2\theta_2 \\[2mm]
a\tan\theta_1 + b\tan\theta_2 = k
\end{array}\right\}
\Rightarrow
\left.\begin{array}{c}
\dfrac{\sin\theta_1}{v_1} = \lambda \\[2mm]
\dfrac{\sin\theta_2}{v_2} = \lambda \\[2mm]
a\tan\theta_1 + b\tan\theta_2 = k
\end{array}\right\}
$$

The first two equations imply that $\dfrac{1}{\lambda} = \dfrac{v_1}{\sin\theta_1} = \dfrac{v_2}{\sin\theta_2}$, and the result follows.

51. False. Take $f(x,y) = x^2 + y^2$ and $g(x,y) = x + y - 1 = 0$. Then $\left(\frac{1}{2}, \frac{1}{2}\right)$ gives a constrained minimum of f subject to $g(x,y) = 0$. But it does not give an unconstrained extremum of f, which in fact has a minimum at $(0,0)$.

Chapter 13   Review   ET 12

Concept Review

1. a. rule; (x,y) **b.** dependent; independent; range   **c.** $\{(x,y,z) \mid z = f(x,y),\,(x,y)\in D\}$

3. $L; L; (a,b)$

5. a. $f(a,b)$ **b.** R

7. a. $\displaystyle\lim_{h\to 0}\frac{f(x+h,y)-f(x,y)}{h}$; $y = b$; $(a,b,f(a,b))$; x; constant; $y = b$ **b.** $y; x$

9. a. $f_x\,dx + f_y\,dy$ **b.** dz **c.** $\Delta x;\ \Delta y;\ 0;\ 0$

    d. $f_x(a,b)\,\Delta x + f_y(a,b)\,\Delta y + \varepsilon_1\,\Delta x + \varepsilon_2\,\Delta y$; $\varepsilon_1 \to 0$; $\varepsilon_2 \to 0$; $(0,0)$

11. a. $\dfrac{\partial w}{\partial x}\dfrac{dx}{dt} + \dfrac{\partial w}{\partial y}\dfrac{dy}{dt}$ **b.** $\dfrac{\partial w}{\partial x}\dfrac{\partial x}{\partial u} + \dfrac{\partial w}{\partial y}\dfrac{\partial y}{\partial u}$ **c.** $-\dfrac{F_x(x,y)}{F_y(x,y)}$; $F_y(x,y)\neq 0$

    d. $-\dfrac{F_x(x,y,z)}{F_z(x,y,z)}$; $-\dfrac{F_y(x,y,z)}{F_z(x,y,z)}$; $F_z(x,y,z)\neq 0$

13. a. $|\nabla f(x,y)|$; $\nabla f(x,y)$ **b.** $-|\nabla f(x,y)|$; $-\nabla f(x,y)$

15. a. relative maximum   **b.** absolute minimum   **c.** does not exist; 0

    d. critical point   **e.** Second Derivative Test

17. a. constrained   **b.** $\lambda\nabla g(x,y)$; multiplier

    c. $\lambda\nabla g(x,y)$; $f(x,y)$; critical points; maximum; minimum

Review Exercises

1. $D = \left\{ (x, y) \mid 0 < x^2 + y^2 \le 9 \right\}$

3. We require that $-1 \le x \le 1$, so
$$D = \{(x, y) \mid -1 \le x \le 1\}.$$

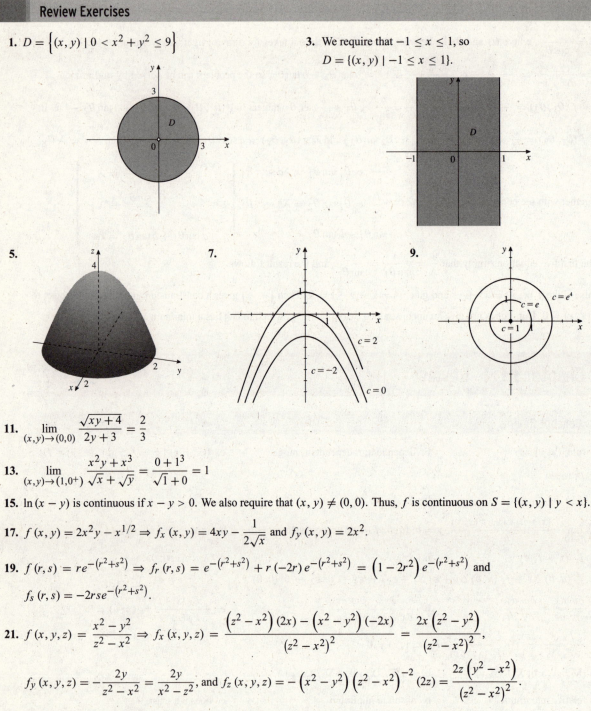

5.

7.

9.

11. $\displaystyle \lim_{(x,y) \to (0,0)} \frac{\sqrt{xy + 4}}{2y + 3} = \frac{2}{3}$

13. $\displaystyle \lim_{(x,y) \to (1,0^+)} \frac{x^2 y + x^3}{\sqrt{x} + \sqrt{y}} = \frac{0 + 1^3}{\sqrt{1} + 0} = 1$

15. $\ln(x - y)$ is continuous if $x - y > 0$. We also require that $(x, y) \ne (0, 0)$. Thus, f is continuous on $S = \{(x, y) \mid y < x\}$.

17. $f(x, y) = 2x^2 y - x^{1/2} \Rightarrow f_x(x, y) = 4xy - \dfrac{1}{2\sqrt{x}}$ and $f_y(x, y) = 2x^2$.

19. $f(r, s) = re^{-(r^2 + s^2)} \Rightarrow f_r(r, s) = e^{-(r^2 + s^2)} + r(-2r)e^{-(r^2 + s^2)} = \left(1 - 2r^2\right)e^{-(r^2 + s^2)}$ and

$f_s(r, s) = -2rse^{-(r^2 + s^2)}$.

21. $f(x, y, z) = \dfrac{x^2 - y^2}{z^2 - x^2} \Rightarrow f_x(x, y, z) = \dfrac{\left(z^2 - x^2\right)(2x) - \left(x^2 - y^2\right)(-2x)}{\left(z^2 - x^2\right)^2} = \dfrac{2x\left(z^2 - y^2\right)}{\left(z^2 - x^2\right)^2}$,

$f_y(x, y, z) = -\dfrac{2y}{z^2 - x^2} = \dfrac{2y}{x^2 - z^2}$, and $f_z(x, y, z) = -\left(x^2 - y^2\right)\left(z^2 - x^2\right)^{-2}(2z) = \dfrac{2z\left(y^2 - x^2\right)}{\left(z^2 - x^2\right)^2}$.

23. $f(x, y) = x^4 - 2x^2 y^3 + y^2 - 2 \Rightarrow f_x(x, y) = 4x^3 - 4xy^3$ and $f_y(x, y) = -6x^2 y^2 + 2y$,

so $f_{xx}(x, y) = \dfrac{\partial}{\partial x}\left(4x^3 - 4xy^3\right) = 12x^2 - 4y^3 = 4\left(3x^2 - y^3\right)$,

$f_{xy}(x, y) = \dfrac{\partial}{\partial y}\left(4x^3 - 4xy^3\right) = -12xy^2 = \dfrac{\partial}{\partial x}\left(-6x^2 y^2 + 2y\right) = f_{yx}(x, y)$, and

$f_{yy}(x, y) = \dfrac{\partial}{\partial y}\left(-6x^2 y^2 + 2y\right) = -12x^2 y + 2$.

25. $f(x, y, z) = x^2yz^3 \Rightarrow f_x(x, y, z) = 2xyz^3$, $f_y(x, y, z) = x^2z^3$, and $f_z(x, y, z) = 3x^2yz^2$, so $f_{xx}(x, y, z) = 2yz^3$, $f_{yy}(x, y, z) = 0$, $f_{zz}(x, y, z) = 6x^2yz$, $f_{xy}(x, y, z) = f_{yx}(x, y, z) = 2xz^3$, $f_{xz}(x, y, z) = f_{zx}(x, y, z) = 6xyz^2$, and $f_{yz}(x, y, z) = f_{zy}(x, y, z) = 3x^2z^2$.

27. $u = \sqrt{x^2 + y^2 + z^2} \Rightarrow \dfrac{\partial u}{\partial x} = x\left(x^2 + y^2 + z^2\right)^{-1/2}$, $\dfrac{\partial u}{\partial y} = y\left(x^2 + y^2 + z^2\right)^{-1/2}$, and $\dfrac{\partial u}{\partial z} = z\left(x^2 + y^2 + z^2\right)^{-1/2}$,

so $\dfrac{\partial^2 u}{\partial x^2} = \dfrac{y^2 + z^2}{\left(x^2 + y^2 + z^2\right)^{3/2}}$, $\dfrac{\partial^2 u}{\partial y^2} = \dfrac{x^2 + z^2}{\left(x^2 + y^2 + z^2\right)^{3/2}}$, and $\dfrac{\partial^2 u}{\partial z^2} = \dfrac{x^2 + y^2}{\left(x^2 + y^2 + z^2\right)^{3/2}}$. Thus,

$\dfrac{\partial^2 u}{\partial x^2} + \dfrac{\partial^2 u}{\partial y^2} + \dfrac{\partial^2 u}{\partial z^2} = \dfrac{2\left(x^2 + y^2 + z^2\right)}{\left(x^2 + y^2 + z^2\right)^{3/2}} = \dfrac{2}{\sqrt{x^2 + y^2 + z^2}} = \dfrac{2}{u}$, as was to be shown.

29. $u = 2z^2 - x^2 - y^2 \Rightarrow u_x = -2x$, $u_{xx} = -2$, $u_y = -2y$, $u_{yy} = -2$, $u_z = 4z$, and $u_{zz} = 4$, so $u_{xx} + u_{yy} + u_{zz} = -2 - 2 + 4 = 0$.

31. $dz = \dfrac{\partial}{\partial x}\left(x^2 \tan^{-1} y^3\right) dx + \dfrac{\partial}{\partial y}\left(x^2 \tan^{-1} y^3\right) dy = 2x \tan^{-1} y^3\, dx + \dfrac{3x^2 y^2}{1 + y^6}\, dy$

33. Let $w = f(x, y, z) = x^2\sqrt{y^2 + z^3}$. Then $f_x = 2x\sqrt{y^2 + z^3}$, $f_y = \dfrac{x^2 y}{\sqrt{y^2 + z^3}}$, and $f_z = \dfrac{3x^2 z^2}{2\sqrt{y^2 + z^3}}$. Suppose (x, y, z)

changes from $(2, 2, 3)$ to $(2.01, 1.98, 3.02)$, so that $dx = 0.01$, $dy = -0.02$, and $dz = 0.02$. Then

$\Delta w = f(2.01, 1.98, 3.02) - f(2, 2, 3) \approx dw = f_x(2, 2, 3)\, dx + f_y(2, 2, 3)\, dy + f_z(2, 2, 3)\, dz$

$= 2 \cdot 2\sqrt{4 + 27}\, (0.01) + \dfrac{4 \cdot 2}{\sqrt{4 + 27}}\, (-0.02) + \dfrac{3 \cdot 4 \cdot 9}{\sqrt{4 + 27}}\, (0.02) = \dfrac{81\sqrt{31}}{775} \approx 0.5819$

so $(2.01)^2\sqrt{(1.98)^2 + (3.02)^3} = f(2.01, 1.98, 3.02) \approx f(2, 2, 3) + dw = 4\sqrt{31} + \dfrac{81\sqrt{31}}{775} = \dfrac{3181\sqrt{31}}{775} \approx 22.853$.

35. No. If such a function f existed, then $\nabla f = \dfrac{\partial f}{\partial x}\mathbf{i} + \dfrac{\partial f}{\partial y}\mathbf{j} = -y\mathbf{i} + x\mathbf{j}$, which is equivalent to $\dfrac{\partial f}{\partial x} = -y$ and $\dfrac{\partial f}{\partial y} = x$. Then

$\dfrac{\partial^2 f}{\partial y\, \partial x} = \dfrac{\partial}{\partial y}(-y) = -1$ and $\dfrac{\partial^2 f}{\partial x\, \partial y} = \dfrac{\partial}{\partial x}(x) = 1$. Both f_{xy} and f_{yx} are continuous, but $f_{xy} \neq f_{yx}$, a contradiction.

Thus, no such function exists.

37. $\dfrac{dz}{dt} = \dfrac{\partial z}{\partial x}\dfrac{dx}{dt} + \dfrac{\partial z}{\partial y}\dfrac{dy}{dt} = \dfrac{\partial}{\partial x}\left(x^2 y - y^{1/2}\right)\dfrac{d}{dt}\left(e^{2t}\right) + \dfrac{\partial}{\partial y}\left(x^2 y - y^{1/2}\right)\dfrac{d}{dt}(\cos t)$

$= (2xy)\left(2e^{2t}\right) + \left(x^2 - \dfrac{1}{2\sqrt{y}}\right)(-\sin t) = 4xye^{2t} + \left(\dfrac{1}{2\sqrt{y}} - x^2\right)\sin t$

39. Let $F(x, y) = x^3 - 3x^2 y + 2xy^2 + 2y^3 - 9 = 0$. Then $\dfrac{dy}{dx} = -\dfrac{F_x}{F_y} = -\dfrac{3x^2 - 6xy + 2y^2}{-3x^2 + 4xy + 6y^2} = \dfrac{3x^2 - 6xy + 2y^2}{3x^2 - 4xy - 6y^2}$.

41. $\nabla f(1, 2) = f_x(1, 2)\mathbf{i} + f_y(1, 2)\mathbf{j} = \left[\dfrac{x}{\sqrt{x^2 + y^2}}\right]_{(1,2)}\mathbf{i} + \left[\dfrac{y}{\sqrt{x^2 + y^2}}\right]_{(1,2)}\mathbf{j} = \dfrac{\sqrt{5}}{5}(\mathbf{i} + 2\mathbf{j})$

43. $\nabla f(2, 1, -3) = \left(f_x\mathbf{i} + f_y\mathbf{j} + f_z\mathbf{k}\right)\big|_{(2,1,-3)} = \left(y^2 + 2xz\right)\mathbf{i} + \left(2xy - z^2\right)\mathbf{j} + \left(-2yz + x^2\right)\mathbf{k}\big|_{(2,1,-3)}$

$= -11\mathbf{i} - 5\mathbf{j} + 10\mathbf{k}$

45. $\mathbf{u} = \dfrac{\mathbf{v}}{|\mathbf{v}|} = \dfrac{3\mathbf{i} - 4\mathbf{j}}{\sqrt{9 + 16}} = \dfrac{3}{5}\mathbf{i} - \dfrac{4}{5}\mathbf{j}$. $\nabla f = \dfrac{\partial}{\partial x}\left(x^3 y^2 - xy^3\right)\mathbf{i} + \dfrac{\partial}{\partial y}\left(x^3 y^2 - xy^3\right)\mathbf{j} = \left(3x^2 y^2 - y^3\right)\mathbf{i} + \left(2x^3 y - 3xy^2\right)\mathbf{j}$,

so $D_{\mathbf{u}} f(2, -1) = \nabla f(2, -1) \cdot \mathbf{u} = [3(4)(1) - (-1)]\left(\dfrac{3}{5}\right) + [2(8)(-1) - 3(2)(1)]\left(-\dfrac{4}{5}\right) = \dfrac{127}{5}$.

47. $\mathbf{u} = \dfrac{\mathbf{v}}{|\mathbf{v}|} = \dfrac{\mathbf{i} - 2\mathbf{j} + 2\mathbf{k}}{\sqrt{1 + 4 + 4}} = \frac{1}{3}\mathbf{i} - \frac{2}{3}\mathbf{j} + \frac{2}{3}\mathbf{k}$.

$\nabla f = \dfrac{\partial}{\partial x}\left(x\sqrt{y^2 + z^2}\right)\mathbf{i} + \dfrac{\partial}{\partial y}\left(x\sqrt{y^2 + z^2}\right)\mathbf{j} + \dfrac{\partial}{\partial z}\left(x\sqrt{y^2 + z^2}\right)\mathbf{k} = \sqrt{y^2 + z^2}\,\mathbf{i} + \dfrac{xy}{\sqrt{y^2 + z^2}}\mathbf{j} + \dfrac{xz}{\sqrt{y^2 + z^2}}\mathbf{k}$. Then

$D_{\mathbf{u}}f(2, 3, 4) = \sqrt{9 + 16}\left(\frac{1}{3}\right) + \dfrac{2 \cdot 3}{\sqrt{9 + 16}}\left(-\frac{2}{3}\right) + \dfrac{2 \cdot 4}{\sqrt{9 + 16}}\left(\frac{2}{3}\right) = \frac{29}{15}$.

49. $\nabla f(x, y) = f_x(x, y)\mathbf{i} + f_y(x, y)\mathbf{j} = \dfrac{\partial}{\partial x}\left(x^{1/2} + xy^2\right)\mathbf{i} + \dfrac{\partial}{\partial y}\left(x^{1/2} + xy^2\right)\mathbf{j} = \left(\dfrac{1}{2\sqrt{x}} + y^2\right)\mathbf{i} + 2xy\mathbf{j}$. At the point

$(4, 1)$, f is increasing most rapidly in the direction of $\nabla f(4, 1) = \left(\dfrac{1}{2\sqrt{4}} + 1^2\right)\mathbf{i} + 2 \cdot 4\mathbf{j} = \frac{5}{4}\mathbf{i} + 8\mathbf{j}$ and its maximum rate

of increase is $|\nabla f(4, 1)| = \sqrt{\left(\frac{5}{4}\right)^2 + 8^2} = \dfrac{\sqrt{1049}}{4}$.

51. Let $F(x, y, z) = 2x^2 + 4y^2 + 9z^2 - 27$. Then $\nabla F(x, y, z) = 4x\mathbf{i} + 8y\mathbf{j} + 18z\mathbf{k}$, so

$\nabla F(1, 2, 1) = 4\mathbf{i} + 16\mathbf{j} + 18\mathbf{k} = 2(2\mathbf{i} + 8\mathbf{j} + 9\mathbf{k})$. Thus, an equation of the tangent plane is

$2(x - 1) + 8(y - 2) + 9(z - 1) = 0 \Rightarrow 2x + 8y + 9z = 27$, and equations of the normal line are $\dfrac{x - 1}{2} = \dfrac{y - 2}{8} = \dfrac{z - 1}{9}$.

53. Let $F(x, y, z) = x^2 + 3xy^2 - z$. Then $\nabla F(x, y, z) = \left(2x + 3y^2\right)\mathbf{i} + 6xy\mathbf{j} - \mathbf{k}$, so $\nabla F(3, 1, 18) = 9\mathbf{i} + 18\mathbf{j} - \mathbf{k}$. Thus,

an equation of the tangent plane is $9(x - 3) + 18(y - 1) - (z - 18) = 0 \Rightarrow 9x + 18y - z = 27$, and equations of the

normal line are $\dfrac{x - 3}{9} = \dfrac{y - 1}{18} = \dfrac{z - 18}{-1}$.

55. a.

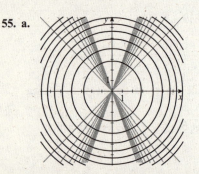

b. We need to show that ∇f and ∇g are orthogonal at each point (x, y) in the plane where the curves intersect. But

$$\nabla f(x, y) \cdot \nabla g(x, y) = (2x\mathbf{i} + 2y\mathbf{j})\left(-\dfrac{2y^2}{x^3}\mathbf{i} + \dfrac{2y}{x^2}\mathbf{j}\right) = -\dfrac{4y^2}{x^2} + \dfrac{4y^2}{x^2} = 0$$

and the result follows.

57. $f(x, y) = x^2 + xy + y^2 - 5x + 8y + 5 \Rightarrow f_x(x, y) = 2x + y - 5$ and $f_y(x, y) = x + 2y + 8$.

Setting $f_x(x, y) = f_y(x, y) = 0$, we find that the sole critical point of f is $(6, -7)$. Next, we calculate

$D(x, y) = f_{xx}(x, y)f_{yy}(x, y) - f_{xy}^2(x, y) = 2 \cdot 2 - 1 = 3$, so since $D(6, -7) = 3 > 0$ and $f_{xx}(6, -7) = 2 > 0$, we

see that $(6, -7)$ gives a relative minimum of f with value $f(6, -7) = -38$.

59. $f(x, y) = x^3 - 3xy + y^2 \Rightarrow f_x(x, y) = 3x^2 - 3y = 3\left(x^2 - y\right)$ and $f_y(x, y) = -3x + 2y$.

We set $\left.\begin{array}{ll} f_x(x, y) = 3\left(x^2 - y\right) = 0 & (1) \\ f_y(x, y) = -3x + 2y = 0 & (2) \end{array}\right\}$ From (1), $y = x^2$. Substituting this into (2) gives

$-3x + 2x^2 = x(-3 + 2x) = 0 \Rightarrow x = 0$ or $x = \frac{3}{2}$, so $(0, 0)$ and $\left(\frac{3}{2}, \frac{9}{4}\right)$ are critical points of f.

$D(x, y) = f_{xx}(x, y)f_{yy}(x, y) - f_{xy}^2(x, y) = (6x)(2) - (-3)^2 = 3(4x - 3)$.

For $(0, 0)$: $D(0, 0) = -9 < 0$, and so $(0, 0, 0)$ is a saddle point.

For $\left(\frac{3}{2}, \frac{9}{4}\right)$: $D\left(\frac{3}{2}, \frac{9}{4}\right) = 9 > 0$ and $f_{xx}\left(\frac{3}{2}, \frac{9}{4}\right) = 9 > 0$, so f has a relative minimum at $\left(\frac{3}{2}, \frac{9}{4}\right)$ with value

$f\left(\frac{3}{2}, \frac{9}{4}\right) = -\frac{27}{16}$.

61. $f(x, y) = x^2 + xy^2 - y^3 \Rightarrow f_x(x, y) = 2x + y^2$ and $f_y(x, y) = 2xy - 3y^2$. We set

$$\left. \begin{array}{l} f_x(x, y) = \quad 2x + y^2 = 0 \\ f_y(x, y) = 2xy - 3y^2 = 0 \end{array} \right\}$$ Solving the system gives $(0, 0)$ as the critical point of f in D because $\left(-\frac{9}{2}, -3\right)$ does

not lie in D, and $f(0, 0) = 0$.

On ℓ_1: $x = 1$, so $g(y) = f(1, y) = 1 + y^2 - y^3$ for $0 \le y \le 2$.

$g'(y) = 2y - 3y^2 = y(2 - 3y) = 0 \Rightarrow y = 0$ or $\frac{2}{3}$.

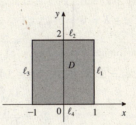

y	0	$\frac{2}{3}$	2
$g(y)$	1	$\frac{31}{27}$	-3

From the table, we see that g has a minimum value of -3 and a maximum

value of $\frac{31}{27}$.

On ℓ_2: $y = 2$, so $h(x) = f(x, 2) = x^2 + 4x - 8$ for $-1 \le x \le 1$. $h'(x) = 2x + 4 = 0 \Rightarrow x = -2$, so h has no critical

number in $(-1, 1)$. $h(-1) = -11$ and $h(1) = -3$ are the minimum and maximum values of h on $[-1, 1]$.

On ℓ_3: $x = -1$, so $s(y) = f(-1, y) = 1 - y^2 - y^3$ for $0 \le y \le 2$. $s'(y) = -2y - 3y^2 = -y(2 + 3y) = 0 \Rightarrow y = 0$ or

$-\frac{2}{3}$, so s has no critical number in $(0, 2)$. We see that $s(0) = 1$ and $s(2) = -11$ are the maximum and minimum values of

g on $[0, 2]$.

On ℓ_4: $y = 0$, so $t(x) = f(x, 0) = x^2$ for $-1 \le x \le 1$, which has a minimum value of 0 and a maximum value of 1.

We see that the absolute minimum value of f on D is -11 attained at $(-1, 2)$ and the absolute maximum value is $\frac{31}{27}$

attained at $\left(1, \frac{2}{3}\right)$.

63. $f(x, y) = xy^2 \Rightarrow \nabla f(x, y) = f_x(x, y)\mathbf{i} + f_y(x, y)\mathbf{j} = y^2\mathbf{i} + 2xy\mathbf{j}$ and $g(x, y) = x^2 + y^2 - 4 \Rightarrow$

$\nabla g(x, y) = g_x(x, y)\mathbf{i} + g_y(x, y)\mathbf{j} = 2x\mathbf{i} + 2y\mathbf{j}$. $\nabla f = \lambda \nabla g$ together with the constraint equation gives

$$\left. \begin{array}{ll} y^2 = 2\lambda x & (1) \\ 2xy = 2\lambda y & (2) \\ x^2 + y^2 = 4 & (3) \end{array} \right\}$$ Note that $x \ne 0$; otherwise, from (1), $y = 0$, which contradicts (3). Thus, $\lambda = \dfrac{y^2}{2x}$.

Substituting this into (2) gives $x = \dfrac{y^2}{2x} \Rightarrow y^2 = 2x^2$, $y \ne 0$. Substituting this into (3) gives $3x^2 = 4 \Leftrightarrow x = \pm\frac{2\sqrt{3}}{3}$. The

corresponding values of y are $y = \pm\frac{2\sqrt{6}}{3}$. If $y = 0$, then $x = \pm 2$.

(x, y)	$(-2, 0)$	$(2, 0)$	$\left(-\frac{2\sqrt{3}}{3}, \frac{2\sqrt{6}}{3}\right)$	$\left(-\frac{2\sqrt{3}}{3}, -\frac{2\sqrt{6}}{3}\right)$	$\left(\frac{2\sqrt{3}}{3}, -\frac{2\sqrt{6}}{3}\right)$	$\left(\frac{2\sqrt{3}}{3}, \frac{2\sqrt{6}}{3}\right)$
$f(x, y)$	0	0	$-\frac{16\sqrt{3}}{9}$	$-\frac{16\sqrt{3}}{9}$	$\frac{16\sqrt{3}}{9}$	$\frac{16\sqrt{3}}{9}$

From the table, we see that f has a minimum value of $-\frac{16\sqrt{3}}{9}$ and a maximum value of $\frac{16\sqrt{3}}{9}$.

65. $f(x, y, z) = xy + yz + xz \Rightarrow \nabla f(x, y, z) = f_x(x, y, z)\mathbf{i} + f_y(x, y, z)\mathbf{j} + f_z(x, y, z)\mathbf{k} = (y + z)\mathbf{i} + (x + z)\mathbf{j} + (x + y)\mathbf{k}$

and $g(x, y, z) = x + 2y + 3z - 1 \Rightarrow \nabla g(x, y, z) = g_x(x, y, z)\mathbf{i} + g_y(x, y, z)\mathbf{j} + g_z(x, y, z)\mathbf{k} = \mathbf{i} + 2\mathbf{j} + 3\mathbf{k}$.

$\nabla f = \lambda \nabla g$ together with the constraint equation gives $\left. \begin{array}{ll} y + z = \lambda & (1) \\ x + z = 2\lambda & (2) \\ x + y = 3\lambda & (3) \\ x + 2y + 3z = 1 \end{array} \right\}$ Solving the system, we find

$\lambda = \frac{1}{4}$, $x = \frac{1}{2}$, $y = \frac{1}{4}$, and $z = 0$. We see that $f\left(\frac{1}{2}, \frac{1}{4}, 0\right) = \frac{1}{8}$ is the maximum value of f.

67. $f(x, y) = Ax^2 + Bxy + Cy^2 + Dx + Ey + F \Rightarrow f_x(x, y) = 2Ax + By + D$ and $f_y(x, y) = Bx + 2Cy + E = 0$.

If f has a relative extremum at (x_0, y_0), then the critical point (x_0, y_0) must satisfy the system $\left.\begin{array}{l} f_x = 0 \\ f_y = 0 \end{array}\right\} \Leftrightarrow$

$\left.\begin{array}{l} 2Ax + By + D = 0 \\ Bx + 2Cy + E = 0 \end{array}\right\}$.

69. We want to minimize $D = f(x, y, z) = (x - 3)^2 + y^2 + z^2$ subject to the constraint $g(x, y, z) = \frac{1}{4}x^2 + \frac{1}{25}y^2 - z = 0$.

$\nabla f(x, y, z) = 2(x - 3)\mathbf{i} + 2y\mathbf{j} + 2z\mathbf{k}$ and $\nabla g(x, y, z) = \frac{1}{2}x\mathbf{i} + \frac{2}{25}y\mathbf{j} - \mathbf{k}$, so $\nabla f = \lambda \nabla g$ together with the constraint

equation leads to $\left.\begin{array}{ll} 2x - 6 = \frac{1}{2}\lambda x & (1) \\ 2y = \frac{2}{25}\lambda y & (2) \\ 2z = -\lambda & (3) \\ \frac{1}{4}x^2 + \frac{1}{25}y^2 = z & (4) \end{array}\right\}$ From (2), we see that $y = 0$ or $\lambda = 25$. Substituting $\lambda = 25$ into

(3), we obtain $z = -\frac{25}{2}$, but this violates (4). So we must have $y = 0$, and the system becomes $\left.\begin{array}{l} 4x - 12 = \lambda x \\ z = -\frac{1}{2}\lambda \\ x^2 - 4z = 0 \end{array}\right\} \Rightarrow$

$x^3 + 8x - 24 = 0 \Rightarrow x = 2$. Thus, the required point is $(2, 0, 1)$.

71. True. If $\mathbf{u} = \mathbf{i}$, then $D_{\mathbf{u}}f(a, b) = D_{\mathbf{i}}f(a, b) = f_x(a, b)$.

Challenge Problems

1. We require that $\left.\begin{array}{l} 36 - 4x^2 - 9y^2 \geq 0 \\ x^2 - 2x + y^2 > 0 \\ 4x^2 + 16x + 4y^2 + 15 > 0 \end{array}\right\} \Rightarrow \left.\begin{array}{l} \frac{1}{9}x^2 + \frac{1}{4}y^2 \leq 1 \\ (x - 1)^2 + y^2 > 1 \\ (x + 2)^2 + y^2 > \frac{1}{4} \end{array}\right\}$

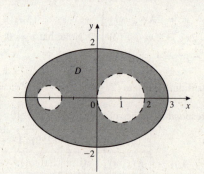

$D = \left\{(x, y, z) \;\middle|\; \begin{array}{l} 4x^2 + 9y^2 \leq 36, \; x^2 - 2x + y^2 > 0, \\ \text{and } 4x^2 + 16y + 4y^2 > -15 \end{array}\right\}$.

3. a. $D_{\mathbf{u}}f(x, y) = (f_x\mathbf{i} + f_y\mathbf{j}) \cdot (u_1\mathbf{i} + u_2\mathbf{j}) = u_1 f_x + u_2 f_y$, so

$D_{\mathbf{u}}^2 f(x, y) = D_{\mathbf{u}}(D_{\mathbf{u}}f) = \left[\frac{\partial}{\partial x}(u_1 f_x + u_2 f_y)\mathbf{i} + \frac{\partial}{\partial y}(u_1 f_x + u_2 f_y)\mathbf{j}\right] \cdot (u_1\mathbf{i} + u_2\mathbf{j})$

$= \left[(u_1 f_{xx} + u_2 f_{yx})\mathbf{i} + (u_1 f_{xy} + u_2 f_{yy})\mathbf{j}\right] \cdot (u_1\mathbf{i} + u_2\mathbf{j}) = f_{xx}u_1^2 + 2f_{xy}u_1 u_2 + f_{yy}u_2^2$

Note that $f_{xy} = f_{yx}$ because, by assumption, f has continuous second partial derivatives in x and y.

b. Here $\mathbf{u} = \dfrac{\mathbf{v}}{|\mathbf{v}|} = \dfrac{2\mathbf{i} - 3\mathbf{j}}{\sqrt{4 + 9}} = \dfrac{2\sqrt{13}}{13}\mathbf{i} - \dfrac{3\sqrt{13}}{13}\mathbf{j}$. $f_x(x, y) = \dfrac{\partial}{\partial x}(xy^2 + e^{xy}) = y^2 + ye^{xy}$ and $f_y(x, y) = 2xy + xe^{xy}$,

so $f_{xx}(x, y) = y^2 e^{xy}$, $f_{xy}(x, y) = 2y + (1 + xy)e^{xy}$, and $f_{yy}(x, y) = 2x + x^2 e^{xy}$. Thus,

$D_{\mathbf{u}}^2 f(1, 0) = f_{xx}(1, 0)u_1^2 + 2f_{xy}(1, 0)u_1 u_2 + f_{yy}(1, 0)u_2^2 = 0 + 2(1)\left(\dfrac{2\sqrt{13}}{13}\right)\left(-\dfrac{3\sqrt{13}}{13}\right) + 3\left(-\dfrac{3\sqrt{13}}{13}\right)^2 = \dfrac{15}{13}$.

5. The area of triangle ABC is given by the sum of the areas of $\triangle COB$, $\triangle CAO$, and $\triangle ABO$. Thus,

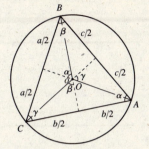

$$A = 2\left[\left(\tfrac{1}{2} \cdot \tfrac{a}{2} \cdot R\cos\alpha\right) + \left(\tfrac{1}{2} \cdot \tfrac{b}{2} \cdot R\cos\beta\right) + \left(\tfrac{1}{2} \cdot \tfrac{c}{2} \cdot R\cos\gamma\right)\right]$$

$$= \tfrac{1}{2}R\left(a\cos\alpha + b\cos\beta + c\cos\gamma\right)$$

and so $\dfrac{\partial f}{\partial a} = \dfrac{\partial A}{\partial a} = \tfrac{1}{2}R\cos\alpha$.

7. $\mathbf{r}(t) = \left\langle t, t^2, t^3\right\rangle \Rightarrow \mathbf{r}'(t) = \left\langle 1, 2t, 3t^2\right\rangle$ and $\mathbf{r}'(1) = \langle 1, 2, 3\rangle \Rightarrow \mathbf{u} = \frac{\sqrt{14}}{14}\langle 1, 2, 3\rangle$, the unit vector in the

desired direction. Next, $\nabla f(x, y, z) = f_x(x, y, z)\mathbf{i} + f_y(x, y, z)\mathbf{j} + f_z(x, y, z)\mathbf{k} = 2x\mathbf{i} + \cos z\,\mathbf{j} - y\sin z\,\mathbf{k}$,

and so $\nabla f\left(1, 2, \frac{\pi}{4}\right) = 2\mathbf{i} + \frac{\sqrt{2}}{2}\mathbf{j} - \sqrt{2}\mathbf{k}$. Therefore,

$$D_{\mathbf{u}}f\left(1, 2, \tfrac{\pi}{4}\right) = \nabla f\left(1, 2, \tfrac{\pi}{4}\right) \cdot \mathbf{u} = \tfrac{\sqrt{14}}{14}\left(2\mathbf{i} + \tfrac{\sqrt{2}}{2}\mathbf{j} - \sqrt{2}\mathbf{k}\right) \cdot (\mathbf{i} + 2\mathbf{j} + 3\mathbf{k}) = \tfrac{\sqrt{14}}{14}\left(2 + \sqrt{2} - 3\sqrt{2}\right) = \tfrac{\sqrt{14}}{7}\left(1 - \sqrt{2}\right).$$

9. Recall that $x = r\cos\theta$, $y = r\sin\theta$, $z = z$. Thus, $\dfrac{\partial u}{\partial r} = \dfrac{\partial u}{\partial x}\dfrac{\partial x}{\partial r} + \dfrac{\partial u}{\partial y}\dfrac{\partial y}{\partial r} + \dfrac{\partial u}{\partial z}\dfrac{\partial z}{\partial r} = \cos\theta\dfrac{\partial u}{\partial x} + \sin\theta\dfrac{\partial u}{\partial y} \Rightarrow$

$$\frac{\partial^2 u}{\partial r^2} = \frac{\partial}{\partial x}\left(\cos\theta\frac{\partial u}{\partial x} + \sin\theta\frac{\partial u}{\partial y}\right)\frac{\partial x}{\partial r} + \frac{\partial}{\partial y}\left(\cos\theta\frac{\partial u}{\partial x} + \sin\theta\frac{\partial u}{\partial y}\right)\frac{\partial y}{\partial r}$$

$$= \left(\cos\theta\frac{\partial^2 u}{\partial x^2} + \sin\theta\frac{\partial^2 u}{\partial x\,\partial y}\right)\cos\theta + \left(\cos\theta\frac{\partial^2 u}{\partial y\,\partial x} + \sin\theta\frac{\partial^2 u}{\partial y^2}\right)\sin\theta$$

$$= \cos^2\theta\frac{\partial^2 u}{\partial x^2} + 2\sin\theta\cos\theta\frac{\partial^2 u}{\partial x\,\partial y} + \sin^2\theta\frac{\partial^2 u}{\partial y^2} \quad\text{since } \frac{\partial^2 u}{\partial x\,\partial y} = \frac{\partial^2 u}{\partial y\,\partial x}$$

$$\frac{\partial u}{\partial \theta} = \frac{\partial u}{\partial x}\frac{\partial x}{\partial \theta} + \frac{\partial u}{\partial y}\frac{\partial y}{\partial \theta} + \frac{\partial u}{\partial z}\frac{\partial z}{\partial \theta} = \frac{\partial u}{\partial x}(-r\sin\theta) + \frac{\partial u}{\partial y}(r\cos\theta) = \frac{\partial u}{\partial y}r\cos\theta - \frac{\partial u}{\partial x}r\sin\theta \Rightarrow$$

$$\frac{\partial^2 u}{\partial \theta^2} = r\left\{\cos\theta\left[\frac{\partial}{\partial x}\left(\frac{\partial u}{\partial y}\right)\frac{\partial x}{\partial \theta} + \frac{\partial}{\partial y}\left(\frac{\partial u}{\partial y}\right)\frac{\partial y}{\partial \theta}\right] - \frac{\partial u}{\partial y}\sin\theta\right\}$$

$$-r\left\{\sin\theta\left[\frac{\partial}{\partial x}\left(\frac{\partial u}{\partial x}\right)\frac{\partial x}{\partial \theta} + \frac{\partial}{\partial y}\left(\frac{\partial u}{\partial x}\right)\frac{\partial y}{\partial \theta}\right] + \frac{\partial u}{\partial x}\cos\theta\right\}$$

$$= r\left\{\cos\theta\left[\frac{\partial^2 u}{\partial x\,\partial y}(-r\sin\theta) + \frac{\partial^2 u}{\partial y^2}(r\cos\theta)\right] - \frac{\partial u}{\partial y}\sin\theta\right\}$$

$$-r\left\{\sin\theta\left[\frac{\partial^2 u}{\partial x^2}(-r\sin\theta) + \frac{\partial^2 u}{\partial y\,\partial x}(r\cos\theta)\right] + \frac{\partial u}{\partial x}\cos\theta\right\}$$

$$= -\frac{\partial^2 u}{\partial x\,\partial y}\left(2r^2\sin\theta\cos\theta\right) + \frac{\partial^2 u}{\partial y^2}\left(r^2\cos^2\theta\right) + \frac{\partial^2 u}{\partial x^2}\left(r^2\sin^2\theta\right) - \frac{\partial u}{\partial y}(r\sin\theta) - \frac{\partial u}{\partial x}(r\cos\theta)$$

Therefore,

$$\frac{\partial^2 u}{\partial r^2} + \frac{1}{r}\frac{\partial u}{\partial r} + \frac{1}{r^2}\frac{\partial^2 u}{\partial \theta^2} + \frac{\partial^2 u}{\partial z^2}$$

$$= \cos^2\theta\frac{\partial^2 u}{\partial x^2} + 2\sin\theta\cos\theta\frac{\partial^2 u}{\partial x\,\partial y} + \sin^2\theta\frac{\partial^2 u}{\partial y^2} + \frac{1}{r}\left(\cos\theta\frac{\partial u}{\partial x} + \sin\theta\frac{\partial u}{\partial y}\right)$$

$$+ \frac{1}{r^2}\cdot r^2\left(\sin^2\theta\frac{\partial^2 u}{\partial x^2} - 2\sin\theta\cos\theta\frac{\partial^2 u}{\partial x\,\partial y} + \cos^2\theta\frac{\partial^2 u}{\partial y^2} - \frac{\sin\theta}{r}\frac{\partial u}{\partial y} - \frac{\cos\theta}{r}\frac{\partial u}{\partial x}\right) + \frac{\partial^2 u}{\partial z^2}$$

$$= \frac{\partial^2 u}{\partial x^2}\left(\sin^2\theta + \cos^2\theta\right) + \frac{\partial^2 u}{\partial y^2}\left(\sin^2\theta + \cos^2\theta\right) + \frac{\partial^2 u}{\partial z^2} = \frac{\partial^2 u}{\partial x^2} + \frac{\partial^2 u}{\partial y^2} + \frac{\partial^2 u}{\partial z^2} = 0$$

14 Multiple Integrals

14.1 Concept Questions

1. a.

$x \backslash y$	1	$\frac{3}{2}$	2	$\frac{5}{2}$	3	$\frac{7}{2}$	4
0	2	3	4	5	6	7	8
$\frac{1}{4}$	$\frac{9}{4}$	$\frac{13}{4}$	$\frac{17}{4}$	$\frac{21}{4}$	$\frac{25}{4}$	$\frac{29}{4}$	$\frac{33}{4}$
$\frac{1}{2}$	$\frac{5}{2}$	$\frac{7}{2}$	$\frac{9}{2}$	$\frac{11}{2}$	$\frac{13}{2}$	$\frac{15}{2}$	$\frac{17}{2}$
$\frac{3}{4}$	$\frac{11}{4}$	$\frac{15}{4}$	$\frac{19}{4}$	$\frac{23}{4}$	$\frac{27}{4}$	$\frac{31}{4}$	$\frac{35}{4}$
1	3	4	5	6	7	8	9

b. $\Delta A = (\Delta x)(\Delta y) = \left(\frac{1-0}{2}\right)\left(\frac{4-1}{3}\right) = \frac{1}{2}$. We find that

$$V \approx \sum_{i=1}^{2} \sum_{j=1}^{3} f\left(x_{ij}^*, y_{ij}^*\right) \Delta A$$

$$= \left[f(0,1) + f(0,2) + f(0,3) + f\left(\frac{1}{2}, 1\right) + f\left(\frac{1}{2}, 2\right) + f\left(\frac{1}{2}, 3\right) \right] \Delta A$$

$$= \frac{1}{2}\left(2 + 4 + 6 + \frac{5}{2} + \frac{9}{2} + \frac{13}{2}\right) = \frac{51}{4}$$

c. $V \approx \sum_{i=1}^{2} \sum_{j=1}^{3} f\left(x_{ij}^*, y_{ij}^*\right) \Delta A$

$$= \left[f\left(\frac{1}{4}, \frac{3}{2}\right) + f\left(\frac{1}{4}, \frac{5}{2}\right) + f\left(\frac{1}{4}, \frac{7}{2}\right) + f\left(\frac{3}{4}, \frac{3}{2}\right) + f\left(\frac{3}{4}, \frac{5}{2}\right) + f\left(\frac{3}{4}, \frac{7}{2}\right) \right] \Delta A$$

$$= \frac{1}{2}\left(\frac{13}{4} + \frac{21}{4} + \frac{29}{4} + \frac{15}{4} + \frac{23}{4} + \frac{31}{4}\right) = \frac{33}{2}$$

14.1 Double Integrals

1. $f(x, y) = 8 - 2x^2 - y^2$ and $\Delta A = (\Delta x)(\Delta y) = \left(\frac{1-0}{2}\right)\left(\frac{2-0}{2}\right) = \frac{1}{2}$, so

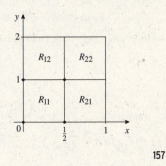

$$V \approx \sum_{i=1}^{2} \sum_{j=1}^{2} f\left(x_{ij}^*, y_{ij}^*\right) \Delta A$$

$$= \left[f(0,0) + f(0,1) + f\left(\frac{1}{2}, 0\right) + f\left(\frac{1}{2}, 1\right) \right] \cdot \frac{1}{2}$$

$$= \frac{1}{2}\left\{ 8 + (8-1) + \left[8 - 2\left(\frac{1}{2}\right)^2 \right] + \left[8 - 2\left(\frac{1}{2}\right)^2 - 1^2 \right] \right\}$$

$$= \frac{1}{2}\left(8 + 7 + \frac{15}{2} + \frac{13}{2}\right) = \frac{29}{2}$$

157

3. $f(x, y) = 8 - 2x^2 - y^2$ and $\Delta A = (\Delta x)(\Delta y) = \left(\frac{1-0}{2}\right)\left(\frac{2-0}{2}\right) = \frac{1}{2}$, so

$$V \approx \sum_{i=1}^{2} \sum_{j=1}^{2} f\left(x_{ij}^*, y_{ij}^*\right) \Delta A$$

$$= \left[f\left(\tfrac{1}{2}, 0\right) + f\left(\tfrac{1}{2}, 1\right) + f(1, 0) + f(1, 1)\right] \cdot \tfrac{1}{2}$$

$$= \tfrac{1}{2}\left\{\left[8 - 2\left(\tfrac{1}{2}\right)^2\right] + \left[8 - 2\left(\tfrac{1}{2}\right)^2 - 1^2\right] + (8 - 2) + (8 - 2 - 1)\right\}$$

$$= \tfrac{1}{2}\left(\tfrac{15}{2} + \tfrac{13}{2} + 6 + 5\right) = \tfrac{25}{2}$$

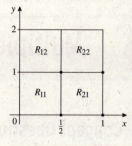

5. $f(x, y) = 2x + 3y$ and $\Delta A = (\Delta x)(\Delta y) = \left(\frac{1-0}{2}\right)\left(\frac{3-0}{3}\right) = \frac{1}{2}$, so

$$V \approx \sum_{i=1}^{2} \sum_{j=1}^{3} f\left(x_{ij}^*, y_{ij}^*\right) \Delta A$$

$$= \left[f(0, 0) + f(0, 1) + f(0, 2) + f\left(\tfrac{1}{2}, 0\right) + f\left(\tfrac{1}{2}, 1\right) + f\left(\tfrac{1}{2}, 2\right)\right] \cdot \tfrac{1}{2}$$

$$= \tfrac{1}{2}\left[0 + 3 + 6 + 1 + (1 + 3) + (1 + 6)\right]$$

$$= \tfrac{21}{2}$$

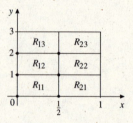

7. $f(x, y) = x^2 + 2y^2$ and $\Delta A = (\Delta x)(\Delta y) = \left(\frac{3+1}{4}\right)\left(\frac{4-0}{4}\right) = 1$, so

$$V \approx \sum_{i=1}^{4} \sum_{j=1}^{4} f\left(x_{ij}^*, y_{ij}^*\right) \Delta A$$

$$= \left[f\left(-\tfrac{1}{2}, \tfrac{1}{2}\right) + f\left(-\tfrac{1}{2}, \tfrac{3}{2}\right) + f\left(-\tfrac{1}{2}, \tfrac{5}{2}\right) + f\left(-\tfrac{1}{2}, \tfrac{7}{2}\right)\right.$$

$$+ f\left(\tfrac{1}{2}, \tfrac{1}{2}\right) + f\left(\tfrac{1}{2}, \tfrac{3}{2}\right) + f\left(\tfrac{1}{2}, \tfrac{5}{2}\right) + f\left(\tfrac{1}{2}, \tfrac{7}{2}\right)$$

$$+ f\left(\tfrac{3}{2}, \tfrac{1}{2}\right) + f\left(\tfrac{3}{2}, \tfrac{3}{2}\right) + f\left(\tfrac{3}{2}, \tfrac{5}{2}\right) + f\left(\tfrac{3}{2}, \tfrac{7}{2}\right)$$

$$\left. + f\left(\tfrac{5}{2}, \tfrac{1}{2}\right) + f\left(\tfrac{5}{2}, \tfrac{3}{2}\right) + f\left(\tfrac{5}{2}, \tfrac{5}{2}\right) + f\left(\tfrac{5}{2}, \tfrac{7}{2}\right)\right] \cdot 1$$

$$= \left(\tfrac{1}{4} + \tfrac{1}{2}\right) + \left(\tfrac{1}{4} + \tfrac{9}{2}\right) + \left(\tfrac{1}{4} + \tfrac{25}{2}\right) + \left(\tfrac{1}{4} + \tfrac{49}{2}\right) + \left(\tfrac{1}{4} + \tfrac{1}{2}\right) + \left(\tfrac{1}{4} + \tfrac{9}{2}\right) + \left(\tfrac{1}{4} + \tfrac{25}{2}\right) + \left(\tfrac{1}{4} + \tfrac{49}{2}\right)$$

$$+ \left(\tfrac{9}{4} + \tfrac{1}{2}\right) + \left(\tfrac{9}{4} + \tfrac{9}{2}\right) + \left(\tfrac{9}{4} + \tfrac{25}{2}\right) + \left(\tfrac{9}{4} + \tfrac{49}{2}\right) + \left(\tfrac{25}{4} + \tfrac{1}{2}\right) + \left(\tfrac{25}{4} + \tfrac{9}{2}\right) + \left(\tfrac{25}{4} + \tfrac{25}{2}\right) + \left(\tfrac{25}{4} + \tfrac{49}{2}\right)$$

$$= 204$$

9. $\Delta A = (\Delta x)(\Delta y) = \left(\frac{5-0}{5}\right)\left(\frac{3-0}{3}\right) = 1$, so

$$V \approx \sum_{i=1}^{5} \sum_{j=1}^{3} f\left(x_{ij}^*, y_{ij}^*\right) \Delta A$$

$$= \left[F_D(x_1, y_1) + F_D(x_1, y_2) + F_D(x_1, y_3) + F_D(x_2, y_1) + F_D(x_2, y_2) + F_D(x_2, y_3) + F_D(x_3, y_1) + F_D(x_3, y_2)\right.$$

$$\left. + F_D(x_3, y_3) + F_D(x_4, y_1) + F_D(x_4, y_2) + F_D(x_4, y_3) + F_D(x_5, y_1) + F_D(x_5, y_2) + F_D(x_5, y_3)\right] \Delta A$$

$$= \left[0 + 1 + 0 + (-2) + (-1) + 3 + (-1) + 0 + 2 + 1 + 2 + 3 + 0 + 3 + 4\right] = 15$$

11. We estimate $f\left(x_{11}^*, y_{11}^*\right) = 25$, $f\left(x_{12}^*, y_{12}^*\right) = 50$, $f\left(x_{21}^*, y_{21}^*\right) = 17$, and $f\left(x_{22}^*, y_{22}^*\right) = 35$, so

$$\iint_R f(x, y)\, dA \approx \sum_{i=1}^{2} \sum_{j=1}^{2} f\left(x_{ij}^*, y_{ij}^*\right) \Delta A \approx (27 + 50 + 17 + 35) \cdot 1 = 129.$$

13. $\iint_R 2\,dA$ represents the volume of the rectangular box with base $R = [-1, 3] \times [2, 5]$ and height 2, so $\iint_R 2\,dA = 2\,[3 - (-1)]\,(5 - 2) = 24$.

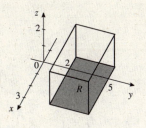

15. $\iint_R (6 - 2y)\,dA$ represents the volume of the solid shown. Its base is $R = [0, 4] \times [0, 2]$ so, calculating the area of trapezoidal cross-sections parallel to the yz-plane and multiplying by the length in the x-direction,
$$\iint_R (6 - 2y)\,dA = \tfrac{1}{2}\,(6 + 2)\,(2)\,(4) = 32.$$

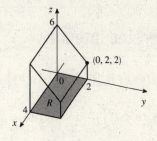

17. The solid is the wedge bounded above by the cylinder $z = 4 - x^2$ and below by the triangular base $R = \{(x, y) \mid 0 \le y \le x, 0 \le x \le 2\}$.

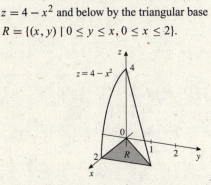

19. $\displaystyle\lim_{m,n\to\infty} \sum_{i=1}^{m} \sum_{j=1}^{n} \left(3 - 2x_{ij}^* + y_{ij}^*\right)\,\Delta A = \iint_R (3 - 2x + y)\,dA$

21. $\iint_R \sqrt{1 + x^2 + y^2}\,dA \approx 1.28079$

23. Observe that $-|f(x, y)| \le f(x, y) \le |f(x, y)|$ for every $(x, y) \in D$. Thus, by Property 4 of Theorem 1, we have $-\iint_D |f(x, y)|\,dA \le \iint_D f(x, y)\,dA \le \iint_D |f(x, y)|\,dA \Leftrightarrow \left|\iint_D f(x, y)\,dA\right| \le \iint_D |f(x, y)|\,dA$.

25. Observe that $0 < e^{-1} \le e^{-x} \le 1$ for all x in $[0, 1]$ and $0 < \cos y \le 1$ for all $0 \le y \le 1$, so $0 \le e^{-x}\cos y \le 1$ for all $(x, y) \in R$. Thus, by Property 4 of Theorem 1, we have $\iint_R 0\,dA \le \iint_R e^{-x}\cos y\,dA \le \iint_R 1\,dA \Leftrightarrow 0 \le \iint_R e^{-x}\cos y\,dA \le 1$.

27. True. $\iint_D [2f(x, y) - 3g(x, y)]\,dA = \iint_D 2f(x, y)\,dA - \iint_D 3g(x, y)\,dA$ by Property 1 of Theorem 1. Continuing, this is equal to $2\iint_D f(x, y)\,dA - 3\iint_D g(x, y)\,dA$ by Property 2 of the same theorem.

29. True. Observe that if (x, y) is a point in R, then $x^2 + xy + y^2 + 1 \ge 1$, so $\sqrt{x^2 + xy + y^2 + 1} \ge 1$. Also, $\cos\left(x^2 + y^2\right) < 1$, so $\dfrac{\sqrt{x^2 + xy + y^2 + 1}}{\cos\left(x^2 + y^2\right)} > 1$ for $(x, y) \in R$. Therefore, by Property 4 of Theorem 1,
$$\iint_R \frac{\sqrt{x^2 + xy + y^2 + 1}}{\cos\left(x^2 + y^2\right)}\,dA \ge \iint_R 1\,dA = \pi \cdot 1^2 = \pi.$$

14.2   Concept Questions   ET 13.2

1. a. See page 1158 (1154 in ET).
   b. See page 1159 (1155 in ET).
   c. See pages 1159 and 1160 (1155 and 1156 in ET).

14.2   Iterated Integrals   ET 13.2

1. $\int_0^1 \int_0^2 (x+2y)\,dy\,dx = \int_0^1 \left[\int_0^2 (x+2y)\,dy\right] dx = \int_0^1 \left[xy+y^2\right]_{y=0}^{y=2} dx = \int_0^1 (2x+4)\,dx = \left(x^2+4x\right)\Big|_0^1 = 5$

3. $\int_0^2 \int_1^4 y\sqrt{x}\,dy\,dx = \int_0^2 \left[\int_1^4 yx^{1/2}\,dy\right] dx = \int_0^2 \left[\tfrac{1}{2}y^2 x^{1/2}\right]_{y=1}^{y=4} dx = \int_0^2 \tfrac{15}{2}x^{1/2}\,dx = 5x^{3/2}\Big|_0^2 = 10\sqrt{2}$

5. $\int_0^\pi \int_0^\pi \cos(x+y)\,dy\,dx = \int_0^\pi \left[\int_0^\pi \cos(x+y)\,dy\right] dx = \int_0^\pi \left[\sin(x+y)\right]_{y=0}^{y=\pi} dx = \int_0^\pi \left[\sin(x+\pi)-\sin x\right] dx$

$\qquad = -2\int_0^\pi \sin x\,dx = 2\cos x|_0^\pi = -4$

7. $\int_0^4 \int_0^{\sqrt{x}} 2xy\,dy\,dx = \int_0^4 \left[\int_0^{x^{1/2}} 2xy\,dy\right] dx = \int_0^4 \left[xy^2\right]_{y=0}^{y=x^{1/2}} dx = \int_0^4 x^2\,dx = \tfrac{1}{3}x^3\Big|_0^4 = \tfrac{64}{3}$

9. $\int_0^1 \int_0^{\sqrt{1-y^2}} x\,dx\,dy = \int_0^1 \left[\int_0^{\sqrt{1-y^2}} x\,dx\right] dy = \int_0^1 \left[\tfrac{1}{2}x^2\right]_{x=0}^{x=\sqrt{1-y^2}} dy = \int_0^1 \tfrac{1}{2}\left(1-y^2\right) dy = \tfrac{1}{2}\left(y-\tfrac{1}{3}y^3\right)\Big|_0^1 = \tfrac{1}{3}$

11. $\int_{-1}^1 \int_x^{2x} e^{x+y}\,dy\,dx = \int_{-1}^1 \left[\int_x^{2x} e^{x+y}\,dy\right] dx = \int_{-1}^1 \left[e^{x+y}\right]_{y=x}^{y=2x} dx = \int_{-1}^1 \left(e^{3x}-e^{2x}\right) dx = \left(\tfrac{1}{3}e^{3x}-\tfrac{1}{2}e^{2x}\right)\Big|_{-1}^1$

$\qquad = \left(\tfrac{1}{3}e^3 - \tfrac{1}{2}e^2\right) - \left(\tfrac{1}{3}e^{-3} - \tfrac{1}{2}e^{-2}\right) = \dfrac{2e^6 - 3e^5 + 3e - 2}{6e^3}$

13. $\iint_R \left(x+y^2\right) dA = \int_{-1}^2 \int_0^1 \left(x+y^2\right) dx\,dy = \int_{-1}^2 \left[\tfrac{1}{2}x^2 + xy^2\right]_{x=0}^{x=1} dy = \int_{-1}^2 \left(\tfrac{1}{2}+y^2\right) dy = \left(\tfrac{1}{2}y + \tfrac{1}{3}y^3\right)\Big|_{-1}^2 = \tfrac{9}{2}$

15. $\iint_R (x\cos y + y\sin x)\,dA = \int_0^{\pi/4} \int_0^{\pi/2} (x\cos y + y\sin x)\,dx\,dy = \int_0^{\pi/4} \left[\tfrac{1}{2}x^2\cos y - y\cos x\right]_{x=0}^{x=\pi/2} dy$

$\qquad = \int_0^{\pi/4} \left(\tfrac{1}{8}\pi^2 \cos y + y\right) dy = \left(\tfrac{1}{8}\pi^2 \sin y + \tfrac{1}{2}y^2\right)\Big|_0^{\pi/4} = \tfrac{1}{32}\pi^2\left(2\sqrt{2}+1\right)$

17. $\iint_R (x+2y)\,dA = \int_0^1 \int_0^x (x+2y)\,dy\,dx$

$\qquad = \int_0^1 \left[xy+y^2\right]_{y=0}^{y=x} dx = 2\int_0^1 x^2\,dx$

$\qquad = \tfrac{2}{3}x^3\Big|_0^1 = \tfrac{2}{3}$

19. $\iint_R \left(x^3+2y\right) dA = \int_0^2 \int_{x^2}^{2x} \left(x^3+2y\right) dy\,dx$

$\qquad = \int_0^2 \left[x^3 y + y^2\right]_{y=x^2}^{y=2x} dx$

$\qquad = \int_0^2 \left(x^4 + 4x^2 - x^5\right) dx$

$\qquad = \left(\tfrac{1}{5}x^5 + \tfrac{4}{3}x^3 - \tfrac{1}{6}x^6\right)\Big|_0^2 = \tfrac{32}{5}$

21. $\iint_R (1 + 2x + 2y)\, dA = \int_0^1 \int_y^{2y} (1 + 2x + 2y)\, dx\, dy$

$$= \int_0^1 \left[x + x^2 + 2xy \right]_{x=y}^{x=2y} dy$$

$$= \int_0^1 \left(5y^2 + y \right) dy$$

$$= \left(\tfrac{5}{3}y^3 + \tfrac{1}{2}y^2 \right)\Big|_0^1 = \tfrac{13}{6}$$

23. $\iint_R x \cos y\, dA = \int_0^{\pi/2} \int_0^{\sin y} x \cos y\, dx\, dy$

$$= \int_0^{\pi/2} \left[\tfrac{1}{2}x^2 \cos y \right]_{x=0}^{x=\sin y} dy$$

$$= \int_0^{\pi/2} \tfrac{1}{2} \sin^2 y \cos y\, dy$$

$$= \tfrac{1}{6} \sin^3 y \Big|_0^{\pi/2} = \tfrac{1}{6}$$

25. $\iint_R x^2 y\, dA = \int_1^2 \int_x^{2x} x^2 y\, dy\, dx = \int_1^2 \left[\tfrac{1}{2}x^2 y^2 \right]_{y=x}^{y=2x} dx$

$$= \tfrac{1}{2} \int_1^2 3x^4\, dx = \tfrac{3}{10}x^5 \Big|_1^2 = \tfrac{93}{10}$$

27. $\iint_R (\sin x - y)\, dA = \int_0^{\pi/2} \int_0^{\cos x} (\sin x - y)\, dy\, dx$

$$= \int_0^{\pi/2} \left[y \sin x - \tfrac{1}{2}y^2 \right]_{y=0}^{y=\cos x} dx$$

$$= \int_0^{\pi/2} \left(\cos x \sin x - \tfrac{1}{2} \cos^2 x \right) dx$$

$$= \int_0^{\pi/2} \left(\cos x \sin x - \tfrac{1}{4} - \tfrac{1}{4} \cos 2x \right) dx$$

$$= \left(\tfrac{1}{2} \sin^2 x - \tfrac{1}{4}x - \tfrac{1}{8} \sin 2x \right)\Big|_0^{\pi/2} = \tfrac{4-\pi}{8}$$

29. $\iint_R 4x^3\, dA = \int_{-1}^2 \int_{(x-1)^2}^{-x+3} 4x^3\, dy\, dx$

$$= \int_{-1}^2 \left[4x^3 y \right]_{y=(x-1)^2}^{y=-x+3} dx$$

$$= 4 \int_{-1}^2 \left(-x^5 + x^4 + 2x^3 \right) dx$$

$$= 4 \left(-\tfrac{1}{6}x^6 + \tfrac{1}{5}x^5 + \tfrac{1}{2}x^4 \right)\Big|_{-1}^2 = \tfrac{72}{5}$$

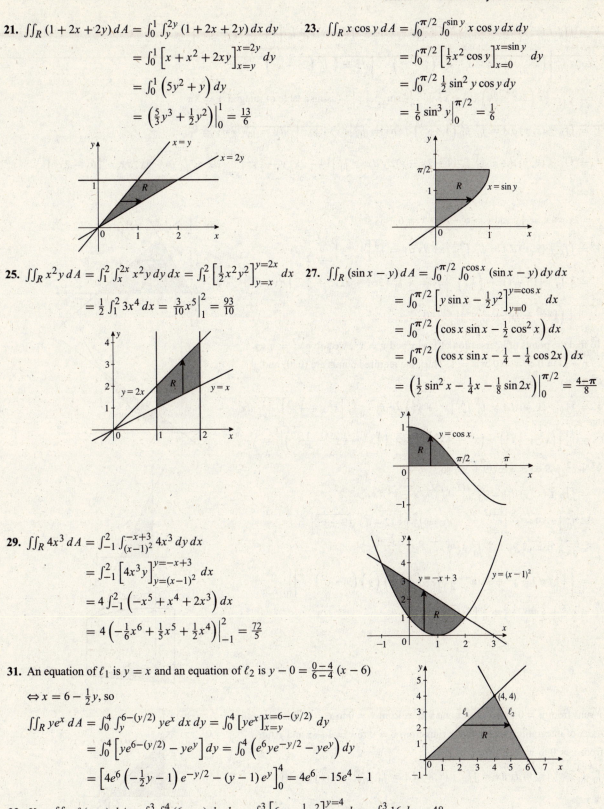

31. An equation of ℓ_1 is $y = x$ and an equation of ℓ_2 is $y - 0 = \tfrac{0-4}{6-4}(x-6)$

$\Leftrightarrow x = 6 - \tfrac{1}{2}y$, so

$$\iint_R ye^x\, dA = \int_0^4 \int_y^{6-(y/2)} ye^x\, dx\, dy = \int_0^4 \left[ye^x \right]_{x=y}^{x=6-(y/2)} dy$$

$$= \int_0^4 \left[ye^{6-(y/2)} - ye^y \right] dy = \int_0^4 \left(e^6 ye^{-y/2} - ye^y \right) dy$$

$$= \left[4e^6 \left(-\tfrac{1}{2}y - 1 \right) e^{-y/2} - (y-1) e^y \right]_0^4 = 4e^6 - 15e^4 - 1$$

33. $V = \iint_R f(x, y)\, dA = \int_0^3 \int_0^4 (6 - y)\, dy\, dx = \int_0^3 \left[6y - \tfrac{1}{2}y^2 \right]_{y=0}^{y=4} dx = \int_0^3 16\, dx = 48$

35. $V = \iint_R f(x, y)\, dA = 4 \int_0^2 \int_0^{\sqrt{4-x^2}} \left(4 - x^2 - y^2\right) dy\, dx = 4 \int_0^2 \left[4y - x^2 y - \tfrac{1}{3} y^3\right]_{y=0}^{y=\sqrt{4-x^2}} dx$

$= 4 \int_0^2 \left[\left(4 - x^2\right)\sqrt{4 - x^2} - \tfrac{1}{3}\left(4 - x^2\right)^{3/2}\right] dx = \tfrac{8}{3} \int_0^2 \left(2^2 - x^2\right)^{3/2} dx$

$= \tfrac{8}{3}\left[-\tfrac{1}{8} x \left(2x^2 - 20\right)\sqrt{4 - x^2} + \tfrac{3 \cdot 16}{8} \sin^{-1} \tfrac{1}{2} x\right]_0^2 \text{ (using a table of integrals)} = 8\pi$

37. $V = \iint_R f(x, y)\, dA = \int_0^4 \int_0^2 \left(4 - x^2\right) dx\, dy = \int_0^4 \left[4x - \tfrac{1}{3} x^3\right]_0^2 dy = \int_0^4 \tfrac{16}{3}\, dy = \tfrac{64}{3}$

39. $V = \iint_R f(x, y)\, dA = \int_0^1 \int_0^2 (4 - 2x - y)\, dy\, dx = \int_0^1 \left[(4 - 2x) y - \tfrac{1}{2} y^2\right]_{y=0}^{y=2} dx = \int_0^1 (6 - 4x)\, dx = \left(6x - 2x^2\right)\Big|_0^1$

$= 4$

41. $y = 2x \Rightarrow x = \tfrac{1}{2} y$ and $y = 6 - x \Rightarrow x = 6 - y$, so

$V = \iint_R f(x, y)\, dA = \int_0^4 \int_{y/2}^{6-y} xy\, dx\, dy = \int_0^4 \left[\tfrac{1}{2} x^2 y\right]_{x=y/2}^{x=6-y} dy$

$= \tfrac{1}{2} \int_0^4 y\left[(6 - y)^2 - \left(\tfrac{1}{2} y\right)^2\right] dy = \tfrac{1}{2} \int_0^4 \left(36y - 12y^2 + \tfrac{3}{4} y^3\right) dy$

$= \tfrac{1}{2} \left(18y^2 - 4y^3 + \tfrac{3}{16} y^4\right)\Big|_0^4 = 40$

43. To find the points of intersection of $y = x$ and $y = x^2$, we solve $x^2 = x \Leftrightarrow$
$x(x - 1) = 0 \Leftrightarrow x = 0$ or $x = 1$. Thus, the required points are $(0, 0)$ and
$(1, 1)$, so

$V = \iint_R f(x, y)\, dA = \int_0^1 \int_{x^2}^x \left(x^2 + y^2\right) dy\, dx = \int_0^1 \left[x^2 y + \tfrac{1}{3} y^3\right]_{y=x^2}^{y=x} dx$

$= \int_0^1 \left[\left(x^3 + \tfrac{1}{3} x^3\right) - \left(x^4 + \tfrac{1}{3} x^6\right)\right] dx = \left(\tfrac{1}{3} x^4 - \tfrac{1}{5} x^5 - \tfrac{1}{21} x^7\right)\Big|_0^1 = \tfrac{3}{35}$

45. $y^2 + z^2 = 9 \Leftrightarrow z = \pm\sqrt{9 - y^2}$, so

$V = \iint_R f(x, y)\, dA = \int_0^2 \int_0^{(2-y)/2} \sqrt{9 - y^2}\, dx\, dy$

$= \int_0^2 \left[x\sqrt{9 - y^2}\right]_{x=0}^{x=(2-y)/2} dy = \int_0^2 \left[\tfrac{1}{2}(2 - y)\sqrt{9 - y^2}\right] dy$

$= \int_0^2 \sqrt{9 - y^2}\, dy - \tfrac{1}{2} \int_0^2 y\sqrt{9 - y^2}\, dy$

$= \left\{\tfrac{1}{2}\left[y\sqrt{9 - y^2} + 9 \sin^{-1} \tfrac{1}{3} y\right] - \tfrac{1}{2}\left(-\tfrac{1}{2}\right)\left(\tfrac{2}{3}\right)\left(9 - y^2\right)^{3/2}\right\}\Big|_0^2$

$= \sqrt{5} + \tfrac{9}{2} \sin^{-1} \tfrac{2}{3} + \tfrac{1}{6}\left(5\sqrt{5} - 27\right)$

47. y runs from $y = 0$ to $y = 1 - x$ and x runs from $x = 0$ to $x = 1$. Reversing the
order of integration, we see that x runs from $x = 0$ to $x = 1 - y$ and y runs
from $y = 0$ to $y = 1$, so

$\int_0^1 \int_0^{1-x} f(x, y)\, dy\, dx = \iint_R f(x, y)\, dA = \int_0^1 \int_0^{1-y} f(x, y)\, dx\, dy.$

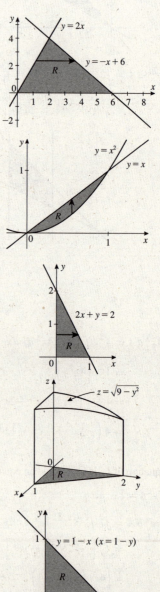

49. x runs from $x = y^2$ to $x = \sqrt[3]{y}$ and y runs from $y = 0$ to $y = 1$. Reversing the

order of integration, we see that y runs from $y = x^3$ to $y = \sqrt{x}$ and x runs from

$x = 0$ to $x = 1$, so

$$\int_0^1 \int_{y^2}^{\sqrt[3]{y}} f(x, y)\, dx\, dy = \iint_R f(x, y)\, dA = \int_0^1 \int_{x^3}^{\sqrt{x}} f(x, y)\, dy\, dx.$$

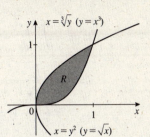

51. x runs from $x = y^2 - 4$ to $x = \frac{3}{2}y - \frac{3}{2}$. To find the points of intersection of the

graphs of $x = y^2 - 4$ and $x = \frac{3}{2}y - \frac{3}{2}$, we solve $y^2 - 4 = \frac{3}{2}y - \frac{3}{2}$ $\Rightarrow$

$2y^2 - 3y - 5 = 0 \Rightarrow (2y - 5)(y + 1) = 0 \Rightarrow y = -1$ or $\frac{5}{2}$, so the points of

intersection are $(-3, -1)$ and $\left(\frac{9}{4}, \frac{5}{2}\right)$. Reversing the order of integration, we

have

$$\int_{-1}^{5/2} \int_{y^2-4}^{(3/2)y-(3/2)} f(x, y)\, dx\, dy = \iint_R f(x, y)\, dA$$

$$= \int_{-4}^{-3} \int_{-\sqrt{x+4}}^{\sqrt{x+4}} f(x, y)\, dy\, dx + \int_{-3}^{9/4} \int_{(2/3)x+1}^{\sqrt{x+4}} f(x, y)\, dy\, dx$$

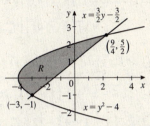

53. y runs from $y = 0$ to $y = \ln x$ and x runs from $x = 1$ to $x = e$. To reverse the

order of integration, observe that x runs from $x = e^y$ to $x = e$ while y runs

from $y = 0$ to $y = 1$. Thus,

$$\int_1^e \int_0^{\ln x} f(x, y)\, dy\, dx = \iint_R f(x, y)\, dA = \int_0^1 \int_{e^y}^e f(x, y)\, dx\, dy.$$

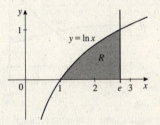

55. $\displaystyle\int_0^1 \int_{2y}^2 e^{-x^2}\, dx\, dy = \int_0^2 \int_0^{x/2} e^{-x^2}\, dy\, dx = \int_0^2 \left[y e^{-x^2} \right]_{y=0}^{y=x/2} dx$

$\qquad = \frac{1}{2} \int_0^2 x e^{-x^2}\, dx = -\frac{1}{4} e^{-x^2} \Big|_0^2 = \frac{1}{4}\left(1 - e^{-4}\right)$

$\qquad = \dfrac{e^4 - 1}{4e^4}$

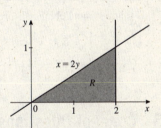

57. $\displaystyle\int_0^4 \int_{\sqrt{x}}^2 \sin y^3\, dy\, dx = \int_0^2 \int_0^{y^2} \sin y^3\, dx\, dy = \int_0^2 \left[x \sin y^3 \right]_{x=0}^{x=y^2} dy$

$\qquad = \int_0^2 y^2 \sin y^3\, dy = -\frac{1}{3} \cos y^3 \Big|_0^2$

$\qquad = \dfrac{1 - \cos 8}{3}$

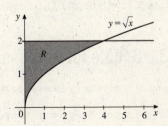

59. $\int_0^4 \int_{\sqrt{y}}^2 \frac{1}{\sqrt{x^3+1}}\, dx\, dy = \int_0^2 \int_0^{x^2} \frac{1}{\sqrt{x^3+1}}\, dy\, dx = \int_0^2 \left[\frac{y}{\sqrt{x^3+1}}\right]_{y=0}^{y=x^2}\, dx$

$$= \int_0^2 \frac{x^2}{\sqrt{x^3+1}}\, dx = \frac{2}{3}\left(x^3+1\right)^{1/2}\Big|_0^2$$

$$= \frac{2}{3}(3-1) = \frac{4}{3}$$

[Graph in upper right: curve $y = x^2$ ($x = \sqrt{y}$) with region R shaded, axes marked to 4 on y and 2 on x.]

61. $\iint_R f(x,y)\, dA = \iint_R g(x)\, h(y)\, dA = \int_a^b \int_c^d g(x)\, h(y)\, dy\, dx = \int_a^b \left[\int_c^d g(x)\, h(y)\, dy\right] dx = \int_a^b \left[g(x) \int_c^d h(y)\, dy\right] dx$

$$= \left[\int_c^d h(y)\, dy\right]\left[\int_a^b g(x)\, dx\right] = \left[\int_a^b g(x)\, dx\right]\left[\int_c^d h(y)\, dy\right]$$

63. $M = \iint_R \delta(x,y)\, dA = \iint_R (1+y)\, dA = \int_{-2}^2 \int_0^{\sqrt{4-x^2}} (1+y)\, dy\, dx = \int_{-2}^2 \left[y + \frac{1}{2}y^2\right]_{y=0}^{y=\sqrt{4-x^2}}\, dx$

$$= \int_{-2}^2 \left[\sqrt{4-x^2} + \frac{1}{2}\left(4-x^2\right)\right] dx = \left[\frac{1}{2}x\sqrt{4-x^2} + 2\sin^{-1}\left(\frac{1}{2}x\right) + 2x - \frac{1}{6}x^3\right]_{-2}^2$$

$$= \left(2\sin^{-1}1 + 4 - \frac{4}{3}\right) - \left[2\sin^{-1}(-1) - 4 + \frac{4}{3}\right] = 2\pi + \frac{16}{3} = \frac{2}{3}(3\pi + 8) \text{ slugs}$$

65. The average population density inside R is $\dfrac{43{,}329}{20} \approx 2166$ people per square mile.

67. a.

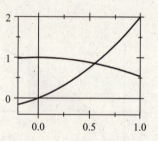

b. Using a CAS, we find the x-coordinate of the point of intersection to be approximately 0.550.

c. $\iint_R x\, dA = \int_0^{0.55} \int_{x^2+x}^{\cos x} x\, dy\, dx = \int_0^{0.55} [xy]_{y=x^2+x}^{y=\cos x}\, dx$

$$= \int_0^{0.55} \left[x\cos x - x\left(x^2 + x\right)\right] dx$$

$$= \left(\cos x + x\sin x - \frac{1}{4}x^4 - \frac{1}{3}x^3\right)\Big|_0^{0.55} \approx 0.062$$

69. Using a calculator or computer, we find $\int_0^1 \int_0^2 x^2 y^3 \cos(x+y)\, dy\, dx \approx -0.8784$.

71. Using a calculator or computer, we find $\int_0^1 \int_0^{1-x} \sqrt{1 + x^2 + y^3}\, dy\, dx \approx 0.5610$.

73. True. This follows from Fubini's Theorem.

75. True. $\int_a^b \left[\int_0^{f(x)} dy\right] dx = \int_a^b [y]_{y=0}^{f(x)}\, dx = \int_a^b f(x)\, dx$. The last integral gives the area under the graph of f on $[a, b]$.

77. False: $\int_0^2 \int_{-1}^1 x\cos\left(y^2\right) dx\, dy = \left[\int_{-1}^1 x\, dx\right]\left[\int_0^2 \cos\left(y^2\right) dy\right] = 0 \cdot \int_0^2 \cos\left(y^2\right) dy = 0$.

14.3   Concept Questions   ET 13.3

1. a. See page 1168 (1164 in ET).

  b. See page 1169 (1165 in ET).

14.3   Double Integrals in Polar Coordinates   ET 13.3

1. The integral is more easily found using rectangular coordinates: $\iint_R f(x,y)\, dA = \int_0^3 \int_0^{-(2/3)x+2} f(x,y)\, dy\, dx$.

3. The integral is more easily found using polar coordinates: $\iint_R f(x,y)\, dA = \int_{-\pi/4}^{\pi/4} \int_0^{\sqrt{2}} f(r\cos\theta, r\sin\theta)\, r\, dr\, d\theta$.

5. r runs from $r = 1$ to $r = 4$ and θ runs from $\theta = 0$ to $\theta = \pi$.

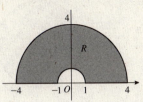

7. r runs from $r = 0$ to $r = 2\sqrt{2}$ and θ runs from $\theta = \frac{\pi}{4}$ to $\theta = \frac{\pi}{2}$.

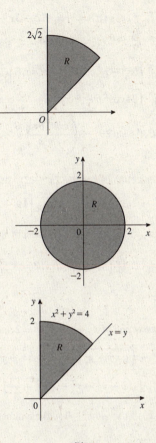

9. $\iint_R 3y\, dA = \int_0^{2\pi} \int_0^2 (3r\sin\theta)\, r\, dr\, d\theta = 8\,(3) \int_0^{2\pi} \left[\frac{1}{3} r^3 \sin\theta \right]_{r=0}^{r=2} d\theta$

$= 24 \cdot \frac{1}{3} \int_0^{2\pi} \sin\theta\, d\theta = -8\cos\theta \big|_0^{2\pi} = 0$

11. $\iint_R xy\, dA = \int_{\pi/4}^{\pi/2} \int_0^2 (r\cos\theta)(r\sin\theta)\, r\, dr\, d\theta$

$= \int_{\pi/4}^{\pi/2} \left[\frac{1}{4} r^4 \sin\theta\cos\theta\right]_{r=0}^{r=2} d\theta$

$= 4 \int_{\pi/4}^{\pi/2} \sin\theta\cos\theta\, d\theta = 2\sin^2\theta \Big|_{\pi/4}^{\pi/2} = 2\left(1 - \frac{1}{2}\right) = 1$

13. $\iint_R \frac{y^2}{x^2 + y^2}\, dA = \int_0^{2\pi} \int_1^{\sqrt{2}} \frac{r^2 \sin^2\theta}{r^2}\, r\, dr\, d\theta = \int_0^{2\pi} \left[\frac{1}{2} r^2 \sin^2\theta\right]_{r=1}^{r=\sqrt{2}} d\theta$

$= \frac{1}{2} \int_0^{2\pi} \sin^2\theta\, d\theta = \frac{1}{2} \int_0^{2\pi} \frac{1}{2}(1 - \cos 2\theta)\, d\theta$

$= \frac{1}{4}\left(\theta - \frac{1}{2}\sin 2\theta\right)\Big|_0^{2\pi} = \frac{\pi}{2}$

15. $x^2 + y^2 = 2x \Rightarrow (x-1)^2 + y^2 = 1$ is an equation of the circle with radius 1 centered at $(1, 0)$. Its equation in polar coordinates is $r = 2\cos\theta$, so

$\iint_R y\, dA = \int_{\pi/4}^{\pi/2} \int_0^{2\cos\theta} (r\sin\theta)\, r\, dr\, d\theta = \int_{\pi/4}^{\pi/2} \left[\frac{1}{3} r^3 \sin\theta\right]_{r=0}^{r=2\cos\theta} d\theta$

$= \frac{8}{3} \int_{\pi/4}^{\pi/2} \cos^3\theta\sin\theta\, d\theta = \frac{8}{3}\left(-\frac{1}{4}\right)\cos^4\theta \Big|_{\pi/4}^{\pi/2} = \frac{1}{6}$

17. By symmetry, $V = 4 \int_0^{\pi/2} \int_0^2 r^2 \cdot r\, dr\, d\theta = 4 \int_0^{\pi/2} \left[\frac{1}{4} r^4\right]_0^2 d\theta = 16 \int_0^{\pi/2} d\theta = 16\theta \big|_0^{\pi/2} = 8\pi$.

19. By symmetry, $V = 4 \int_0^{\pi/2} \int_0^2 r \cdot r\, dr\, d\theta = 4 \int_0^{\pi/2} \left[\frac{1}{3} r^3\right]_0^2 d\theta = \frac{32}{3} \int_0^{\pi/2} d\theta = \frac{16\pi}{3}$.

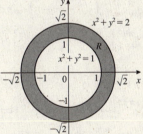

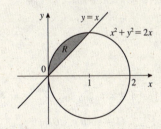

21. $x^2 + y^2 = 2x \Rightarrow (x-1)^2 + y^2 = 1$ is an equation of the circle with radius 1 centered at $(1, 0)$. Its polar equation is $r = 2\cos\theta$, so

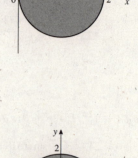

$$V = \iint_R f(x, y)\, dA = \iint_R (12 - 3x - 4y)\, dA$$
$$= \int_0^{\pi} \int_0^{2\cos\theta} (12 - 3r\cos\theta - 4r\sin\theta)\, r\, dr\, d\theta$$
$$= \int_0^{\pi} \left[6r^2 - r^3\cos\theta - \tfrac{4}{3}r^3\sin\theta \right]_{r=0}^{r=2\cos\theta} d\theta$$
$$= \int_0^{\pi} \left(24\cos^2\theta - 8\cos^4\theta - \tfrac{32}{3}\cos^3\theta\sin\theta \right) d\theta$$
$$\left[12\theta + 6\sin 2\theta - 8\left(\frac{\cos^3\theta\sin\theta}{4} + \frac{3}{8}\theta + \frac{3}{16}\sin 2\theta \right) + \tfrac{8}{3}\cos^4\theta \right]_0^{\pi} = 9\pi$$

23. The intersection of $z = 9 - 2x^2 - 2y^2$ and $z = 1$ is the curve $9 - 2x^2 - 2y^2 = 1$ $\Rightarrow x^2 + y^2 = 4$. Thus, $V = \iint_R \left[\left(9 - 2x^2 - 2y^2 \right) - 1 \right] dA$, where $R = \left\{ (x, y) \mid x^2 + y^2 \le 4 \right\}$. Thus, by symmetry,

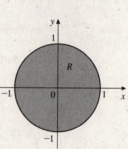

$$V = 8\int_0^{\pi/2} \int_0^2 \left(4 - r^2 \right) r\, dr\, d\theta = 8\int_0^{\pi/2} \left[2r^2 - \tfrac{1}{4}r^4 \right]_0^2 d\theta = 32\int_0^{\pi/2} d\theta = 16\pi.$$

25. The intersection of $x^2 + y^2 + z^2 = 2$ and $z = \sqrt{x^2 + y^2}$ is the curve $x^2 + y^2 + \left(\sqrt{x^2 + y^2} \right)^2 = 2 \Rightarrow x^2 + y^2 = 1$, so $V = \iint_R \left(\sqrt{2 - x^2 - y^2} - \sqrt{x^2 + y^2} \right) dA$, where $R = \left\{ (x, y) \mid x^2 + y^2 \le 1 \right\}$. By symmetry, we have

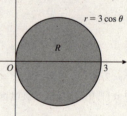

$$V = 4\int_0^{\pi/2} \int_0^1 \left(\sqrt{2 - r^2} - r \right) r\, dr\, d\theta = 4\int_0^{\pi/2} \left[-\tfrac{1}{2}\left(\tfrac{2}{3} \right)\left(2 - r^2 \right)^{3/2} - \tfrac{1}{3}r^3 \right]_0^1 d\theta$$
$$= 4\int_0^{\pi/2} \left[-\tfrac{1}{3} - \tfrac{1}{3} + \tfrac{1}{3}(2)^{3/2} \right] d\theta = \tfrac{8}{3}\left(\sqrt{2} - 1 \right) \int_0^{\pi/2} d\theta = \tfrac{4}{3}\left(\sqrt{2} - 1 \right)\pi$$

27. $A = \iint_R dA = 2\int_0^{\pi/2} \int_0^{3\cos\theta} r\, dr\, d\theta = 2\int_0^{\pi/2} \left[\tfrac{1}{2}r^2 \right]_0^{3\cos\theta} d\theta$
$$= 9\int_0^{\pi/2} \cos^2\theta\, d\theta = \tfrac{9}{2}\int_0^{\pi/2} (1 + \cos 2\theta)\, d\theta$$
$$= \tfrac{9}{2}\left(\theta + \tfrac{1}{2}\sin 2\theta \right)\Big|_0^{\pi/2} = \frac{9\pi}{4}$$

29. $A = \iint_R dA = 2\int_{-\pi/2}^{\pi/2} \int_0^{3-3\sin\theta} r\, dr\, d\theta = 2\int_{-\pi/2}^{\pi/2} \left[\tfrac{1}{2}r^2 \right]_0^{3(1-\sin\theta)} d\theta$
$$= 9\int_{-\pi/2}^{\pi/2} (1 - \sin\theta)^2\, d\theta = 9\int_{-\pi/2}^{\pi/2} \left(1 - 2\sin\theta + \sin^2\theta \right) d\theta$$
$$= 9\left[\theta + 2\cos\theta + \tfrac{1}{2}\left(\theta - \tfrac{1}{2}\sin 2\theta \right) \right]_{-\pi/2}^{\pi/2} = \frac{27\pi}{2}$$

31. First, we find the points of intersection of the two circles by solving

$a = 2a \sin\theta \Rightarrow \sin\theta = \frac{1}{2} \Rightarrow \theta = \frac{\pi}{6}$ or $\frac{5\pi}{6}$, giving the points as $\left(a, \frac{\pi}{6}\right)$ and

$\left(a, \frac{5\pi}{6}\right)$. Thus, by symmetry,

$A = \iint_R dA = 2\int_{\pi/6}^{\pi/2}\int_a^{2a\sin\theta} r\,dr\,d\theta = 2\int_{\pi/6}^{\pi/2}\left[\frac{1}{2}r^2\right]_a^{2a\sin\theta} d\theta$

$= \int_{\pi/6}^{\pi/2}\left(4a^2\sin^2\theta - a^2\right)d\theta = a^2\int_{\pi/6}^{\pi/2}\left(4\sin^2\theta - 1\right)d\theta$

$= a^2\int_{\pi/6}^{\pi/2}(1 - 2\cos2\theta)\,d\theta = a^2\left(\theta - \sin2\theta\right)\Big|_{\pi/6}^{\pi/2} = \frac{2\pi + 3\sqrt{3}}{6}a^2$

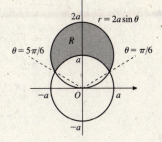

33. $\int_{-2}^{2}\int_0^{\sqrt{4-x^2}} \sqrt{x^2+y^2}\,dy\,dx = \int_0^{\pi}\int_0^2 \sqrt{r^2}r\,dr\,d\theta = \int_0^{\pi}\left[\frac{1}{3}r^3\right]_0^2 d\theta$

$= \frac{8}{3}\int_0^{\pi} d\theta = \frac{8\pi}{3}$

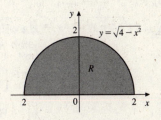

35. $\int_{-1}^{1}\int_0^{\sqrt{1-y^2}} \frac{1}{1+x^2+y^2}\,dx\,dy = \int_{-\pi/2}^{\pi/2}\int_0^1 \frac{1}{1+r^2}r\,dr\,d\theta$

$= \int_{-\pi/2}^{\pi/2}\left[\frac{1}{2}\ln\left(1+r^2\right)\right]_0^1 d\theta$

$= \frac{1}{2}\ln2\int_{-\pi/2}^{\pi/2} d\theta = \frac{\pi}{2}\ln2$

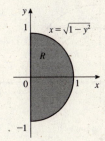

37. $\int_{-2}^{2}\int_0^{\sqrt{4-x^2}} e^{x^2+y^2}\,dy\,dx = \int_0^{\pi}\int_0^2 e^{r^2}r\,dr\,d\theta = \int_0^{\pi}\left[\frac{1}{2}e^{r^2}\right]_0^2 d\theta$

$= \int_0^{\pi}\frac{1}{2}\left(e^4 - 1\right)d\theta = \frac{\pi}{2}\left(e^4 - 1\right)$

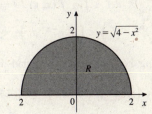

39. Observe that $y = \pm\sqrt{2x-x^2} \Rightarrow x^2 + y^2 = 2x \Rightarrow (x-1)^2 + y^2 = 1$. In

polar coordinates, an equation is $r = 2\cos\theta, 0 \le \theta \le \pi$, so

$\int_0^2\int_{-\sqrt{2x-x^2}}^{\sqrt{2x-x^2}} x\,dy\,dx = \int_0^{\pi}\int_0^{2\cos\theta} (r\cos\theta)r\,dr\,d\theta$

$= \int_0^{\pi}\left[\frac{1}{3}r^3\cos\theta\right]_{r=0}^{r=2\cos\theta} d\theta = \frac{8}{3}\int_0^{\pi}\cos^4\theta\,d\theta$

$= \frac{8}{3}\left[\frac{1}{4}\cos^3\theta\sin\theta + \frac{3}{8}\theta + \frac{3}{16}\sin2\theta\right]_0^{\pi} = \pi$

41. $\int_0^{\sqrt{2}} \int_0^x xy \, dy \, dx + \int_{\sqrt{2}}^2 \int_0^{\sqrt{4-x^2}} xy \, dy \, dx$

$$= \iint_{R_1} xy \, dA + \iint_{R_2} xy \, dA = \iint_{R_1 \cup R_2} xy \, dA$$

$$= \int_0^{\pi/4} \int_0^2 (r \cos\theta)(r \sin\theta) r \, dr \, d\theta$$

$$= \int_0^{\pi/4} \left[\tfrac{1}{4} r^4 \cos\theta \sin\theta \right]_{r=0}^{r=2} d\theta$$

$$= 4 \int_0^{\pi/4} \sin\theta \cos\theta \, d\theta = 2 \sin^2\theta \Big|_0^{\pi/4} = 1$$

43. a. Since $y = 1$ can be expressed in polar form as $r \sin\theta = 1 \Leftrightarrow$
$r = \csc\theta$, we find

$$\iint_R f(x, y) \, dA = \int_{\pi/4}^{3\pi/4} \int_0^{\csc\theta} f(r \cos\theta, r \sin\theta) r \, dr \, d\theta.$$

b. $\int_0^1 \int_{-y}^y \sqrt{x^2 + y^2} \, dx \, dy = \iint_R \sqrt{x^2 + y^2} \, dA$

$$= \int_{\pi/4}^{3\pi/4} \int_0^{\csc\theta} \sqrt{r^2} \, r \, dr \, d\theta$$

$$= \int_{\pi/4}^{3\pi/4} \left[\tfrac{1}{3} r^3 \right]_0^{\csc\theta} d\theta = \tfrac{1}{3} \int_{\pi/4}^{3\pi/4} \csc^3\theta \, d\theta$$

$$= \tfrac{1}{3} \left[-\tfrac{1}{2} \csc\theta \cot\theta + \tfrac{1}{2} \ln|\csc\theta - \cot\theta| \right]_{\pi/4}^{3\pi/4} = \tfrac{1}{3} \left[\sqrt{2} + \ln\left(\sqrt{2} + 1\right) \right]$$

45. a.

b. If $f(r, \theta) = e^{-r^2}$, then

$$\iint_{R_1} f(r, \theta) \, dA = \int_0^{\pi/2} \int_0^a e^{-r^2} r \, dr \, d\theta = \int_0^{\pi/2} \left[-\tfrac{1}{2} e^{-r^2} \right]_0^a d\theta$$

$$= \tfrac{1}{2} \left(1 - e^{-a^2} \right) \int_0^{\pi/2} d\theta = \tfrac{\pi}{4} \left(1 - e^{-a^2} \right)$$

and $\iint_{R_3} f(r, \theta) \, dA = \int_0^{\pi/2} \int_0^{\sqrt{2}a} e^{-r^2} r \, dr \, d\theta = \tfrac{\pi}{4} \left(1 - e^{-2a^2} \right)$.

c. $\iint_{R_2} f(x, y) \, dA = \int_0^a \int_0^a e^{-x^2 - y^2} \, dx \, dy = \left[\int_0^a e^{-x^2} \, dx \right] \left[\int_0^a e^{-y^2} \, dy \right]$

$$= \left(\int_0^a e^{-x^2} \, dx \right)^2$$

Using polar coordinates and referring to part a, we have

$\iint_{R_1} e^{-r^2} \, dA < \iint_{R_2} f(x, y) \, dA = \iint_{R_2} e^{-x^2 - y^2} \, dA = \iint_{R_2} e^{-r^2} \, dA < \iint_{R_3} e^{-r^2} \, dA$, so the result of part a gives

$\tfrac{\pi}{4} \left(1 - e^{-a^2} \right) < \left(\int_0^a e^{-x^2} \, dx \right)^2 < \tfrac{\pi}{4} \left(1 - e^{-2a^2} \right)$.

d. Using the result of part c, we find $\lim_{a \to \infty} \tfrac{\pi}{4} \left(1 - e^{-a^2} \right) < \lim_{a \to \infty} \left(\int_0^a e^{-x^2} \, dx \right)^2 < \lim_{a \to \infty} \tfrac{\pi}{4} \left(1 - e^{-2a^2} \right) \Rightarrow$

$\tfrac{\pi}{4} \le \left(\lim_{a \to \infty} \int_0^a e^{-x^2} \, dx \right)^2 \le \tfrac{\pi}{4} \Leftrightarrow \left(\int_0^\infty e^{-x^2} \, dx \right)^2 = \tfrac{\pi}{4} \Rightarrow \int_0^\infty e^{-x^2} \, dx = \tfrac{\sqrt{\pi}}{2}$. Next, consider

$I = \int_{-\infty}^\infty e^{-x^2/2} \, dx = 2 \int_0^\infty e^{-x^2/2} \, dx$. Let $x = \sqrt{2} u$, so $dx = \sqrt{2} \, du$, $x = 0 \Rightarrow u = 0$, and $x \to \infty \Rightarrow u \to \infty$.

Thus, $I = 2 \int_0^\infty e^{-u^2} \left(\sqrt{2} \, du \right) = 2\sqrt{2} \int_0^\infty e^{-u^2} \, du = 2\sqrt{2} \left(\tfrac{\sqrt{\pi}}{2} \right) = \sqrt{2\pi}$.

47. Let $u = \sqrt{x} \Rightarrow u^2 = x$, so $dx = 2u \, du$, $x = 0 \Rightarrow u = 0$, and $x \to \infty \Rightarrow u \to \infty$. Then

$$\int_0^\infty \frac{e^{-x}}{\sqrt{x}} \, dx = \int_0^\infty \frac{e^{-u^2}}{u} (2u \, du) = 2 \int_0^\infty e^{-u^2} \, du = 2 \left(\tfrac{\sqrt{\pi}}{2} \right) = \sqrt{\pi}.$$

49. True. If $f(r \cos\theta, r \sin\theta) = 1$, then $\int_\alpha^\beta \int_0^{g(\theta)} f(r \cos\theta, r \sin\theta) r \, dr \, d\theta = \int_\alpha^\beta \int_0^{g(\theta)} r \, dr \, d\theta = \iint_R dA = A(R)$.

14.4 Concept Questions   ET 13.4

1. a. See page 1176 (1172 in ET).

  b. See page 1178 (1174 in ET).

  c. See page 1178 (1174 in ET).

3. See page 1180 (1176 in ET).

14.4 Applications of Double Integrals   ET 13.4

1. $m = \iint_R \rho(x, y)\, dA = \int_0^3 \int_0^2 y\, dy\, dx = \int_0^3 \left[\frac{1}{2}y^2\right]_0^2 dx = \int_0^3 2\, dx = 6,$

  $M_x = \iint_R y\rho(x, y)\, dA = \int_0^3 \int_0^2 y \cdot y\, dy\, dx = \int_0^3 \left[\frac{1}{3}y^3\right]_0^2 dx = \int_0^3 \frac{8}{3}\, dx = 8,$

  and

  $M_y = \iint_R x\rho(x, y)\, dA = \int_0^3 \int_0^2 xy\, dy\, dx = \int_0^3 \left[\frac{1}{2}xy^2\right]_{y=0}^{y=2} dx$

  $= \int_0^3 2x\, dx = x^2\Big|_0^3 = 9$

  so $(\overline{x}, \overline{y}) = \left(\frac{M_y}{m}, \frac{M_x}{m}\right) = \left(\frac{9}{6}, \frac{8}{6}\right) = \left(\frac{3}{2}, \frac{4}{3}\right).$

3. The line passing through $(0, 0)$ and $(2, 1)$ has equation $y = \frac{1}{2}x \Leftrightarrow x = 2y$ and

  the line passing through $(2, 1)$ and $(4, 0)$ has equation $y - 0 = \frac{0-1}{4-2}(x - 4) \Rightarrow$

  $y = -\frac{1}{2}x + 2 \Leftrightarrow x = 4 - 2y.$

  $m = \iint_R \rho(x, y)\, dA = \int_0^1 \int_{2y}^{4-2y} x\, dx\, dy = \int_0^1 \left[\frac{1}{2}x^2\right]_{2y}^{4-2y} dy$

  $= \int_0^1 8(1 - y)\, dy = 8\left(y - \frac{1}{2}y^2\right)\Big|_0^1 = 4$

  $M_x = \iint_R y\rho(x, y)\, dA = \int_0^1 \int_{2y}^{4-2y} xy\, dx\, dy = \int_0^1 \left[\frac{1}{2}x^2 y\right]_{x=2y}^{x=4-2y} dy = 8\int_0^1 y(1 - y)\, dy = 8\left(\frac{1}{2}y^2 - \frac{1}{3}y^3\right)\Big|_0^1 = \frac{4}{3},$

  and

  $M_y = \iint_R x\rho(x, y)\, dA = \int_0^1 \int_{2y}^{4-2y} x^2\, dx\, dy = \int_0^1 \left[\frac{1}{3}x^3\right]_{2y}^{4-2y} dy = \frac{8}{3}\int_0^1 \left(8 - 12y + 6y^2 - 2y^3\right) dy$

  $= \frac{8}{3}\left(8y - 6y^2 + 2y^3 - \frac{1}{2}y^4\right)\Big|_0^1 = \frac{28}{3}$

  so $(\overline{x}, \overline{y}) = \left(\frac{M_y}{m}, \frac{M_x}{m}\right) = \left(\frac{7}{3}, \frac{1}{3}\right).$

5. $m = \iint_R \rho(x, y)\, dA = \int_0^4 \int_0^{\sqrt{x}} xy\, dy\, dx = \int_0^4 \left[\frac{1}{2}xy^2\right]_{y=0}^{y=\sqrt{x}} dx$

$\qquad = \int_0^4 \frac{1}{2}x^2\, dx = \frac{1}{6}x^3 \Big|_0^4 = \frac{32}{3}$

$\qquad M_x = \iint_R y\rho(x, y)\, dA = \int_0^4 \int_0^{\sqrt{x}} xy^2\, dy\, dx = \int_0^4 \left[\frac{1}{3}xy^3\right]_{y=0}^{y=\sqrt{x}} dx$

$\qquad = \int_0^4 \frac{1}{3}x^{5/2}\, dx = \frac{2}{21}x^{7/2}\Big|_0^4 = \frac{256}{21}$

and

$M_y = \iint_R x\rho(x, y)\, dA = \int_0^4 \int_0^{\sqrt{x}} x^2 y\, dy\, dx = \int_0^4 \left[\frac{1}{2}x^2 y^2\right]_{y=0}^{y=\sqrt{x}} dx = \int_0^4 \frac{1}{2}x^3\, dx = \frac{1}{8}x^4\Big|_0^4 = 32$, so

$(\bar{x}, \bar{y}) = \left(\frac{M_y}{m}, \frac{M_x}{m}\right) = \left(3, \frac{8}{7}\right)$.

7. $m = \iint_R \rho(x, y)\, dA = \int_0^1 \int_1^{e^x} 2xy\, dy\, dx = \int_0^1 \left[xy^2\right]_{y=1}^{y=e^x} dx$

$\qquad = \int_0^1 xe^{2x}\, dx = \left(\frac{1}{2}xe^{2x} - \frac{1}{4}e^{2x}\right)\Big|_0^1 = \frac{1}{4}\left(e^2 + 1\right)$

$\qquad M_x = \iint_R y\rho(x, y)\, dA = \int_0^1 \int_0^{e^x} 2xy^2\, dy\, dx = 2\int_0^1 \left[\frac{1}{3}xy^3\right]_{y=0}^{y=e^x} dx$

$\qquad = \frac{2}{3}\int_0^1 xe^{3x}\, dx = \frac{2}{3}\left(\frac{1}{3}xe^{3x} - \frac{1}{9}e^{3x}\right)\Big|_0^1 = \frac{2}{27}\left(2e^3 + 1\right)$

and

$M_y = \iint_R x\rho(x, y)\, dA = \int_0^1 \int_0^{e^x} 2x^2 y\, dy\, dx = 2\int_0^1 \left[\frac{1}{2}x^2 y^2\right]_{y=0}^{y=e^x} dx = \int_0^1 x^2 e^{2x}\, dx$

$\qquad = \left[e^{2x}\left(\frac{1}{2}x^2 - \frac{1}{2}x + \frac{1}{4}\right)\right]_0^1 = \frac{1}{4}\left(e^2 - 1\right)$

so $(\bar{x}, \bar{y}) = \left(\frac{M_y}{m}, \frac{M_x}{m}\right) = \left(\frac{e^2 - 1}{4} \cdot \frac{4}{e^2 + 1}, \frac{2\left(2e^3 + 1\right)}{27} \cdot \frac{4}{e^2 + 1}\right) = \left(\frac{e^2 - 1}{e^2 + 1}, \frac{8\left(2e^3 + 1\right)}{27\left(e^2 + 1\right)}\right) \approx (0.76, 1.45)$.

9. $m = \iint_R \rho(x, y)\, dA = \int_0^\pi \int_0^{\sin x} y\, dy\, dx = \int_0^\pi \left[\frac{1}{2}y^2\right]_0^{\sin x} dx$

$\qquad = \frac{1}{2}\int_0^\pi \sin^2 x\, dx = \frac{1}{4}\left(x - \frac{1}{2}\sin 2x\right)\Big|_0^\pi = \frac{\pi}{4}$

$\qquad M_x = \iint_R y\rho(x, y)\, dA = \int_0^\pi \int_0^{\sin x} y^2\, dy\, dx = \int_0^\pi \left[\frac{1}{3}y^3\right]_0^{\sin x} dx$

$\qquad = \frac{1}{3}\int_0^\pi \sin^3 x\, dx = \frac{1}{3}\left[-\frac{1}{3}\sin^2 x \cos x - \frac{2}{3}\cos x\right]_0^\pi = \frac{4}{9}$

and

$M_y = \iint_R x\rho(x, y)\, dA = \int_0^\pi \int_0^{\sin x} xy\, dy\, dx = \int_0^\pi \left[\frac{1}{2}xy^2\right]_{y=0}^{y=\sin x} dx = \frac{1}{2}\int_0^\pi x\sin^2 x\, dx$

$\qquad = \frac{1}{2}\left[\frac{1}{4}x^2 - \frac{1}{2}x\sin x\cos x + \frac{1}{4}\sin^2 x\right]_0^\pi = \frac{\pi^2}{8}$

so $(\bar{x}, \bar{y}) = \left(\frac{M_y}{m}, \frac{M_x}{m}\right) = \left(\frac{\pi}{2}, \frac{16}{9\pi}\right)$. Note that we could also have found $\bar{x}$ using symmetry.

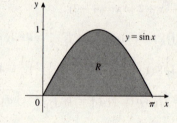

11. $m = \iint_R \rho\,(r, \theta)\, dA = \int_{-\pi/2}^{\pi/2} \int_0^{2\cos\theta} r \cdot r\, dr\, d\theta = \int_{-\pi/2}^{\pi/2} \left[\frac{1}{3}r^3\right]_0^{2\cos\theta} d\theta$

$\qquad = \int_{-\pi/2}^{\pi/2} \frac{8}{3}\cos^3\theta\, d\theta = \frac{8}{3}\left(\frac{1}{3}\sin\theta\cos^2\theta + \frac{2}{3}\sin\theta\right)\Big|_{-\pi/2}^{\pi/2}$

$\qquad = \frac{32}{9}$

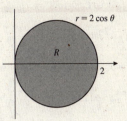

$M_x = 0$ because ρ does not depend on θ and the region is symmetric with respect to the polar axis.

$M_y = \iint_R x\rho\,(x, y)\, dA = \int_{-\pi/2}^{\pi/2} \int_0^{2\cos\theta} (r\cos\theta)r \cdot r\, dr\, d\theta$

$\qquad = \int_{-\pi/2}^{\pi/2} \left[\frac{1}{4}r^4\cos\theta\right]_{r=0}^{r=2\cos\theta} d\theta = 4\int_{-\pi/2}^{\pi/2} \cos^5\theta\, d\theta$

$\qquad = 4\left[\frac{1}{5}\sin\theta\cos^4\theta + \frac{4}{15}\sin\theta\cos^3\theta + \frac{8}{15}\sin\theta\right]_{-\pi/2}^{\pi/2} = \frac{64}{15}$

so $(\overline{x}, \overline{y}) = \left(\frac{M_y}{m}, \frac{M_x}{m}\right) = \left(\frac{6}{5}, 0\right)$.

13. $Q = \iint_R \sigma\,(x, y)\, dA = \int_0^1 \int_0^3 \left(2x^2 + y^3\right) dx\, dy = \int_0^1 \left[\frac{2}{3}x^3 + xy^3\right]_{x=0}^{x=3} dy = \int_0^1 \left(18 + 3y^3\right) dy = \left(18y + \frac{3}{4}y^4\right)\Big|_0^1$

$\qquad = \frac{75}{4}$ coulombs .

15. $T_{\text{av}} = \dfrac{\iint_R T\,(x, y)\, dA}{\iint_R dA} = \dfrac{1}{16\pi}\int_0^{2\pi} \int_0^4 400\cos\left(0.1\sqrt{r^2}\right) r\, dr\, d\theta = \dfrac{25}{\pi}\int_0^{2\pi} \int_0^4 \cos(0.1r)\, r\, dr\, d\theta$

$\qquad = \frac{25}{\pi}\int_0^{2\pi} [10(10\cos 0.1r + 4\sin 0.1r)]_0^4\, d\theta = \frac{250}{\pi}\int_0^{2\pi} (10\cos 0.4 + 4\sin 0.4 - 10)\, d\theta \approx 384.14^\circ\text{F}$

17. $m = \iint_R \rho\,(x, y)\, dA = \rho\iint_R dA = \rho ab$,

$I_x = \iint_R y^2\rho\,(x, y)\, dA = \rho\int_0^a \int_0^b y^2\, dy\, dx = \rho\int_0^a \left[\frac{1}{3}y^3\right]_0^b dx$

$\qquad = \frac{1}{3}\rho\int_0^a b^3\, dx = \frac{1}{3}\rho ab^3$

$I_y = \iint_R x^2\rho\,(x, y)\, dA = \rho\int_0^a \int_0^b x^2\, dy\, dx = \rho\int_0^a \left[x^2 y\right]_{y=0}^{y=b} dx$

$\qquad = \rho\int_0^a bx^2\, dx = \frac{1}{3}\rho a^3 b$

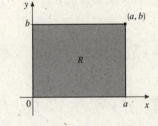

Thus, $I_0 = I_x + I_y = \frac{1}{3}\rho ab\left(a^2 + b^2\right)$, $\overline{\overline{x}} = \sqrt{\dfrac{I_y}{m}} = \sqrt{\dfrac{\rho a^3 b}{3\rho ab}} = \sqrt{\frac{1}{3}a^2} = \frac{\sqrt{3}}{3}a$, and

$\overline{\overline{y}} = \sqrt{\dfrac{I_x}{m}} = \sqrt{\dfrac{\rho ab^3}{3\rho ab}} = \sqrt{\frac{1}{3}b^2} = \frac{\sqrt{3}}{3}b$.

19. $m = \iint_D \rho\,(x, y)\, dA = \rho\iint_D dA = \frac{\pi}{2}\rho R^2$,

$I_x = \iint_D y^2\rho\, dA = \rho\int_0^\pi \int_0^R r^2\sin^2\theta\, r\, dr\, d\theta = \rho\int_0^\pi \left[\frac{1}{4}r^4\sin^2\theta\right]_{r=0}^{r=R} d\theta$

$\qquad = \frac{1}{4}\rho R^4\int_0^\pi \sin^2\theta\, d\theta = \frac{1}{4}\rho R^4\left[\frac{1}{2}\theta - \frac{1}{4}\sin 2\theta\right]_0^\pi = \frac{1}{8}\pi\rho R^4$

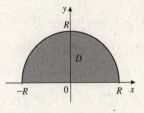

$I_y = \iint_D x^2\rho\, dA = \rho\int_0^\pi \int_0^R r^2\cos^2\theta\, r\, dr\, d\theta = \rho\int_0^\pi \left[\frac{1}{4}r^4\cos^2\theta\right]_{r=0}^{r=R} d\theta$

$\qquad = \frac{1}{4}\rho R^4\int_0^\pi \cos^2\theta\, d\theta = \frac{1}{4}\rho R^4\left[\frac{1}{2}\theta + \frac{1}{4}\sin 2\theta\right]_0^\pi = \frac{1}{8}\pi\rho R^4$

Thus, $I_0 = I_x + I_y = \frac{1}{4}\pi\rho R^4$, $\overline{\overline{x}} = \sqrt{\dfrac{I_y}{m}} = \frac{1}{2}R$, and $\overline{\overline{y}} = \sqrt{\dfrac{I_x}{m}} = \frac{1}{2}R$.

21. $I_x = \iint_R y^2 \rho(x, y) \, dA = \int_0^3 \int_0^2 y^2 \cdot y \, dy \, dx = \int_0^3 \left[\frac{1}{4} y^4\right]_0^2 dx = 4 \int_0^3 dx = 12,$

$I_y = \iint_R x^2 \rho(x, y) \, dA = \int_0^3 \int_0^2 x^2 y \, dy \, dx = \int_0^3 \left[\frac{1}{2} x^2 y^2\right]_{y=0}^{y=2} dx = \int_0^3 2x^2 = \frac{2}{3} x^3 \Big|_0^3 = 18,$ and

$I_0 = I_x + I_y = 12 + 18 = 30.$ Using the result of Exercise 1, $m = 6, \bar{\bar{x}} = \sqrt{\frac{I_y}{m}} = \sqrt{3},$ and $\bar{\bar{y}} = \sqrt{\frac{I_x}{m}} = \sqrt{2}.$

23. $I_x = \iint_R y^2 \rho(x, y) \, dA = \int_0^4 \int_0^{\sqrt{x}} y^2 xy \, dy \, dx = \int_0^4 \left[\frac{1}{4} xy^4\right]_{y=0}^{y=\sqrt{x}} dx = \frac{1}{4} \int_0^4 x^3 \, dx = 16,$

$I_y = \iint_R x^2 \rho(x, y) \, dA = \int_0^4 \int_0^{\sqrt{x}} x^2 xy \, dy \, dx = \int_0^4 \left[\frac{1}{2} x^3 y^2\right]_{y=0}^{y=\sqrt{x}} dx = \frac{1}{2} \int_0^4 x^4 \, dx = \frac{512}{5},$ and $I_0 = I_x + I_y = \frac{592}{5}.$

Using the result of Exercise 5, $m = \frac{32}{3}, \bar{\bar{x}} = \sqrt{\frac{I_y}{m}} = \frac{4\sqrt{15}}{5},$ and $\bar{\bar{y}} = \sqrt{\frac{I_x}{m}} = \frac{\sqrt{6}}{2}.$

25. $m = \iint_R \rho(x, y) \, dA.$ Observe that the line $x = 1$ can be represented in polar coordinates by $r \cos\theta = 1 \Leftrightarrow r = \sec\theta,$ so

$m = \int_0^{\pi/4} \int_{\sec\theta}^2 \frac{r \sin\theta}{r \cos\theta} \, r \, dr \, d\theta = \int_0^{\pi/4} \int_{\sec\theta}^2 \tan\theta \, r \, dr \, d\theta = \int_0^{\pi/4} \left[\frac{1}{2} r^2 \tan\theta\right]_{r=\sec\theta}^{r=2} d\theta = \frac{1}{2} \int_0^{\pi/4} \left(4 - \sec^2\theta\right) \tan\theta \, d\theta$

$= \frac{1}{2} \int_0^{\pi/4} \left(4 \tan\theta - \tan\theta \sec^2\theta\right) d\theta = \frac{1}{2} \left(-4 \ln|\cos\theta| - \frac{1}{2} \tan^2\theta\right)\Big|_0^{\pi/4} = \ln 2 - \frac{1}{4}$

27. True: $m = \iint_R \left[\rho_1(x, y) + \rho_2(x, y)\right] dA = \iint_R \rho_1(x, y) \, dA + \iint_R \rho_2(x, y) \, dA.$

29. False. For example, the annulus $R = \left\{(x, y) \mid 1 \leq x^2 + y^2 \leq 2\right\}$ with $\rho = 1$ has center of mass $(0, 0),$ which does not lie in $R.$

14.5 Concept Questions   ET 13.5

1. See page 1184 (1180 in ET).

14.5 Surface Area   ET 13.5

1. $2x + 3y + z = 12 \Rightarrow z = f(x, y) = 12 - 2x - 3y,$ so $f_x(x, y) = -2$ and $f_y(x, y) = -3.$ Thus,

$A = \iint_R \sqrt{f_x^2 + f_y^2 + 1} \, dA = \int_0^2 \int_0^1 \sqrt{(-2)^2 + (-3)^2 + 1} \, dy \, dx = \sqrt{14} \int_0^2 \int_0^1 dy \, dx = 2\sqrt{14}.$

3. $z = f(x, y) = \frac{1}{2} x^2 + y \Rightarrow f_x(x, y) = x$ and $f_y(x, y) = 1,$ so

$A = \iint_R \sqrt{f_x^2 + f_y^2 + 1} \, dA = \int_0^1 \int_0^x \sqrt{x^2 + 2} \, dy \, dx = \int_0^1 \left[\sqrt{x^2 + 2} \, y\right]_{y=0}^{y=x} dx$

$= \int_0^1 x \left(x^2 + 2\right)^{1/2} dx = \frac{1}{2} \cdot \frac{2}{3} \left(x^2 + 2\right)^{3/2}\Big|_0^1 = \frac{1}{3} \left(3\sqrt{3} - 2\sqrt{2}\right)$

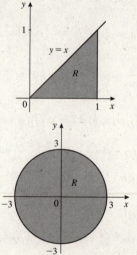

5. Setting $z = 0$ gives $0 = 9 - x^2 - y^2 \Leftrightarrow x^2 + y^2 = 9,$ so we take

$R = \left\{(x, y) \mid x^2 + y^2 \leq 9\right\}.$ Now $z = f(x, y) = 9 - x^2 - y^2 \Rightarrow$

$f_x(x, y) = -2x$ and $f_y(x, y) = -2y,$ so

$A = \iint_R \sqrt{f_x^2 + f_y^2 + 1} \, dA = \iint_R \sqrt{4x^2 + 4y^2 + 1} \, dA$

$= \int_0^{2\pi} \int_0^3 \sqrt{4r^2 + 1} \, r \, dr \, d\theta = \int_0^{2\pi} \left[\frac{1}{8} \cdot \frac{2}{3} \left(4r^2 + 1\right)^{3/2}\right]_0^3 d\theta$

$= \int_0^{2\pi} \frac{1}{12} \left(37^{3/2} - 1\right) d\theta = \frac{2\pi}{12} \left(37^{3/2} - 1\right) = \frac{\pi}{6} \left(37\sqrt{37} - 1\right)$

7. The sphere $x^2 + y^2 + z^2 = 9$ and the plane $z = 2$ intersect along the curve

$x^2 + y^2 = 5$, a circle of radius $\sqrt{5}$ centered at $(0, 0, 2)$. Thus,

$R = \left\{ (x, y) \mid x^2 + y^2 \le 5 \right\}$. Now $z = f(x, y) = \sqrt{9 - x^2 - y^2}$, so

$f_x(x, y) = -\dfrac{x}{\sqrt{9 - x^2 - y^2}}$ and $f_y(x, y) = -\dfrac{y}{\sqrt{9 - x^2 - y^2}}$, so

$A = \iint_R \sqrt{f_x^2 + f_y^2 + 1}\, dA = \iint_R \sqrt{\dfrac{x^2}{9 - x^2 - y^2} + \dfrac{y^2}{9 - x^2 - y^2} + 1}\, dA$

$= \iint_R \dfrac{3}{\sqrt{9 - x^2 - y^2}}\, dA = \int_0^{2\pi} \int_0^{\sqrt{5}} 3 \left(9 - r^2\right)^{-1/2} r\, dr\, d\theta = 3 \int_0^{2\pi} \left[-\tfrac{1}{2} \cdot 2 \left(9 - r^2\right)^{1/2} \right]_0^{\sqrt{5}} d\theta = \int_0^{2\pi} 3\, d\theta = 6\pi$

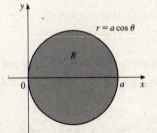

9. Here $x = h(y, z) = yz \Rightarrow h_y = z$ and $h_z = y$. $R = \left\{ (y, z) \mid y^2 + z^2 \le 16 \right\}$, so

$A = \iint_R \sqrt{h_y^2 + h_z^2 + 1}\, dA = \iint_R \sqrt{z^2 + y^2 + 1}\, dA = \int_0^{2\pi} \int_0^4 \sqrt{r^2 + 1}\, r\, dr\, d\theta = \int_0^{2\pi} \left[\tfrac{1}{2} \cdot \tfrac{2}{3} \left(r^2 + 1\right)^{3/2} \right]_0^4 d\theta$

$= \int_0^{2\pi} \tfrac{1}{3} \left(17\sqrt{17} - 1\right) d\theta = \tfrac{2\pi}{3} \left(17\sqrt{17} - 1\right)$

11. The sphere $x^2 + y^2 + z^2 = 8$ and the cone $z^2 = x^2 + y^2$ intersect along the curve $x^2 + y^2 + \left(x^2 + y^2\right) = 8 \Rightarrow$

$x^2 + y^2 = 4$, so $R = \left\{ (x, y) \mid x^2 + y^2 \le 4 \right\}$. Here $z = \pm\sqrt{8 - x^2 - y^2}$ and the function describing the upper hemisphere

is $z = f(x, y) = \sqrt{8 - x^2 - y^2} \Rightarrow f_x(x, y) = -\dfrac{x}{\sqrt{8 - x^2 - y^2}}$ and $f_y(x, y) = -\dfrac{y}{\sqrt{8 - x^2 - y^2}}$. Thus, by symmetry,

the required area is

$A = 2 \iint_R \sqrt{f_x^2 + f_y^2 + 1}\, dA = 2 \iint_R \sqrt{\dfrac{x^2}{8 - x^2 - y^2} + \dfrac{y^2}{8 - x^2 - y^2} + 1}\, dA = 2 \iint_R \dfrac{2\sqrt{2}}{\sqrt{8 - x^2 - y^2}}\, dA$

$= 4\sqrt{2} \int_0^{2\pi} \int_0^2 \dfrac{1}{\sqrt{8 - r^2}}\, r\, dr\, d\theta = 4\sqrt{2} \int_0^{2\pi} \left[-\tfrac{1}{2} \cdot 2\sqrt{8 - r^2} \right]_0^2 d\theta = 4\sqrt{2} \int_0^{2\pi} \left(2\sqrt{2} - 2\right) d\theta = 16\pi \left(2 - \sqrt{2}\right)$

13. $x^2 - ax + y^2 = 0 \Rightarrow \left(x - \tfrac{1}{2}a\right)^2 + y^2 = \tfrac{1}{4}a^2$, so

$R = \left\{ (x, y) \mid \left(x - \tfrac{1}{2}a\right)^2 + y^2 \le \tfrac{1}{4}a^2 \right\}$. Also, $z = \pm\sqrt{a^2 - x^2 - y^2}$ and the

upper hemisphere is the graph of $z = f(x, y) = \sqrt{a^2 - x^2 - y^2} \Rightarrow$

$f_x(x, y) = -\dfrac{x}{\sqrt{a^2 - x^2 - y^2}}$ and $f_y(x, y) = -\dfrac{y}{\sqrt{a^2 - x^2 - y^2}}$. Thus, by

symmetry,

$A = 2 \iint_R \sqrt{f_x^2 + f_y^2 + 1}\, dA = 2 \iint_R \sqrt{\dfrac{x^2}{a^2 - x^2 - y^2} + \dfrac{y^2}{a^2 - x^2 - y^2} + 1}\, dA = 2a \iint_R \dfrac{1}{\sqrt{a^2 - x^2 - y^2}}\, dA$

$= 2a \int_0^{\pi} \int_0^{a\cos\theta} \left(a^2 - r^2\right)^{-1/2} r\, dr\, d\theta \text{ (in polar coordinates)} = 2a \int_0^{\pi} \left[-\tfrac{1}{2} \cdot 2 \left(a^2 - r^2\right)^{1/2} \right]_0^{a\cos\theta} d\theta$

$= 2a \int_0^{\pi} \left[a - \sqrt{a^2 \left(1 - \cos^2\theta\right)} \right] d\theta = 2a \int_0^{\pi} a \left(1 - \sin\theta\right) d\theta = 2a^2 \left(\theta + \cos\theta\right)\Big|_0^{\pi} = 2a^2 \left(\pi - 2\right)$

15. From $ax + by + cz = d$, we find $z = f(x, y) = \frac{1}{c}(d - ax - by) \Rightarrow f_x(x, y) = -\frac{a}{c}$ and $f_y(x, y) = -\frac{b}{c}$. Thus,

$$A = \iint_R \sqrt{f_x^2 + f_y^2 + 1}\, dA = \iint_R \sqrt{\frac{a^2}{c^2} + \frac{b^2}{c^2} + 1}\, dA = \iint_R \frac{\sqrt{a^2 + b^2 + c^2}}{c}\, dA = \frac{1}{c}\sqrt{a^2 + b^2 + c^2} \iint_R dA$$

$$= \frac{1}{c}\sqrt{a^2 + b^2 + c^2}\, A(R), \text{ where } A(R) \text{ is the area of } R.$$

17. $z = f(x, y) = x^2 + y^2 \Rightarrow f_x(x, y) = 2x$ and $f_y(x, y) = 2y$, so

$$A = \iint_R \sqrt{f_x^2 + f_y^2 + 1}\, dA = \int_0^2 \int_0^2 \sqrt{4x^2 + 4y^2 + 1}\, dy\, dx \approx 13.0046.$$

19. $z = f(x, y) = e^{-x^2 - y^2} \Rightarrow f_x(x, y) = -2xe^{-x^2-y^2}$ and $f_y(x, y) = -2ye^{-x^2-y^2}$, so

$$A = \iint_R \sqrt{f_x^2 + f_y^2 + 1}\, dA = \iint_R \sqrt{4x^2 e^{-2(x^2+y^2)} + 4y^2 e^{-2(x^2+y^2)} + 1}\, dA = \int_0^{2\pi} \int_0^2 \frac{\sqrt{4r^2 + e^{2r^2}}}{e^{r^2}} r\, dr\, d\theta$$

$$\approx 13.9783$$

21. $f(x, y) = 3x^2 y^2 \Rightarrow f_x(x, y) = 6xy^2$ and $f_y(x, y) = 6x^2 y$, so

$$A = \iint_R \sqrt{f_x^2 + f_y^2 + 1}\, dA = \int_{-1}^1 \int_{-1}^1 \sqrt{36x^2 y^4 + 36x^4 y^2 + 1}\, dy\, dx.$$

23. $f(x, y) = \frac{1}{2x + 3y} \Rightarrow f_x(x, y) = -\frac{2}{(2x + 3y)^2}$ and

$f_y(x, y) = -\frac{3}{(2x + 3y)^2}$, so

$$A = \iint_R \sqrt{f_x^2 + f_y^2 + 1}\, dA = \iint_R \sqrt{\frac{4}{(2x + 3y)^4} + \frac{9}{(2x + 3y)^4} + 1}\, dA$$

$$= \int_0^2 \int_0^x \frac{\sqrt{13 + (2x + 3y)^4}}{(2x + 3y)^2}\, dy\, dx$$

25. True. $\iint_R \sqrt{f_x^2 + f_y^2 + 1}\, dA$ for $f(x, y) = \sqrt{4 - x^2 - y^2}$ gives the surface area of the upper hemisphere of radius 2

centered at the origin. The area is $\frac{1}{2}\left(4\pi r^2\right) = \frac{1}{2}(4\pi)\left(2^2\right) = 8\pi$.

14.6   Concept Questions   ET 13.6

1. a. See page 1189 (1185 in ET).
   b. See page 1189 (1185 in ET).

3. a. See pages 1190 and 1192 (1186 and 1188 in ET).
   b. See pages 1191 and 1192 (1187 and 1188 in ET).

14.6   Triple Integrals   ET 13.6

1. a. $\iiint_B f(x, y, z)\, dV = \int_0^3 \int_0^1 \int_0^2 (x + y + z)\, dx\, dy\, dz = \int_0^3 \int_0^1 \left[\frac{1}{2}x^2 + yx + zx\right]_{x=0}^{x=2} dy\, dz$

$$= \int_0^3 \int_0^1 (2 + 2y + 2z)\, dy\, dz = 2\int_0^3 \left[y + \frac{1}{2}y^2 + yz\right]_{y=0}^{y=1} dz = 2\int_0^3 \left(\frac{3}{2} + z\right) dz$$

$$= 2\left(\frac{3}{2}z + \frac{1}{2}z^2\right)\Big|_0^3 = 18$$

b. $\iiint_B f(x,y,z)\,dV = \int_0^2 \int_0^1 \int_0^3 (x+y+z)\,dz\,dy\,dx = \int_0^2 \int_0^1 \left[xz + yz + \frac{1}{2}z^2\right]_{z=0}^{z=3} dy\,dx$

$\qquad = \int_0^2 \int_0^1 \left(3x + 3y + \frac{9}{2}\right) dy\,dx = 3\int_0^2 \left[xy + \frac{1}{2}y^2 + \frac{3}{2}y\right]_{y=0}^{y=1} dx = 3\int_0^2 (x+2)\,dx$

$\qquad = 3\left(\frac{1}{2}x^2 + 2x\right)\Big|_0^2 = 18$

3. a. $\iiint_B f(x,y,z)\,dV = \int_0^2 \int_{-1}^1 \int_0^3 \left(xy^2 + yz^2\right) dz\,dy\,dx = \int_0^2 \int_{-1}^1 \left[xy^2 z + \frac{1}{3}yz^3\right]_{z=0}^{z=3} dy\,dx$

$\qquad = \int_0^2 \int_{-1}^1 \left(3xy^2 + 9y\right) dy\,dx = \int_0^2 \left[xy^3 + \frac{9}{2}y^2\right]_{y=-1}^{y=1} dx = \int_0^2 2x\,dx = x^2\Big|_0^2 = 4$

b. $\iiint_B f(x,y,z)\,dV = \int_0^3 \int_{-1}^1 \int_0^2 \left(xy^2 + yz^2\right) dx\,dy\,dz = \int_0^3 \int_{-1}^1 \left[\frac{1}{2}x^2 y^2 + xyz^2\right]_{x=0}^{x=2} dy\,dz$

$\qquad = \int_0^3 \int_{-1}^1 \left(2y^2 + 2yz^2\right) dy\,dz = \int_0^3 \left[\frac{2}{3}y^3 + y^2 z^2\right]_{y=-1}^{y=1} dz = \int_0^3 \frac{4}{3}\,dz = \frac{4}{3}z\Big|_0^3 = 4$

5. $\int_0^1 \int_0^x \int_0^{x+y} x\,dz\,dy\,dx = \int_0^1 \int_0^x [xz]_{z=0}^{z=x+y} dy\,dx = \int_0^1 \int_0^x \left(x^2 + xy\right) dy\,dx = \int_0^1 \left[x^2 y + \frac{1}{2}xy^2\right]_{y=0}^{y=x} dx = \int_0^1 \frac{3}{2}x^3\,dx$

$\qquad = \frac{3}{8}x^4\Big|_0^1 = \frac{3}{8}$

7. $\int_0^{\pi/2} \int_1^2 \int_0^{\sqrt{1-z}} y\cos x\,dy\,dz\,dx = \int_0^{\pi/2} \int_1^2 \left[\frac{1}{2}y^2 \cos x\right]_{y=0}^{y=\sqrt{1-z}} dz\,dx = \int_0^{\pi/2} \int_1^2 \frac{1}{2}(1-z)\cos x\,dz\,dx$

$\qquad = \int_0^{\pi/2} \left[\frac{1}{2}\left(z - \frac{1}{2}z^2\right)\cos x\right]_{z=1}^{z=2} dx = \int_0^{\pi/2} \left(-\frac{1}{4}\cos x\right) dx = -\frac{1}{4}\sin x\Big|_0^{\pi/2} = -\frac{1}{4}$

9. $\int_0^4 \int_0^1 \int_0^x 2\sqrt{y}e^{-x^2}\,dz\,dx\,dy = \int_0^4 \int_0^1 \left[2z\sqrt{y}e^{-x^2}\right]_{z=0}^{z=x} dx\,dy = \int_0^4 \int_0^1 2\sqrt{y}xe^{-x^2}\,dx\,dy = \int_0^4 \left[-\sqrt{y}e^{-x^2}\right]_{x=0}^{x=1} dy$

$\qquad = \int_0^4 \left(-e^{-1} + 1\right) y^{1/2}\,dy = \left(1 - \frac{1}{e}\right)\left(\frac{2}{3}y^{3/2}\right)\Big|_0^4 = \frac{16(e-1)}{3e}$

11. $6x + 3y + 4z = 12 \Rightarrow z = \frac{1}{4}(12 - 6x - 3y)$, so

$\iiint_T f(x,y,z)\,dV = \int_0^2 \int_0^{4-2x} \int_0^{(12-6x-3y)/4} f(x,y,z)\,dz\,dy\,dx$

$\qquad = \int_0^4 \int_0^{(4-y)/2} \int_0^{(12-6x-3y)/4} f(x,y,z)\,dz\,dx\,dy$

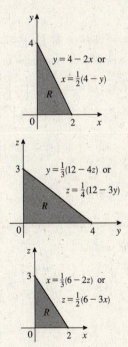

$y = 4 - 2x$ or

$x = \frac{1}{2}(4 - y)$

$6x + 3y + 4z = 12 \Rightarrow x = \frac{1}{6}(12 - 3y - 4z)$, so

$\iiint_T f(x,y,z)\,dV = \int_0^4 \int_0^{(12-3y)/4} \int_0^{(12-3y-4z)/6} f(x,y,z)\,dx\,dz\,dy$

$\qquad = \int_0^3 \int_0^{(12-4z)/3} \int_0^{(12-3y-4z)/6} f(x,y,z)\,dx\,dy\,dz$

$y = \frac{1}{3}(12 - 4z)$ or

$z = \frac{1}{4}(12 - 3y)$

$6x + 3y + 4z = 12 \Rightarrow y = \frac{1}{3}(12 - 6x - 4z)$, so

$\iiint_T f(x,y,z)\,dV = \int_0^2 \int_0^{(6-3x)/2} \int_0^{(12-6x-4z)/3} f(x,y,z)\,dy\,dz\,dx$

$\qquad = \int_0^3 \int_0^{(6-2z)/3} \int_0^{(12-6x-4z)/3} f(x,y,z)\,dy\,dx\,dz$

$x = \frac{1}{3}(6 - 2z)$ or

$z = \frac{1}{2}(6 - 3x)$

13. $x + z = 1 \Rightarrow z = 1 - x$, so

$$\iiint_T f(x,y,z)\, dV = \int_0^1 \int_{-\sqrt{x}}^{\sqrt{x}} \int_0^{1-x} f(x,y,z)\, dz\, dy\, dx$$

$$= \int_{-1}^1 \int_{y^2}^1 \int_0^{1-x} f(x,y,z)\, dz\, dx\, dy$$

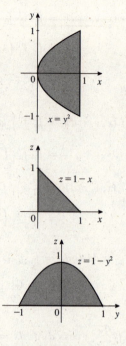

$$\iiint_T f(x,y,z)\, dV = \int_0^1 \int_0^{1-x} \int_{-\sqrt{x}}^{\sqrt{x}} f(x,y,z)\, dy\, dz\, dx$$

$$= \int_0^1 \int_0^{1-z} \int_{-\sqrt{x}}^{\sqrt{x}} f(x,y,z)\, dy\, dx\, dz$$

The intersection of $x + z = 1$ and $x = y^2$ is $1 - z = y^2 \Leftrightarrow z = 1 - y^2$, so

$$\iiint_T f(x,y,z)\, dV = \int_{-1}^1 \int_0^{1-y^2} \int_{y^2}^{1-z} f(x,y,z)\, dx\, dz\, dy$$

$$= \int_0^1 \int_{-\sqrt{1-z}}^{\sqrt{1-z}} \int_{y^2}^{1-z} f(x,y,z)\, dx\, dy\, dz$$

15. $x + y + z = 1 \Rightarrow z = 1 - x - y$, so

$$\iiint_T f(x,y,z)\, dV = \int_0^1 \int_0^{1-x} \int_0^{1-x-y} x\, dz\, dy\, dx$$

$$= \int_0^1 \int_0^{1-x} [xz]_{z=0}^{z=1-x-y}\, dy\, dx = \int_0^1 \int_0^{1-x} x(1-x-y)\, dy\, dx$$

$$= \int_0^1 \left[xy - x^2 y - \tfrac{1}{2}xy^2 \right]_{y=0}^{y=1-x}\, dx = \int_0^1 \left[\left(x - x^2\right)(1-x) - \tfrac{1}{2}x(1-x)^2 \right] dx$$

$$= \int_0^1 \left(\tfrac{1}{2}x^3 - x^2 + \tfrac{1}{2}x \right) dx = \left(\tfrac{1}{8}x^4 - \tfrac{1}{3}x^3 + \tfrac{1}{4}x^2 \right)\Big|_0^1 = \tfrac{1}{24}$$

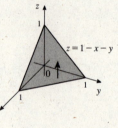

17. $\iiint_T f(x,y,z)\, dV = \int_0^1 \int_{x^3}^x \int_0^{2x} 2z\, dz\, dy\, dx$, where we have viewed T as z-simple. The projection onto the xy-plane is shown. Thus,

$$\iiint_T f(x,y,z)\, dV = \int_0^1 \int_{x^3}^x \left[z^2 \right]_0^{2x}\, dy\, dx = \int_0^1 \int_{x^3}^x 4x^2\, dy\, dx$$

$$= \int_0^1 \left[4x^2 y \right]_{y=x^3}^{y=x}\, dx = \int_0^1 \left(4x^3 - 4x^5 \right) dx$$

$$= \left(x^4 - \tfrac{2}{3}x^6 \right)\Big|_0^1 = \tfrac{1}{3}$$

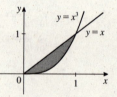

19. $\iiint_T f(x,y,z)\, dV = \iint_R \left(\int_{x^2+z^2}^4 y\, dy \right) dA$, where

$R = \left\{ (x,z) \mid x^2 + z^2 \le 4 \right\}$. Using polar coordinates $x = r\cos\theta$ and $z = r\sin\theta$, we find

$$\iiint_T f(x,y,z)\, dV = \int_0^{2\pi} \int_0^2 \left[\tfrac{1}{2}y^2 \right]_{r^2}^4 r\, dr\, d\theta$$

$$= \int_0^{2\pi} \int_0^2 \left(8r - \tfrac{1}{2}r^5 \right) dr\, d\theta$$

$$= \int_0^{2\pi} \left[4r^2 - \tfrac{1}{12}r^6 \right]_0^2\, d\theta = \int_0^{2\pi} \tfrac{32}{3}\, d\theta = \tfrac{64\pi}{3}$$

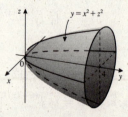

21. In the first octant, $x^2 + z^2 = 4 \Rightarrow z = \sqrt{4-x^2}$, so

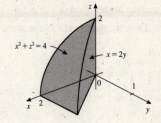

$$\iiint_T f(x,y,z)\,dV = \int_0^1 \int_{2y}^2 \int_0^{\sqrt{4-x^2}} z\,dz\,dx\,dy = \int_0^1 \int_{2y}^2 \left[\tfrac{1}{2}z^2\right]_0^{\sqrt{4-x^2}} dx\,dy$$

$$= \tfrac{1}{2}\int_0^1 \int_{2y}^2 \left(4 - x^2\right) dx\,dy = \tfrac{1}{2}\int_0^1 \left[4x - \tfrac{1}{3}x^3\right]_{2y}^2 dy$$

$$= \tfrac{1}{2}\int_0^1 \left(\tfrac{16}{3} - 8y + \tfrac{8}{3}y^3\right) dy$$

$$= 4\left(\tfrac{2}{3}y - \tfrac{1}{2}y^2 + \tfrac{1}{12}y^4\right)\Big|_0^1 = 1$$

23. $V = \int_0^2 \int_0^{(6-3x)/2} \int_0^{6-3x-2y} dz\,dy\,dx = \int_0^2 \int_0^{(6-3x)/2} [z]_0^{6-3x-2y}\,dy\,dx$

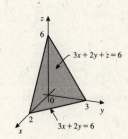

$$= \int_0^2 \int_0^{(6-3x)/2} (6 - 3x - 2y)\,dy\,dx = \int_0^2 \left[6y - 3xy - y^2\right]_{y=0}^{y=(6-3x)/2} dx$$

$$= \int_0^2 \left(\tfrac{9}{4}x^2 - 9x + 9\right) dx = \left(\tfrac{3}{4}x^3 - \tfrac{9}{2}x^2 + 9x\right)\Big|_0^2 = 6$$

25. $V = \int_{-2}^2 \int_0^{4-y^2} \int_0^{4-x} dz\,dx\,dy = \int_{-2}^2 \int_0^{4-y^2} [z]_0^{4-x}\,dx\,dy$

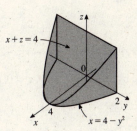

$$= \int_{-2}^2 \int_0^{4-y^2} (4 - x)\,dx\,dy = \int_{-2}^2 \left[4x - \tfrac{1}{2}x^2\right]_0^{4-y^2} dy$$

$$= \int_{-2}^2 \left(8 - \tfrac{1}{2}y^4\right) dy = \left(8y - \tfrac{1}{10}y^5\right)\Big|_{-2}^2 = \tfrac{128}{5}$$

27. The intersection of the paraboloids $z = x^2 + y^2$ and $z = 8 - x^2 - y^2$ is the

curve $x^2 + y^2 = 8 - x^2 - y^2 \Leftrightarrow x^2 + y^2 = 4$, a circle. Thus,

$$V = \iiint_T dV = \iint_R \left(\int_{x^2+y^2}^{8-x^2-y^2} dz\right) dA, \text{ where } R = \left\{(x,y) \mid x^2 + y^2 \le 4\right\}.$$

Using polar coordinates and symmetry, we find

$$V = 4\int_0^{\pi/2} \int_0^2 \left(\int_{r^2}^{8-r^2} dz\right) r\,dr\,d\theta = 4\int_0^{\pi/2} \int_0^2 [z]_{r^2}^{8-r^2}\,r\,dr\,d\theta$$

$$= 4\int_0^{\pi/2} \int_0^2 r\left(8 - 2r^2\right) dr\,d\theta = 8\int_0^{\pi/2} \left[2r^2 - \tfrac{1}{4}r^4\right]_0^2 d\theta$$

$$= 8\int_0^{\pi/2} 4\,d\theta = 16\pi$$

29. First, we find an equation of the plane containing $A(1, 0, 0)$, $B(0, 3, 0)$, and

$C(0, 0, 2)$. A normal vector to the plane is

$\mathbf{n} = \overrightarrow{AB} \times \overrightarrow{AC}$

$= \begin{vmatrix} \mathbf{i} & \mathbf{j} & \mathbf{k} \\ -1 & 3 & 0 \\ -1 & 0 & 2 \end{vmatrix} = 6\mathbf{i} + 2\mathbf{j} + 3\mathbf{k}$

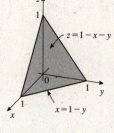

Thus, an equation of the plane is $6(x - 0) + 2(y - 0) + 3(z - 2) = 0 \Leftrightarrow 6x + 2y + 3z = 6 \Rightarrow$

$z = f(x, y) = \frac{1}{3}(6 - 6x - 2y)$, and so

$V = \int_0^1 \int_0^{3-3x} \int_0^{(6-6x-2y)/3} dz\, dy\, dx = \int_0^1 \int_0^{3-3x} [z]_0^{(6-6x-2y)/3} dy\, dx = \frac{1}{3} \int_0^1 \left[6y - 6xy - y^2 \right]_{y=0}^{y=3-3x} dx$

$= \frac{1}{3} \int_0^1 9(1-x)^2 dx = 3(-1)\left(\frac{1}{3}\right)(1-x)^3 \Big|_0^1 = 1$

31. $T = \{(x, y, z) \mid 0 \le x \le 1 - y, 0 \le y \le 1, 0 \le z \le 1 - x - y\}$. We see that

the solid is bounded by the plane $z = 1 - x - y$ and the coordinate planes.

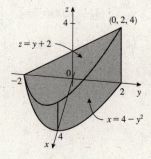

33. $T = \left\{(x, y, z) \mid 0 \le x \le 4 - y^2, -2 \le y \le 2, 0 \le z \le y + 2\right\}$. The solid is

bounded by the planes $x = 0$, $z = 0$, and $z = y + 2$ and the cylinder $x = 4 - y^2$.

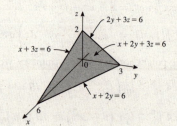

35. $\iiint_T f(x, y, z)\, dV = \int_0^6 \int_0^{(6-x)/2} \int_0^{(6-x-2y)/3} f(x, y, z)\, dz\, dy\, dx$

$= \int_0^6 \int_0^{(6-x)/3} \int_0^{(6-x-3z)/2} f(x, y, z)\, dy\, dz\, dx$

$= \int_0^3 \int_0^{6-2y} \int_0^{(6-x-2y)/3} f(x, y, z)\, dz\, dx\, dy$

$= \int_0^3 \int_0^{(6-2y)/3} \int_0^{6-2y-3z} f(x, y, z)\, dx\, dz\, dy$

$= \int_0^2 \int_0^{6-3z} \int_0^{(6-x-3z)/2} f(x, y, z)\, dy\, dx\, dz$

$= \int_0^2 \int_0^{(6-3z)/2} \int_0^{6-2y-3z} f(x, y, z)\, dx\, dy\, dz$

37. $\iiint_T f(x,y,z)\,dV = \int_{-1}^{1} \int_{-\sqrt{1-x^2}}^{\sqrt{1-x^2}} \int_0^2 f(x,y,z)\,dz\,dy\,dx$

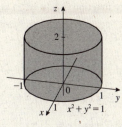

$= \int_{-1}^{1} \int_{-\sqrt{1-y^2}}^{\sqrt{1-y^2}} \int_0^2 f(x,y,z)\,dz\,dx\,dy$

$= \int_{-1}^{1} \int_0^2 \int_{-\sqrt{1-x^2}}^{\sqrt{1-x^2}} f(x,y,z)\,dy\,dz\,dx$

$= \int_0^2 \int_{-1}^{1} \int_{-\sqrt{1-x^2}}^{\sqrt{1-x^2}} f(x,y,z)\,dy\,dx\,dz$

$= \int_{-1}^{1} \int_0^2 \int_{-\sqrt{1-y^2}}^{\sqrt{1-y^2}} f(x,y,z)\,dx\,dz\,dy$

$= \int_0^2 \int_{-1}^{1} \int_{-\sqrt{1-y^2}}^{\sqrt{1-y^2}} f(x,y,z)\,dx\,dy\,dz$

39. a. Here $\Delta x = \frac{4-0}{2} = 2$, $\Delta y = \frac{4-0}{2} = 2$, and $\Delta z = \frac{4-0}{2} = 2$, so $\Delta V = (\Delta x)(\Delta y)(\Delta z) = 8$. Also, $x_0 = 0$, $x_1 = 2$,

$x_3 = 4$, $y_0 = 0$, $y_1 = 2$, $y_3 = 4$, $z_0 = 0$, $z_1 = 2$, and $z_3 = 4$. Thus,

$\iiint_B (x+y+z)\,dV \approx \sum_{i=1}^{2} \sum_{j=1}^{2} \sum_{k=1}^{2} f\left(x_{ijk}^*, y_{ijk}^*, z_{ijk}^*\right) \Delta V$

$= \left[f(1,1,1) + f(1,1,3) + f(1,3,1) + f(1,3,3) + f(3,1,1) + f(3,1,3) + f(3,3,1) + f(3,3,3) \right](8)$

$= (3+5+5+7+5+7+7+9)(8) = 384$

b. $\iiint_B f(x,y,z)\,dV = \int_0^4 \int_0^4 \int_0^4 (x+y+z)\,dx\,dy\,dz = \int_0^4 \int_0^4 \left[\frac{1}{2}x^2 + xy + xz\right]_{x=0}^{x=4} dy\,dz = \int_0^4 \int_0^4 (8 + 4y + 4z)\,dy\,dz$

$= \int_0^4 \left[8y + 2y^2 + 4yz\right]_{y=0}^{y=4} dz = \int_0^4 (64 + 16z)\,dz = \left(64z + 8z^2\right)\Big|_0^4 = 384$

41. Using a CAS, we find $\displaystyle \int_{-1}^{1} \int_0^2 \int_1^2 \frac{\cos xy}{\sqrt{1+xyz^2}}\,dx\,dy\,dz \approx 0.4439$.

43. $\rho(x,y,z) = kx$, where k is a positive constant, so

$m = \iiint_T \rho(x,y,z)\,dV = \int_0^1 \int_0^{1-x} \int_0^{1-x-y} kx\,dz\,dy\,dx$

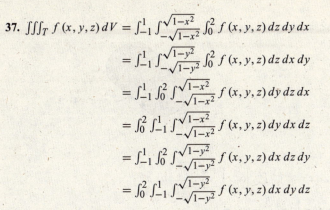

$= k \int_0^1 \int_0^{1-x} [xz]_{z=0}^{z=1-x-y}\,dy\,dx = k \int_0^1 \int_0^{1-x} \left(x - x^2 - xy\right)\,dy\,dx$

$= k \int_0^1 \left[xy - x^2y - \frac{1}{2}xy^2\right]_{y=0}^{y=1-x}\,dx = k \int_0^1 \left(\frac{1}{2}x - x^2 + \frac{1}{2}x^3\right)\,dx$

$= k \left(\frac{1}{4}x^2 - \frac{1}{3}x^3 + \frac{1}{8}x^4\right)\Big|_0^1 = \frac{1}{24}k$

$M_{yz} = \iiint_T x\rho(x,y,z)\,dV = \int_0^1 \int_0^{1-x} \int_0^{1-x-y} kx^2\,dz\,dy\,dx = k \int_0^1 \int_0^{1-x} \left(x^2 - x^3 - x^2y\right)\,dy\,dx$

$= k \int_0^1 \left[x^2y - x^3y - \frac{1}{2}x^2y^2\right]_{y=0}^{y=1-x}\,dx = k \int_0^1 \left(\frac{1}{2}x^2 - x^3 + \frac{1}{2}x^4\right)\,dx = k \left(\frac{1}{6}x^3 - \frac{1}{4}x^4 + \frac{1}{10}x^5\right)\Big|_0^1 = \frac{1}{60}k$

$M_{xz} = \iiint_T y\rho(x,y,z)\,dV = \int_0^1 \int_0^{1-x} \int_0^{1-x-y} kxy\,dz\,dy\,dx = k \int_0^1 \int_0^{1-x} [xyz]_{z=0}^{z=1-x-y}\,dy\,dx$

$= k \int_0^1 \int_0^{1-x} \left(xy - x^2y - xy^2\right)\,dy\,dx = k \int_0^1 \left[\frac{1}{2}\left(x - x^2\right)y^2 - \frac{1}{3}xy^3\right]_{y=0}^{y=1-x}\,dx = \frac{1}{6}k \int_0^1 \left(x - 3x^2 + 3x^3 - x^4\right)\,dx$

$= \frac{1}{6}k \left(\frac{1}{2}x^2 - x^3 + \frac{3}{4}x^4 - \frac{1}{5}x^5\right)\Big|_0^1 = \frac{1}{120}k$

$M_{xy} = \iiint_T z\rho(x,y,z)\,dV = \int_0^1 \int_0^{1-x} \int_0^{1-x-y} kxz\,dz\,dy\,dx = k \int_0^1 \int_0^{1-x} \left[\frac{1}{2}xz^2\right]_{z=0}^{z=1-x-y}\,dy\,dx$

$= \frac{1}{2}k \int_0^1 \int_0^{1-x} x\left(1 - 2x - 2y + 2xy + x^2 + y^2\right)\,dy\,dx = \frac{1}{6}k \int_0^1 \left(x - 3x^2 + 3x^3 - x^4\right)\,dx$

$= \frac{1}{6}k \left(\frac{1}{2}x^2 - x^3 + \frac{3}{4}x^4 - \frac{1}{5}x^5\right)\Big|_0^1 = \frac{1}{120}k$

Thus, $(\bar{x}, \bar{y}, \bar{z}) = \left(\dfrac{M_{yz}}{m}, \dfrac{M_{xz}}{m}, \dfrac{M_{xy}}{m}\right) = \left(\dfrac{2}{5}, \dfrac{1}{5}, \dfrac{1}{5}\right)$.

45. The density function is $\rho(x, y, z) = kz$, where k is a positive constant, so

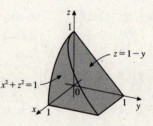

$m = \iiint_T \rho(x, y, z)\, dV = \int_0^3 \int_{-2}^2 \int_{-\sqrt{4-y^2}}^{\sqrt{4-y^2}} kx\, dz\, dy\, dx$

$\quad = k \int_0^3 \int_{-2}^2 [xz]_{z=-\sqrt{4-y^2}}^{z=\sqrt{4-y^2}}\, dy\, dx$

$\quad = k \int_0^3 \int_{-2}^2 2x\sqrt{4-y^2}\, dy\, dx$

$\quad = k \int_0^3 2\,(2\pi)\, x\, dx = 2k\pi x^2 \Big|_0^3 = 18k\pi$

$M_{yz} = \iiint_T x\rho(x, y, z)\, dV = \int_0^3 \int_{-2}^2 \int_{-\sqrt{4-y^2}}^{\sqrt{4-y^2}} kx^2\, dz\, dy\, dx = k \int_0^3 \int_{-2}^2 [x^2 z]_{z=-\sqrt{4-y^2}}^{z=\sqrt{4-y^2}}\, dy\, dx$

$\quad = k \int_0^3 \int_{-2}^2 2x^2 \sqrt{4-y^2}\, dy\, dx = 2k \int_0^3 (2\pi) x^2\, dx = 4k\pi \left[\tfrac{1}{3}x^3\right]_0^3 = 36k\pi$

$M_{xz} = \iiint_T y\rho(x, y, z)\, dV = \int_0^3 \int_{-2}^2 \int_{-\sqrt{4-y^2}}^{\sqrt{4-y^2}} kxy\, dz\, dy\, dx = k \int_0^3 \int_{-2}^2 [xyz]_{z=-\sqrt{4-y^2}}^{z=\sqrt{4-y^2}}\, dy\, dx$

$\quad = k \int_0^3 \int_{-2}^2 2xy\sqrt{4-y^2}\, dy\, dx = k \int_0^3 \left[2x\left(-\tfrac{1}{2}\right)\left(\tfrac{2}{3}\right)\left(4-y^2\right)^{3/2}\right]_{y=-2}^{y=2}\, dx = 0$

$M_{xy} = \iiint_T z\rho(x, y, z)\, dV = \int_0^3 \int_{-2}^2 \int_{-\sqrt{4-y^2}}^{\sqrt{4-y^2}} kxz\, dz\, dy\, dx = k \int_0^3 \int_{-2}^2 \left[\tfrac{1}{2}xz^2\right]_{z=-\sqrt{4-y^2}}^{z=\sqrt{4-y^2}}\, dy\, dx = 0$. Thus,

$(\overline{x}, \overline{y}, \overline{z}) = \left(\dfrac{M_{yz}}{m}, \dfrac{M_{xz}}{m}, \dfrac{M_{xy}}{m}\right) = (2, 0, 0)$.

47. $m = \iiint_T \rho(x, y, z)\, dV$

$\quad = \int_0^1 \int_0^{1-y} \int_0^{\sqrt{1-z^2}} \left(xy + z^2\right) dx\, dz\, dy$

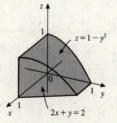

49. $m = \iiint_T \rho(x, y, z)\, dV$

$\quad = \int_0^1 \int_0^{(2-y)/2} \int_0^{1-y^2} \sqrt{x^2 + y^2 + z^2}\, dz\, dx\, dy$

viewing T as z-simple.

51. $I_x = \iiint_T \left(y^2 + z^2\right) \rho\left(x, y, z\right) dV = \int_0^1 \int_0^1 \int_0^1 k \left(y^2 + z^2\right) dz\, dy\, dx$

$= k \int_0^1 \int_0^1 \left[y^2 z + \tfrac{1}{3} z^3\right]_{z=0}^{z=1} dy\, dx = k \int_0^1 \int_0^1 \left(y^2 + \tfrac{1}{3}\right) dy\, dx$

$= k \int_0^1 \left[\tfrac{1}{3} y^3 + \tfrac{1}{3} y\right]_0^1 dx = k \int_0^1 \tfrac{2}{3}\, dx = \tfrac{2}{3} k$

$I_y = \iiint_T \left(x^2 + z^2\right) \rho\left(x, y, z\right) dV = \int_0^1 \int_0^1 \int_0^1 k \left(x^2 + z^2\right) dz\, dy\, dx$

$= k \int_0^1 \int_0^1 \left[x^2 z + \tfrac{1}{3} z^3\right]_{z=0}^{z=1} dy\, dx = k \int_0^1 \int_0^1 \left(x^2 + \tfrac{1}{3}\right) dy\, dx$

$= k \int_0^1 \left[x^2 y + \tfrac{1}{3} y\right]_{y=0}^{y=1} dx = k \int_0^1 \left(x^2 + \tfrac{1}{3}\right) dx = k \left(\tfrac{1}{3} x^3 + \tfrac{1}{3} x\right)\big|_0^1 = \tfrac{2}{3} k$

$I_z = \iiint_T \left(x^2 + y^2\right) \rho\left(x, y, z\right) dV = \int_0^1 \int_0^1 \int_0^1 k \left(x^2 + y^2\right) dz\, dy\, dx = k \int_0^1 \int_0^1 \left[\left(x^2 + y^2\right) z\right]_{z=0}^{z=1} dy\, dx$

$= k \int_0^1 \int_0^1 \left(x^2 + y^2\right) dy\, dx = k \int_0^1 \left[x^2 y + \tfrac{1}{3} y^3\right]_{y=0}^{y=1} dx = k \int_0^1 \left(x^2 + \tfrac{1}{3}\right) dx = k \left(\tfrac{1}{3} x^3 + \tfrac{1}{3} x\right)\big|_0^1 = \tfrac{2}{3} k$

53. $I_x = \iiint_T \left(y^2 + z^2\right) \rho\left(x, y, z\right) dV = \int_0^1 \int_0^{1-x} \int_0^{1-x-y} x \left(y^2 + z^2\right) dz\, dy\, dx$

$= \int_0^1 \int_0^{1-x} \left[xy^2 z + \tfrac{1}{3} x z^3\right]_{z=0}^{z=1-x-y} dy\, dx$

$= -\tfrac{1}{3} \int_0^1 \int_0^{1-x} \left(x^4 + 3x^3 y + 6x^2 y^2 + 4xy^3 - 3x^3\right.$

$\left. \qquad\qquad - 6xy^2 - 6x^2 y + 3x^2 + 3xy - x\right) dy\, dx$

$= \tfrac{1}{6} \int_0^1 \left(x^5 - 4x^4 + 6x^3 - 4x^2 + x\right) dx = \tfrac{1}{180}$

$I_y = \iiint_T \left(x^2 + z^2\right) \rho\left(x, y, z\right) dV = \int_0^1 \int_0^{1-x} \int_0^{1-x-y} x \left(x^2 + z^2\right) dz\, dy\, dx$

$= -\tfrac{1}{3} \int_0^1 \int_0^{1-x} \left(4x^4 + 6x^3 y + 3x^2 y^2 + xy^3 - 6x^3 - 6x^2 y - 3xy^2 + 3x^2 + 3xy - x\right) dy\, dx$

$= \tfrac{1}{12} \int_0^1 \left(7x^5 - 16x^4 + 12x^3 - 4x^2 + x\right) dx = \tfrac{1}{90}$

$I_z = \iiint_T \left(x^2 + y^2\right) \rho\left(x, y, z\right) dV = \int_0^1 \int_0^{1-x} \int_0^{1-x-y} x \left(x^2 + y^2\right) dz\, dy\, dx$

$= \int_0^1 \int_0^{1-x} x \left(x^2 + y^2\right) (1 - x - y)\, dy\, dx = \int_0^1 \int_0^{1-x} \left(x^3 - x^4 - x^3 y + xy^2 - x^2 y^2 - xy^3\right) dy\, dx$

$= \tfrac{1}{12} \int_0^1 \left(7x^5 - 16x^4 + 12x^3 - 4x^2 + x\right) dx = \tfrac{1}{90}$

55. $\iiint_T f\left(x, y, z\right) dV = \int_0^1 \int_0^2 \int_0^3 (x + y + z)\, dz\, dy\, dx = \int_0^1 \int_0^2 \left[xz + yz + \tfrac{1}{2} z^2\right]_{z=0}^{z=3} dy\, dx = \int_0^1 \int_0^2 \left(3x + 3y + \tfrac{9}{2}\right) dy\, dx$

$= \int_0^1 \left[3xy + \tfrac{3}{2} y^2 + \tfrac{9}{2} y\right]_{y=0}^{y=2} dx = \int_0^1 (6x + 15)\, dx = 18$

The volume of the box is $V(T) = 1 \cdot 2 \cdot 3 = 6$, so $f_{av} = \tfrac{1}{V(T)} \iiint_T f\left(x, y, z\right) dV = \tfrac{18}{6} = 3$.

57. The region under consideration is

$$T = \left\{ (x, y, z) \mid x^2 + y^2 + z^2 \le 4, x \ge 0, y \ge 0, z \ge 0 \right\}.$$

$$\iiint_T f(x, y, z)\,dV = \int_0^2 \int_0^{\sqrt{4-x^2}} \int_0^{\sqrt{4-x^2-y^2}} xyz\,dz\,dy\,dx$$

$$= \int_0^2 \int_0^{\sqrt{4-x^2}} \left[\tfrac{1}{2}xyz^2\right]_{z=0}^{z=\sqrt{4-x^2-y^2}} dy\,dx$$

$$= \tfrac{1}{2} \int_0^2 \int_0^{\sqrt{4-x^2}} xy\left(4 - x^2 - y^2\right) dy\,dx$$

$$= \tfrac{1}{2} \int_0^2 \left[2xy^2 - \tfrac{1}{2}x^3 y^2 - \tfrac{1}{4}xy^4\right]_{y=0}^{y=\sqrt{4-x^2}} dx$$

$$= \tfrac{1}{2} \int_0^2 \left[\left(2x - \tfrac{1}{2}x^3\right)\left(4 - x^2\right) - \tfrac{1}{4}x\left(4 - x^2\right)^2\right] dx = \int_0^2 \left(2x - x^3 + \tfrac{1}{8}x^5\right) dx$$

$$= \left(x^2 - \tfrac{1}{4}x^4 + \tfrac{1}{48}x^6\right)\Big|_0^2 = \tfrac{4}{3}$$

Since $V = \tfrac{1}{8} \cdot \tfrac{4}{3}\pi\left(2^3\right) = \tfrac{4\pi}{3}$, we have $f_{av} = \tfrac{4/3}{4\pi/3} = \tfrac{1}{\pi}$.

59. The integrand $f(x, y, z) = \left(1 - 2x^2 - 3y^2 - z^2\right)^{1/3}$ is defined in all of three-space. To maximize the value of $\iiint_T f(x, y, z)\,dV$, we want the largest region T such that $f(x, y, z) \ge 0$ for all (x, y, z) in T. Now $f(x, y, z) \ge 0 \Leftrightarrow$ $1 - 2x^2 - 3y^2 - z^2 \ge 0 \Leftrightarrow 2x^2 + 3y^2 + z^2 \le 1$. This means we should take $T = \left\{ (x, y, z) \mid 2x^2 + 3y^2 + z^2 \le 1 \right\}$, a region bounded by an ellipsoid.

61. True. The integrand $f(x, y, z) = \sqrt{x^2 + y^2 + z^2} > 0$ on B.

63. True. The integral is taken over the set $T = \{(x, y, z) \mid 1 \le x \le 2, 1 \le y \le 3, 1 \le z \le 4\}$.
On this set, $f(x, y, z) = \sqrt{1 + x^2 + y^2 + z^2}$ satisfies $2 \le \sqrt{1 + x^2 + y^2 + z^2} \le \sqrt{30}$, so $2 \cdot 6 \le \int_1^2 \int_1^3 \int_1^4 \sqrt{1 + x^2 + y^2 + z^2}\,dz\,dy\,dx \le 6\sqrt{30}$.

14.7 Concept Questions   ET 13.7

1. See page 1201 (1197 in ET).

3. a. $dV = r\,dz\,dr\,d\theta$
   b. $dV = \rho^2 \sin\phi\,d\rho\,d\phi\,d\theta$

14.7 Triple Integrals in Cylindrical and Spherical Coordinates   ET 13.7

1. $\int_0^{\pi/2} \int_0^3 \int_0^{r^2} r\,dz\,dr\,d\theta = \int_0^{\pi/2} \int_0^3 [rz]_{z=0}^{z=r^2}\,dr\,d\theta = \int_0^{\pi/2} \int_0^3 r^3\,dr\,d\theta$

$$= \int_0^{\pi/2} \left[\tfrac{1}{4}r^4\right]_0^3 = \int_0^{\pi/2} \tfrac{81}{4}\,d\theta = \tfrac{81\pi}{8}$$

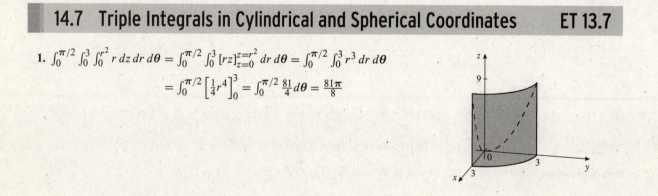

3. $\int_0^{2\pi} \int_0^{\pi/2} \int_0^2 \rho^2 \sin\phi \, d\rho \, d\phi \, d\theta = \int_0^{2\pi} \int_0^{\pi/2} \left[\frac{1}{3}\rho^3 \sin\phi\right]_{\rho=0}^{\rho=2} d\phi \, d\theta$

$$= \int_0^{2\pi} \int_0^{\pi/2} \frac{8}{3} \sin\phi \, d\phi \, d\theta$$

$$= \int_0^{2\pi} \left[-\frac{8}{3}\cos\phi\right]_0^{\pi/2} d\theta = \int_0^{2\pi} \frac{8}{3} \, d\theta$$

$$= \frac{8}{3}(2\pi) = \frac{16\pi}{3}$$

5. In cylindrical coordinates, $T = \{(r, \theta, z) \mid 0 \le \theta \le 2\pi, 0 \le r \le 1, 1 \le z \le 3\}$ and $R = \{(r, \theta) \mid 0 \le \theta \le 2\pi, 0 \le r \le 1\}$.

Also, $\sqrt{x^2 + y^2} = \sqrt{r^2} = r$, so

$\iiint_T \sqrt{x^2 + y^2} \, dV = \int_0^{2\pi} \int_0^1 \int_1^3 r \cdot r \, dz \, dr \, d\theta = \int_0^{2\pi} \int_0^1 \left[r^2 z\right]_{z=1}^{z=3} dr \, d\theta = \int_0^{2\pi} \int_0^1 2r^2 \, dr \, d\theta = \int_0^{2\pi} \left[\frac{2}{3}r^3\right]_0^1 d\theta$

$$= \int_0^{2\pi} \frac{2}{3} \, d\theta = \frac{4\pi}{3}$$

7. In cylindrical coordinates, $T = \left\{(r, \theta, z) \mid 0 \le \theta \le \frac{\pi}{2}, 0 \le r \le 2, 0 \le z \le 4 - r^2\right\}$, so

$\iiint_T y \, dV = \int_0^{\pi/2} \int_0^2 \int_0^{4-r^2} (r\sin\theta) r \, dz \, dr \, d\theta = \int_0^{\pi/2} \int_0^2 \left[z\left(r^2 \sin\theta\right)\right]_{z=0}^{z=4-r^2} dr \, d\theta = \int_0^{\pi/2} \int_0^2 r^2 \left(4 - r^2\right) \sin\theta \, dr \, d\theta$

$$= \int_0^{\pi/2} \left[\left(\frac{4}{3}r^3 - \frac{1}{5}r^5\right)\sin\theta\right]_{r=0}^{r=2} d\theta = \int_0^{2\pi} \frac{64}{15} \sin\theta \, d\theta = -\frac{64}{15}\cos\theta\Big|_0^{\pi/2} = \frac{64}{15}$$

9. In cylindrical coordinates, $T = \{(r, \theta, z) \mid 0 \le \theta \le 2\pi, 0 \le r \le 4, 0 \le z \le 4 - r\}$, so

$\iiint_T \left(x^2 + y^2\right) dV = \int_0^{2\pi} \int_0^4 \int_0^{4-r} r^2 \cdot r \, dz \, dr \, d\theta = \int_0^{2\pi} \int_0^4 \left[r^3 z\right]_{z=0}^{z=4-r} dr \, d\theta = \int_0^{2\pi} \int_0^4 \left(4r^3 - r^4\right) dr \, d\theta$

$$= \int_0^{2\pi} \left[r^4 - \frac{1}{5}r^5\right]_0^4 d\theta = \int_0^{2\pi} \frac{256}{5} \, d\theta = \frac{512\pi}{5}$$

11. To find the intersection of $x^2 + y^2 + z^2 = 9$ and $8z = x^2 + y^2$, we solve the

two equations simultaneously to obtain $8z + z^2 = 9 \Leftrightarrow z^2 + 8z - 9 = 0 \Leftrightarrow$

$(z + 9)(z - 1) = 0 \Rightarrow z = 1$, since $z > 0$. The intersection is the circle

$x^2 + y^2 = 8 = \left(2\sqrt{2}\right)^2$. Thus,

$T = \left\{(r, \theta, z) \mid 0 \le \theta \le 2\pi, 0 \le r \le 2\sqrt{2}, \frac{1}{8}r^2 \le z \le \sqrt{9 - r^2}\right\}$, and using

symmetry, we have

$V = \iiint_T dV = 4 \int_0^{\pi/2} \int_0^{2\sqrt{2}} \int_{r^2/8}^{\sqrt{9-r^2}} r \, dz \, dr \, d\theta = 4 \int_0^{\pi/2} \int_0^{2\sqrt{2}} [rz]_{z=r^2/8}^{z=\sqrt{9-r^2}} dr \, d\theta$

$$= 4 \int_0^{\pi/2} \int_0^{2\sqrt{2}} \left(r\sqrt{9 - r^2} - \frac{1}{8}r^3\right) dr \, d\theta = 4 \int_0^{\pi/2} \left[-\frac{1}{2} \cdot \frac{2}{3}\left(9 - r^2\right)^{3/2} - \frac{1}{32}r^4\right]_0^{2\sqrt{2}} d\theta = 4 \int_0^{\pi/2} \frac{20}{3} \, d\theta = \frac{40\pi}{3}$$

13. The solid can be described by

$T = \{(r, \theta, z) \mid 0 \le \theta \le 2\pi, 0 \le r \le 2, 0 \le z \le 3\}$. Also, $\rho(x, y, z) = kz$,

where k is a positive constant. Thus,

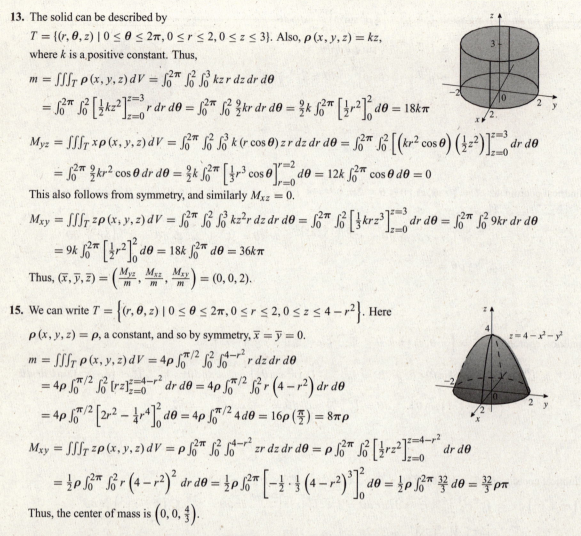

$m = \iiint_T \rho(x, y, z)\, dV = \int_0^{2\pi} \int_0^2 \int_0^3 kz\, r\, dz\, dr\, d\theta$

$= \int_0^{2\pi} \int_0^2 \left[\frac{1}{2}kz^2\right]_{z=0}^{z=3} r\, dr\, d\theta = \int_0^{2\pi} \int_0^2 \frac{9}{2}kr\, dr\, d\theta = \frac{9}{2}k \int_0^{2\pi} \left[\frac{1}{2}r^2\right]_0^2 d\theta = 18k\pi$

$M_{yz} = \iiint_T x\rho(x, y, z)\, dV = \int_0^{2\pi} \int_0^2 \int_0^3 k(r\cos\theta)\, z\, r\, dz\, dr\, d\theta = \int_0^{2\pi} \int_0^2 \left[(kr^2\cos\theta)\left(\frac{1}{2}z^2\right)\right]_{z=0}^{z=3} dr\, d\theta$

$= \int_0^{2\pi} \frac{9}{2}kr^2\cos\theta\, dr\, d\theta = \frac{9}{2}k \int_0^{2\pi} \left[\frac{1}{3}r^3\cos\theta\right]_{r=0}^{r=2} d\theta = 12k \int_0^{2\pi} \cos\theta\, d\theta = 0$

This also follows from symmetry, and similarly $M_{xz} = 0$.

$M_{xy} = \iiint_T z\rho(x, y, z)\, dV = \int_0^{2\pi} \int_0^2 \int_0^3 kz^2 r\, dz\, dr\, d\theta = \int_0^{2\pi} \int_0^2 \left[\frac{1}{3}krz^3\right]_{z=0}^{z=3} dr\, d\theta = \int_0^{2\pi} \int_0^2 9kr\, dr\, d\theta$

$= 9k \int_0^{2\pi} \left[\frac{1}{2}r^2\right]_0^2 d\theta = 18k \int_0^{2\pi} d\theta = 36k\pi$

Thus, $(\bar{x}, \bar{y}, \bar{z}) = \left(\frac{M_{yz}}{m}, \frac{M_{xz}}{m}, \frac{M_{xy}}{m}\right) = (0, 0, 2)$.

15. We can write $T = \left\{(r, \theta, z) \mid 0 \le \theta \le 2\pi, 0 \le r \le 2, 0 \le z \le 4 - r^2\right\}$. Here

$\rho(x, y, z) = \rho$, a constant, and so by symmetry, $\bar{x} = \bar{y} = 0$.

$m = \iiint_T \rho(x, y, z)\, dV = 4\rho \int_0^{\pi/2} \int_0^2 \int_0^{4-r^2} r\, dz\, dr\, d\theta$

$= 4\rho \int_0^{\pi/2} \int_0^2 [rz]_{z=0}^{z=4-r^2} dr\, d\theta = 4\rho \int_0^{\pi/2} \int_0^2 r\left(4 - r^2\right) dr\, d\theta$

$= 4\rho \int_0^{\pi/2} \left[2r^2 - \frac{1}{4}r^4\right]_0^2 d\theta = 4\rho \int_0^{\pi/2} 4\, d\theta = 16\rho\left(\frac{\pi}{2}\right) = 8\pi\rho$

$M_{xy} = \iiint_T z\rho(x, y, z)\, dV = \rho \int_0^{2\pi} \int_0^2 \int_0^{4-r^2} zr\, dz\, dr\, d\theta = \rho \int_0^{2\pi} \int_0^2 \left[\frac{1}{2}rz^2\right]_{z=0}^{z=4-r^2} dr\, d\theta$

$= \frac{1}{2}\rho \int_0^{2\pi} \int_0^2 r\left(4 - r^2\right)^2 dr\, d\theta = \frac{1}{2}\rho \int_0^{2\pi} \left[-\frac{1}{2} \cdot \frac{1}{3}\left(4 - r^2\right)^3\right]_0^2 d\theta = \frac{1}{2}\rho \int_0^{2\pi} \frac{32}{3} d\theta = \frac{32}{3}\rho\pi$

Thus, the center of mass is $\left(0, 0, \frac{4}{3}\right)$.

17. To find the curve of intersection of $z = \sqrt{x^2 + y^2}$ and $z = x^2 + y^2$, we solve the equation $z^2 = z \Leftrightarrow z(z - 1) = 0 \Leftrightarrow$

$z = 0$ or $z = 1$, that is, the point $(0, 0, 0)$ and the circle with radius 1 in the plane $z = 1$ centered at $(0, 0, 1)$. The solid of

interest can be written as $T = \left\{(r, \theta, z) \mid 0 \le \theta \le 2\pi, 0 \le r \le 1, r^2 \le z \le r\right\}$. Thus,

$I_z = \iiint_T \left(x^2 + y^2\right)\rho(x, y, z)\, dV = \rho \int_0^{2\pi} \int_0^1 \int_{r^2}^r r^2 \cdot r\, dz\, dr\, d\theta = \rho \int_0^{2\pi} \int_0^1 \left[r^3 z\right]_{z=r^2}^{z=r} dr\, d\theta$

$= \rho \int_0^{2\pi} \int_0^1 \left(r^4 - r^5\right) dr\, d\theta = \rho \int_0^{2\pi} \left[\frac{1}{5}r^5 - \frac{1}{6}r^6\right]_0^1 d\theta = \rho \int_0^{2\pi} \frac{1}{30} d\theta = \frac{1}{15}\pi\rho$

19. $\iiint_B \sqrt{x^2 + y^2 + z^2}\, dV = 8 \int_0^{\pi/2} \int_0^{\pi/2} \int_0^1 \sqrt{\rho^2}\, \rho^2 \sin\phi\, d\rho\, d\phi\, d\theta = 8 \int_0^{\pi/2} \int_0^{\pi/2} \left[\frac{1}{4}\rho^4 \sin\phi\right]_{\rho=0}^{\rho=1} d\phi\, d\theta$

$= 8 \int_0^{\pi/2} \int_0^{\pi/2} \frac{1}{4} \sin\phi\, d\phi\, d\theta = 2 \int_0^{\pi/2} [-\cos\phi]_0^{\pi/2} d\theta = 2 \int_0^{\pi/2} d\theta = \pi$

21. $\iiint_T y\, dV = \int_0^{2\pi} \int_0^{\pi/2} \int_0^1 (\rho\sin\phi\sin\theta)\, \rho^2 \sin\phi\, d\rho\, d\phi\, d\theta = \int_0^{2\pi} \int_0^{\pi/2} \left[\frac{1}{4}\rho^4 \sin^2\phi\sin\theta\right]_{\rho=0}^{\rho=1} d\phi\, d\theta$

$= \int_0^{2\pi} \int_0^{\pi/2} \frac{1}{4} \sin^2\phi\sin\theta\, d\phi\, d\theta = \frac{1}{4} \int_0^{2\pi} \int_0^{\pi/2} \left(\frac{1}{2} - \frac{1}{2}\cos 2\phi\right) \sin\theta\, d\phi\, d\theta$

$= \frac{1}{4} \int_0^{2\pi} \left[\frac{1}{2}\phi - \frac{1}{4}\sin 2\phi\right]_{\phi=0}^{\phi=\pi/2} \sin\theta\, d\theta = \frac{\pi}{16} \int_0^{2\pi} \sin\theta\, d\theta = 0$

23. The solid in question can be described as

$T = \{(\rho, \theta, \phi) \mid 0 \le \rho \le 2, 0 \le \phi \le \frac{\pi}{4}, 0 \le \theta \le 2\pi\}$, so

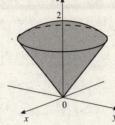

$$\iiint_T xz\, dV = \int_0^{2\pi} \int_0^{\pi/4} \int_0^2 (\rho \sin\phi \cos\theta)(\rho\cos\phi)\rho^2 \sin\phi\, d\rho\, d\phi\, d\theta$$

$$= \int_0^{2\pi} \int_0^{\pi/4} \left[\tfrac{1}{5}\rho^5 \sin^2\phi \cos\phi \cos\theta \right]_{\rho=0}^{\rho=2} d\phi\, d\theta$$

$$= \int_0^{2\pi} \int_0^{\pi/4} \tfrac{32}{5} \sin^2\phi \cos\phi \cos\theta\, d\phi\, d\theta$$

$$= \tfrac{32}{5} \int_0^{2\pi} \left[\tfrac{1}{3} \sin^3\phi \cos\theta \right]_{\phi=0}^{\phi=\pi/4} d\theta = \tfrac{32}{5} \int_0^{2\pi} \tfrac{\sqrt{2}}{12} \cos\theta\, d\theta = 0$$

25. The solid can be described as

$T = \{(\rho, \theta, \phi) \mid 0 \le \rho \le \sec\phi, 0 \le \phi \le \frac{\pi}{4}, 0 \le \theta \le 2\pi\}$. By symmetry, the

volume of T is

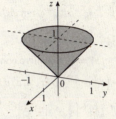

$$V = \iiint_T dV = 4\int_0^{\pi/2} \int_0^{\pi/4} \int_0^{\sec\phi} \rho^2 \sin\phi\, d\rho\, d\phi\, d\theta$$

$$= 4\int_0^{\pi/2} \int_0^{\pi/4} \left[\tfrac{1}{3}\rho^3 \sin\phi \right]_{\rho=0}^{\rho=\sec\phi} d\phi\, d\theta = \tfrac{4}{3} \int_0^{\pi/2} \int_0^{\pi/4} \sec^3\phi \sin\phi\, d\phi\, d\theta$$

$$= \tfrac{4}{3} \int_0^{\pi/2} \int_0^{\pi/4} \tan\phi \sec^2\phi\, d\phi\, d\theta = \tfrac{4}{3} \int_0^{\pi/2} \left[\tfrac{1}{2}\tan^2\phi \right]_0^{\pi/4} d\theta = \tfrac{2}{3} \int_0^{\pi/2} d\theta = \tfrac{\pi}{3}$$

27. The projection of the solid onto the yz-plane is shown in the figure.   Let

$T = \{(\rho, \theta, \phi) \mid 0 \le \rho \le 1, \frac{\pi}{4} \le \phi \le \frac{\pi}{2}, 0 \le \theta \le 2\pi\}$. Then the volume of T is

$$V = \iiint_T dV = \int_0^{2\pi} \int_{\pi/4}^{\pi/2} \int_0^1 \rho^2 \sin\phi\, d\rho\, d\phi\, d\theta = \int_0^{2\pi} \int_{\pi/4}^{\pi/2} \left[\tfrac{1}{3}\rho^3 \sin\phi \right]_{\rho=0}^{\rho=1} d\phi\, d\theta = \tfrac{1}{3} \int_0^{2\pi} \int_{\pi/4}^{\pi/2} \sin\phi\, d\phi\, d\theta$$

$$= \tfrac{1}{3} \int_0^{2\pi} [-\cos\phi]_{\pi/4}^{\pi/2}\, d\theta = \tfrac{1}{3} \int_0^{2\pi} \tfrac{\sqrt{2}}{2}\, d\theta = \tfrac{\sqrt{2}}{6}(2\pi) = \tfrac{\sqrt{2}\pi}{3}$$

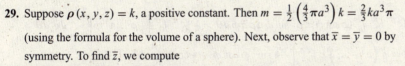

29. Suppose $\rho(x, y, z) = k$, a positive constant. Then $m = \tfrac{1}{2}\left(\tfrac{4}{3}\pi a^3\right)k = \tfrac{2}{3}ka^3\pi$

(using the formula for the volume of a sphere). Next, observe that $\bar{x} = \bar{y} = 0$ by

symmetry. To find $\bar{z}$, we compute

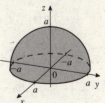

$$M_{xy} = \iiint_T kz\, dV = k\int_0^{2\pi} \int_0^{\pi/2} \int_0^a (\rho\cos\phi)\rho^2 \sin\phi\, d\rho\, d\phi\, d\theta$$

$$= k\int_0^{2\pi} \int_0^{\pi/2} \left[\tfrac{1}{4}\rho^4 \cos\phi \sin\phi \right]_{\rho=0}^{\rho=a} d\phi\, d\theta = k\int_0^{2\pi} \int_0^{\pi/2} \tfrac{1}{4}a^4 \cos\phi \sin\phi\, d\phi\, d\theta$$

$$= \tfrac{1}{4}ka^4 \int_0^{2\pi} \left[\tfrac{1}{2}\sin^2\phi \right]_0^{\pi/2} d\theta = \tfrac{1}{8}ka^4 \int_0^{2\pi} d\theta = \tfrac{1}{4}ka^4\pi$$

Thus, $(\bar{x}, \bar{y}, \bar{z}) = \left(\dfrac{M_{yz}}{m}, \dfrac{M_{xz}}{m}, \dfrac{M_{xy}}{m}\right) = \left(0, 0, \tfrac{3}{8}a\right)$.

31. Here $\rho(x, y, z) = kz$, where k is a positive constant. Then

$$m = \iiint_T \rho(x, y, z)\, dV = \int_0^{2\pi} \int_0^{\pi/2} \int_0^a k\,(\rho\cos\phi)\,\rho^2 \sin\phi\,d\rho\,d\phi\,d\theta$$

$$= k \int_0^{2\pi} \int_0^{\pi/2} \left[\tfrac{1}{4}\rho^4 \cos\phi \sin\phi\right]_{\rho=0}^{\rho=a} d\phi\,d\theta$$

$$= \tfrac{1}{4}ka^4 \int_0^{2\pi} \left[\tfrac{1}{2}\sin^2\phi\right]_0^{\pi/2} d\theta = \tfrac{1}{8}ka^4 \int_0^{2\pi} d\theta = \tfrac{1}{4}ka^4\pi$$

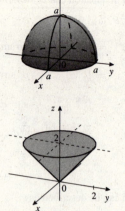

33. Here $\rho(x, y, z) = k\left(x^2 + y^2 + z^2\right)$, where k is a positive constant. Then

$$m = \iiint_T \rho(x, y, z)\, dV = \int_0^{2\pi} \int_0^{\pi/4} \int_0^{2\sec\phi} \left(k\rho^2\right)\rho^2 \sin\phi\,d\rho\,d\phi\,d\theta$$

$$= k \int_0^{2\pi} \int_0^{\pi/4} \left[\tfrac{1}{5}\rho^5 \sin\phi\right]_{\rho=0}^{\rho=2\sec\phi} d\phi\,d\theta = k \int_0^{2\pi} \int_0^{\pi/4} \tfrac{32}{5} \sec^5\phi \sin\phi\,d\phi\,d\theta$$

$$= \tfrac{32}{5}k \int_0^{2\pi} \int_0^{\pi/4} \sec^4\phi \tan\phi\,d\phi\,d\theta = \tfrac{32}{5}k \int_0^{2\pi} \left[\tfrac{1}{4}\sec^4\phi\right]_0^{\pi/4} d\theta$$

$$= \tfrac{24}{5}k \int_0^{2\pi} d\theta = \tfrac{48}{5}k\pi$$

35. Here $\rho(x, y, z) = k$, a positive constant.

$$I_z = \iiint_T \left(x^2 + y^2\right)\rho(x, y, z)\, dV = \int_0^{2\pi} \int_0^{\pi/6} \int_0^{4\cos\phi} k\,(\rho\sin\phi)^2\,\rho^2 \sin\phi\,d\rho\,d\phi\,d\theta$$

$$= k \int_0^{2\pi} \int_0^{\pi/6} \left[\tfrac{1}{5}\rho^5 \sin^3\phi\right]_{\rho=0}^{\rho=4\cos\phi} d\phi\,d\theta = \tfrac{1024}{5}k \int_0^{2\pi} \int_0^{\pi/6} \cos^5\phi \sin^3\phi\,d\phi\,d\theta$$

$$= \tfrac{1024}{5}k \int_0^{2\pi} \int_0^{\pi/6} \left(\sin^3\phi\right)\left(1 - 2\sin^2\phi + \sin^4\phi\right)\cos\phi\,d\phi\,d\theta$$

$$= \tfrac{1024}{5}k \int_0^{2\pi} \left[\tfrac{1}{4}\sin^4\phi - \tfrac{1}{3}\sin^6\phi + \tfrac{1}{8}\sin^8\phi\right]_0^{\pi/6} d\theta = \tfrac{67}{15}k\pi$$

37. Here $\rho(x, y, z) = k$, a positive constant. The required moment of inertia is given by I_x or I_y. We find

$$I_x = \iiint_T \left(y^2 + z^2\right)\rho(x, y, z)\, dV$$

$$= k \int_0^{2\pi} \int_0^{\pi/2} \int_0^a \left[(\rho\sin\phi\sin\theta)^2 + (\rho\cos\phi)^2\right]\rho^2 \sin\phi\,d\rho\,d\phi\,d\theta$$

$$= k \int_0^{2\pi} \int_0^{\pi/2} \left[\tfrac{1}{5}\rho^5 \left(\sin^2\phi\sin^2\theta + \cos^2\phi\right)\sin\phi\right]_{\rho=0}^{\rho=a} d\phi\,d\theta$$

$$= \tfrac{1}{5}ka^5 \int_0^{2\pi} \int_0^{\pi/2} \left(\sin^3\phi\sin^2\theta + \cos^2\phi\sin\phi\right) d\phi\,d\theta$$

$$= \tfrac{1}{5}ka^5 \int_0^{2\pi} \left[\sin^2\theta\left(-\tfrac{2}{3}\cos\phi - \tfrac{1}{3}\sin^2\phi\cos\phi\right) - \tfrac{1}{3}\cos^3\phi\right]_{\phi=0}^{\phi=\pi/2} d\theta = \tfrac{1}{5}ka^5 \int_0^{2\pi} \left(\tfrac{2}{3}\sin^2\theta + \tfrac{1}{3}\right) d\theta$$

$$= \tfrac{1}{5}ka^5 \int_0^{2\pi} \left[\tfrac{2}{3}\cdot\tfrac{1}{2}(1 - \cos 2\theta) + \tfrac{1}{3}\right] d\theta = \tfrac{1}{5}ka^5 \left[\tfrac{2}{3}\theta - \tfrac{1}{6}\sin 2\theta\right]_0^{2\pi} = \tfrac{4}{15}ka^5\pi$$

39. Place the center of the solid at the origin. By symmetry, the moment of inertia is I_z. Let k denote the density of the solid. Then

$$I_z = \iiint_T \left(x^2 + y^2\right)\rho(x, y, z)\, dV = k \int_0^{2\pi} \int_0^{\pi} \int_a^b (\rho\sin\phi)^2\,\rho^2 \sin\phi\,d\rho\,d\phi\,d\theta = k \int_0^{2\pi} \int_0^{\pi} \left[\tfrac{1}{5}\rho^5 \sin^3\phi\right]_{\rho=a}^{\rho=b} d\phi\,d\theta$$

$$= \tfrac{1}{5}\left(b^5 - a^5\right)k \int_0^{2\pi} \int_0^{\pi} \sin^3\phi\,d\phi\,d\theta = \tfrac{1}{5}\left(b^5 - a^5\right)k \int_0^{2\pi} \left[-\tfrac{1}{3}\sin^2\phi\cos\phi - \tfrac{2}{3}\cos\phi\right]_0^{\pi} d\theta$$

$$= \tfrac{1}{5}\left(b^5 - a^5\right)k \int_0^{2\pi} \tfrac{4}{3}\,d\theta = \tfrac{1}{5}\left(b^5 - a^5\right)k \cdot \tfrac{4}{3}\cdot 2\pi = \frac{8\left(b^5 - a^5\right)\pi k}{15}$$

Now $m = \iiint_T k\,dV = \tfrac{4}{3}k\left(b^3 - a^3\right)\pi$ (that is, k times the volume of the solid), so

$$k = \tfrac{3}{4}m\left[\left(b^3 - a^3\right)\pi\right]^{-1} = \frac{3m}{4\left(b^3 - a^3\right)\pi}, \text{ and thus } I = \frac{8\left(b^5 - a^5\right)\pi}{15} \cdot \frac{3m}{4\left(b^3 - a^3\right)\pi} = \frac{2m}{5}\left(\frac{b^5 - a^5}{b^3 - a^3}\right).$$

41. $\int_{-1}^{1} \int_{0}^{\sqrt{1-x^2}} \int_{0}^{\sqrt{4-x^2-y^2}} z \, dz \, dy \, dx = \int_{0}^{\pi} \int_{0}^{1} \int_{0}^{\sqrt{4-r^2}} zr \, dz \, dr \, d\theta$

$= \int_{0}^{\pi} \int_{0}^{1} \left[\frac{1}{2} z^2 r \right]_{z=0}^{z=\sqrt{4-r^2}} dr \, d\theta$

$= \int_{0}^{\pi} \int_{0}^{1} \frac{1}{2} r \left(4 - r^2 \right) dr \, d\theta$

$= \int_{0}^{\pi} \left[r^2 - \frac{1}{8} r^4 \right]_{0}^{1} d\theta = \int_{0}^{\pi} \frac{7}{8} d\theta = \frac{7\pi}{8}$

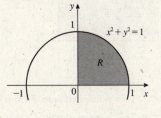

Projection of T onto the xy-plane

43. The intersection of the cone $z = \sqrt{x^2 + y^2}$ and the upper hemisphere $z = \sqrt{2 - x^2 - y^2}$ is found by solving $\sqrt{x^2 + y^2} = \sqrt{2 - x^2 - y^2} \Leftrightarrow x^2 + y^2 = 1$. The projection of this circle onto the xy-plane is shown.

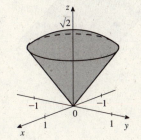

Thus,

$\int_{0}^{1} \int_{0}^{\sqrt{1-x^2}} \int_{\sqrt{x^2+y^2}}^{\sqrt{2-x^2-y^2}} \left(x^2 + y^2 + z^2 \right)^{3/2} dz \, dy \, dx = \int_{0}^{\pi/2} \int_{0}^{\pi/4} \int_{0}^{\sqrt{2}} \left(\rho^2 \right)^{3/2} \rho^2 \sin\phi \, d\rho \, d\phi \, d\theta$

$= \int_{0}^{\pi/2} \int_{0}^{\pi/4} \left[\frac{1}{6} \rho^6 \sin\phi \right]_{\rho=0}^{\rho=\sqrt{2}} d\phi \, d\theta = \frac{4}{3} \int_{0}^{\pi/2} \int_{0}^{\pi/4} \sin\phi \, d\phi \, d\theta$

$= \frac{4}{3} \int_{0}^{\pi/2} \left[-\cos\phi \right]_{0}^{\pi/4} d\theta = \frac{4}{3} \left(1 - \frac{\sqrt{2}}{2} \right) \int_{0}^{\pi/2} d\theta = \frac{1}{3} \left(2 - \sqrt{2} \right) \pi$

45. Making use of symmetry, we have

$\iiint_{T} f(x, y, z) \, dV = 8 \int_{0}^{3} \int_{0}^{\sqrt{9-x^2}} \int_{0}^{\sqrt{9-x^2-y^2}} 20 \left(x^2 + y^2 + z^2 \right) dz \, dy \, dx = 160 \int_{0}^{\pi/2} \int_{0}^{\pi/2} \int_{0}^{3} \rho^2 \cdot \rho^2 \sin\phi \, d\rho \, d\phi \, d\theta$

$= 160 \int_{0}^{\pi/2} \int_{0}^{\pi/2} \left[\frac{1}{5} \rho^5 \sin\phi \right]_{\rho=0}^{\rho=3} d\phi \, d\theta = \frac{160 \cdot 243}{5} \int_{0}^{\pi/2} \int_{0}^{\pi/2} \sin\phi \, d\phi \, d\theta = 7776 \int_{0}^{\pi/2} \left[-\cos\phi \right]_{0}^{\pi/2} d\theta$

$= 7776 \int_{0}^{\pi/2} d\theta = 7776 \cdot \frac{\pi}{2} = 3888\pi$

The volume of the sphere is $V = \frac{4}{3} \pi \cdot 3^3 = 36\pi$, so the average temperature is

$f_{av} = \frac{1}{V} \iiint_{T} f(x, y, z) \, dV = \frac{3888\pi}{36\pi} = 108°\text{F}$.

47. True. The triple integral represents the volume of a hemisphere of radius 2, and so it is given by $V = \frac{1}{2} \cdot \frac{4}{3} \pi \cdot 2^3 = \frac{16\pi}{3}$.

49. True: $I_z = \iiint_{T} \left(x^2 + y^2 \right) k \, dV = k \iiint_{T} \left[(\rho \sin\phi \cos\theta)^2 + (\rho \sin\phi \sin\theta)^2 \right] dV = k \iiint_{T} \rho^2 \sin^2\phi \, dV$

14.8 Concept Questions   ET 13.8

1. a. See page 1211 (1207 in ET).

   b. See page 1215 (1211 in ET).

14.8   Change of Variables in Multiple Integrals   ET 13.8

1. Here $T : x = u - v, y = v$.

On $S_1: v = 0, 0 \le u \le 2$. This is mapped onto $T(S_1): y = 0, 0 \le x \le 2$.

On $S_2: u = 2, 0 \le v \le 1$. This is mapped onto $T(S_2): x = 2 - y, 0 \le y \le 1$.

On $S_3: v = 1, 0 \le u \le 2$. This is mapped onto $T(S_3): y = 1, -1 \le x \le 1$.

On $S_4: u = 0, 0 \le v \le 1$. This is mapped onto $T(S_4): x = -y, 0 \le y \le 1$.

The region R is the parallelogram shown.

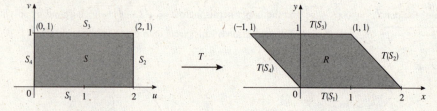

3. Here $T : x = u + 2v, y = 2v$.

On $S_1: v = u, 0 \le u \le 1$. This is mapped onto $T(S_1): y = \frac{2}{3}x, 0 \le x \le 3$.

On $S_2: v = 1, 0 \le u \le 1$. This is mapped onto $T(S_2): y = 2, 2 \le x \le 3$.

On $S_3: u = 0, 0 \le v \le 1$. This is mapped onto $T(S_3): y = x, 0 \le x \le 2$.

Thus, R is the triangular region shown.

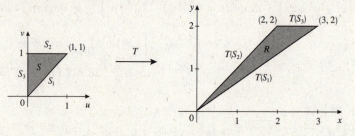

5. Here $T : x = u^2 - v^2, y = 2uv$.

The segment $S_1: 0 \le u \le 1, v = 0$ is mapped onto the line segment $T(S_1): y = 0, 0 \le x \le 1$.

Next, $x^2 + y^2 = \left(u^2 - v^2\right)^2 + (2uv)^2 = u^4 - 2u^2v^2 + v^4 + 4u^2v^2 = u^4 + 2u^2v^2 + v^4 = \left(u^2 + v^2\right)^2$, showing that the

circle $u^2 + v^2 = 1$ is mapped onto the circle $x^2 + y^2 = 1$. Thus, the quarter-circle $S_2: u^2 + v^2 = 1, 0 \le u \le 1, 0 \le v \le 1$

is mapped onto the half-circle $T(S_2): x^2 + y^2 = 1, 0 \le y \le 1$.

The segment $S_3: u = 0, 0 \le v \le 1$ is mapped onto the line segment $T(S_3): y = 0, -1 \le x \le 0$. Thus, R is the half-disk

shown.

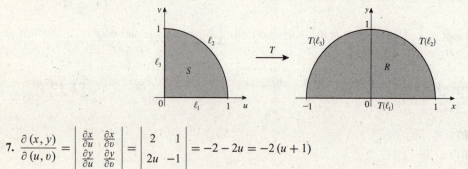

7. $\dfrac{\partial(x, y)}{\partial(u, v)} = \begin{vmatrix} \frac{\partial x}{\partial u} & \frac{\partial x}{\partial v} \\ \frac{\partial y}{\partial u} & \frac{\partial y}{\partial v} \end{vmatrix} = \begin{vmatrix} 2 & 1 \\ 2u & -1 \end{vmatrix} = -2 - 2u = -2(u + 1)$

9. $\dfrac{\partial (x,y)}{\partial (u,v)} = \begin{vmatrix} \frac{\partial x}{\partial u} & \frac{\partial x}{\partial v} \\ \frac{\partial y}{\partial u} & \frac{\partial y}{\partial v} \end{vmatrix} = \begin{vmatrix} e^u \cos 2v & -2e^u \sin 2v \\ e^u \sin 2v & 2e^u \cos 2v \end{vmatrix} = 2e^{2u} \left(\cos^2 2v + \sin^2 2v \right) = 2e^{2u}$

11. $\dfrac{\partial (x,y,z)}{\partial (u,v,w)} = \begin{vmatrix} \frac{\partial x}{\partial u} & \frac{\partial x}{\partial v} & \frac{\partial x}{\partial w} \\ \frac{\partial y}{\partial u} & \frac{\partial y}{\partial v} & \frac{\partial y}{\partial w} \\ \frac{\partial z}{\partial u} & \frac{\partial z}{\partial v} & \frac{\partial z}{\partial w} \end{vmatrix} = \begin{vmatrix} 1 & 1 & 1 \\ 1 & -1 & 1 \\ 1 & -2 & 3 \end{vmatrix} = -4$

13. To find T^{-1}, we solve the system of equations of $T : x = u + 2v,\ y = v - 2u$ for u and v, obtaining

$T^{-1} : u = \frac{1}{5}(x - 2y),\ v = \frac{1}{5}(2x + y)$. Using this transformation, we obtain the region $S = T^{-1}(R)$. Next, we find

$J = \dfrac{\partial (x,y)}{\partial (u,v)} = \begin{vmatrix} \frac{\partial x}{\partial u} & \frac{\partial x}{\partial v} \\ \frac{\partial y}{\partial u} & \frac{\partial y}{\partial v} \end{vmatrix} = \begin{vmatrix} 1 & 2 \\ -2 & 1 \end{vmatrix} = 5$. Thus,

$\iint_R (x + y)\, dA = \iint_S [(u + 2v) + (v - 2u)] \left| \dfrac{\partial (x,y)}{\partial (u,v)} \right| du\, dv = 5 \int_0^2 \int_0^3 (3v - u)\, du\, dv$

$\qquad = 5 \int_0^2 \left[3uv - \frac{1}{2}u^2 \right]_{u=0}^{u=3} dv = 5 \int_0^2 \left(9v - \frac{9}{2} \right) dv = 5 \left(\frac{9}{2}v^2 - \frac{9}{2}v \right) \Big|_0^2 = 45$

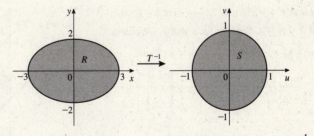

15. Here $T : x = 3u,\ y = 2v$. Then $4x^2 + 9y^2 = 36$

$\Rightarrow 4(3u)^2 + 9(2v)^2 = 36 \Leftrightarrow u^2 + v^2 = 1$, the

circle with radius 1 centered at the origin of the

uv-plane. Next, we find

$J = \dfrac{\partial (x,y)}{\partial (u,v)} = \begin{vmatrix} \frac{\partial x}{\partial u} & \frac{\partial x}{\partial v} \\ \frac{\partial y}{\partial u} & \frac{\partial y}{\partial v} \end{vmatrix} = \begin{vmatrix} 3 & 0 \\ 0 & 2 \end{vmatrix} = 6$, so

$\iint_R 2xy\, dA = 2 \iint_S (3u)(2v) \left| \dfrac{\partial (x,y)}{\partial (u,v)} \right| du\, dv = 12 \cdot 6 \int_0^{2\pi} \int_0^1 (r \cos \theta)(r \sin \theta)\, r\, dr\, d\theta = 72 \int_0^{2\pi} \left[\frac{1}{4}r^4 \cos \theta \sin \theta \right]_{r=0}^{r=1} d\theta$

$\qquad = 72 \int_0^{2\pi} \frac{1}{4} \cos \theta \sin \theta\, d\theta = -9 \sin^2 \theta \Big|_0^{2\pi} = 0$

17. Here the transformation is $T : x = 2u, y = 3v$. Now $\frac{1}{4}x^2 + \frac{1}{9}y^2 \le 1 \Rightarrow u^2 + v^2 \le 1$, so the elliptical region R is mapped onto the circular region S.

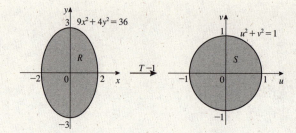

Next, we find $J = \dfrac{\partial(x, y)}{\partial(u, v)} = \begin{vmatrix} \frac{\partial x}{\partial u} & \frac{\partial x}{\partial v} \\ \frac{\partial y}{\partial u} & \frac{\partial y}{\partial v} \end{vmatrix} = \begin{vmatrix} 2 & 0 \\ 0 & 3 \end{vmatrix} = 6$, and so

$$\iint_R \sqrt{1 - \tfrac{1}{4}x^2 - \tfrac{1}{9}y^2}\, dA = \iint_S \sqrt{1 - (u^2 + v^2)}\, \left|\frac{\partial(x,y)}{\partial(u,v)}\right|\, du\, dv = 6\int_0^{2\pi}\int_0^1 \sqrt{1 - r^2}\, r\, dr\, d\theta$$

$$= 6\int_0^{2\pi}\left[-\tfrac{1}{2}\cdot\tfrac{2}{3}\left(1 - r^2\right)^{3/2}\right]_0^1 d\theta = 2\int_0^{2\pi} d\theta = 4\pi$$

19. Using the result of Exercise 5, we see that T^{-1} maps the semicircular region R onto the quarter of the unit disk lying in the first quadrant of the uv-plane; that is, $T^{-1}(R) = S = \left\{(u, v) \mid u^2 + v^2 \le 1, u \ge 0, v \ge 0\right\}$. Next, we find

$$J = \frac{\partial(x, y)}{\partial(u, v)} = \begin{vmatrix} \frac{\partial x}{\partial u} & \frac{\partial x}{\partial v} \\ \frac{\partial y}{\partial u} & \frac{\partial y}{\partial v} \end{vmatrix} = \begin{vmatrix} 2u & -2v \\ 2v & 2u \end{vmatrix} = 4\left(u^2 + v^2\right), \text{ and so}$$

$$\iint_R \frac{1}{\sqrt{x^2 + y^2}}\, dA = \iint_S \frac{1}{\sqrt{(u^2 - v^2)^2 + (2uv)^2}}\, \left|\frac{\partial(x, y)}{\partial(u, v)}\right|\, du\, dv = \iint_S \frac{4\left(u^2 + v^2\right)}{\sqrt{(u^2 + v^2)^2}}\, du\, dv$$

$$= \iint_S 4\, du\, dv = 4\,(\text{area of } S) = 4\left(\tfrac{1}{4}\pi \cdot 1^2\right) = \pi$$

21. Let $u = x + y$ and $v = 2x - y$. Then $-1 \le u \le 3$ and $0 \le v \le 4$. Solving for x and y, we obtain $x = \frac{1}{3}(u + v)$ and $y = \frac{1}{3}(2u - v)$. Thus,

$$J = \frac{\partial(x, y)}{\partial(u, v)} = \begin{vmatrix} \frac{\partial x}{\partial u} & \frac{\partial x}{\partial v} \\ \frac{\partial y}{\partial u} & \frac{\partial y}{\partial v} \end{vmatrix} = \begin{vmatrix} \frac{1}{3} & \frac{1}{3} \\ \frac{2}{3} & -\frac{1}{3} \end{vmatrix} = -\frac{1}{3},$$

so

$$\iint_R (2x + y)\, dA = \int_{-1}^3\int_0^4 \left[\tfrac{2}{3}(u + v) + \tfrac{1}{3}(2u - v)\right]\left|-\tfrac{1}{3}\right|\, dv\, du = \tfrac{1}{9}\int_{-1}^3\int_0^4 (4u + v)\, dv\, du$$

$$= \tfrac{1}{9}\int_{-1}^3 \left[4uv + \tfrac{1}{2}v^2\right]_{v=0}^{v=4} du = \tfrac{1}{9}\int_{-1}^3 (16u + 8)\, du = \tfrac{8}{9}\left(u^2 + u\right)\Big|_{-1}^3 = \tfrac{32}{3}$$

23. Let $T : u = x - y, v = x + y$. Then $T^{-1} : x = \frac{1}{2}(u + v), y = \frac{1}{2}(v - u)$. The triangular region R is mapped onto the triangular region S.

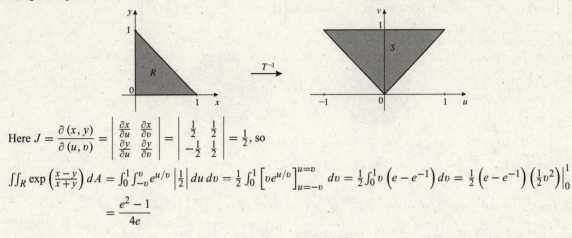

Here $J = \dfrac{\partial(x, y)}{\partial(u, v)} = \begin{vmatrix} \frac{\partial x}{\partial u} & \frac{\partial x}{\partial v} \\ \frac{\partial y}{\partial u} & \frac{\partial y}{\partial v} \end{vmatrix} = \begin{vmatrix} \frac{1}{2} & \frac{1}{2} \\ -\frac{1}{2} & \frac{1}{2} \end{vmatrix} = \frac{1}{2}$, so

$$\iint_R \exp\left(\frac{x-y}{x+y}\right) dA = \int_0^1 \int_{-v}^{v} e^{u/v} \left|\tfrac{1}{2}\right| du\, dv = \tfrac{1}{2} \int_0^1 \left[v e^{u/v}\right]_{u=-v}^{u=v} dv = \tfrac{1}{2} \int_0^1 v \left(e - e^{-1}\right) dv = \tfrac{1}{2}\left(e - e^{-1}\right)\left(\tfrac{1}{2}v^2\right)\Big|_0^1$$

$$= \frac{e^2 - 1}{4e}$$

25. Let $T : x = au, y = bv$, so $T^{-1} : u = \dfrac{x}{a}, v = \dfrac{b}{y}$. The region

$$R = \left\{(x, y) \ \Big| \ \frac{x^2}{a^2} + \frac{y^2}{b^2} \le 1, 0 \le x \le a, 0 \le y \le b\right\} \text{ is}$$

mapped onto the region

$$S = \left\{(u, v) \mid u^2 + v^2 \le 1, 0 \le u \le 1, 0 \le v \le 1\right\}.$$

Now $J = \dfrac{\partial(x, y)}{\partial(u, v)} = \begin{vmatrix} \frac{\partial x}{\partial u} & \frac{\partial x}{\partial v} \\ \frac{\partial y}{\partial u} & \frac{\partial y}{\partial v} \end{vmatrix} = \begin{vmatrix} a & 0 \\ 0 & b \end{vmatrix} = ab$, so

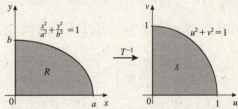

$$\iint_R xy\, dA = \int_0^{\pi/2} \int_0^1 (ar\cos\theta)(br\sin\theta)\,|ab|\,r\, dr\, d\theta = a^2 b^2 \int_0^{\pi/2} \left[\tfrac{1}{4}r^4 \cos\theta \sin\theta\right]_0^1 d\theta = \tfrac{1}{4}a^2 b^2 \left[-\tfrac{1}{2}\cos^2\theta\right]_0^{\pi/2} = \tfrac{1}{8}a^2 b^2.$$

27. Let $T : x = au, y = bv, z = cw$. Then $E = \left\{(x, y, z) \ \Big| \ \dfrac{x^2}{a^2} + \dfrac{y^2}{b^2} + \dfrac{z^2}{c^2} \le 1\right\}$ is mapped by T^{-1} onto

$$S = \left\{(u, v, w) \mid u^2 + v^2 + w^2 \le 1\right\} \text{ and } J = \dfrac{\partial(x, y, z)}{\partial(u, v, w)} = \begin{vmatrix} \frac{\partial x}{\partial u} & \frac{\partial x}{\partial v} & \frac{\partial x}{\partial w} \\ \frac{\partial y}{\partial u} & \frac{\partial y}{\partial v} & \frac{\partial y}{\partial w} \\ \frac{\partial z}{\partial u} & \frac{\partial z}{\partial v} & \frac{\partial z}{\partial w} \end{vmatrix} = \begin{vmatrix} a & 0 & 0 \\ 0 & b & 0 \\ 0 & 0 & c \end{vmatrix} = abc, \text{ so}$$

$$V = \iiint_E dV = \iiint_S |J|\, dV = abc \iiint_S dV = \tfrac{4}{3}\pi abc.$$

29. We complete the square in x: $x^2 + y^2 - ax \le 0 \Rightarrow \left(x - \frac{1}{2}a\right)^2 + y^2 \le \left(\frac{1}{2}a\right)^2$. Let $T : x = u + \frac{1}{2}a, y = v$. Then

$$R = \left\{(x,y) \mid x^2 + y^2 - ax \le 0\right\} \text{ is mapped by } T^{-1} \text{ onto } S = \left\{(u,v) \ \Big| \ u^2 + v^2 \le \left(\frac{1}{2}a\right)^2\right\}.$$

$$J = \frac{\partial(x,y)}{\partial(u,v)} = \begin{vmatrix} \frac{\partial x}{\partial u} & \frac{\partial x}{\partial v} \\ \frac{\partial y}{\partial u} & \frac{\partial y}{\partial v} \end{vmatrix} = \begin{vmatrix} 1 & 0 \\ 0 & 1 \end{vmatrix} = 1, \text{ so}$$

$$I_x = \iint_R y^2 \rho(x,y)\,dA = \rho \iint_S v^2\,du\,dv = \rho \int_0^{2\pi} \int_0^{a/2} r^2 \sin^2\theta\, r\,dr\,d\theta = \rho \int_0^{2\pi} \left[\frac{1}{4}r^4 \sin^2\theta\right]_0^{a/2} d\theta$$

$$= \frac{1}{64}\rho a^4 \int_0^{2\pi} \sin^2\theta\,d\theta = \frac{1}{64}\rho a^4 \int_0^{2\pi} \frac{1}{2}(1 - \cos 2\theta)\,d\theta = \frac{1}{64}\rho a^4 \left[\frac{1}{2}\theta - \frac{1}{4}\sin 2\theta\right]_0^{2\pi} = \frac{1}{64}\pi\rho a^4$$

31. We use the transformation $T : x = r\cos\theta, y = r\sin\theta, z = z$, so

$$J = \frac{\partial(x,y,z)}{\partial(r,\theta,z)} = \begin{vmatrix} \frac{\partial x}{\partial r} & \frac{\partial x}{\partial \theta} & \frac{\partial x}{\partial z} \\ \frac{\partial y}{\partial r} & \frac{\partial y}{\partial \theta} & \frac{\partial y}{\partial z} \\ \frac{\partial z}{\partial r} & \frac{\partial z}{\partial \theta} & \frac{\partial z}{\partial z} \end{vmatrix} = \begin{vmatrix} \cos\theta & -r\sin\theta & 0 \\ \sin\theta & r\cos\theta & 0 \\ 0 & 0 & 1 \end{vmatrix} = r. \text{ Thus,}$$

$$\iiint_R f(x,y,z)\,dV = \iiint_S f(r\cos\theta, r\sin\theta, z)\, r\,dz\,dr\,d\theta.$$

33. False. See Exercise 6. Here $\iint_R \left(x^2 + y^2\right)dx\,dy = \int_0^{\pi/2} \int_1^2 r^2 r\,dr\,d\theta = \int_0^{\pi/2} \left[\frac{1}{4}r^4\right]_1^2 d\theta = \frac{15}{8}\pi$. Next,

$$J = \frac{\partial(x,y)}{\partial(u,v)} = \begin{vmatrix} \frac{\partial x}{\partial u} & \frac{\partial x}{\partial v} \\ \frac{\partial y}{\partial u} & \frac{\partial y}{\partial v} \end{vmatrix} = \begin{vmatrix} \cos v & -u\sin v \\ \sin v & u\cos v \end{vmatrix} = u, \text{ so}$$

$$\iint_S \left(u^2 + v^2\right)\left|\frac{\partial(x,y)}{\partial(u,v)}\right|du\,dv = \int_0^{\pi/2} \int_1^2 \left(u^2 + v^2\right) u\,du\,dv = \int_0^{\pi/2} \left[\frac{1}{4}u^4 + \frac{1}{2}u^2 v^2\right]_1^2 dv = \int_0^{\pi/2} \left(\frac{15}{4} + \frac{3}{2}v^2\right)dv$$

$$= \left(\frac{15}{4}v + \frac{1}{2}v^3\right)\Big|_0^{\pi/2} = \frac{1}{16}\pi\left(30 + \pi^2\right)$$

Thus, $\iint_R \left(x^2 + y^2\right)dx\,dy \ne \iint_S \left(u^2 + v^2\right)\left|\frac{\partial(x,y)}{\partial(u,v)}\right|du\,dv$.

Chapter 14   Review   ET 13

Concept Review

1. a. $\sum_{i=1}^{m} \sum_{j=1}^{n} f\left(x_{ij}^*, y_{ij}^*\right) \Delta A$

   b. $\lim_{m,n \to \infty} \sum_{i=1}^{m} \sum_{j=1}^{n} f\left(x_{ij}^*, y_{ij}^*\right) \Delta A;\ \left(x_{ij}^*, y_{ij}^*\right)$

   c. volume; $z = f(x,y)$

   d. $\lim_{m,n \to \infty} \sum_{i=1}^{m} \sum_{j=1}^{n} f_D\left(x_{ij}^*, y_{ij}^*\right) \Delta A;\ f(x,y);\ 0$

3. a. $\int_a^b \int_c^d f(x,y)\,dy\,dx;\ \int_c^d \int_a^b f(x,y)\,dx\,dy$

   b. iterated

5. a. $\{(r,\theta) \mid a \le r \le b; \alpha \le \theta \le \beta\}$

   b. $\int_\alpha^\beta \int_a^b f(r\cos\theta, r\sin\theta)\, r\,dr\,d\theta$

   c. $\{(r,\theta) \mid \alpha \le \theta \le \beta, g_1(\theta) \le r \le g_2(\theta)\}$

   d. $\int_\alpha^\beta \int_{g_1(\theta)}^{g_2(\theta)} f(r\cos\theta, r\sin\theta)\, r\,dr\,d\theta$

7. a. $\iint_R \sqrt{(f_x)^2 + (f_y)^2 + 1}\, dA$ **b.** $\iint_R \sqrt{(g_x)^2 + (g_z)^2 + 1}\, dA$ **c.** $\iint_R \sqrt{(h_y)^2 + (h_z)^2 + 1}\, dA$

9. a. order; $\int_p^q \int_c^d \int_a^b f(x, y, z)\, dx\, dy\, dz$ **b.** $\iint_R \left[\int_{k_1(x,y)}^{k_2(x,y)} f(x, y, z)\, dz \right] dA$

11. a. $\int_\alpha^\beta \int_{g_1(\theta)}^{g_2(\theta)} \int_{h_1(r\cos\theta, r\sin\theta)}^{h_2(r\cos\theta, r\sin\theta)} f(r\cos\theta, r\sin\theta, z)\, r\, dz\, dr\, d\theta$

    b. $\int_\alpha^\beta \int_c^d \int_a^b f(\rho\sin\phi\cos\theta, \rho\sin\phi\sin\theta, \rho\cos\phi)\, \rho^2 \sin\phi\, d\rho\, d\phi\, d\theta$

    c. $\int_\alpha^\beta \int_c^d \int_{h_1(\phi,\theta)}^{h_2(\phi,\theta)} f(\rho\sin\phi\cos\theta, \rho\sin\phi\sin\theta, \rho\cos\phi)\, \rho^2 \sin\phi\, d\rho\, d\phi\, d\theta$

Review Exercises

1. $\int_0^2 \int_{-1}^2 \left(2x + 3xy^2\right) dx\, dy = \int_0^2 \left[x^2 + \frac{3}{2}x^2 y^2\right]_{x=-1}^{x=2} dy = \int_0^2 \left(3 + \frac{9}{2}y^2\right) dy = \left(3y + \frac{3}{2}y^3\right)\Big|_0^2 = 18$

3. $\int_0^1 \int_x^{\sqrt{x}} (2x + 3y)\, dy\, dx = \int_0^1 \left[2xy + \frac{3}{2}y^2\right]_{y=x}^{y=\sqrt{x}} dx = \int_0^1 \left(2x^{3/2} + \frac{3}{2}x - \frac{7}{2}x^2\right) dx = \left(\frac{4}{5}x^{5/2} + \frac{3}{4}x^2 - \frac{7}{6}x^3\right)\Big|_0^1 = \frac{23}{60}$

5. $\int_0^2 \int_y^2 \frac{1}{4 + y^2}\, dx\, dy = \int_0^2 \left[\frac{x}{4 + y^2}\right]_{x=y}^{x=2} dy = \int_0^2 \left(\frac{2}{4 + y^2} - \frac{y}{4 + y^2}\right) dy = \left[\tan^{-1}\frac{y}{2} - \frac{1}{2}\ln\left(4 + y^2\right)\right]_0^2 = \frac{\pi}{4} - \frac{\ln 2}{2}$

7. $\int_0^2 \int_0^{\sqrt{z}} \int_0^x (x + 2z)\, dy\, dx\, dz = \int_0^2 \int_0^{\sqrt{z}} \left[(x + 2z)\,y\right]_{y=0}^{y=x} dx\, dz = \int_0^2 \int_0^{\sqrt{z}} \left(x^2 + 2xz\right) dx\, dz = \int_0^2 \left[\frac{1}{3}x^3 + x^2 z\right]_{x=0}^{x=\sqrt{z}} dz$

    $= \int_0^2 \left(\frac{1}{3}z^{3/2} + z^2\right) dz = \left(\frac{2}{15}z^{5/2} + \frac{1}{3}z^3\right)\Big|_0^2 = \frac{8}{15}\left(5 + \sqrt{2}\right)$

9.

11.

13. $\int_0^1 \int_y^1 \sin x^2\, dx\, dy = \int_0^1 \int_0^x \sin x^2\, dy\, dx$

    $= \int_0^1 \left[y \sin x^2\right]_{y=0}^{y=x} dx$

    $= \int_0^1 x \sin x^2\, dx$

    $= -\frac{1}{2}\cos x^2\Big|_0^1 = \frac{1}{2}(1 - \cos 1)$

15. $\iint_R \left(x^2 + 3y^2\right) dA = \int_0^2 \int_{-1}^1 \left(x^2 + 3y^2\right) dx\, dy = \int_0^2 \left[\frac{1}{3}x^3 + 3xy^2\right]_{x=-1}^{x=1} dy = \int_0^2 \left(\frac{2}{3} + 6y^2\right) dy = \left(\frac{2}{3}y + 2y^3\right)\Big|_0^2 = \frac{52}{3}$

17. $\iint_R y\, dA = \int_{-1}^3 \int_{y^2}^{2y+3} y\, dx\, dy = \int_{-1}^3 \left[xy\right]_{x=y^2}^{x=2y+3} dy$

    $= \int_{-1}^3 \left(2y^2 + 3y - y^3\right) dy$

    $= \left(\frac{2}{3}y^3 + \frac{3}{2}y^2 - \frac{1}{4}y^4\right)\Big|_{-1}^3 = \frac{32}{3}$

19. $\iint_R x\, dA = \int_0^3 \int_0^{(2\sqrt{9-x^2})/3} x\, dy\, dx = \int_0^3 [xy]_{y=0}^{y=(2\sqrt{9-x^2})/3}\, dx = \int_0^3 \frac{2}{3}x\left(9-x^2\right)^{1/2}\, dx$

$\qquad = \left(\frac{2}{3}\right)\left(-\frac{1}{2}\right)\left(\frac{2}{3}\right)\left(9-x^2\right)^{3/2}\Big|_0^3 = 6$

21. $\iiint_T xy\, dV = \int_0^1 \int_0^{x^2} \int_0^{x+y} xy\, dz\, dy\, dx = \int_0^1 \int_0^{x^2} [xyz]_{z=0}^{z=x+y}\, dy\, dx = \int_0^1 \int_0^{x^2} \left(x^2 y + xy^2\right)\, dy\, dx$

$\qquad = \int_0^1 \left[\frac{1}{2}x^2 y^2 + \frac{1}{3}xy^3\right]_{y=0}^{y=x^2}\, dx = \int_0^1 \left(\frac{1}{2}x^6 + \frac{1}{3}x^7\right)\, dx = \left(\frac{1}{14}x^7 + \frac{1}{24}x^8\right)\Big|_0^1 = \frac{19}{168}$

23. $\iiint_T xyz\, dV = \int_0^{2\pi} \int_0^{\pi/2} \int_0^1 (\rho\sin\phi\cos\theta)(\rho\sin\phi\sin\theta)(\rho\cos\phi)\,\rho^2 \sin\phi\, d\rho\, d\phi\, d\theta$

$\qquad = \int_0^{2\pi} \int_0^{\pi/2} \left[\left(\sin^3\phi\cos\phi\cos\theta\sin\theta\right)\left(\frac{1}{6}\rho^6\right)\right]_{\rho=0}^{\rho=1}\, d\phi\, d\theta = \frac{1}{6}\int_0^{2\pi} \int_0^{\pi/2} \sin^3\phi\cos\phi\cos\theta\sin\theta\, d\phi\, d\theta$

$\qquad = \frac{1}{6}\int_0^{2\pi}\left[\frac{1}{4}\sin^4\phi\cos\theta\sin\theta\right]_{\phi=0}^{\phi=\pi/2}\, d\theta = \frac{1}{24}\int_0^{2\pi}\cos\theta\sin\theta\, d\theta = \frac{1}{48}\sin^2\theta\Big|_0^{2\pi} = 0$

25. $\iiint_T x^2 z\, dV = \int_0^\pi \int_0^1 \int_0^{1-r^2} (r\cos\theta)^2 (r\sin\theta)\, r\, dy\, dr\, d\theta$

$\qquad = \int_0^\pi \int_0^1 \left[\left(r^4\cos^2\theta\sin\theta\right) y\right]_{y=0}^{y=1-r^2}\, dr\, d\theta$

$\qquad = \int_0^\pi \int_0^1 \left(\cos^2\theta\sin\theta\right)\left(r^4 - r^6\right)\, dr\, d\theta$

$\qquad = \int_0^\pi \left[\left(\cos^2\theta\sin\theta\right)\left(\frac{1}{5}r^5 - \frac{1}{7}r^7\right)\right]_{r=0}^{r=1}\, d\theta$

$\qquad = \frac{2}{35}\int_0^\pi \cos^2\theta\sin\theta\, d\theta = \left(\frac{2}{35}\right)\left(-\frac{1}{3}\right)\cos^3\theta\Big|_0^\pi = \frac{4}{105}$

27. $V = \iint_R xy^2\, dA = \int_0^1 \int_1^2 xy^2\, dy\, dx = \int_0^1 \left[\frac{1}{3}xy^3\right]_{y=1}^{y=2}\, dx = \int_0^1 \frac{7}{3}x\, dx = \frac{7}{6}x^2\Big|_0^1 = \frac{7}{6}$

29. By symmetry, we have $V = 4\int_0^{\pi/2} \int_0^1 r^2 \cdot r\, dr\, d\theta = 4\int_0^{\pi/2} \left[\frac{1}{4}r^4\right]_0^1\, d\theta = \int_0^{\pi/2} d\theta = \frac{\pi}{2}$.

31. $V = \iint_R z\, dA = \int_0^{2\pi} \int_0^1 e^{-r^2} r\, dr\, d\theta = \int_0^{2\pi}\left[-\frac{1}{2}e^{-r^2}\right]_0^1\, d\theta = \frac{1}{2}\left(1 - e^{-1}\right)\int_0^{2\pi} d\theta = \pi\left(1 - \frac{1}{e}\right) = \frac{\pi(e-1)}{e}$

33. $m = \iint_R \rho(x,y)\, dA = \int_0^1 \int_{x^3}^x y\, dy\, dx = \int_0^1 \left[\frac{1}{2}y^2\right]_{x^3}^x\, dx$

$\qquad = \frac{1}{2}\int_0^1 \left(x^2 - x^6\right)\, dx = \frac{1}{2}\left(\frac{1}{3}x^3 - \frac{1}{7}x^7\right)\Big|_0^1 = \frac{2}{21}$

$M_x = \iint_R y\rho(x,y)\, dA = \int_0^1 \int_{x^3}^x y^2\, dy\, dx = \int_0^1 \left[\frac{1}{3}y^3\right]_{x^3}^x\, dx$

$\qquad = \frac{1}{3}\int_0^1 \left(x^3 - x^9\right)\, dx = \frac{1}{3}\left(\frac{1}{4}x^4 - \frac{1}{10}x^{10}\right)\Big|_0^1 = \frac{1}{20}$

$M_y = \iint_R x\rho(x,y)\, dA = \int_0^1 \int_{x^3}^x xy\, dy\, dx = \int_0^1 \left[\frac{1}{2}xy^2\right]_{y=x^3}^{y=x}\, dx = \frac{1}{2}\int_0^1 \left(x^3 - x^7\right)\, dx = \frac{1}{2}\left(\frac{1}{4}x^4 - \frac{1}{8}x^8\right)\Big|_0^1 = \frac{1}{16}$.

Thus, $(\overline{x}, \overline{y}) = \left(\frac{M_y}{m}, \frac{M_x}{m}\right) = \left(\frac{21}{32}, \frac{21}{40}\right)$.

35. $m = \iint_R \rho(x,y)\, dA = \int_0^{\pi/2} \int_0^1 \sqrt{r^2}\, r\, dr\, d\theta = \int_0^{\pi/2}\left[\frac{1}{3}r^3\right]_0^1\, d\theta = \frac{1}{3}\int_0^{\pi/2} d\theta$

$\qquad = \frac{\pi}{6}$

$M_x = \iint_R y\rho(x,y)\, dA = \int_0^{\pi/2} \int_0^1 (r\sin\theta)\sqrt{r^2}\, r\, dr\, d\theta$

$\qquad = \int_0^{\pi/2}\left[\frac{1}{4}r^4 \sin\theta\right]_{r=0}^{r=1}\, d\theta = \frac{1}{4}\int_0^{\pi/2}\sin\theta\, d\theta = -\frac{1}{4}\cos\theta\Big|_0^{\pi/2} = \frac{1}{4}$

and $M_y = \frac{1}{4}$, so $(\overline{x}, \overline{y}) = \left(\frac{M_y}{m}, \frac{M_x}{m}\right) = \left(\frac{3}{2\pi}, \frac{3}{2\pi}\right)$.

37. $I_x = \iint_R y^2 \rho(x,y)\,dA = \int_0^1 \int_0^y y^2 \left(x^2 + y^2\right) dx\,dy = \int_0^1 \left[\frac{1}{3}x^3 y^2 + xy^4\right]_{x=0}^{x=y} dy$

$\quad = \int_0^1 \frac{4}{3} y^5\,dy = \frac{4}{3}\left(\frac{1}{6}y^6\right)\Big|_0^1 = \frac{2}{9}$

$\quad I_y = \iint_R x^2 \rho(x,y)\,dA = \int_0^1 \int_0^y x^2\left(x^2 + y^2\right) dx\,dy = \int_0^1 \left[\frac{1}{5}x^5 + \frac{1}{3}x^3 y^2\right]_{x=0}^{x=y} dy$

$\quad = \int_0^1 \frac{8}{15}y^5\,dy = \frac{4}{45}$

$\quad I_0 = I_x + I_y = \frac{2}{9} + \frac{4}{45} = \frac{14}{45}$

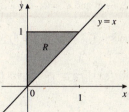

39. $2x + 3y + z = 6 \Rightarrow z = f(x,y) = 6 - 2x - 3y$, so $f_x = -2$, $f_y = -3$, and

$\quad A = \iint_R \sqrt{f_x^2 + f_y^2 + 1}\,dA = \int_0^3 \int_0^{(6-2x)/3} \sqrt{4 + 9 + 1}\,dy\,dx$

$\quad = \sqrt{14} \int_0^3 [y]_0^{(6-2x)/3}\,dx = \frac{2\sqrt{14}}{3} \int_0^3 (3 - x)\,dx$

$\quad = \frac{2\sqrt{14}}{3}\left(3x - \frac{1}{2}x^2\right)\Big|_0^3 = 3\sqrt{14}$

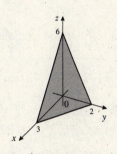

41. Using cylindrical coordinates, we find

$\quad \int_0^2 \int_0^{\sqrt{4-x^2}} \int_0^1 \left(x^2 + y^2\right)^{3/2} dz\,dy\,dx = \int_0^{\pi/2} \int_0^2 \int_0^1 \left(r^2\right)^{3/2} r\,dz\,dr\,d\theta$

$\quad\quad = \int_0^{\pi/2} \int_0^2 \left[r^4 z\right]_{z=0}^{z=1} dr\,d\theta$

$\quad\quad = \int_0^{\pi/2} \int_0^2 r^4\,dr\,d\theta = \int_0^{\pi/2} \left[\frac{1}{5}r^5\right]_0^2 d\theta$

$\quad\quad = \frac{32}{5} \int_0^{\pi/2} d\theta = \frac{16\pi}{5}$

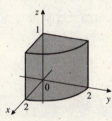

43. $\iiint_T f(x,y,z)\,dV = \int_0^3 \int_0^{(6-2x)/3} \int_0^{6-2x-3y} f(x,y,z)\,dz\,dy\,dx$

$\quad = \int_0^2 \int_0^{(6-3y)/2} \int_0^{6-2x-3y} f(x,y,z)\,dz\,dx\,dy$

$\quad = \int_0^3 \int_0^{6-2x} \int_0^{(6-2x-z)/3} f(x,y,z)\,dy\,dz\,dx$

$\quad = \int_0^6 \int_0^{(6-z)/2} \int_0^{(6-2x-z)/3} f(x,y,z)\,dy\,dx\,dz$

$\quad = \int_0^2 \int_0^{6-3y} \int_0^{(6-3y-z)/2} f(x,y,z)\,dx\,dz\,dy$

$\quad = \int_0^6 \int_0^{(6-z)/3} \int_0^{(6-3y-z)/2} f(x,y,z)\,dx\,dy\,dz$

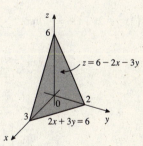

45. $J = \dfrac{\partial(x,y,z)}{\partial(u,v,w)} = \begin{vmatrix} \frac{\partial x}{\partial u} & \frac{\partial x}{\partial v} & \frac{\partial x}{\partial w} \\[4pt] \frac{\partial y}{\partial u} & \frac{\partial y}{\partial v} & \frac{\partial y}{\partial w} \\[4pt] \frac{\partial z}{\partial u} & \frac{\partial z}{\partial v} & \frac{\partial z}{\partial w} \end{vmatrix} = \begin{vmatrix} 1 & 0 & 2w \\ 4u & 1 & 0 \\ 2u & -2v & 2 \end{vmatrix} = 2 + 2w(-8uv - 2u) = 2 - 16uvw - 4uw$

47. Let $u = x - y$ and $v = x + y$. Solving for x and y in terms of u and v gives the desired

transformation $T : x = \frac{1}{2}(u + v)$, $y = \frac{1}{2}(v - u)$. $J = \dfrac{\partial(x, y)}{\partial(u, v)} = \begin{vmatrix} \dfrac{\partial x}{\partial u} & \dfrac{\partial x}{\partial v} \\ \dfrac{\partial y}{\partial u} & \dfrac{\partial y}{\partial v} \end{vmatrix} = \begin{vmatrix} \frac{1}{2} & \frac{1}{2} \\ -\frac{1}{2} & \frac{1}{2} \end{vmatrix} = \frac{1}{2}$, so

$\iint_R e^{(x-y)/(x+y)} dA = \int_0^2 \int_0^v e^{u/v} \left| \frac{1}{2} \right| du \, dv = \frac{1}{2} \int_0^2 \left[v e^{u/v} \right]_{u=0}^{u=v} dv = \frac{1}{2} \int_0^2 (e - 1) v \, dv = \frac{1}{2}(e - 1) \left(\frac{1}{2} v^2 \right) \Big|_0^2 = e - 1.$

49. True. If $R = [-2, 3] \times [0, 1]$, then $\int_0^1 \int_{-2}^3 (x + \cos xy) \, dx \, dy = \iint_R (x + \cos xy) \, dA = \int_{-2}^3 \int_0^1 (x + \cos xy) \, dy \, dx.$

51. False. Take $f(x, y) = x$ and $D = [-1, 2] \times [0, 2]$. Then

$\iint_D f(x, y) \, dA = \int_0^2 \int_{-1}^2 x \, dx \, dy = \int_0^2 \left[\frac{1}{2} x^2 \right]_{-1}^2 dy = \frac{3}{2} \int_0^2 dy = 3 > 0$, but $f(-0.5, 0) = -0.5 < 0$ and so
$f(x, y) < 0$ for at least one value of (x, y) in D.

53. True. Since $1 \leq \sqrt{x} \leq \sqrt{3} < \sqrt{4} = 2$ and $\cos^2 xy \leq 1$, we have $\sqrt{x} + \cos^2 xy < 3$, so
$\int_0^1 \int_1^3 \left(\sqrt{x} + \cos^2 xy \right) dx \, dy \leq \int_0^1 \int_1^3 3 \, dx \, dy = 3 \cdot 1 \cdot 2 = 6.$

Challenge Problems

1. a. Let $\Delta x = \dfrac{2}{m}$ and $\Delta y = \dfrac{1}{n}$. We use $\left(x_{ij}^*, y_{ij}^* \right) = \left(\dfrac{2i}{m}, \dfrac{j}{n} \right)$. The figure

shows the case where $m = 4$ and $n = 2$. Thus,

$\iint_R \left(3x^2 + 2y \right) dA = \lim_{m,n \to \infty} \sum_{i=1}^m \sum_{j=1}^n \left[3 \left(\frac{2i}{m} \right)^2 + 2 \left(\frac{j}{n} \right) \right] \left(\frac{2}{m} \right) \left(\frac{1}{n} \right)$

$= \lim_{m,n \to \infty} \sum_{i=1}^m \left[\sum_{j=1}^n \left(\frac{12i^2}{m^2} + \frac{2j}{n} \right) \right] \left(\frac{2}{mn} \right)$

$= \lim_{m,n \to \infty} \sum_{i=1}^m \left[\frac{12n}{m^2} i^2 + \frac{2n(n+1)}{2n} \right] \left(\frac{2}{mn} \right)$

$= \lim_{m,n \to \infty} \left[\frac{12n}{m^2} \cdot \frac{m(m+1)(2m+1)}{6} + \frac{2n(n+1)}{2n} \cdot m \right] \left(\frac{2}{mn} \right)$

$= \lim_{m,n \to \infty} \left[\frac{24}{m^3} \cdot \frac{m(m+1)(2m+1)}{6} + \frac{2(n+1)}{n} \right]$

$= \lim_{m,n \to \infty} \left[4 \left(1 + \frac{1}{m} \right) \left(2 + \frac{1}{m} \right) + 2 \left(1 + \frac{1}{n} \right) \right] = 8 + 2 = 10$

b. $\iint_R \left(3x^2 + 2y \right) dA = \int_0^2 \int_0^1 \left(3x^2 + 2y \right) dy \, dx = \int_0^2 \left[3x^2 y + y^2 \right]_{y=0}^{y=1} dx = \int_0^2 \left(3x^2 + 1 \right) dx = \left(x^3 + x \right) \Big|_0^2 = 10$

3. Let us denote the slope of L_1 and L_2 by m. Then their equations are
$y = mx + c$ and $y = mx + d$ respectively, and

$$A = \iint_R dA = \int_a^b \int_{mx+c}^{mx+d} 1 \, dy \, dx = \int_a^b [y]_{mx+c}^{mx+d} \, dx$$

$$= \int_a^b (d - c) \, dx = (b - a)(d - c)$$

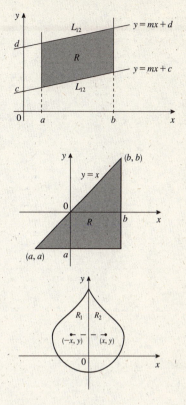

5. Answers will vary.

7. a. $\int_a^b \left[\int_a^x f(x, y) \, dy \right] dx = \iint_R f(x, y) \, dA = \int_a^b \left[\int_y^b f(x, y) \, dx \right] dy$

  b. $\int_0^1 \left[\int_y^1 \sin x^2 \, dx \right] dy = \int_0^1 \left[\int_0^x \sin x^2 \, dy \right] dx = \int_0^1 \left[y \sin x^2 \right]_{y=0}^{y=x} dx$

$$= \int_0^1 x \sin x^2 \, dx = -\tfrac{1}{2} \cos x^2 \Big|_0^1 = -\tfrac{1}{2} \cos 1 + \tfrac{1}{2}$$

$$= \tfrac{1}{2} (1 - \cos 1)$$

9. a. The region $R = R_1 \cup R_2$. Observe that because of symmetry, each point
(x, y) in R_2 corresponds to exactly one point $(-x, y)$ in R_1, so

$$\iint_{R_2} f(x, y) \, dA = \iint_{R_1} f(-x, y) \, dA$$

$$= \iint_{R_1} [-f(x, y)] \, dA \quad [\text{since } f(-x, y) = -f(x, y)]$$

$$= -\iint_{R_1} f(x, y) \, dA$$

Thus, since R_1 and R_2 do not overlap,
$$\iint_R f(x, y) \, dA = \iint_{R_1} f(x, y) \, dA + \iint_{R_2} f(x, y) \, dA = \iint_{R_1} f(x, y) \, dA - \iint_{R_1} f(x, y) \, dA = 0.$$

  b. Without loss of generality, we can place the lamina so that the line L is on the y-axis. Then R is symmetric with respect to the y-axis. Next, observe that the function $f(x, y) = \rho x$ satisfies $f(-x, y) = -\rho x = -f(x, y)$. Thus, using the result of part a, we have $M_y = \iint_R x \rho(x, y) \, dA = \rho \iint_R x \, dA = 0$, and so $\bar{x} = 0$. In other words, the centroid lies on the y-axis.

11. a. Let P be a partition of the rectangle $[a, b] \times [c, d]$ into mn subrectangles. The mass of the piece of plate occupying subrectangle R_{ij} is $m_{ij} = k \Delta A$, where $\Delta A = \Delta x \, \Delta y$ with $\Delta x = \frac{a}{m}$ and $\Delta y = \frac{b}{n}$. The kinetic energy of this piece of plate is $\Delta E_{ij} = \tfrac{1}{2} m_{ij} v_{ij}^2$, where $v_{ij} = \sqrt{x_{ij}^2 + y_{ij}^2}$ is its velocity. Summing, we see that the total energy of the plate is

$$E \approx \sum_{i=1}^m \sum_{j=1}^n \Delta E_{ij} = \sum_{i=1}^m \sum_{j=1}^n k\omega^2 \left(x_{ij}^2 + y_{ij}^2 \right) \Delta x \, \Delta y. \text{ Letting } m, n \to \infty, \text{ we see that}$$

$$E = k\omega^2 \int_0^b \int_0^a \left(x^2 + y^2 \right) dx \, dy = k\omega^2 \int_0^b \left[\tfrac{1}{3} x^3 + x y^2 \right]_0^a dy = k\omega^2 \int_0^b \left(\tfrac{1}{3} a^3 + a y^2 \right) dy = k\omega^2 \left(\tfrac{1}{3} a^3 y + \tfrac{1}{3} a y^3 \right) \Big|_0^b$$

$$= k\omega^2 \left(\tfrac{1}{3} a^3 b + \tfrac{1}{3} a b^3 \right) = \tfrac{1}{3} k\omega^2 a b \left(a^2 + b^2 \right) = \tfrac{1}{3} \left(a^2 + b^2 \right) M \omega^2 \text{ where } M = kab.$$

  b. If we let $I = \tfrac{2}{3} \left(a^2 + b^2 \right) M$, then $E = \tfrac{1}{2} I \omega^2$, where $I = \tfrac{2}{3} \left(a^2 + b^2 \right) M$.

15 Vector Analysis ET 14

15.1 Concept Questions ET 14.1

1. a. See pages 1224 and 1225 (1220 and 1221 in ET).

   b. Answers will vary.

15.1 Vector Fields ET 14.1

1. The answer is b. Note that the vectors of $\mathbf{F}(x, y) = y\mathbf{i}$ are parallel to the x-axis and their lengths increase as we move away from the x-axis.

3. The answer is c. $\mathbf{F}(x, y) = -\dfrac{y}{x^2 + y^2}\mathbf{i} + \dfrac{x}{x^2 + y^2}\mathbf{j}$ is a spin vector field similar to that of Example 2, but the lengths of its vectors decrease as we move away from the origin.

5. The answer is e. Observe that $|\mathbf{F}(x, y)| = 1$ and $\mathbf{F}(x, y) = -\dfrac{1}{\sqrt{x^2 + y^2}}\langle x, y\rangle$, showing that all the vectors of $\mathbf{F}$ have unit length and point toward the origin.

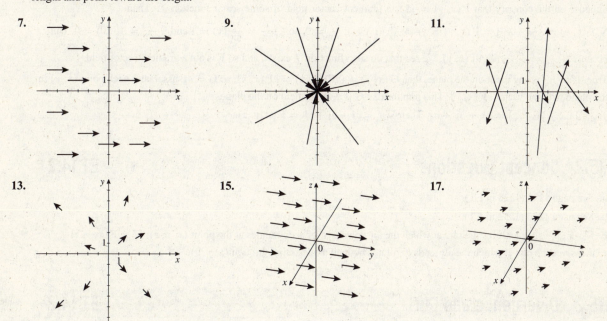

7.   **9.**   **11.**

13.   **15.**   **17.**

19. The vector field $\mathbf{F}(x, y, z) = \mathbf{i} + \mathbf{j} + 2\mathbf{k}$ consists of a family of vectors having constant length $\sqrt{6}$ and the same direction as the vector $\mathbf{i} + \mathbf{j} + 2\mathbf{k}$, so it matches plot c.

21. The vectors in the vector field $\mathbf{F}(x, y, z) = -x\mathbf{i} - y\mathbf{j} - z\mathbf{k} = -(x\mathbf{i} + y\mathbf{j} + z\mathbf{k})$ point toward the origin, and so the vector field matches plot a.

23.

25.

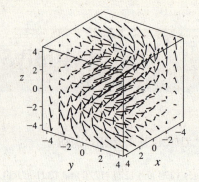

27. $\mathbf{F}(x, y) = \nabla f(x, y) = \frac{\partial}{\partial x}\left(x^2 y - y^3\right)\mathbf{i} + \frac{\partial}{\partial y}\left(x^2 y - y^3\right)\mathbf{j} = 2xy\mathbf{i} + \left(x^2 - 3y^2\right)\mathbf{j}$

29. $\mathbf{F}(x, y, z) = \nabla f(x, y, z) = \frac{\partial}{\partial x}(xyz)\mathbf{i} + \frac{\partial}{\partial y}(xyz)\mathbf{j} + \frac{\partial}{\partial z}(xyz)\mathbf{k} = yz\mathbf{i} + xz\mathbf{j} + xy\mathbf{k}$

31. $\mathbf{F}(x, y, z) = \nabla f(x, y, z) = \frac{\partial}{\partial x}\left[y\ln(x+z)\right]\mathbf{i} + \frac{\partial}{\partial y}\left[y\ln(x+z)\right]\mathbf{j} + \frac{\partial}{\partial z}\left[y\ln(x+z)\right]\mathbf{k}$

$$= \frac{y}{x+z}\mathbf{i} + \ln(x+z)\mathbf{j} + \frac{y}{x+z}\mathbf{k}$$

33. a. The velocity of the particle at $t = 2$ is given by $\mathbf{v} = \mathbf{V}(1, 3, 2) = \left(2x\mathbf{i} + (x + 3y)\mathbf{j} + z^2\mathbf{k}\right)\Big|_{(1,3,2)} = 2\mathbf{i} + 10\mathbf{j} + 4\mathbf{k}$.

   b. At $t = 2$, the position of the particle is given by $\mathbf{r}(2) = \mathbf{i} + 3\mathbf{j} + 2\mathbf{k}$. The displacement of the particle from $t = 2$ to $t = 2.01$ is approximately $\Delta\mathbf{s} \approx \mathbf{v}\Delta t = 0.02\mathbf{i} + 0.1\mathbf{j} + 0.04\mathbf{k}$, so its approximate location at $t = 2.01$ is given by $\mathbf{r}(2.01) \approx \mathbf{r}(2) + \Delta\mathbf{s} \approx (\mathbf{i} + 3\mathbf{j} + 2\mathbf{k}) + (0.02\mathbf{i} + 0.1\mathbf{j} + 0.04\mathbf{k}) = 1.02\mathbf{i} + 3.1\mathbf{j} + 2.04\mathbf{k}$. Thus, it is near the point $(1.02, 3.1, 2.04)$.

35. Suppose on the contrary that $\mathbf{F}(x, y) = y\mathbf{i}$ is a gradient vector field of some scalar function f. Then

$\nabla f(x, y) = \frac{\partial f}{\partial x}\mathbf{i} + \frac{\partial f}{\partial y}\mathbf{j} = \mathbf{F}(x, y) = y\mathbf{i} \Rightarrow \frac{\partial f}{\partial x} = y$ and $\frac{\partial f}{\partial y} = 0$, so $\frac{\partial^2 f}{\partial y\,\partial x} = \frac{\partial f}{\partial y}(y) = 1$ and $\frac{\partial^2 f}{\partial x\,\partial y} = \frac{\partial f}{\partial x}(0) = 0$. But

because $\frac{\partial^2 f}{\partial y\,\partial x} \neq \frac{\partial^2 f}{\partial x\,\partial y}$ for all (x, y), we see that no such function f exists, and so $\mathbf{F}$ is not a gradient vector field.

37. True. If (x, y) is any point on the plane, then $\mathbf{G}(x, y) = (c\mathbf{F})(x, y) = c\mathbf{F}(x, y)$, so $\mathbf{G} = c\mathbf{F}$ associates each point (x, y) in the plane with the vector $c\mathbf{F}(x, y)$. This implies that $\mathbf{G}$ is a vector field on the plane.

39. True. Take $f(x, y, z) = ax + by + cz$. Then $\nabla f = a\mathbf{i} + b\mathbf{j} + c\mathbf{k} = \mathbf{F}(x, y, z)$.

15.2 Concept Questions   ET 14.2

1. a. See page 1232 (1228 in ET).

   b. See page 1236 (1232 in ET).

   c. $\nabla \cdot \mathbf{F}(x, y, z)$ measures the rate at which the air flows past or accumulates at the point (x, y, z); $\nabla \times \mathbf{F}(x, y, z)$ measures the rate at which the air tends to rotate about an axis passing through (x, y, z).

15.2 Divergence and Curl   ET 14.2

1. a. If $P(x, y, z)$ is any point in the domain of $\mathbf{F}$, then the flow entering a neighborhood of P is matched exactly by the flow exiting the neighborhood, so div $\mathbf{F} = 0$.

   b. div $\mathbf{F} = \frac{\partial}{\partial x}\left(\frac{x}{|x|}\right) + \frac{\partial}{\partial y}(0) + \frac{\partial}{\partial z}(0) = 0$

   c. A paddle wheel placed at any point will not rotate, so curl $\mathbf{F} = \mathbf{0}$.

d. $\text{curl } \mathbf{F} = \begin{vmatrix} \mathbf{i} & \mathbf{j} & \mathbf{k} \\ \frac{\partial}{\partial x} & \frac{\partial}{\partial y} & \frac{\partial}{\partial z} \\ \frac{x}{|x|} & 0 & 0 \end{vmatrix} = 0\mathbf{i} + \frac{\partial}{\partial z}\left(\frac{x}{|x|}\right)\mathbf{j} - \frac{\partial}{\partial y}\left(\frac{x}{|x|}\right)\mathbf{k} = 0\mathbf{i} + 0\mathbf{j} - 0\mathbf{k} = \mathbf{0}$

3. a. First observe that $|\mathbf{F}(x, y, z)| = 1$, so each vector has the same (unit) length. Consider the neighborhood shown. Since there are more arrows emanating from the larger upper arc AB than entering the smaller arc, we see that the flow out of the neighborhood is greater than the flow into it. Thus, $\text{div } \mathbf{F} > 0$.

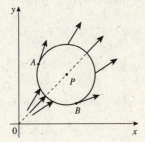

b. $\text{div } \mathbf{F} = \frac{\partial}{\partial x}\left(\frac{x}{\sqrt{x^2 + y^2}}\right) + \frac{\partial}{\partial y}\left(\frac{y}{\sqrt{x^2 + y^2}}\right) + \frac{\partial}{\partial z}(0)$

$= \frac{\left(x^2 + y^2\right)^{1/2} - x\left(\frac{1}{2}\right)\left(x^2 + y^2\right)^{-1/2}(2x)}{x^2 + y^2} + \frac{\left(x^2 + y^2\right)^{1/2} - y\left(\frac{1}{2}\right)\left(x^2 + y^2\right)^{-1/2}(2y)}{x^2 + y^2} = \frac{y^2 + x^2}{\left(x^2 + y^2\right)^{3/2}}$

$= \frac{1}{\sqrt{x^2 + y^2}} > 0$ for $(x, y) \neq (0, 0)$

c. The line passing through the origin and P (see the diagram in part a) divides the neighborhood into two half-circles. The arrows in the upper left half of the circle tend to rotate the paddle wheel in a clockwise direction, but because these are balanced by an equal number of arrows in the bottom right half that tend to rotate the paddle wheel counterclockwise, the wheel will not rotate, and so $\text{curl } \mathbf{F} = \mathbf{0}$.

d. $\text{curl } \mathbf{F} = \begin{vmatrix} \mathbf{i} & \mathbf{j} & \mathbf{k} \\ \frac{\partial}{\partial x} & \frac{\partial}{\partial y} & \frac{\partial}{\partial z} \\ \frac{x}{\sqrt{x^2+y^2}} & \frac{y}{\sqrt{x^2+y^2}} & 0 \end{vmatrix}$

$= -\frac{\partial}{\partial z}\left(\frac{y}{\sqrt{x^2 + y^2}}\right)\mathbf{i} + \frac{\partial}{\partial z}\left(\frac{x}{\sqrt{x^2 + y^2}}\right)\mathbf{j} + \left[\frac{\partial}{\partial x}\left(\frac{y}{\sqrt{x^2 + y^2}}\right) - \frac{\partial}{\partial y}\left(\frac{x}{\sqrt{x^2 + y^2}}\right)\right]\mathbf{k}$

$= 0\mathbf{i} + 0\mathbf{j} + \left[\left(-\frac{1}{2}\right)y\left(x^2 + y^2\right)^{-3/2}(2x) - \left(-\frac{1}{2}\right)x\left(x^2 + y^2\right)^{-3/2}(2y)\right]\mathbf{k} = 0\mathbf{i} + 0\mathbf{j} + 0\mathbf{k} = \mathbf{0}$

5. a. $\text{div } \mathbf{F} = \frac{\partial}{\partial x}(yz) + \frac{\partial}{\partial y}(xz) + \frac{\partial}{\partial z}(xy) = 0 + 0 + 0 = 0$

b. $\text{curl } \mathbf{F} = \begin{vmatrix} \mathbf{i} & \mathbf{j} & \mathbf{k} \\ \frac{\partial}{\partial x} & \frac{\partial}{\partial y} & \frac{\partial}{\partial z} \\ yz & xz & xy \end{vmatrix} = (x - x)\mathbf{i} - (y - y)\mathbf{j} + (z - z)\mathbf{k} = \mathbf{0}$

7. a. $\text{div } \mathbf{F} = \frac{\partial}{\partial x}\left(x^2 y^3\right) + \frac{\partial}{\partial y}(0) + \frac{\partial}{\partial z}\left(xz^2\right) = 2xy^3 + 2xz = 2x\left(y^3 + z\right)$

b. $\text{curl } \mathbf{F} = \begin{vmatrix} \mathbf{i} & \mathbf{j} & \mathbf{k} \\ \frac{\partial}{\partial x} & \frac{\partial}{\partial y} & \frac{\partial}{\partial z} \\ x^2 y^3 & 0 & xz^2 \end{vmatrix} = 0\mathbf{i} - \left(z^2 - 0\right)\mathbf{j} + \left(0 - 3x^2 y^2\right)\mathbf{k} = -z^2\mathbf{j} - 3x^2 y^2\mathbf{k}$

9. a. $\operatorname{div} \mathbf{F} = \dfrac{\partial}{\partial x}(\sin x) + \dfrac{\partial}{\partial y}(x\cos y) + \dfrac{\partial}{\partial z}(\sin z) = \cos x - x\sin y + \cos z$

b. $\operatorname{curl}\mathbf{F} = \begin{vmatrix} \mathbf{i} & \mathbf{j} & \mathbf{k} \\ \dfrac{\partial}{\partial x} & \dfrac{\partial}{\partial y} & \dfrac{\partial}{\partial z} \\ \sin x & x\cos y & \sin z \end{vmatrix} = 0\mathbf{i} - 0\mathbf{j} + (\cos y - 0)\,\mathbf{k} = \cos y\,\mathbf{k}$

11. a. $\operatorname{div}\mathbf{F} = \dfrac{\partial}{\partial x}\left(e^{-x}\cos y\right) + \dfrac{\partial}{\partial y}\left(e^{-x}\sin y\right) + \dfrac{\partial}{\partial z}(\ln z) = -e^{-x}\cos y + e^{-x}\cos y + \dfrac{1}{z} = \dfrac{1}{z}$

b. $\operatorname{curl}\mathbf{F} = \begin{vmatrix} \mathbf{i} & \mathbf{j} & \mathbf{k} \\ \dfrac{\partial}{\partial x} & \dfrac{\partial}{\partial y} & \dfrac{\partial}{\partial z} \\ e^{-x}\cos y & e^{-x}\sin y & \ln z \end{vmatrix} = 0\mathbf{i} - 0\mathbf{j} + \left(-e^{-x}\sin y + e^{-x}\sin y\right)\mathbf{k} = \mathbf{0}$

13. a. No   **b.** No   **c.** Yes, a vector field   **d.** No

15. a. Yes, a vector field   **b.** Yes, a scalar field   **c.** No   **d.** Yes, a vector field

17. $\operatorname{div}\mathbf{F} = \dfrac{\partial}{\partial x}f(y,z) + \dfrac{\partial}{\partial y}g(x,z) + \dfrac{\partial}{\partial z}h(x,y) = 0 + 0 + 0 = 0$, so the vector field $\mathbf{F}$ is incompressible.

19. Let $\mathbf{F}(x,y,z) = P_1\mathbf{i} + Q_1\mathbf{j} + R_1\mathbf{k}$ and $\mathbf{G}(x,y,z) = P_2\mathbf{i} + Q_2\mathbf{j} + R_2\mathbf{k}$. Then

$\operatorname{div}(\mathbf{F}+\mathbf{G}) = \dfrac{\partial}{\partial x}(P_1+P_2) + \dfrac{\partial}{\partial y}(Q_1+Q_2) + \dfrac{\partial}{\partial z}(R_1+R_2)$

$= \left(\dfrac{\partial P_1}{\partial x} + \dfrac{\partial P_2}{\partial x}\right) + \left(\dfrac{\partial Q_1}{\partial y} + \dfrac{\partial Q_2}{\partial y}\right) + \left(\dfrac{\partial R_1}{\partial z} + \dfrac{\partial R_2}{\partial z}\right)$

$= \left(\dfrac{\partial P_1}{\partial x} + \dfrac{\partial Q_1}{\partial y} + \dfrac{\partial R_1}{\partial z}\right) + \left(\dfrac{\partial P_2}{\partial x} + \dfrac{\partial Q_2}{\partial y} + \dfrac{\partial R_2}{\partial z}\right) = \operatorname{div}\mathbf{F} + \operatorname{div}\mathbf{G}$

21. Let $\mathbf{F}(x,y,z) = P\mathbf{i} + Q\mathbf{j} + R\mathbf{k}$. Then

$\operatorname{curl}(f\mathbf{F}) = \begin{vmatrix} \mathbf{i} & \mathbf{j} & \mathbf{k} \\ \dfrac{\partial}{\partial x} & \dfrac{\partial}{\partial y} & \dfrac{\partial}{\partial z} \\ fP & fQ & fR \end{vmatrix} = \left[\dfrac{\partial}{\partial y}(fR) - \dfrac{\partial}{\partial z}(fQ)\right]\mathbf{i} - \left[\dfrac{\partial}{\partial x}(fR) - \dfrac{\partial}{\partial z}(fP)\right]\mathbf{j} + \left[\dfrac{\partial}{\partial x}(fQ) - \dfrac{\partial}{\partial y}(fP)\right]\mathbf{k}$

$= \left(f\dfrac{\partial R}{\partial y} + R\dfrac{\partial f}{\partial y} - f\dfrac{\partial Q}{\partial z} - Q\dfrac{\partial f}{\partial z}\right)\mathbf{i} - \left(f\dfrac{\partial R}{\partial x} + R\dfrac{\partial f}{\partial x} - f\dfrac{\partial P}{\partial z} - P\dfrac{\partial f}{\partial z}\right)\mathbf{j} + \left(f\dfrac{\partial Q}{\partial x} + Q\dfrac{\partial f}{\partial x} - f\dfrac{\partial P}{\partial y} - P\dfrac{\partial f}{\partial y}\right)\mathbf{k}$

$= f\left[\left(\dfrac{\partial R}{\partial y} - \dfrac{\partial Q}{\partial z}\right)\mathbf{i} - \left(\dfrac{\partial R}{\partial x} - \dfrac{\partial P}{\partial z}\right)\mathbf{j} + \left(\dfrac{\partial Q}{\partial x} - \dfrac{\partial P}{\partial y}\right)\mathbf{k}\right] + \left[\left(R\dfrac{\partial f}{\partial y} - Q\dfrac{\partial f}{\partial z}\right)\mathbf{i} - \left(R\dfrac{\partial f}{\partial x} - P\dfrac{\partial f}{\partial z}\right)\mathbf{j} + \left(Q\dfrac{\partial f}{\partial x} - P\dfrac{\partial f}{\partial y}\right)\mathbf{k}\right]$

$= f\begin{vmatrix} \mathbf{i} & \mathbf{j} & \mathbf{k} \\ \dfrac{\partial}{\partial x} & \dfrac{\partial}{\partial y} & \dfrac{\partial}{\partial z} \\ P & Q & R \end{vmatrix} + \begin{vmatrix} \mathbf{i} & \mathbf{j} & \mathbf{k} \\ \dfrac{\partial f}{\partial x} & \dfrac{\partial f}{\partial y} & \dfrac{\partial f}{\partial z} \\ P & Q & R \end{vmatrix} = f\operatorname{curl}\mathbf{F} + (\nabla f)\times\mathbf{F}$

23. Let $\mathbf{F}(x,y,z) = P_1\mathbf{i} + Q_1\mathbf{j} + R_1\mathbf{k}$ and $\mathbf{G}(x,y,z) = P_2\mathbf{i} + Q_2\mathbf{j} + R_2\mathbf{k}$. Then

$\operatorname{div}(\mathbf{F}\times\mathbf{G}) = \nabla\cdot(\mathbf{F}\times\mathbf{G}) = \begin{vmatrix} \dfrac{\partial}{\partial x} & \dfrac{\partial}{\partial y} & \dfrac{\partial}{\partial z} \\ P_1 & Q_1 & R_1 \\ P_2 & Q_2 & R_2 \end{vmatrix} = \dfrac{\partial}{\partial x}(Q_1R_2 - R_1Q_2) - \dfrac{\partial}{\partial y}(P_1R_2 - R_1P_2) + \dfrac{\partial}{\partial z}(P_1Q_2 - Q_1P_2)$

$= \left(R_2\dfrac{\partial Q_1}{\partial x} + Q_1\dfrac{\partial R_2}{\partial x}\right) - \left(R_1\dfrac{\partial Q_2}{\partial x} + Q_2\dfrac{\partial R_1}{\partial x}\right) - \left(R_2\dfrac{\partial P_1}{\partial y} + P_1\dfrac{\partial R_2}{\partial y}\right)$

$\qquad + \left(P_2\dfrac{\partial R_1}{\partial y} - R_1\dfrac{\partial P_2}{\partial y}\right) + \left(Q_2\dfrac{\partial P_1}{\partial z} + P_1\dfrac{\partial R_2}{\partial z}\right) - \left(P_2\dfrac{\partial Q_1}{\partial z} + Q_1\dfrac{\partial P_2}{\partial z}\right)$

$= \left[P_2\left(\dfrac{\partial R_1}{\partial y} - \dfrac{\partial Q_1}{\partial z}\right) + Q_2\left(\dfrac{\partial P_1}{\partial z} - \dfrac{\partial R_1}{\partial x}\right) + R_2\left(\dfrac{\partial Q_1}{\partial x} - \dfrac{\partial P_1}{\partial y}\right)\right]$

$\qquad - \left[P_1\left(\dfrac{\partial R_2}{\partial y} - \dfrac{\partial Q_2}{\partial z}\right) + Q_1\left(\dfrac{\partial P_2}{\partial z} - \dfrac{\partial R_2}{\partial x}\right) + R_1\left(\dfrac{\partial Q_2}{\partial x} - \dfrac{\partial P_2}{\partial y}\right)\right]$

$= \mathbf{G}\cdot\operatorname{curl}\mathbf{F} - \mathbf{F}\cdot\operatorname{curl}\mathbf{G}$

25. Let $\mathbf{F} = P\mathbf{i} + Q\mathbf{j} + R\mathbf{k}$. Then $\nabla \times \mathbf{F} = \left(\frac{\partial R}{\partial y} - \frac{\partial Q}{\partial z}\right)\mathbf{i} - \left(\frac{\partial R}{\partial x} - \frac{\partial P}{\partial z}\right)\mathbf{j} + \left(\frac{\partial Q}{\partial x} - \frac{\partial P}{\partial y}\right)\mathbf{k}$, so

$$\nabla \times (\nabla \times \mathbf{F}) = \begin{vmatrix} \mathbf{i} & \mathbf{j} & \mathbf{k} \\ \frac{\partial}{\partial x} & \frac{\partial}{\partial y} & \frac{\partial}{\partial z} \\ \frac{\partial R}{\partial y} - \frac{\partial Q}{\partial z} & \frac{\partial P}{\partial z} - \frac{\partial R}{\partial x} & \frac{\partial Q}{\partial x} - \frac{\partial P}{\partial y} \end{vmatrix}$$

$$= \left(\frac{\partial^2 Q}{\partial y\,\partial x} - \frac{\partial^2 P}{\partial y^2} - \frac{\partial^2 P}{\partial z^2} + \frac{\partial^2 R}{\partial z\,\partial x}\right)\mathbf{i} - \left(\frac{\partial^2 Q}{\partial x^2} - \frac{\partial^2 P}{\partial x\,\partial y} - \frac{\partial^2 R}{\partial z\,\partial y} + \frac{\partial^2 Q}{\partial z^2}\right)\mathbf{j} + \left(\frac{\partial^2 P}{\partial x\,\partial z} - \frac{\partial^2 R}{\partial x^2} - \frac{\partial^2 R}{\partial y^2} + \frac{\partial^2 Q}{\partial y\,\partial z}\right)\mathbf{k}$$

On the other hand,

$$\nabla(\nabla \cdot \mathbf{F}) - \nabla^2 \mathbf{F} = \nabla\left(\frac{\partial P}{\partial x} + \frac{\partial Q}{\partial y} + \frac{\partial R}{\partial z}\right) - \left(\frac{\partial^2}{\partial x^2} + \frac{\partial^2}{\partial y^2} + \frac{\partial^2}{\partial z^2}\right)(P\mathbf{i} + Q\mathbf{j} + R\mathbf{k})$$

$$= \left[\left(\frac{\partial^2 P}{\partial x^2} + \frac{\partial^2 Q}{\partial x\,\partial y} + \frac{\partial^2 R}{\partial x\,\partial z}\right)\mathbf{i} + \left(\frac{\partial^2 P}{\partial y\,\partial x} + \frac{\partial^2 Q}{\partial y^2} + \frac{\partial^2 R}{\partial y\,\partial z}\right)\mathbf{j} + \left(\frac{\partial^2 P}{\partial z\,\partial x} + \frac{\partial^2 Q}{\partial z\,\partial y} + \frac{\partial^2 R}{\partial z^2}\right)\mathbf{k}\right]$$

$$\quad - \left[\left(\frac{\partial^2 P}{\partial x^2} + \frac{\partial^2 P}{\partial y^2} + \frac{\partial^2 P}{\partial z^2}\right)\mathbf{i} + \left(\frac{\partial^2 Q}{\partial x^2} + \frac{\partial^2 Q}{\partial y^2} + \frac{\partial^2 Q}{\partial z^2}\right)\mathbf{j} + \left(\frac{\partial^2 R}{\partial x^2} + \frac{\partial^2 R}{\partial y^2} + \frac{\partial^2 R}{\partial z^2}\right)\mathbf{k}\right]$$

$$= \left(\frac{\partial^2 Q}{\partial x\,\partial y} + \frac{\partial^2 R}{\partial x\,\partial z} - \frac{\partial^2 P}{\partial y^2} - \frac{\partial^2 P}{\partial z^2}\right)\mathbf{i} + \left(\frac{\partial^2 P}{\partial y\,\partial x} + \frac{\partial^2 R}{\partial y\,\partial z} - \frac{\partial^2 Q}{\partial x^2} - \frac{\partial^2 Q}{\partial z^2}\right)\mathbf{j} + \left(\frac{\partial^2 P}{\partial z\,\partial x} + \frac{\partial^2 Q}{\partial z\,\partial y} - \frac{\partial^2 R}{\partial x^2} - \frac{\partial^2 R}{\partial y^2}\right)\mathbf{k}$$

Since the second partial derivatives of P, Q, and R are continuous, the mixed derivatives are equal and so
$\nabla \times (\nabla \times \mathbf{F}) = \nabla(\nabla \cdot \mathbf{F}) - \nabla^2 \mathbf{F}$.

27. Suppose on the contrary that there is a vector field $\mathbf{F}$ such that $\operatorname{curl} \mathbf{F} = xy\mathbf{i} - yz\mathbf{j} + xy\mathbf{k}$. Then
$\operatorname{div}(\operatorname{curl} \mathbf{F}) = \frac{\partial}{\partial x}(xy) + \frac{\partial}{\partial y}(-yz) + \frac{\partial}{\partial z}(xy) = y - z \neq 0$, but this contradicts the result of Example 9. Therefore, there is
no vector field with the given property.

29. Suppose on the contrary that $\mathbf{F}$ is a gradient vector field. Then $\mathbf{F} = \nabla f$ for some scalar field. Also, using the result of
Exercise 22, we must have $\operatorname{curl} \mathbf{F} = \operatorname{curl} \nabla f = \mathbf{0}$. But

$$\operatorname{curl} \mathbf{F} = \begin{vmatrix} \mathbf{i} & \mathbf{j} & \mathbf{k} \\ \frac{\partial}{\partial x} & \frac{\partial}{\partial y} & \frac{\partial}{\partial z} \\ y\cos x & x\sin y & 0 \end{vmatrix} = -\frac{\partial}{\partial z}(x\sin y)\mathbf{i} - \left[-\frac{\partial}{\partial z}(y\cos x)\right]\mathbf{j} + \left[\frac{\partial}{\partial x}(x\sin y) - \frac{\partial}{\partial y}(y\cos x)\right]\mathbf{k}$$

$$= 0\mathbf{i} + 0\mathbf{j} + (\sin y - \cos x)\mathbf{k} = (\sin y - \cos x)\mathbf{k} \neq \mathbf{0},$$

and this contradiction establishes the result.

31. $r = |\mathbf{r}| = \sqrt{x^2 + y^2 + z^2}$, so

$$\nabla r = \frac{\partial}{\partial x}\left(x^2 + y^2 + z^2\right)^{1/2}\mathbf{i} + \frac{\partial}{\partial y}\left(x^2 + y^2 + z^2\right)^{1/2}\mathbf{j} + \frac{\partial}{\partial z}\left(x^2 + y^2 + z^2\right)^{1/2}\mathbf{k}$$

$$= \frac{x}{\sqrt{x^2 + y^2 + z^2}}\mathbf{i} + \frac{y}{\sqrt{x^2 + y^2 + z^2}}\mathbf{j} + \frac{z}{\sqrt{x^2 + y^2 + z^2}}\mathbf{k} = \frac{\mathbf{r}}{|\mathbf{r}|} = \frac{\mathbf{r}}{r}$$

33. $r = |\mathbf{r}| = \sqrt{x^2 + y^2 + z^2} \Rightarrow \ln r = \ln\left(x^2 + y^2 + z^2\right)^{1/2} = \frac{1}{2}\ln\left(x^2 + y^2 + z^2\right)$, so

$$\nabla(\ln r) = \frac{\partial}{\partial x}\left[\frac{1}{2}\ln\left(x^2 + y^2 + z^2\right)\right]\mathbf{i} + \frac{\partial}{\partial y}\left[\frac{1}{2}\ln\left(x^2 + y^2 + z^2\right)\right]\mathbf{j} + \frac{\partial}{\partial z}\left[\frac{1}{2}\ln\left(x^2 + y^2 + z^2\right)\right]\mathbf{k}$$

$$= \frac{1}{2}\cdot\frac{2x}{x^2 + y^2 + z^2}\mathbf{i} + \frac{1}{2}\cdot\frac{2y}{x^2 + y^2 + z^2}\mathbf{j} + \frac{1}{2}\cdot\frac{2z}{x^2 + y^2 + z^2}\mathbf{k} = \frac{1}{x^2 + y^2 + z^2}(x\mathbf{i} + y\mathbf{j} + z\mathbf{k}) = \frac{\mathbf{r}}{r^2}$$

35. $\nabla f = \frac{\partial f}{\partial x}\mathbf{i} + \frac{\partial f}{\partial y}\mathbf{j} + \frac{\partial f}{\partial z}\mathbf{k}$, so $\nabla \cdot (\nabla f) = \frac{\partial}{\partial x}\left(\frac{\partial f}{\partial x}\right) + \frac{\partial}{\partial y}\left(\frac{\partial f}{\partial y}\right) + \frac{\partial}{\partial z}\left(\frac{\partial f}{\partial z}\right) = \frac{\partial^2 f}{\partial x^2} + \frac{\partial^2 f}{\partial y^2} + \frac{\partial^2 f}{\partial z^2} = \nabla^2 f$.

37. $r = |\mathbf{r}| = \sqrt{x^2 + y^2 + z^2}$, so

$$\nabla^2 r^3 = \nabla^2 \left(x^2 + y^2 + z^2\right)^{3/2} = \frac{\partial^2}{\partial x^2}\left(x^2 + y^2 + z^2\right)^{3/2} + \frac{\partial^2}{\partial y^2}\left(x^2 + y^2 + z^2\right)^{3/2} + \frac{\partial^2}{\partial z^2}\left(x^2 + y^2 + z^2\right)^{3/2}$$

$$= \frac{\partial}{\partial x}\left[\frac{\partial}{\partial x}\left(x^2 + y^2 + z^2\right)^{3/2}\right] + \frac{\partial}{\partial y}\left[\frac{\partial}{\partial y}\left(x^2 + y^2 + z^2\right)^{3/2}\right] + \frac{\partial}{\partial z}\left[\frac{\partial}{\partial z}\left(x^2 + y^2 + z^2\right)^{3/2}\right]$$

$$= \frac{\partial}{\partial x}\left[3x\left(x^2 + y^2 + z^2\right)^{1/2}\right] + \frac{\partial}{\partial y}\left[3y\left(x^2 + y^2 + z^2\right)^{1/2}\right] + \frac{\partial}{\partial z}\left[3z\left(x^2 + y^2 + z^2\right)^{1/2}\right]$$

$$= 3\left[\left(x^2 + y^2 + z^2\right)^{1/2} + \frac{x^2}{\left(x^2 + y^2 + z^2\right)^{1/2}}\right] + 3\left[\left(x^2 + y^2 + z^2\right)^{1/2} + \frac{y^2}{\left(x^2 + y^2 + z^2\right)^{1/2}}\right]$$

$$+ 3\left[\left(x^2 + y^2 + z^2\right)^{1/2} + \frac{z^2}{\left(x^2 + y^2 + z^2\right)^{1/2}}\right]$$

$$= 3\left[\frac{2x^2 + y^2 + z^2}{\sqrt{x^2 + y^2 + z^2}}\right] + 3\left[\frac{x^2 + 2y^2 + z^2}{\sqrt{x^2 + y^2 + z^2}}\right] + 3\left[\frac{x^2 + y^2 + 2z^2}{\sqrt{x^2 + y^2 + z^2}}\right] = \frac{12\left(x^2 + y^2 + z^2\right)}{\sqrt{x^2 + y^2 + z^2}}$$

$$= 12\sqrt{x^2 + y^2 + z^2} = 12r$$

39. a. We have $v = \omega R$, so $|\mathbf{v}| = |\mathbf{w}|\, R$. However, $R = |\mathbf{r}| \sin\theta$, so $|\mathbf{v}| = |\mathbf{w}|\,|\mathbf{r}| \sin\theta = |\mathbf{w} \times \mathbf{r}|$, and thus we have $\mathbf{v} = \mathbf{w} \times \mathbf{r}$.

b. $\mathbf{w} = \omega\mathbf{k}$ and $\mathbf{r} = R\cos\omega t\,\mathbf{i} + R\sin\omega t\,\mathbf{j} + h\mathbf{k}$, so

$$\mathbf{v} = \mathbf{w} \times \mathbf{r} = \begin{vmatrix} \mathbf{i} & \mathbf{j} & \mathbf{k} \\ 0 & 0 & \omega \\ R\cos\omega t & R\sin\omega t & h \end{vmatrix} = -R\omega\sin\omega t\,\mathbf{i} + R\omega\cos\omega t\,\mathbf{j} = -\omega y\mathbf{i} + \omega x\mathbf{j}.$$

c. $\operatorname{curl}\mathbf{v} = \begin{vmatrix} \mathbf{i} & \mathbf{j} & \mathbf{k} \\ \frac{\partial}{\partial x} & \frac{\partial}{\partial y} & \frac{\partial}{\partial z} \\ -\omega y & \omega x & 0 \end{vmatrix} = 0\mathbf{i} + 0\mathbf{j} + (\omega + \omega)\,\mathbf{k} = 2\omega\mathbf{k} = 2\mathbf{w}$

41. False. Take $\mathbf{F} = y\mathbf{i} + z\mathbf{j} - x\mathbf{k}$. Then $\mathbf{F}$ is nonconstant, but $\operatorname{div}\mathbf{F} = \frac{\partial}{\partial x}\,(y) + \frac{\partial}{\partial y}\,(z) + \frac{\partial}{\partial z}\,(-x) = 0$.

43. False. See Example 2, parts a and c.

45. False. Take $\mathbf{F}(x, y, z) = y\mathbf{i}$. Then the streamlines of $\mathbf{F}$ are straight lines, but $\operatorname{curl}\mathbf{F} = \begin{vmatrix} \mathbf{i} & \mathbf{j} & \mathbf{k} \\ \frac{\partial}{\partial x} & \frac{\partial}{\partial y} & \frac{\partial}{\partial z} \\ y & 0 & 0 \end{vmatrix} = 0\mathbf{i} + 0\mathbf{j} - \mathbf{k} \neq \mathbf{0}$.

47. False. Take $\mathbf{F}(x, y, z) = \mathbf{i} + 2\mathbf{j} + 3\mathbf{k}$.

15.3 Concept Questions   ET 14.3

1. a. See page 1242 and 1243 (1238 and 1239 in ET).

   b. See page 1243 (1239 in ET).

3. a. See page 1250 (1246 in ET).

   b. It represents the work done by the force $\mathbf{F}$ in moving a particle along the curve $C : \mathbf{r}(t)$ from $t = a$ to $t = b$.

15.3   Line Integrals   ET 14.3

1. $\mathbf{r}(t) = 3t\mathbf{i} + 4t\mathbf{j} \Rightarrow \mathbf{r}'(t) = 3\mathbf{i} + 4\mathbf{j} \Rightarrow |\mathbf{r}'(t)| = \sqrt{3^2 + 4^2} = 5$, so $\int_C (x+y)\,ds = \int_0^1 (3t + 4t)(5)\,dt = 35\int_0^1 t\,dt = \frac{35}{2}$.

3. Here $x(t) = 2t$ and $y(t) = t^3$, so $|\mathbf{r}'(t)| = \sqrt{2^2 + (3t^2)^2} = \sqrt{9t^4 + 4}$ and

$$\int_C y\,ds = \int_0^1 t^3 \sqrt{9t^4 + 4}\,dt = \frac{1}{36}\left(\frac{2}{3}\right)\left(9t^4 + 4\right)^{3/2}\Big|_0^1 = \frac{13\sqrt{13} - 8}{54}.$$

5. We use the parametrization $C : x(t) = 2\cos t,\, y = 2\sin t,\, 0 \le t \le \pi$, so

$x'(t) = -2\sin t,\, y'(t) = 2\cos t$, and $|\mathbf{r}'(t)| = \sqrt{(-2\sin t)^2 + (2\cos t)^2} = 2$.

Thus,

$$\int_C \left(xy^2 + yx^2\right) ds = \int_0^\pi \left[(2\cos t)(2\sin t)^2 + (2\sin t)(2\cos t)^2\right](2\,dt)$$

$$= 16\int_0^\pi \left(\sin^2 t \cos t + \cos^2 t \sin t\right) dt = 16\left(\frac{1}{3}\sin^3 t - \frac{1}{3}\cos^3 t\right)\Big|_0^\pi = \frac{32}{3}$$

7. We use the parametrization $C : r(t) = (-2 + 3t)\mathbf{i} + (-1 + 4t)\mathbf{j},\, 0 \le t \le 1$, so $|\mathbf{r}'(t)| = \sqrt{3^2 + 4^2} = 5$. Thus,

$$\int_C 2xy\,ds = 2\int_0^1 (-2 + 3t)(-1 + 4t)(5)\,dt = 10\int_0^1 \left(12t^2 - 11t + 2\right) dt = 10\left(4t^3 - \frac{11}{2}t^2 + 2t\right)\Big|_0^1 = 5.$$

9. Here $C : \mathbf{r}(t) = (-1 + 2t)\mathbf{i} + (1 + 3t)\mathbf{j},\, 0 \le t \le 1$, so $x(t) = -1 + 2t,\, y = 1 + 3t$, and $dx = 2\,dt$. Thus,

$$\int_C \left(x + 3y^2\right) dx = \int_0^1 \left[(-1 + 2t) + 3(1 + 3t)^2\right](2)\,dt = 2\int_0^1 \left(27t^2 + 20t + 2\right) dt = 2\left(9t^3 + 10t^2 + 2t\right)\Big|_0^1 = 42.$$

11. We have $C_1 : x(t) = 1 + 2t,\, y(t) = 2 + 2t,\, 0 \le t \le 1 \Rightarrow x'(t) = 2$ and

$y'(t) = 2$; and $C_2 : x(t) = 3 + t,\, y(t) = 4 - 4t,\, 0 \le t \le 1 \Rightarrow x'(t) = 1$ and

$y'(t) = -4$. Thus,

$\int_C xy\,dx + (x + y)\,dy$

$$= \int_{C_1} xy\,dx + (x + y)\,dy + \int_{C_2} xy\,dx + (x + y)\,dy$$

$$= \int_0^1 \{(1 + 2t)(2 + 2t)(2) + [(1 + 2t) + (2 + 2t)](2)\}\,dt$$

$$\quad + \int_0^1 \{(3 + t)(4 - 4t)(1) + [(3 + t) + (4 - 4t)](-4)\}\,dt$$

$$= \int_0^1 \left(4t^2 + 24t - 6\right) dt = \left(\frac{4}{3}t^3 + 12t^2 - 6t\right)\Big|_0^1 = \frac{22}{3}$$

13. We parametrize $C_1 : x(t) = t,\, y(t) = 4 - t^2,\, -2 \le t \le 0 \Rightarrow$

$x'(t) = 1,\, y'(t) = -2t$ and $C_2 : x(t) = 2t,\, y(t) = 4 - 4t,\, 0 \le t \le 1 \Rightarrow$

$x'(t) = 2,\, y'(t) = -4$. Thus,

$\int_C y\,dx + x\,dy = \int_{C_1} y\,dx + x\,dy + \int_{C_2} y\,dx + x\,dy$

$$= \int_{-2}^0 \left[\left(4 - t^2\right)(1) + t(-2t)\right] dt + \int_0^1 [(4 - 4t)(2) + (2t)(-4)]\,dt$$

$$= \int_{-2}^0 \left(4 - 3t^2\right) dt + 8\int_0^1 (1 - 2t)\,dt = \left(4t - t^3\right)\Big|_{-2}^0 + 8\left(t - t^2\right)\Big|_0^1$$

$$= 0 + 0 = 0$$

15. $\mathbf{r}(t) = (1 + t)\mathbf{i} + 2t\mathbf{j} + (1 - t)\mathbf{k} \Rightarrow \mathbf{r}'(t) = \mathbf{i} + 2\mathbf{j} - \mathbf{k}$, so $|\mathbf{r}'(t)| = \sqrt{1^2 + 2^2 + (-1)^2} = \sqrt{6}$. Thus,

$$\int_C xyz\,ds = \int_0^1 (1 + t)(2t)(1 - t)\sqrt{6}\,dt = 2\sqrt{6}\int_0^1 \left(t - t^3\right) dt = 2\sqrt{6}\left(\frac{1}{2}t^2 - \frac{1}{4}t^4\right)\Big|_0^1 = \frac{\sqrt{6}}{2}.$$

17. $\mathbf{r}(t) = \cos 2t\,\mathbf{i} + \sin 2t\,\mathbf{j} + 3t\mathbf{k} \Rightarrow \mathbf{r}'(t) = -2\sin 2t\,\mathbf{i} + 2\cos 2t\,\mathbf{j} + 3\mathbf{k}$, so $|\mathbf{r}'(t)| = \sqrt{(-2\sin 2t)^2 + (2\cos 2t)^2 + 3^2} = \sqrt{13}$.

Thus, $\int_C xy^2\,ds = \int_0^{\pi/2} (\cos 2t)(\sin 2t)^2 \sqrt{13}\,dt = \sqrt{13} \int_0^{\pi/2} \sin^2 2t \cos 2t\,dt = \sqrt{13} \left(\frac{1}{2}\right)\left(\frac{1}{3}\right) \sin^3 2t \Big|_0^{\pi/2} = 0$.

19. $\mathbf{r}(t) = e^t\mathbf{i} + e^{-t}\mathbf{j} + 2e^{2t}\mathbf{k} \Rightarrow x(t) = e^t$, $y(t) = e^{-t}$, and $z(t) = 2e^{2t}$. Thus, $x'(t) = e^t$, $y'(t) = -e^{-t}$, and $z'(t) = 4e^{2t}$ and

$$\int_C (x+y)\,dx + xy\,dy + y\,dz = \int_0^1 \left[(e^t + e^{-t})e^t + e^t(e^{-t})(-e^{-t}) + e^{-t}\left(4e^{2t}\right)\right] dt = \int_0^1 \left(e^{2t} + 1 - e^{-t} + 4e^t\right) dt$$

$$= \left(\tfrac{1}{2}e^{2t} + t + e^{-t} + 4e^t\right)\Big|_0^1 = \tfrac{1}{2}e^2 + 4e + \tfrac{1}{e} - \tfrac{9}{2}$$

21. $C = C_1 \cup C_2$ where $C_1 : x(t) = t, y(t) = t, z(t) = 0, 0 \le t \le 1 \Rightarrow$
$x'(t) = 1, y'(t) = 1$, and $z'(t) = 0$; and
$C_2 : x(t) = 1 + t, y(t) = 1 + 2t, z(t) = 5t, 0 \le t \le 1 \Rightarrow x'(t) = 1$,
$y'(t) = 2$, and $z'(t) = 5$. Thus,

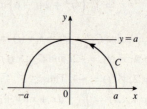

$$\int_C xy\,dx - yz\,dy + x^2\,dz = \int_{C_1} xy\,dx - yz\,dy + x^2\,dz$$

$$+ \int_{C_2} xy\,dx - yz\,dy + x^2\,dz$$

$$= \int_0^1 \left[(t)(t) - (t)(0) + \left(t^2\right)(0)\right] dt$$

$$+ \int_0^1 \left[(1+t)(1+2t) - (1+2t)(5t)(2) + (1+t)^2(5)\right] dt$$

$$= \int_0^1 \left(-12t^2 + 3t + 6\right) dt = \left(-4t^3 + \tfrac{3}{2}t^2 + 6t\right)\Big|_0^1 = \tfrac{7}{2}$$

23. Place the curve within a coordinate system as shown. Then

$C : x(t) = a\cos t, y(t) = a\sin t, 0 \le t \le \pi \Rightarrow x'(t) = -a\sin t$ and
$y'(t) = a\cos t$. The mass of the wire is $m = \pi ka$, and by symmetry,
$\bar{x} = \frac{M_y}{m} = 0$. Next,

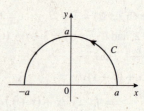

$$\bar{y} = \frac{M_x}{m} = \frac{1}{\pi ka}\int_C ky\,ds = \frac{1}{\pi a}\int_0^\pi (a\sin t)\sqrt{(-a\sin t)^2 + (a\cos t)^2}\,dt$$

$$= \frac{a^2}{\pi a}\int_0^\pi \sin t\,dt = \frac{a}{\pi}\left[-\cos t\right]_0^\pi = \frac{2a}{\pi}$$

Thus, the mass of the wire is $m = \pi ka$ and its center of mass is $\left(0, \frac{2a}{\pi}\right)$.

25. Here $C : x(t) = a\cos t, y(t) = a\sin t, 0 \le t \le \pi \Rightarrow$
$x'(t) = -a\sin t, y'(t) = a\cos t$. Also, $\rho(x,y) = k(a - y)$, so

$$m = \int_C \rho\,ds = k\int_0^\pi (a - a\sin t)\sqrt{(-a\sin t)^2 + (a\cos t)^2}\,dt$$

$$= ka^2 \int_0^\pi (1 - \sin t)\,dt = ka^2 (t + \cos t)\Big|_0^\pi = ka^2 (\pi - 2)$$

By symmetry, $\bar{x} = \frac{M_y}{m} = 0$. Next,

$$\bar{y} = \frac{M_x}{m} = \frac{1}{ka^2(\pi - 2)}\int_C \rho y\,ds = \frac{1}{ka^2(\pi - 2)}\int_0^\pi k(a - a\sin t)(a\sin t)\sqrt{a^2}\,dt$$

$$= \frac{ka^3}{ka^2(\pi - 2)}\int_0^\pi \left(\sin t - \sin^2 t\right) dt = \frac{a}{\pi - 2}\int_0^\pi \left(\sin t - \frac{1 - \cos 2t}{2}\right) dt$$

$$= \frac{a}{\pi - 2}\left(-\cos t - \frac{t}{2} + \frac{\sin 2t}{4}\right)\Big|_0^\pi = \frac{a(4 - \pi)}{2(\pi - 2)}$$

Thus, the center of mass of the wire is $\left(0, \frac{a(4 - \pi)}{2(\pi - 2)}\right)$.

27. $x = \cos^3 t$ and $y = \sin^3 t \Rightarrow x' = -3\cos^2 t \sin t$ and $y' = 3\sin^2 t \cos t$, so

$$m = \int_C k\,ds = k\int_0^{\pi/2} \sqrt{\left(-3\cos^2 t \sin t\right)^2 + \left(3\sin^2 t \cos t\right)^2}\,dt$$

$$= 3k\int_0^{\pi/2}\sqrt{\cos^2 t \sin^2 t}\,dt = 3k\int_0^{\pi/2}\cos t \sin t\,dt$$

$$= \tfrac{3}{2}k\sin^2 t \Big|_0^{\pi/2} = \tfrac{3}{2}k,$$

$$\overline{x} = \frac{M_y}{m} = \frac{2}{3k}\int_C \rho x\,ds = \frac{2}{3k}\int_0^{\pi/2} k\left(\cos^3 t\right) 3\cos t \sin t\,dt$$

$$= 2\int_0^{\pi/2}\sin t \cos^4 t\,dt = -\tfrac{2}{5}\cos^5 t\Big|_0^{\pi/2} = \tfrac{2}{5}$$

By symmetry, $\overline{y} = \frac{M_x}{m} = \frac{2}{5}$, so the center of mass is $\left(\frac{2}{5}, \frac{2}{5}\right)$.

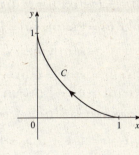

29. a. In being moved along the upper semicircle from $(-2, 0)$ to $(2, 0)$, the particle is moving *against* the force field, and so we expect the work done by the force on the particle to be *negative*.

b. The path of the particle is the semicircle parametrized by

$C : x(t) = -2\cos t, \ y(t) = 2\sin t, \ 0 \le t \le \pi$. We find $x'(t) = 2\sin t$,

$y'(t) = 2\cos t$, and

$\mathbf{F} = (x - y)\mathbf{i} + (x + y)\mathbf{j} = (-2\cos t - 2\sin t)\mathbf{i} + (-2\cos t + 2\sin t)\mathbf{j}$,

so $W = \int_C \mathbf{F}\cdot d\mathbf{r} = \int_0^\pi \mathbf{F}\cdot \mathbf{r}'(t)\,dt$, where $\mathbf{r}(t) = -2\cos t\,\mathbf{i} + 2\sin t\,\mathbf{j}$.

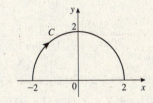

Continuing,

$$W = \int_0^\pi \left[-2(\cos t + \sin t)\mathbf{i} - 2(\cos t - \sin t)\mathbf{j}\right]\cdot(2\sin t\,\mathbf{i} + 2\cos t\,\mathbf{j})\,dt = -4\int_0^\pi\left(\sin^2 t + \cos^2 t\right)dt = -4\int_0^\pi dt$$

$$= -4\pi$$

31. $\mathbf{r}(t) = t^2\mathbf{i} + t^3\mathbf{j} \Rightarrow \mathbf{r}'(t) = 2t\mathbf{i} + 3t^2\mathbf{j}$ and $\mathbf{F}(t) = \left[\left(t^2\right)^2 + \left(t^3\right)^2\right]\mathbf{i} + \left(t^2\right)\left(t^3\right)\mathbf{j} = \left(t^4 + t^6\right)\mathbf{i} + t^5\mathbf{j}$, so

$$W = \int_C \mathbf{F}\cdot d\mathbf{r} = \int_C \mathbf{F}\cdot\mathbf{r}'(t)\,dt = \int_0^1\left[\left(t^4 + t^6\right)\mathbf{i} + t^5\mathbf{j}\right]\cdot\left(2t\mathbf{i} + 3t^2\mathbf{j}\right)dt = \int_0^1\left(2t^5 + 5t^7\right)dt = \left(\tfrac{1}{3}t^6 + \tfrac{5}{8}t^8\right)\Big|_0^1 = \tfrac{23}{24}.$$

33. A parametric representation of C is $\mathbf{r}(t) = t\mathbf{i} + t^2\mathbf{j}$, $-1 \le t \le 2 \Rightarrow \mathbf{r}'(t) = \mathbf{i} + 2t\mathbf{j}$. Also, $\mathbf{F} = te^{t^2}\mathbf{i} + t^2\mathbf{j}$, so

$$W = \int_C \mathbf{F}\cdot d\mathbf{r} = \int_C \mathbf{F}\cdot\mathbf{r}'(t)\,dt = \int_{-1}^2\left(te^{t^2}\mathbf{i} + t^2\mathbf{j}\right)\cdot(\mathbf{i} + 2t\mathbf{j})\,dt = \int_{-1}^2\left(te^{t^2} + 2t^3\right)dt = \left(\tfrac{1}{2}e^{t^2} + \tfrac{1}{2}t^4\right)\Big|_{-1}^2$$

$$= \tfrac{1}{2}\left(e^4 - e + 15\right)$$

35. $\mathbf{r}(t) = t\mathbf{i} + t^2\mathbf{j} + t^3\mathbf{k} \Rightarrow \mathbf{r}'(t) = \mathbf{i} + 2t\mathbf{j} + 3t^2\mathbf{k}$. Also, $\mathbf{F} = x^2\mathbf{i} + y^2\mathbf{j} + z^2\mathbf{k} = (t)^2\mathbf{i} + \left(t^2\right)^2\mathbf{j} + \left(t^3\right)^2\mathbf{k} = t^2\mathbf{i} + t^4\mathbf{j} + t^6\mathbf{k}$.

Thus,

$$W = \int_C \mathbf{F}\cdot d\mathbf{r} = \int_C \mathbf{F}\cdot\mathbf{r}'(t)\,dt = \int_0^1\left(t^2\mathbf{i} + t^4\mathbf{j} + t^6\mathbf{k}\right)\cdot\left(\mathbf{i} + 2t\mathbf{j} + 3t^2\mathbf{k}\right)dt = \int_0^1\left(t^2 + 2t^5 + 3t^8\right)dt$$

$$= \left(\tfrac{1}{3}t^3 + \tfrac{1}{3}t^6 + \tfrac{1}{3}t^9\right)\Big|_0^1 = 1$$

37. Here $C : \mathbf{r}(t) = 5\cos t\,\mathbf{i} + 5\sin t\,\mathbf{j} + \frac{16}{\pi}t\mathbf{k}$, $0 \le t \le \frac{\pi}{2} \Rightarrow \mathbf{r}'(t) = -5\sin t\,\mathbf{i} + 5\cos t\,\mathbf{j} + \frac{16}{\pi}\mathbf{k}$.

The girl weighs 90 lbs and so $\mathbf{F} = -90\mathbf{k}$. Therefore, the work done by her against gravity is

$$W = \int_C \mathbf{F}\cdot d\mathbf{r} = \int_C \mathbf{F}\cdot\mathbf{r}'(t)\,dt = \int_0^{\pi/2}(-90\mathbf{k})\cdot\left(-5\sin t\,\mathbf{i} + 5\cos t\,\mathbf{j} + \tfrac{16}{\pi}\mathbf{k}\right)dt = -\frac{1440}{\pi}\int_0^{\pi/2}dt = -720 \text{ ft-lb}.$$

39. a. $\mathbf{F}(\mathbf{r}(t)) = \frac{d}{dt}[m(t)\mathbf{v}(t)] = m'(t)\mathbf{v}(t) + m(t)\mathbf{v}'(t)$, so

$$\mathbf{F}(\mathbf{r}(t)) \cdot \mathbf{r}'(t) = \mathbf{F}(\mathbf{r}(t)) \cdot \mathbf{v}(t) = [m'(t)\mathbf{v}(t) + m(t)\mathbf{v}'(t)] \cdot \mathbf{v}(t) = m'(t)[\mathbf{v}(t) \cdot \mathbf{v}(t)] + m(t)[\mathbf{v}'(t) \cdot \mathbf{v}(t)]$$

$$= m'(t)|\mathbf{v}(t)|^2 + m(t)[\mathbf{v}'(t) \cdot \mathbf{v}(t)]$$

but $\mathbf{v}(t) \cdot \mathbf{v}(t) = |\mathbf{v}(t)|^2 = v^2(t) \Rightarrow \mathbf{v}'(t) \cdot \mathbf{v}(t) + \mathbf{v}(t) \cdot \mathbf{v}'(t) = 2\mathbf{v}(t)\mathbf{v}'(t) = 2\mathbf{v}'(t)\mathbf{v}(t) \Rightarrow \mathbf{v}'(t) \cdot \mathbf{v}(t) = v(t)v'(t)$.

Thus, $\mathbf{F}(\mathbf{r}(t)) \cdot \mathbf{r}'(t) = m'(t)v^2(t) + m(t)v(t)v'(t)$, as was to be shown.

b. If m is constant, then $m'(t) = \frac{d}{dt}(m) = 0$ and so $\mathbf{F}(\mathbf{r}(t)) \cdot \mathbf{r}'(t) = mv(t)v'(t)$. Therefore,

$$W = \int_C \mathbf{F} \cdot d\mathbf{r} = \int_C \mathbf{F} \cdot \mathbf{r}'(t)\,dt = \int_a^b mv(t)\frac{dv}{dt}\,dt = m\int_a^b v(t)\,dv = \frac{1}{2}mv^2(t)\Big|_a^b = \frac{1}{2}m\left[v^2(b) - v^2(a)\right].$$

41. a. i. Here $C : \mathbf{r}(t) = (2-2t)\mathbf{i} + (1+4t)\mathbf{j} + 5t\mathbf{k}, 0 \le t \le 1 \Rightarrow \mathbf{r}'(t) = -2\mathbf{i} + 4\mathbf{j} + 5\mathbf{k}$. Thus,

$$W = \int_C \mathbf{F} \cdot d\mathbf{r} = q\int_C \mathbf{E} \cdot \mathbf{r}'(t)\,dt = \frac{qQ}{4\pi\varepsilon_0}\int \frac{\mathbf{r}(t) \cdot \mathbf{r}'(t)}{|\mathbf{r}(t)|^3}\,dt$$

$$= \frac{qQ}{4\pi\varepsilon_0}\int_0^1 \frac{[(2-2t)\mathbf{i} + (1+4t)\mathbf{j} + 5t\mathbf{k}] \cdot (-2\mathbf{i} + 4\mathbf{j} + 5\mathbf{k})}{\left[(2-2t)^2 + (1+4t)^2 + (5t)^2\right]^{3/2}}\,dt = \frac{9\sqrt{5}qQ}{20\pi\varepsilon_0}\int_0^1 \frac{t}{(1+9t^2)^{3/2}}\,dt$$

$$= -\frac{9\sqrt{5}qQ}{20\pi\varepsilon_0}\left[\frac{1}{9\sqrt{1+9t^2}}\right]_0^1 = \frac{\sqrt{2}\left(\sqrt{10}-1\right)qQ}{40\pi\varepsilon_0} = \frac{\left(2\sqrt{5}-\sqrt{2}\right)qQ}{40\pi\varepsilon_0}$$

ii. Here $C = C_1 \cup C_2 \cup C_3$, where $C_1 : \mathbf{r}(t) = 2\mathbf{i} + (1+4t)\mathbf{j}, 0 \le t \le 1 \Rightarrow$
$\mathbf{r}'(t) = 4\mathbf{j}, C_2 : \mathbf{r}(t) = (2-2t)\mathbf{i} + 5\mathbf{j}, 0 \le t \le 1 \Rightarrow \mathbf{r}'(t) = -2\mathbf{i}$, and
$C_3 : \mathbf{r}(t) = 5\mathbf{j} + 5t\mathbf{k}, 0 \le t \le 1 \Rightarrow \mathbf{r}'(t) = 5\mathbf{k}$. Thus,

$$W = \int_C \mathbf{F} \cdot d\mathbf{r}$$

$$= q\int_{C_1}\mathbf{E} \cdot \mathbf{r}'(t)\,dt + q\int_{C_2}\mathbf{E} \cdot \mathbf{r}'(t)\,dt + q\int_{C_3}\mathbf{E} \cdot \mathbf{r}'(t)\,dt$$

$$= \frac{qQ}{4\pi\varepsilon_0}\left\{\int_0^1 \frac{[2\mathbf{i} + (1+4t)\mathbf{j}] \cdot (4\mathbf{j})}{\left[2^2 + (1+4t)^2\right]^{3/2}}\,dt + \int_0^1 \frac{[(2-2t)\mathbf{i} + 5\mathbf{j}](-2\mathbf{i})}{\left[(2-2t)^2 + 5^2\right]^{3/2}}\,dt + \int_0^1 \frac{(5\mathbf{j} + 5t\mathbf{k}) \cdot (5\mathbf{k})}{\left[5^2 + (5t)^2\right]^{3/2}}\,dt\right\}$$

$$= \frac{qQ}{4\pi\varepsilon_0}\left\{\int_0^1 \frac{16t+4}{(16t^2+8t+5)^{3/2}}\,dt + \int_0^1 \frac{4t-4}{(4t^2-8t+29)^{3/2}}\,dt + \int_0^1 \frac{25t}{125(t^2+1)^{3/2}}\,dt\right\}$$

$$= -\frac{qQ}{4\pi\varepsilon_0}\left[\frac{1}{\sqrt{16t^2+8t+5}} + \frac{1}{\sqrt{4t^2-8t+29}} + \frac{1}{5\sqrt{t^2+1}}\right]_0^1 = \frac{qQ\left(2\sqrt{5}-\sqrt{2}\right)}{40\pi\varepsilon_0}$$

b. The work done in parts a.i and a.ii are the same.

43.

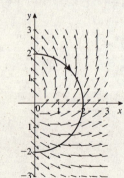

A particle moving along C will move against the vector field and so we expect $\int_C \mathbf{F} \cdot d\mathbf{r}$ to be negative. Here $\mathbf{r}(t) = 2\sin t\,\mathbf{i} + 2\cos t\,\mathbf{j} \Rightarrow$
$\mathbf{F}(x(t), y(t)) = (\sin t - \cos t)\mathbf{i} + (\sin t + \cos t)\mathbf{j}$. Also,
$\mathbf{r}'(t) = 2\cos t\,\mathbf{i} - 2\sin t\,\mathbf{j}$, so

$$\int_C \mathbf{F} \cdot d\mathbf{r} = \int_C \mathbf{F}(\mathbf{r}(t)) \cdot \mathbf{r}'(t)\,dt$$

$$= \int_0^\pi [(\sin t - \cos t)\mathbf{i} + (\sin t + \cos t)\mathbf{j}] \cdot (2\cos t\,\mathbf{i} - 2\sin t\,\mathbf{j})\,dt$$

$$= \int_0^\pi \left(-2\cos^2 t - 2\sin^2 t\right)dt = -2\int_0^\pi dt = -2\pi$$

45. True. The force is perpendicular to the path (or, more accurately, the tangent line to the path) at any point on the path, so the work is 0. We can also prove this analytically as follows. Let $C : \mathbf{r}(t) = a\cos t\,\mathbf{i} + a\sin t\,\mathbf{j}, 0 \le t \le 2\pi$. Then $\mathbf{r}'(t) = -a\sin t\,\mathbf{i} + a\cos t\,\mathbf{j}$, so

$$\int_C \mathbf{F} \cdot d\mathbf{r} = \int_0^{2\pi} \mathbf{F}(\mathbf{r}(t)) \cdot \mathbf{r}'(t)\,dt = \int_0^{2\pi} [(a\cos t)\,\mathbf{i} + (a\sin t)\,\mathbf{j}] \cdot (-a\sin t\,\mathbf{i} + a\cos t\,\mathbf{j})\,dt = \int_0^{2\pi} 0\,dt = 0.$$

47. False. Take $C : \mathbf{r}(t) = t\mathbf{i} + 2t\mathbf{j}, 0 \le t \le 3$. Then $y'(t) = 2$, so $\int_C xy\,dy = \int_0^3 t \cdot 2t \cdot 2\,dt = 4\int_0^3 t^2\,dt = 36$, but

$$\tfrac{1}{2}xy^2\Big|_{t=0}^{t=3} = \tfrac{1}{2}(t)(2t)^2\Big|_0^3 = 2 \cdot 3^3 = 54.$$

15.4 Concept Questions   ET 14.4

1. See page 1257 (1253 in ET).

3. a. See page 1262 (1258 in ET).

   b. See page 1264 (1260 in ET).

15.4 Independence of Path and Conservative Vector Fields   ET 14.4

1. Here $P(x, y) = 4x + 3y$ and $Q(x, y) = 3x - 2y$. Since $\frac{\partial Q}{\partial x} = 3 = \frac{\partial P}{\partial y}$ for all (x, y) in the plane, $\mathbf{F}$ is conservative in the plane, so there exists a function f such that $\frac{\partial f}{\partial x} = 4x + 3y$ and $\frac{\partial f}{\partial y} = 3x - 2y$. Integrating the first equation gives $f(x, y) = 2x^2 + 3xy + g(y)$; then $\frac{\partial f}{\partial y} = 3x + g'(y) = 3x - 2y \Rightarrow g'(y) = -2y \Rightarrow g(y) = -y^2 + C$, where C is a constant. Thus, $f(x, y) = 2x^2 + 3xy - y^2 + C$.

3. Here $P(x, y) = 2x + y^2$ and $Q(x, y) = x^2 + y$. Since $\frac{\partial Q}{\partial x} = 2x$ and $\frac{\partial P}{\partial y} = 2y$, we see that $\frac{\partial Q}{\partial x} \ne \frac{\partial P}{\partial y}$ except along the line $y = x$. Thus, by Theorem 5, $\mathbf{F}$ is not conservative.

5. Here $P(x, y) = y^2\cos x$ and $Q(x, y) = 2y\sin x + 3$. Since $\frac{\partial Q}{\partial x} = 2y\cos x = \frac{\partial P}{\partial y}$ for all (x, y), $\mathbf{F}$ is conservative, so there exists a function f such that $\frac{\partial f}{\partial x} = y^2\cos x$ and $\frac{\partial f}{\partial y} = 2y\sin x + 3$. Integrating the first equation gives $f(x, y) = y^2\sin x + g(y)$; then $\frac{\partial f}{\partial y} = 2y\sin x + g'(y) = 2y\sin x + 3 \Rightarrow g'(y) = 3 \Rightarrow g(y) = 3y + C$, where C is a constant. Thus, $f(x, y) = y^2\sin x + 3y + C$.

7. Here $P(x, y) = e^{-x} - 2y\cos 2x$ and $Q(x, y) = \sin 2x + ye^{-x}$, so $\frac{\partial Q}{\partial x} = 2\cos 2x - ye^{-x}$ and $\frac{\partial P}{\partial y} = -2\cos 2x$. Since $\frac{\partial Q}{\partial x} \ne \frac{\partial P}{\partial y}$, $\mathbf{F}$ is not conservative.

9. Here $P(x, y) = x^2 + \frac{y}{x}$ and $Q(x, y) = y^2 + \ln x$. Since $\frac{\partial Q}{\partial x} = \frac{1}{x} = \frac{\partial P}{\partial y}$ for $\{(x, y) \mid x \ne 0\}$, Theorem 5 tells us that $\mathbf{F}$ is conservative in the each of the half-planes $x < 0$ and $x > 0$, so there exists a function f such that $\frac{\partial f}{\partial x} = x^2 + \frac{y}{x}$ and $\frac{\partial f}{\partial y} = y^2 + \ln x$. Integrating the first equation gives $f(x, y) = \frac{1}{3}x^3 + y\ln x + g(y)$; then $\frac{\partial f}{\partial y} = \ln x + g'(y) = y^2 + \ln x \Rightarrow g'(y) = y^2 \Rightarrow g(y) = \frac{1}{3}y^3 + C$, where C is a constant. Thus, $f(x, y) = \frac{1}{3}x^3 + y\ln x + \frac{1}{3}y^3 + C$.

11. a. Here $P(x, y) = 2y + 1$ and $Q(x, y) = 2x + 3$. Since $\frac{\partial Q}{\partial x} = 2 = \frac{\partial P}{\partial y}$ for all (x, y), $\mathbf{F}$ is conservative, so there exists a function f such that $\frac{\partial f}{\partial x} = 2y + 1$ and $\frac{\partial f}{\partial y} = 2x + 3$. Integrating the first equation gives $f(x, y) = 2xy + x + g(y)$; then $\frac{\partial f}{\partial y} = 2x + g'(y) = 2x + 3 \Rightarrow g'(y) = 3 \Rightarrow g(y) = 3y + C$, where C is a constant. Thus, $f(x, y) = 2xy + x + 3y + C$.

   b. $\int_C \mathbf{F} \cdot d\mathbf{r} = \int_C \nabla f \cdot d\mathbf{r} = f(-1, 1) - f(0, 0) = [2(-1)(1) + (-1) + 3(1)] - 0 = 0$

13. a. Here $P(x, y) = 2xy^2 + 2y$ and $Q(x, y) = 2x^2y + 2x$. Since $\frac{\partial Q}{\partial x} = 4xy + 2 = \frac{\partial P}{\partial y}$ for all (x, y), $\mathbf{F}$ is conservative, so

there exists a function f such that $\frac{\partial f}{\partial x} = 2xy^2 + 2y$ and $\frac{\partial f}{\partial y} = 2x^2y + 2x$. Integrating the first equation gives

$f(x, y) = x^2y^2 + 2xy + g(y)$; then $\frac{\partial f}{\partial y} = 2x^2y + 2x + g'(y) = 2x^2y + 2x \Rightarrow g'(y) = 0 \Rightarrow g(y) = C$, a constant.

Thus, $f(x, y) = x^2y^2 + 2xy + C$.

b. $\int_C \mathbf{F} \cdot d\mathbf{r} = \int_C \nabla f \cdot d\mathbf{r} = f(1, 2) - f(-1, 1) = \left[(1)^2(2)^2 + 2(1)(2)\right] - \left[(-1)^2(1)^2 + 2(-1)(1)\right] = 9$

15. a. Here $P(x, y) = xe^{2y}$ and $Q(x, y) = x^2e^{2y}$. Since $\frac{\partial Q}{\partial x} = 2xe^{2y} = \frac{\partial P}{\partial y}$ for all (x, y), $\mathbf{F}$ is conservative, so there exists a

function f such that $\frac{\partial f}{\partial x} = xe^{2y}$ and $\frac{\partial f}{\partial y} = x^2e^{2y}$. Integrating the first equation gives $f(x, y) = \frac{1}{2}x^2e^{2y} + g(y)$; then

$\frac{\partial f}{\partial y} = x^2e^{2y} + g'(y) = x^2e^{2y} \Rightarrow g'(y) = 0 \Rightarrow g(y) = C$, a constant. Thus, $f(x, y) = \frac{1}{2}x^2e^{2y} + C$.

b. $\int_C \mathbf{F} \cdot d\mathbf{r} = \int_C \nabla f \cdot d\mathbf{r} = f(-1, 1) - f(0, 0) = \frac{1}{2}(-1)^2 e^2 - 0 = \frac{1}{2}e^2$

17. a. Here $P(x, y) = e^x \sin y$ and $Q(x, y) = e^x \cos y + y$. Since $\frac{\partial Q}{\partial x} = e^x \cos y = \frac{\partial P}{\partial y}$ for all (x, y), $\mathbf{F}$ is conservative, so

there exists a function f such that $\frac{\partial f}{\partial x} = e^x \sin y$ and $\frac{\partial f}{\partial y} = e^x \cos y + y$. Integrating the first equation gives

$f(x, y) = e^x \sin y + g(y)$; then $\frac{\partial f}{\partial y} = e^x \cos y + g'(y) = e^x \cos y + y \Rightarrow g'(y) = y \Rightarrow g(y) = \frac{1}{2}y^2 + C$, where C is

a constant. Thus, $f(x, y) = e^x \sin y + \frac{1}{2}y^2 + C$.

b. $\int_C \mathbf{F} \cdot d\mathbf{r} = \int_C \nabla f \cdot d\mathbf{r} = f(0, \pi) - f(0, 0) = e^0 \sin \pi + \frac{1}{2}(\pi)^2 - 0 = \frac{1}{2}\pi^2$

19. Here $P(x, y) = 2xy^2 + \cos y$ and $Q(x, y) = 2x^2y - x \sin y$, so

$\frac{\partial Q}{\partial x} = 4xy - \sin y = \frac{\partial P}{\partial y}$ and $\mathbf{F}$ is conservative. Therefore, $\int_C \mathbf{F} \cdot d\mathbf{r}$ is

independent of path. Let $C_1 : \mathbf{r}(t) = 2t\mathbf{i}, 0 \le t \le 1$. Then $\mathbf{r}'(t) = 2\mathbf{i}$ and on

C_1,

$\mathbf{F} = \left(2xy^2 + \cos y\right)\mathbf{i} + \left(2x^2y - x \sin y\right)\mathbf{j}$

$= [2(2t)(0) + \cos 0]\mathbf{i} + \left[2(2t)^2(0) - 2t(0)\right]\mathbf{j} = \mathbf{i}$

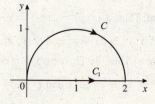

so $\int_C \mathbf{F} \cdot d\mathbf{r} = \int_{C_1} \mathbf{F} \cdot d\mathbf{r} = \int_{C_1} \mathbf{F}(\mathbf{r}(t)) \cdot \mathbf{r}'(t)\, dt = \int_0^1 \mathbf{i} \cdot (2\mathbf{i})\, dt = 2\int_0^1 dt = 2$.

21. Here $P(x, y) = 2\sqrt{y}$ and $Q(x, y) = \dfrac{x}{\sqrt{y}}$, so $\dfrac{\partial Q}{\partial x} = \dfrac{1}{\sqrt{y}}$ and $\dfrac{\partial P}{\partial y} = \dfrac{1}{\sqrt{y}}$. Because $\dfrac{\partial Q}{\partial x} = \dfrac{\partial P}{\partial y}$ for

$\{(x, y) \mid -\infty < x < \infty, y > 0\}$, we see that $\mathbf{F}$ is conservative on that set. Therefore, the work done by $\mathbf{F}$ in moving a

particle along a path from the point $A(1, 1)$ to the point $(2, 9)$ depends only on the endpoints. Now there exists a function f

such that $F = \nabla f$; that is, $\frac{\partial f}{\partial x} = 2y^{1/2}$ and $\frac{\partial f}{\partial y} = xy^{-1/2}$. Integrating the first equation gives $f(x, y) = 2xy^{1/2} + g(y)$,

so $\frac{\partial f}{\partial y} = xy^{-1/2} + g'(y) = xy^{-1/2} \Rightarrow g'(y) = 0 \Rightarrow g(y) = C$, a constant. Thus, $f(x, y) = 2x\sqrt{y} + C$, and so

$W = \int_C \mathbf{F} \cdot d\mathbf{r} = \int_C \nabla f \cdot d\mathbf{r} = f(2, 9) - f(1, 1) = 2(2)\sqrt{9} - 2(1)\sqrt{1} = 10$.

23. Here $\mathbf{F}(x, y, z) = yz\mathbf{i} + xz\mathbf{j} + xyz\mathbf{k}$. Observe that $P = yz$, $Q = xz$, and $R = xyz$ have continuous first-order partial

derivatives. Next, $\operatorname{curl} \mathbf{F} = \begin{vmatrix} \mathbf{i} & \mathbf{j} & \mathbf{k} \\ \frac{\partial}{\partial x} & \frac{\partial}{\partial y} & \frac{\partial}{\partial z} \\ yz & xz & xyz \end{vmatrix} = (xz - x)\mathbf{i} - (yz - y)\mathbf{j} + (z - z)\mathbf{k} = x(z - 1)\mathbf{i} + y(1 - z)\mathbf{j} \ne \mathbf{0}$, so by

Theorem 6, $\mathbf{F}$ is not conservative, and therefore $\int_C yz\, dx + xz\, dy + xyz\, dz = \int_C \mathbf{F} \cdot d\mathbf{r}$ is not independent of path.

25. $\text{curl } \mathbf{F} = \begin{vmatrix} \mathbf{i} & \mathbf{j} & \mathbf{k} \\ \frac{\partial}{\partial x} & \frac{\partial}{\partial y} & \frac{\partial}{\partial z} \\ yz & xz & xy \end{vmatrix} = (x - x)\mathbf{i} - (y - y)\mathbf{j} + (z - z)\mathbf{k} = \mathbf{0}$. Since all the components of $\mathbf{F}$ have continuous first-order

partial derivatives, we see that $\mathbf{F}$ is a conservative vector field, so there exists a function f such that $\frac{\partial f}{\partial x} = yz$ (1),

$\frac{\partial f}{\partial y} = xz$ (2), and $\frac{\partial f}{\partial z} = xy$ (3). Integrating (1), we obtain $f(x, y, z) = xyz + g(y, z)$. Differentiating f with respect to y

and using (2) gives $\frac{\partial f}{\partial y} = xz + \frac{\partial g}{\partial y} = xz \Rightarrow \frac{\partial g}{\partial y} = 0 \Rightarrow g(y, z) = h(z)$. Thus, $f(x, y, z) = xyz + h(z)$, and differentiating

with respect to z and using (3) then gives $\frac{\partial f}{\partial z} = xy + h'(z) = xy \Rightarrow h'(z) = 0 \Rightarrow h(z) = C$, a constant. Thus,

$f(x, y, z) = xyz + C$.

27. $\text{curl } \mathbf{F} = \begin{vmatrix} \mathbf{i} & \mathbf{j} & \mathbf{k} \\ \frac{\partial}{\partial x} & \frac{\partial}{\partial y} & \frac{\partial}{\partial z} \\ 2xy & x^2 + z^2 & xy \end{vmatrix} = (x - 2z)\mathbf{i} - (y - 0)\mathbf{j} + (2x - 2x)\mathbf{k} \neq \mathbf{0}$, so $\mathbf{F}$ is not conservative.

29. $\text{curl } \mathbf{F} = \begin{vmatrix} \mathbf{i} & \mathbf{j} & \mathbf{k} \\ \frac{\partial}{\partial x} & \frac{\partial}{\partial y} & \frac{\partial}{\partial z} \\ e^x \cos z & z \sinh y & \cosh y - e^x \sin z \end{vmatrix} = (\sinh y - \sinh y)\mathbf{i} - (-e^x \sin z + e^x \sin z)\mathbf{j} + (0 - 0)\mathbf{k} = \mathbf{0}$, and since

all the components of $\mathbf{F}$ have continuous partial derivatives, we see that $\mathbf{F}$ is a conservative vector field. Thus, there exists a

function f such that $\frac{\partial f}{\partial x} = e^x \cos z$ (1), $\frac{\partial f}{\partial y} = z \sinh y$ (2), and $\frac{\partial f}{\partial z} = \cosh y - e^x \sin z$ (3). Integrating (1), we obtain

$f(x, y, z) = e^x \cos z + g(y, z)$. Differentiating f with respect to y and using (2) gives $\frac{\partial f}{\partial y} = 0 + \frac{\partial g}{\partial y} = z \sinh y \Rightarrow$

$g(y, z) = z \cosh y + h(z)$. Thus, $f(x, y, z) = e^x \cos z + z \cosh y + h(z)$, and differentiating with respect to z and

using (3) then gives $\frac{\partial f}{\partial z} = -e^x \sin z + \cosh y + h'(z) = \cosh y - e^x \sin z \Rightarrow h'(z) = 0 \Rightarrow h(z) = C$, a constant. Thus,

$f(x, y, z) = e^x \cos z + z \cosh y + C$.

31. $\text{curl } \mathbf{F} = \begin{vmatrix} \mathbf{i} & \mathbf{j} & \mathbf{k} \\ \frac{\partial}{\partial x} & \frac{\partial}{\partial y} & \frac{\partial}{\partial z} \\ z \cos(x + y) & z \sin(x + y) & \cos(x + y) \end{vmatrix}$

$= \left[-\sin(x + y) - \sin(x + y) \right]\mathbf{i} - \left[-\sin(x + y) - \cos(x + y) \right]\mathbf{j} + \left[z \cos(x + y) + z \sin(x + y) \right]\mathbf{k}$

$= -2 \sin(x + y)\mathbf{i} + \left[\sin(x + y) + \cos(x + y) \right]\mathbf{j} + z \left[\cos(x + y) + \sin(x + y) \right]\mathbf{k} \neq \mathbf{0}$

so $\mathbf{F}$ is not conservative.

33. a. $\text{curl } \mathbf{F} = \begin{vmatrix} \mathbf{i} & \mathbf{j} & \mathbf{k} \\ \frac{\partial}{\partial x} & \frac{\partial}{\partial y} & \frac{\partial}{\partial z} \\ yz^2 & xz^2 & 2xyz \end{vmatrix} = (2xz - 2xz)\mathbf{i} - (2yz - 2yz)\mathbf{j} + \left(z^2 - z^2 \right)\mathbf{k} = \mathbf{0}$, so $\mathbf{F}$ is a conservative vector field.

Therefore, there exists a function f such that $\frac{\partial f}{\partial x} = yz^2$ (1), $\frac{\partial f}{\partial y} = xz^2$ (2), and $\frac{\partial f}{\partial z} = 2xyz$ (3). Integrating (1), we

obtain $f(x, y, z) = xyz^2 + g(y, z)$. Differentiating f with respect to y and using (2) gives $\frac{\partial f}{\partial y} = xz^2 + \frac{\partial g}{\partial y} = xz^2 \Rightarrow$

$\frac{\partial g}{\partial y} = 0 \Rightarrow g(y, z) = h(z)$. Thus, $f(x, y, z) = xyz^2 + h(z)$, and differentiating with respect to z and using (3) then

gives $\frac{\partial f}{\partial z} = 2xyz + h'(z) = 2xyz \Rightarrow h'(z) = 0 \Rightarrow h(z) = C$, a constant. Thus, $f(x, y, z) = xyz^2 + C$.

b. $\int_C \mathbf{F} \cdot d\mathbf{r} = f(1, 3, 2) - f(0, 0, 1) = (1)(3)(2)^2 - (0)(0)(1)^2 = 12$

35. a. $\operatorname{curl} \mathbf{F} = \begin{vmatrix} \mathbf{i} & \mathbf{j} & \mathbf{k} \\ \frac{\partial}{\partial x} & \frac{\partial}{\partial y} & \frac{\partial}{\partial z} \\ \cos y & z^2 - x\sin y & 2yz \end{vmatrix} = (2z - 2z)\mathbf{i} - (0 - 0)\mathbf{j} + (-\sin y + \sin y)\mathbf{k} = \mathbf{0}$, so $\mathbf{F}$ is a conservative vector

field. Therefore, there exists a function f such that $\frac{\partial f}{\partial x} = \cos y$ (1), $\frac{\partial f}{\partial y} = z^2 - x\sin y$ (2), and $\frac{\partial f}{\partial z} = 2yz$ (3).

Integrating (1), we obtain $f(x, y, z) = x\cos y + g(y, z)$. Differentiating f with respect to y and using (2) gives

$\frac{\partial f}{\partial y} = -x\sin y + \frac{\partial g}{\partial y} = z^2 - x\sin y \Rightarrow \frac{\partial g}{\partial y} = z^2 \Rightarrow g(y, z) = yz^2 + h(z)$. Thus, $f(x, y, z) = x\cos y + yz^2 + h(z)$,

and differentiating with respect to z and using (3) then gives $\frac{\partial f}{\partial z} = 2yz + h'(z) = 2yz \Rightarrow h'(z) = 0 \Rightarrow h(z) = C$, a

constant. Thus, $f(x, y, z) = x\cos y + yz^2 + C$.

b. $\int_C \mathbf{F} \cdot d\mathbf{r} = f(2, 2\pi, 1) - f(1, 0, 0) = \left[2\cos 2\pi + 2\pi(1)^2\right] - \left[1\cos 0 + 0(0)^2\right] = 2\pi + 1$.

37. $\int_C \left(2xy^2 - 3\right)dx + \left(2x^2y + 1\right)dy = \int \mathbf{F} \cdot d\mathbf{r}$, where $\mathbf{F}(x, y) = \left(2xy^2 - 3\right)\mathbf{i} + \left(2x^2y + 1\right)\mathbf{j}$. Here

$P(x, y) = 2xy^2 - 3$ and $Q(x, y) = 2x^2y + 1$. Since $\frac{\partial Q}{\partial x} = 4xy = \frac{\partial P}{\partial y}$ for all (x, y), we see that $\mathbf{F}$ is a

conservative vector field, so there exists a function f such that $\frac{\partial f}{\partial x} = 2xy^2 - 3$ and $\frac{\partial f}{\partial y} = 2x^2y + 1$. Integrating

the first equation gives $f(x, y) = x^2y^2 - 3x + g(y)$. Differentiating f with respect to y and using the second

equation, we have $\frac{\partial f}{\partial y} = 2x^2y + g'(y) = 2x^2y + 1 \Rightarrow g'(y) = 1 \Rightarrow g(y) = y + C$, where C is a constant, so

$f(x, y) = x^2y^2 - 3x + y + C$. Finally, since $\mathbf{F}$ is conservative, the integral depends only on the endpoints, and we have

$\int_C \mathbf{F} \cdot d\mathbf{r} = \int_C \nabla f \cdot d\mathbf{r} = f(2, 1) - f(0, 0) = \left[(2)^2(1)^2 - 3(2) + 1\right] - 0 = -1$.

39. $\mathbf{E}(x, y, z) = \frac{kQ}{|\mathbf{r}|^3}\mathbf{r} = kQ\left[\frac{x}{(x^2 + y^2 + z^2)^{3/2}}\mathbf{i} + \frac{y}{(x^2 + y^2 + z^2)^{3/2}}\mathbf{j} + \frac{z}{(x^2 + y^2 + z^2)^{3/2}}\mathbf{k}\right]$, so

$\operatorname{curl} \mathbf{E} = kQ \begin{vmatrix} \mathbf{i} & \mathbf{j} & \mathbf{k} \\ \frac{\partial}{\partial x} & \frac{\partial}{\partial y} & \frac{\partial}{\partial z} \\ \dfrac{x}{(x^2 + y^2 + z^2)^{3/2}} & \dfrac{y}{(x^2 + y^2 + z^2)^{3/2}} & \dfrac{z}{(x^2 + y^2 + z^2)^{3/2}} \end{vmatrix}$

$= kQ\left[\frac{-3yz}{(x^2 + y^2 + z^2)^{5/2}} + \frac{3yz}{(x^2 + y^2 + z^2)^{5/2}}\right]\mathbf{i} - kQ\left[\frac{-3xz}{(x^2 + y^2 + z^2)^{5/2}} + \frac{3xz}{(x^2 + y^2 + z^2)^{5/2}}\right]\mathbf{j}$

$+ kQ\left[\frac{-3xy}{(x^2 + y^2 + z^2)^{5/2}} + \frac{3xy}{(x^2 + y^2 + z^2)^{5/2}}\right]\mathbf{k}$

$= \mathbf{0}$

Thus, $\mathbf{E}$ is a conservative vector field, and so there exists a function f such that $\frac{\partial f}{\partial x} = \frac{kQx}{(x^2 + y^2 + z^2)^{3/2}}$ (1),

$\frac{\partial f}{\partial y} = \frac{kQy}{(x^2 + y^2 + z^2)^{3/2}}$ (2), and $\frac{\partial f}{\partial z} = \frac{kQz}{(x^2 + y^2 + z^2)^{3/2}}$ (3). Integrating (1), we obtain

$f(x, y, z) = -\frac{kQ}{\sqrt{x^2 + y^2 + z^2}} + g(y, z)$.

Differentiating with respect to y and using (2), we find $\dfrac{\partial f}{\partial y} = \dfrac{kQy}{\left(x^2 + y^2 + z^2\right)^{3/2}} + \dfrac{\partial g}{\partial y} = \dfrac{kQy}{\left(x^2 + y^2 + z^2\right)^{3/2}} \Rightarrow$

$\dfrac{\partial g}{\partial y} = 0 \Rightarrow g(y, z) = h(z)$, so $f(x, y, z) = -\dfrac{kQ}{\sqrt{x^2 + y^2 + z^2}} + h(z)$. Differentiating with respect to z and using (3),

we have $\dfrac{\partial f}{\partial z} = \dfrac{kQz}{\left(x^2 + y^2 + z^2\right)^{3/2}} + h'(z) = \dfrac{kQz}{\left(x^2 + y^2 + z^2\right)^{3/2}} \Rightarrow h'(z) = 0 \Rightarrow h(z) = C$, a constant. Thus,

$f(x, y, z) = -\dfrac{kQ}{\sqrt{x^2 + y^2 + z^2}} + C$. Finally, because $\mathbf{E}$ is conservative, the work done is

$q \int_C \mathbf{E} \cdot d\mathbf{r} = q \int_C \nabla f \cdot d\mathbf{r} = q\left[f(2, 4, 1) - f(1, 3, 2)\right] = q\left(-\dfrac{kQ}{\sqrt{4 + 16 + 1}} + \dfrac{kQ}{\sqrt{1 + 9 + 4}}\right)$

$= kqQ\left(\dfrac{\sqrt{14}}{14} - \dfrac{\sqrt{21}}{21}\right) \approx 0.05kqQ$

41. a. $\dfrac{\partial Q}{\partial x} = \dfrac{\partial}{\partial x}\left(-\dfrac{x}{x^2 + y^2}\right) = -\dfrac{\left(x^2 + y^2\right)(1) - x(2x)}{\left(x^2 + y^2\right)^2} = \dfrac{x^2 - y^2}{\left(x^2 + y^2\right)^2}$ and

$\dfrac{\partial P}{\partial y} = \dfrac{\partial}{\partial y}\left(\dfrac{y}{x^2 + y^2}\right) = \dfrac{\left(x^2 + y^2\right)(1) - y(2y)}{\left(x^2 + y^2\right)^2} = \dfrac{x^2 - y^2}{\left(x^2 + y^2\right)^2}$, so $\dfrac{\partial Q}{\partial x} = \dfrac{\partial P}{\partial y}$.

b. Let $C_1 : \mathbf{r}_1(t) = \cos t\,\mathbf{i} + \sin t\,\mathbf{j}, 0 \le t \le \pi$ and

$C_2 : \mathbf{r}_2(t) = \cos t\,\mathbf{i} - \sin t\,\mathbf{j}, 0 \le t \le \pi$. Then

$\mathbf{r}_1'(t) = -\sin t\,\mathbf{i} + \cos t\,\mathbf{j}$ and $\mathbf{r}_2'(t) = -\sin t\,\mathbf{i} - \cos t\,\mathbf{j}$, so

$\int_{C_1} \mathbf{F} \cdot d\mathbf{r} = \int_{C_1} \mathbf{F}(\mathbf{r}_1(t)) \cdot \mathbf{r}_1'(t)\, dt$

$= \int_0^\pi (\sin t\,\mathbf{i} - \cos t\,\mathbf{j}) \cdot (-\sin t\,\mathbf{i} + \cos t\,\mathbf{j})\, dt$

$= -\int_0^\pi dt = -\pi$

and $\int_{C_2} \mathbf{F} \cdot d\mathbf{r} = \int_{C_2} \mathbf{F}(\mathbf{r}_2(t)) \cdot \mathbf{r}_2'(t)\, dt = \int_0^\pi (-\sin t\,\mathbf{i} - \cos t\,\mathbf{j}) \cdot (-\sin t\,\mathbf{i} - \cos t\,\mathbf{j})\, dt = \int_0^\pi dt = \pi$. This shows

that $\int_C \mathbf{F} \cdot d\mathbf{r}$ is not independent of path.

c. This result does not contradict Theorem 5; the requirement that P and Q have continuous first-order partial derivatives on R is not satisfied.

43. False. The simple closed curve $C : \mathbf{r}(t) = \frac{1}{2}\cos t\,\mathbf{i} + \frac{1}{2}\sin t\,\mathbf{j}, 0 \le t \le 2\pi$ lies in R but encloses the origin, which is not in R.

45. False. Let $\mathbf{F}(x, y) = (x - y)\,\mathbf{i} + (x + y)\,\mathbf{j}$. Then $\mathbf{F}$ has continuous first-order partial derivatives on the plane. But $\dfrac{\partial Q}{\partial x} = 1$ and $\dfrac{\partial P}{\partial y} = -1$, so $\dfrac{\partial Q}{\partial x} \ne \dfrac{\partial P}{\partial y}$ for all (x, y) on the plane. Thus, $\mathbf{F}$ is not a conservative vector field, and so $\int_C \nabla f \cdot d\mathbf{r}$ is not independent of path; that is, it depends on the curve C and not only on its endpoints.

47. True. The curve $-C$ is the same as C but is traversed in the opposite direction.

15.5   Concept Questions   ET 14.5

1. See page 1268 (1264 in ET).

15.5   Green's Theorem   ET 14.5

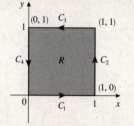

1. **a.** Let $C = C_1 \cup C_2 \cup C_3 \cup C_4$. We have

$C_1 : x = t, y = 0, 0 \leq t \leq 1$, so $\int_{C_1} 2xy\,dx + 3xy^2\,dy = \int_0^1 0\,dt = 0$;

$C_2 : x = 1, y = t, 0 \leq t \leq 1$, so $\int_{C_2} 2xy\,dx + 3xy^2\,dy = \int_0^1 3t^2\,dt = 1$;

$C_3 : x = 1 - t, y = 1, 0 \leq t \leq 1$, so

$\int_{C_3} 2xy\,dx + 3xy^2\,dy = \int_0^1 2(1 - t)(-dt) = 2\int_0^1 (t - 1)\,dt$

$\qquad = (t - 1)^2 \Big|_0^1 = -1$

and $C_4 : x = 0, y = 1 - t, 0 \leq t \leq 1$, so $\int_{C_4} 2xy\,dx + 3xy^2\,dy = \int_0^1 0\,dt = 0$. Therefore,

$\int_C 2xy\,dx + 3xy^2\,dy = \sum_{i=1}^4 \int_{C_i} 2xy\,dx + 3xy^2\,dy = 0 + 1 - 1 + 0 = 0$.

b. Here $P(x, y) = 2xy$ and $Q(x, y) = 3xy^2$, so by Green's Theorem,

$\oint 2xy\,dx + 3xy^2\,dy = \iint_R \left(\frac{\partial Q}{\partial x} - \frac{\partial P}{\partial y} \right) dA = \int_0^1 \int_0^1 \left(3y^2 - 2x \right) dy\,dx = \int_0^1 \left[y^3 - 2xy \right]_{y=0}^{y=1} dx = \int_0^1 (1 - 2x)\,dx$

$\qquad = \left(x - x^2 \right) \Big|_0^1 = 0$

3. **a.** Let $C = C_1 \cup C_2$. We have

$C_1 : x = t, y = t^3, 0 \leq t \leq 1$, so $dx = dt, dy = 3t^2\,dt$ and

$\int_{C_1} y^2\,dx + \left(x^2 + 2xy \right) dy = \int_0^1 \left\{ \left(t^3 \right)^2 + \left[t^2 + 2(t)\left(t^3 \right) \right] \left(3t^2 \right) \right\} dt$

$\qquad = \int_0^1 \left(7t^6 + 3t^4 \right) dt = \left(t^7 + \frac{3}{5}t^5 \right) \Big|_0^1 = \frac{8}{5}$

$C_2 : x = 1 - t, y = 1 - t, 0 \leq t \leq 1$, so $dx = dy = -dt$ and

$\int_{C_2} y^2\,dx + \left(x^2 + 2xy \right) dy = \int_0^1 \left[(1 - t)^2 + (1 - t)^2 + 2(1 - t)(1 - t) \right] (-dt)$

$\qquad = -4 \int_0^1 (1 - t)^2\,dt = (-4)(-1)\left(\frac{1}{3} \right)(1 - t)^3 \Big|_0^1 = -\frac{4}{3}$

Therefore, $\int_C y^2\,dx + \left(x^2 + 2xy \right) dy = \sum_{i=1}^2 \int_{C_i} y^2\,dx + \left(x^2 + 2xy \right) dy = \frac{8}{5} - \frac{4}{3} = \frac{4}{15}$.

b. Here $P(x, y) = y^2$ and $Q(x, y) = x^2 + 2xy$, so by Green's Theorem,

$\oint y^2\,dx + \left(x^2 + 2xy \right) dy = \iint_R \left(\frac{\partial Q}{\partial x} - \frac{\partial P}{\partial y} \right) dA = \int_0^1 \int_{x^3}^x (2x + 2y - 2y)\,dy\,dx = \int_0^1 [2xy]_{y=x^3}^{y=x}\,dx$

$\qquad = 2 \int_0^1 \left(x^2 - x^4 \right) dx = 2 \left(\frac{1}{3}x^3 - \frac{1}{5}x^5 \right) \Big|_0^1 = \frac{4}{15}$

5. Here $P(x, y) = x^3$ and $Q(x, y) = xy$, so $\frac{\partial Q}{\partial x} - \frac{\partial P}{\partial y} = y$ and

$\oint_C x^3\,dx + xy\,dy = \iint_R \left(\frac{\partial Q}{\partial x} - \frac{\partial P}{\partial y} \right) dA = \int_0^1 \int_0^y y\,dx\,dy = \int_0^1 [xy]_{x=0}^{x=y}\,dy$

$\qquad = \int_0^1 y^2\,dy = \frac{1}{3}y^3 \Big|_0^1 = \frac{1}{3}$

7. Here $P(x, y) = x^2 y + x^3$ and $Q(x, y) = 2xy$, so $\frac{\partial Q}{\partial x} - \frac{\partial P}{\partial y} = 2y - x^2$ and

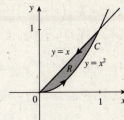

$$\oint_C \left(x^2 y + x^3\right) dx + 2xy \, dy = \iint_R \left(\frac{\partial Q}{\partial x} - \frac{\partial P}{\partial y}\right) dA = \int_0^1 \int_{x^2}^x \left(2y - x^2\right) dy \, dx$$

$$= \int_0^1 \left[y^2 - x^2 y\right]_{y=x^2}^{y=x} dx = \int_0^1 \left(x^2 - x^3\right) dx$$

$$= \left(\tfrac{1}{3}x^3 - \tfrac{1}{4}x^4\right)\Big|_0^1 = \tfrac{1}{12}$$

9. Here $P(x, y) = y^2 + \cos x$ and $Q(x, y) = x - \tan^{-1} y$, so

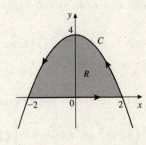

$$\frac{\partial Q}{\partial x} - \frac{\partial P}{\partial y} = 1 - 2y \text{ and}$$

$$\oint_C \left(y^2 + \cos x\right) dx + \left(x - \tan^{-1} y\right) dy = \iint_R \left(\frac{\partial Q}{\partial x} - \frac{\partial P}{\partial y}\right) dA$$

$$= \int_{-2}^2 \int_0^{4-x^2} (1 - 2y) \, dy \, dx = \int_{-2}^2 \left[y - y^2\right]_0^{4-x^2} dx$$

$$= \int_{-2}^2 \left(-12 + 7x^2 - x^4\right) dx = 2 \int_0^2 \left(-12 + 7x^2 - x^4\right) dx$$

$$= 2 \left(-12x + \tfrac{7}{3}x^3 - \tfrac{1}{5}x^5\right)\Big|_0^2 = -\tfrac{352}{15}$$

11. Here $P(x, y) = x^2 - y$ and $Q(x, y) = \sqrt{1 + y^2}$, so $\frac{\partial Q}{\partial x} - \frac{\partial P}{\partial y} = 1$ and

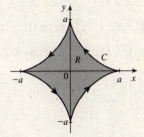

$$\oint_C \left(x^2 - y\right) dx + \sqrt{1 + y^2} \, dy = \iint_R \left(\frac{\partial Q}{\partial x} - \frac{\partial P}{\partial y}\right) dA = \iint_R 1 \, dA$$

$$= \tfrac{3}{8}\pi a^2 \text{ (see Exercise 10.3.49/9.3.49 in ET)}$$

13. Here $P(x, y) = x + e^x \sin y$ and $Q(x, y) = x + e^x \cos y$, so

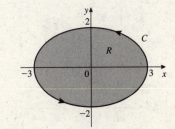

$$\frac{\partial Q}{\partial x} - \frac{\partial P}{\partial y} = 1 + e^x \cos y - e^x \cos y = 1 \text{ and}$$

$$\oint_C \left(x + e^x \sin y\right) dx + \left(x + e^x \cos y\right) dy = \iint_R \left(\frac{\partial Q}{\partial x} - \frac{\partial P}{\partial y}\right) dA = \iint_R 1 \, dA$$

$$= 6\pi \text{ since an ellipse has area } \pi ab.$$

15. Here $P(x, y) = -y$ and $Q(x, y) = x$, so $\frac{\partial Q}{\partial x} - \frac{\partial P}{\partial y} = 1 - (-1) = 2$ and

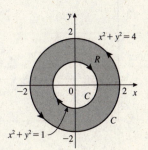

$$\oint_C (-y) \, dx + x \, dy = \iint_R \left(\frac{\partial Q}{\partial x} - \frac{\partial P}{\partial y}\right) dA = \iint_R 2 \, dA = 2(4\pi - \pi) = 6\pi.$$

17. $W = \oint_C \mathbf{F} \cdot d\mathbf{r} = \oint_C \left[\left(x^2 - y^2 \right) \mathbf{i} + 2xy\mathbf{j} \right] \cdot (dx\,\mathbf{i} + dy\,\mathbf{j})$

$\quad = \oint_C \left(x^2 - y^2 \right) dx + 2xy\,dy = \iint_R \left[\frac{\partial}{\partial x} (2xy) - \frac{\partial}{\partial y} \left(x^2 - y^2 \right) \right] dA$

$\quad = \iint_R 4y\,dA = 4 \int_0^1 \int_0^{1-x} y\,dy\,dx = 4 \int_0^1 \left[\frac{1}{2} y^2 \right]_0^{1-x} dx$

$\quad = 2 \int_0^1 (1-x)^2\,dx = -\frac{2}{3} (1-x)^3 \Big|_0^1 = \frac{2}{3}$

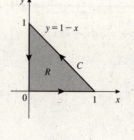

19. A parametrization of the astroid is $C : x = a\cos^3 t,\ y = a\sin^3 t,\ 0 \le t \le 2\pi$, so

$A = \oint_C x\,dy = \int_0^{2\pi} \left(a\cos^3 t \right) \left(3a\sin^2 t\cos t \right) dt = 3a^2 \int_0^{2\pi} \cos^4 t\sin^2 t\,dt$

$\quad = 3a^2 \int_0^{2\pi} \cos^4 t \left(1 - \cos^2 t \right) dt = 3a^2 \int_0^{2\pi} \left(\cos^4 t - \cos^6 t \right) dt$

$\quad = 3a^2 \left\{ \int_0^{2\pi} \cos^4 t\,dt - \left[\frac{1}{6} \cos^5 t\sin t \right]_0^{2\pi} - \frac{5}{6} \int_0^{2\pi} \cos^4 t\,dt \right\}$

$\quad = 3a^2 \left(\frac{1}{6} \right) \int_0^{2\pi} \cos^4 t\,dt = \frac{1}{2} a^2 \left[\left(\frac{1}{4} \cos^3 t\sin t \right) \Big|_0^{2\pi} + \frac{3}{4} \int_0^{2\pi} \cos^2 t\,dt \right]$

$\quad = \frac{3}{8} a^2 \int_0^{2\pi} \cos^2 t\,dt = \frac{3}{8} a^2 \int_0^{2\pi} \frac{1}{2} (1 + \cos 2t)\,dt = \frac{3}{8} a^2 \left(\frac{1}{2} t + \frac{1}{4} \sin 2t \right) \Big|_0^{2\pi} = \frac{3}{8} \pi a^2$

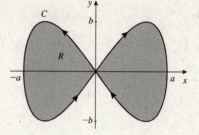

21. Let $C : x = a\sin t,\ y = b\sin 2t,\ \pi \le t \le 2\pi$. Then by symmetry,

$A = 2 \oint_C x\,dy = 2 \int_\pi^{2\pi} (a\sin t)(2b\cos 2t)\,dt$

$\quad = 4ab \int_\pi^{2\pi} \left(2\cos^2 t - 1 \right) \sin t\,dt$

$\quad = 4ab \left(-\frac{2}{3} \cos^3 t + \cos t \right) \Big|_\pi^{2\pi} = \frac{8}{3} ab$

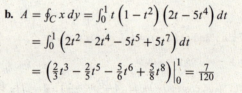

23. a.

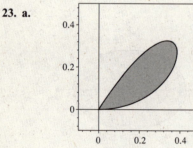

b. $A = \oint_C x\,dy = \int_0^1 t \left(1 - t^2 \right) \left(2t - 5t^4 \right) dt$

$\quad = \int_0^1 \left(2t^2 - 2t^4 - 5t^5 + 5t^7 \right) dt$

$\quad = \left(\frac{2}{3} t^3 - \frac{2}{5} t^5 - \frac{5}{6} t^6 + \frac{5}{8} t^8 \right) \Big|_0^1 = \frac{7}{120}$

25. a.

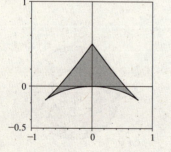

b. Using symmetry,

$A = \oint_C x\,dy = \int_{-1}^1 2t \left(1 - t^2 \right) \left(6t^3 - 2t \right) dt$

$\quad = 8 \int_0^1 \left(4t^4 - t^2 - 3t^6 \right) dt = 8 \left(\frac{4}{5} t^5 - \frac{1}{3} t^3 - \frac{3}{7} t^7 \right) \Big|_0^1 = \frac{32}{105}$

27. Applying Green's Theorem to the multiply-connected region R and denoting its area

by $A(R)$, we have $\int_{C_1} \mathbf{F} \cdot d\mathbf{r} - \int_{C_2} \mathbf{F} \cdot d\mathbf{r} - \int_{C_3} \mathbf{F} \cdot d\mathbf{r} = \iint_R \left(\frac{\partial Q}{\partial x} - \frac{\partial P}{\partial y} \right) dA$, so

$\int_{C_1} \mathbf{F} \cdot d\mathbf{r} = \int_{C_2} \mathbf{F} \cdot d\mathbf{r} + \int_{C_3} \mathbf{F} \cdot d\mathbf{r} + \iint_R \left(\frac{\partial Q}{\partial x} - \frac{\partial P}{\partial y} \right) dA = 2\pi + 3\pi + 6A(R) = 5\pi + 6[8 \cdot 6 - 2\pi] = 288 - 7\pi.$

29. Let C_2 be the path $\mathbf{r}(t) = (1 + 3t)\mathbf{i}, 0 \le t \le 1$. Then $C_1 \cup C_2$ is a closed path and Green's Theorem

gives $\int_{C_1} \mathbf{F} \cdot d\mathbf{r} + \int_{C_2} \mathbf{F} \cdot d\mathbf{r} = \iint_R \left(\frac{\partial Q}{\partial x} - \frac{\partial P}{\partial y} \right) dA$, where $\mathbf{F}(x, y) = \left(x^2 + 2y \right)\mathbf{i} + (3x - \sinh y)\mathbf{j}$

and R is the region bounded by C_1 and C_2. Now on C_2, $x = 1 + 3t$ and $y = 0$, so

$\int_{C_2} \mathbf{F} \cdot d\mathbf{r} = \int_0^1 \mathbf{F}(\mathbf{r}(t)) \cdot \mathbf{r}'(t)\, dt = \int_0^1 (1 + 3t)^2\,(3\,dt) = \frac{1}{3}(1 + 3t)^3 \Big|_0^1 = 21.$ Next,

$\frac{\partial Q}{\partial x} - \frac{\partial P}{\partial y} = 3 - 2 = 1$, so $\iint_R \left(\frac{\partial Q}{\partial x} - \frac{\partial P}{\partial y} \right) dA = \iint_R dA = (2)(5) - 2\left[\frac{1}{2}(1)(1)\right] = 9.$ Finally,

$\int_{C_1} \left(x^2 + 2y \right) dx + (3x - \sinh y)\, dy = \int_{C_1} \mathbf{F} \cdot d\mathbf{r} = -\int_{C_2} \mathbf{F} \cdot d\mathbf{r} + \iint_R \left(\frac{\partial Q}{\partial x} - \frac{\partial P}{\partial y} \right) dA = -21 + 9 = -12.$

31. The required area is that of the quadrilateral $ABCD$ with vertices $A(3, -3)$, $B(5, 5)$, $C(0, 5)$

and $D(-5, 0)$ minus that of a circle with radius 1 so using the result of Exercise 26, we have

$A = \frac{1}{2} \{[(3)(5) - 5(-3)] + [(5)(5) - (0)(5)] + [(0)(0) - (-5)(5)] + [(-5)(-3) - (3)(0)]\} - \pi(1)^2 = 47.5 - \pi.$

33. $A = \frac{1}{2} \{[(0)(0) - (0)(2)] + [(2)(1) - (0)(3)] + [(3)(3) - (1)(1)] + [(1)(1) - (3)(-1)] + [(-1)(0) - (0)(1)]\} = 7$

35. $\bar{x} = \frac{M_y}{m} = \frac{\iint_R x\rho\, dA}{\iint_R \rho\, dA} = \frac{\iint_R x\, dA}{\iint_R dA} = \frac{1}{A} \iint_R x\, dA = \frac{1}{2A} \iint_R 2x\, dA = \frac{1}{2A} \oint_C x^2\, dy$, where we have used Green's

Theorem at the last step. Similarly, $\bar{y} = \frac{M_x}{m} = \frac{\iint_R y\rho\, dA}{\iint_R \rho\, dA} = \frac{1}{A} \iint_R y\, dA = \frac{1}{2A} \iint_R 2y\, dA = -\frac{1}{2A} \oint_C y^2\, dx$

37. $A = 2\int_0^3 \left(9 - x^2 \right) dx = 2\left(9x - \frac{1}{3}x^3 \right) \Big|_0^3 = 36.$ Let $C = C_1 \cup C_2$, where

$C_1 : x = -3 + 6t, y = 0, 0 \le t \le 1$ and $C_2 : x = -t, y = 9 - t^2, -3 \le t \le 3.$

Now $\bar{x} = 0$ by symmetry, and

$\bar{y} = -\frac{1}{2A} \oint_C y^2\, dx = -\frac{1}{72} \left(\int_{C_1} y^2\, dx + \int_{C_2} y^2\, dx \right)$

$= -\frac{1}{72} \left[\int_0^1 0\,(6\,dt) + \int_{-3}^3 \left(9 - t^2 \right)^2 (-dt) \right]$

$= \frac{1}{72}(2) \int_0^3 \left(9 - t^2 \right)^2 dt$ (the integrand is even)

$= \frac{1}{36} \int_0^3 \left(81 - 18t^2 + t^4 \right) dt = \frac{1}{36} \left(81t - 6t^3 + \frac{1}{5}t^5 \right) \Big|_0^3 = \frac{18}{5}$

Thus, the centroid is $(\bar{x}, \bar{y}) = \left(0, \frac{18}{5} \right).$

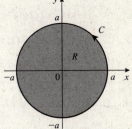

39. Place the lamina so that its center is at the origin. Then

$C : x = a \cos t, y = a \sin t, 0 \le t \le 2\pi$ and the required moment of inertia is

$I_x = -\frac{1}{3}\rho \oint_C y^3\, dx = -\frac{1}{3}\rho \int_0^{2\pi} (a \sin t)^3 (-a \sin t)\, dt = \frac{1}{3}\rho a^4 \int_0^{2\pi} \sin^4 t\, dt$

$= \frac{1}{3}\rho a^4 \left[\left(-\frac{1}{4}\sin^3 t \cos t \right) \Big|_0^{2\pi} + \frac{3}{4} \int_0^{2\pi} \sin^2 t\, dt \right]$

$= \frac{1}{8}\rho a^4 \int_0^{2\pi} (1 - \cos 2t)\, dt = \frac{1}{8}\rho a^4 \left(t - \frac{1}{2}\sin 2t \right) \Big|_0^{2\pi} = \frac{1}{4}\rho a^4 \pi$

41. Using Green's Theorem, we have

$\oint_C (ay + b)\, dx + (cx + d)\, dy = \iint_R \left[\frac{\partial}{\partial x}(cx + d) - \frac{\partial}{\partial y}(ay + b) \right] dA = \iint_R (c - a)\, dA = (c - a)A.$

43. a. Here $C : x = \cos t$, $y = \sin t$, $0 \le t \le 2\pi$, so

$$\int_C P\,dx + Q\,dy = \int \frac{-y}{x^2 + y^2}\,dx + \frac{x}{x^2 + y^2}\,dy$$

$$= \int_0^{2\pi} \left[-\frac{\sin t}{\cos^2 t + \sin^2 t}\,(-\sin t) + \frac{\cos t}{\cos^2 t + \sin^2 t}\,(\cos t) \right] dt$$

$$= \int_0^{2\pi} dt = t\big|_0^{2\pi} = 2\pi \ne 0$$

b. $\dfrac{\partial P}{\partial y} = \dfrac{\partial}{\partial y}\left(-\dfrac{y}{x^2 + y^2}\right) = \dfrac{y^2 - x^2}{(x^2 + y^2)^2}$ and $\dfrac{\partial Q}{\partial x} = \dfrac{\partial}{\partial x}\left(\dfrac{x}{x^2 + y^2}\right) = \dfrac{y^2 - x^2}{(x^2 + y^2)^2}$, so $\dfrac{\partial P}{\partial y} = \dfrac{\partial Q}{\partial x}$.

c. Green's Theorem does not apply because P and Q are not defined at the origin.

45. a. Without loss of generality, assume C is positively oriented. Then

$$\oint_C \left(\cos x + x^3 y\right) dx + \left(x^4 + e^y\right) dy$$

$$= \iint_R \left[\frac{\partial}{\partial x}\left(x^4 + e^y\right) - \frac{\partial}{\partial y}\left(\cos x + x^3 y\right)\right] dA$$

$$= \iint_R \left(4x^3 - x^3\right) dA = \int_{-1}^1 \int_{-1}^1 3x^3\,dx\,dy$$

$$= \int_{-1}^1 \left[\tfrac{3}{4}x^4\right]_{-1}^1 dy = \int_{-1}^1 0\,dy = 0$$

b. The result does not contradict Theorem 15.4.4 because $\dfrac{\partial Q}{\partial x} \ne \dfrac{\partial P}{\partial y} \Rightarrow \mathbf{F}$ is not a conservative vector field, and so for a specific closed path C, the integral $\int_C \mathbf{F} \cdot d\mathbf{r}$ may or may not be equal to zero.

c. $\oint_C \left(\cos x + x^3 y\right) dx + \left(x^4 + e^y\right) dy$

$$= \iint_R \left[\frac{\partial}{\partial x}\left(x^4 + e^y\right) - \frac{\partial}{\partial y}\left(\cos x + x^3 y\right)\right] dA$$

$$= \int_0^1 \int_0^1 3x^3\,dx\,dy = \int_0^1 \left[\tfrac{3}{4}x^4\right]_0^1 dy = \int_0^1 \tfrac{3}{4}\,dy = \tfrac{3}{4}$$

47. Using Green's Theorem, we find

$$\oint_C P(y)\,dx + Q(x)\,dy = \iint_R \left\{\frac{\partial}{\partial x}[Q(x)] - \frac{\partial}{\partial y}[P(y)]\right\} dA$$

$$= \int_{-1}^1 \int_{-1}^1 [Q'(x) - P'(y)]\,dx\,dy$$

$$= \int_{-1}^1 [Q(x) - P'(y)x]_{x=-1}^{x=1}\,dy$$

$$= \int_{-1}^1 \{[Q(1) - P'(y)] - [Q(-1) + P'(y)]\}\,dy$$

$$= \int_{-1}^1 [Q(1) - Q(-1) - 2P'(y)]\,dy = \{[Q(1) - Q(-1)]\,y + 2P(y)\}|_{-1}^1$$

$$= [Q(1) - Q(-1) + 2P(1)] - \{[Q(1) - Q(-1)](-1) + 2P(-1)\}$$

$$= 2[Q(1) - Q(-1)] + 2[P(1) - P(-1)] = 2[Q(1) + P(1)] - 2[Q(-1) + P(-1)]$$

$$= 2[Q(t) + P(t)]_{t=-1}^{t=1}$$

49. False. Using Green's Theorem, we have $\oint_C a\,dx + b\,dy = \iint_R \left[\frac{\partial}{\partial x}(b) - \frac{\partial}{\partial y}(a)\right] dA = \iint_R 0\,dA = 0$.

51. True. Using Green's Theorem, we have

$$W = \oint_C \mathbf{F}\cdot d\mathbf{r} = \oint_C \left(-\tfrac{1}{2}y\right) dx + \left(\tfrac{1}{2}x\right) dy = \iint_R \left[\frac{\partial}{\partial x}\left(\tfrac{1}{2}x\right) - \frac{\partial}{\partial y}\left(-\tfrac{1}{2}y\right)\right] dA = \iint_R \left(\tfrac{1}{2} + \tfrac{1}{2}\right) dA = \iint_R 1\,dA = A(R).$$

15.6 Concept Questions   ET 14.6

1. a. See page 1279 (1275 in ET).
   b. See page 1280 (1276 in ET).

3. $A = \iint_D |\mathbf{r}_u \times \mathbf{r}_v|\, dA$

15.6 Parametric Surfaces   ET 14.6

1. The equation matches graph b. Here $x = 2\cos u$, $y = 2\sin u$, $z = v$, so $x^2 + y^2 = 4$ and the surface is a circular cylinder
with axis parallel to the z-axis.

3. The equation matches graph a. Here $x = u\cos v$, $y = u\sin v$, $z = u^2$, so $x^2 + y^2 = u^2\cos^2 v + u^2\sin^2 v = u^2 = z$; that
is, $x^2 + y^2 = z$, an equation of a paraboloid.

5. $\mathbf{r}(u, v) = (u - v)\mathbf{i} + 3v\mathbf{j} + (u + v)\mathbf{k} \Rightarrow x = u - v$, $y = 3v$, $z = u + v$. The
second equation gives $v = \frac{1}{3}y$, then the other two equations give
$x - z = (u - v) - (u + v) = -2v = -\frac{2}{3}y \Rightarrow 3x + 2y - 3z = 0$, an equation
of a plane.

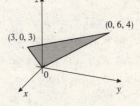

7. $\mathbf{r}(u, v) = 3\sin u\,\mathbf{i} + 2\cos u\,\mathbf{j} + v\mathbf{k}$, $0 \le v \le 2 \Rightarrow$
$x = 3\sin u$, $y = 2\cos u$, $z = v \Rightarrow$
$\left(\frac{1}{3}x\right)^2 + \left(\frac{1}{2}y\right)^2 = \frac{1}{9}x^2 + \frac{1}{4}y^2 = 1$, $0 \le z \le 2$, a cylinder with an elliptical
cross-section and axis the z-axis, bounded below by the plane $z = 0$ and above
by the plane $z = 2$.

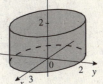

9.

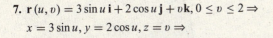

11.

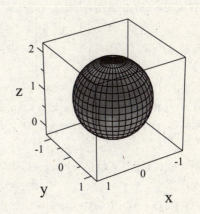

13.

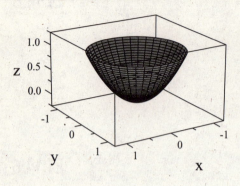

15. $\mathbf{r}(u, v) = \mathbf{r}_0 + u\mathbf{a} + v\mathbf{b}$, where $\mathbf{r}_0 = 2\mathbf{i} + \mathbf{j} - 3\mathbf{k}$, $\mathbf{a} = 2\mathbf{i} + \mathbf{j} - \mathbf{k}$, and $\mathbf{b} = \mathbf{i} - 2\mathbf{j} - \mathbf{k} \Rightarrow$
$\mathbf{r}(u, v) = (2\mathbf{i} + \mathbf{j} - 3\mathbf{k}) + u(2\mathbf{i} + \mathbf{j} - \mathbf{k}) + v(\mathbf{i} - 2\mathbf{j} - \mathbf{k}) = (2 + 2u + v)\mathbf{i} + (1 + u - 2v)\mathbf{j} - (3 + u + v)\mathbf{k}$ with domain
the entire uv-plane.

17. We have $\mathbf{r}(u, v) = \cos v \cos u\,\mathbf{i} + \cos v \sin u\,\mathbf{j} - \sin v\,\mathbf{k}$ with domain $D = \{(u, v) \mid 0 \le u \le 2\pi, 0 \le v \le \pi\}$.

19. We have $\mathbf{r}(u, v) = 2\cos u\,\mathbf{i} + 2\sin u\,\mathbf{j} + v\mathbf{k}$ with domain $D = \{(u, v) \mid 0 \le u \le 2\pi, -1 \le v \le 3\}$.

21. The region inside the cylinder $x^2 + y^2 = 4$ can be described by $x^2 + y^2 = v^2$, $0 \le v \le 2$, so since $z = 9 - \left(x^2 + y^2\right)$, we

have the representation $\mathbf{r}(u, v) = v\cos u\,\mathbf{i} + v\sin u\,\mathbf{j} + \left(9 - v^2\right)\mathbf{k}$ with domain $D = \{(u, v) \mid 0 \le u \le 2\pi, 0 \le v \le 2\}$.

23. $y = f(x) = \sqrt{x}$. Let $x = u$, $y = f(u)\cos v$, and
$z = f(u)\sin v$. Then the required equation is
$\mathbf{r}(u, v) = u\mathbf{i} + \sqrt{u}\cos v\,\mathbf{j} + \sqrt{u}\sin v\,\mathbf{k}$ with domain
$D = \{(u, v) \mid 0 \le u \le 4, 0 \le v \le 2\pi\}$.

25. $x = g(y) = 9 - y^2$. Let $y = u$, $x = g(u)\cos v$, and
$z = g(u)\sin v$. Then the required equation is
$\mathbf{r}(u, v) = \left(9 - u^2\right)\cos v\,\mathbf{i} + u\mathbf{j} + \left(9 - u^2\right)\sin v\,\mathbf{k}$ with
domain $D = \{(u, v) \mid 0 \le u \le 3, 0 \le v \le 2\pi\}$.

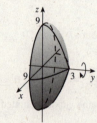

27. $\mathbf{r}(u, v) = (u + v)\mathbf{i} + (u - v)\mathbf{j} + v^2\mathbf{k} \Rightarrow \mathbf{r}_u(u, v) = \mathbf{i} + \mathbf{j}$ and $\mathbf{r}_v = \mathbf{i} - \mathbf{j} + 2v\mathbf{k}$. Observe that the point
$(2, 0, 1)$ on the surface corresponds to $(u, v) = (1, 1)$, so a normal vector to the required tangent plane is

$$\mathbf{n} = \mathbf{r}_u(1, 1) \times \mathbf{r}_v(1, 1) = \begin{vmatrix} \mathbf{i} & \mathbf{j} & \mathbf{k} \\ 1 & 1 & 0 \\ 1 & -1 & 2 \end{vmatrix} = 2\mathbf{i} - 2\mathbf{j} - 2\mathbf{k}.$$ Therefore, an equation of the tangent plane is

$2(x - 2) - 2(y - 0) - 2(z - 1) = 0 \Rightarrow x - y - z = 1$.

29. $\mathbf{r}(u, v) = u\cos v\,\mathbf{i} + 2u\sin v\,\mathbf{j} + u^2\mathbf{k} \Rightarrow \mathbf{r}_u(u, v) = \cos v\,\mathbf{i} + 2\sin v\,\mathbf{j} + 2u\mathbf{k}$ and $\mathbf{r}_v(u, v) = -u\sin v\,\mathbf{i} + 2u\cos v\,\mathbf{j}$.
The point $(u, v) = (1, \pi)$ gives the point $(-1, 0, 1)$ on the surface, so a normal vector to the required tangent

plane is $\mathbf{n} = \mathbf{r}_u(1, \pi) \times \mathbf{r}_v(1, \pi) = \begin{vmatrix} \mathbf{i} & \mathbf{j} & \mathbf{k} \\ -1 & 0 & 2 \\ 0 & -2 & 0 \end{vmatrix} = 4\mathbf{i} + 2\mathbf{k}$. Therefore, an equation of the tangent plane is

$4(x + 1) + 2(z - 1) = 0 \Rightarrow 2x + z = -1$.

31. $\mathbf{r}(u, v) = ue^v\mathbf{i} + ve^u\mathbf{j} + uv\mathbf{k} \Rightarrow \mathbf{r}_u(u, v) = e^v\mathbf{i} + ve^u\mathbf{j} + v\mathbf{k}$ and $\mathbf{r}_v(u, v) = ue^v\mathbf{i} + e^u\mathbf{j} + u\mathbf{k}$. The point
$(u, v) = (0, \ln 2)$ gives the point $(0, \ln 2, 0)$ on the surface, so a normal vector to the required tangent plane is

$$\mathbf{n} = \mathbf{r}_u(0, \ln 2) \times \mathbf{r}_v(0, \ln 2) = \begin{vmatrix} \mathbf{i} & \mathbf{j} & \mathbf{k} \\ 2 & \ln 2 & \ln 2 \\ 0 & 1 & 0 \end{vmatrix} = -\ln 2\,\mathbf{i} + 2\mathbf{k}.$$ Therefore, an equation of the tangent plane is

$(-\ln 2)(x - 0) + 2(z - 0) = 0 \Rightarrow (\ln 2)\,x - 2z = 0.$

33. $\mathbf{r}(u, v) = (u + 2v - 1)\mathbf{i} + (2u + 3v + 1)\mathbf{j} + (u + v + 2)\mathbf{k}$, so $\mathbf{r}_u(u, v) = \mathbf{i} + 2\mathbf{j} + \mathbf{k}$ and $\mathbf{r}_v(u, v) = 2\mathbf{i} + 3\mathbf{j} + \mathbf{k}$.

Thus, $\mathbf{r}_u \times \mathbf{r}_v = \begin{vmatrix} \mathbf{i} & \mathbf{j} & \mathbf{k} \\ 1 & 2 & 1 \\ 2 & 3 & 1 \end{vmatrix} = -\mathbf{i} + \mathbf{j} - \mathbf{k}$ and $|\mathbf{r}_u \times \mathbf{r}_v| = \sqrt{(-1)^2 + 1^2 + (-1)^2} = \sqrt{3}$, so the required area is

$A = \iint_D |\mathbf{r}_u \times \mathbf{r}_v|\, du\, dv = \int_0^1 \int_0^2 \sqrt{3}\, dv\, du = \int_0^1 \left[\sqrt{3}v\right]_0^2 du = 2\sqrt{3}u\,\Big|_0^1 = 2\sqrt{3}.$

35. A vector representation of the surface is $\mathbf{r}(u, v) = u\cos v\,\mathbf{i} + u\sin v\,\mathbf{j} + (8 - 2u\cos v - 3u\sin v)\mathbf{k}$, $0 \le u \le 2$, $0 \le v \le 2\pi$,
so $\mathbf{r}_u(u, v) = \cos v\,\mathbf{i} + \sin v\,\mathbf{j} + (-2\cos v - 3\sin v)\mathbf{k}$ and $\mathbf{r}_v(u, v) = -u\sin v\,\mathbf{i} + u\cos v\,\mathbf{j} + (2u\sin v - 3u\cos v)\mathbf{k}$.

Thus, $\mathbf{r}_u \times \mathbf{r}_v = \begin{vmatrix} \mathbf{i} & \mathbf{j} & \mathbf{k} \\ \cos v & \sin v & -(2\cos v + 3\sin v) \\ -u\sin v & u\cos v & 2u\sin v - 3u\cos v \end{vmatrix} = 2u\mathbf{i} + 3u\mathbf{j} + u\mathbf{k}$

and $|\mathbf{r}_u \times \mathbf{r}_v| = \sqrt{(2u)^2 + (3u)^2 + u^2} = \sqrt{14}\,u$, so the required area is
$A = \iint_D |\mathbf{r}_u \times \mathbf{r}_v|\, du\, dv = \int_0^{2\pi} \int_0^2 \sqrt{14}\,u\, du\, dv = \int_0^{2\pi} \left[\sqrt{14}\left(\tfrac{1}{2}u^2\right)\right]_0^2 dv = 2\sqrt{14} \int_0^{2\pi} dv = 4\sqrt{14}\pi.$

37. $\mathbf{r}(u, v) = u\cos v\,\mathbf{i} + u\sin v\,\mathbf{j} + u\mathbf{k}$, so $\mathbf{r}_u(u, v) = \cos v\,\mathbf{i} + \sin v\,\mathbf{j} + \mathbf{k}$ and $\mathbf{r}_v(u, v) = -u\sin v\,\mathbf{i} + u\cos v\,\mathbf{j}$. Thus,

$\mathbf{r}_u \times \mathbf{r}_v = \begin{vmatrix} \mathbf{i} & \mathbf{j} & \mathbf{k} \\ \cos v & \sin v & 1 \\ -u\sin v & u\cos v & 0 \end{vmatrix} = -u\cos v\,\mathbf{i} - u\sin v\,\mathbf{j} + u\mathbf{k}$ and $|\mathbf{r}_u \times \mathbf{r}_v| = \sqrt{u^2\cos^2 v + u^2\sin^2 v + u^2} = \sqrt{2}\,u$, so

the required area is $A = \iint_D |\mathbf{r}_u \times \mathbf{r}_v|\, du\, dv = \int_0^{\pi/2} \int_1^2 \sqrt{2}\,u\, du\, dv = \int_0^{\pi/2} \left[\tfrac{\sqrt{2}}{2}u^2\right]_1^2 dv = \int_0^{\pi/2} \tfrac{3\sqrt{2}}{2}\, dv = \tfrac{3\sqrt{2}\pi}{4}.$

39. $\mathbf{r}(u, v) = \sin u\cos v\,\mathbf{i} + \sin u\sin v\,\mathbf{j} + u\mathbf{k}$, $0 \le u \le \pi$, $0 \le v \le 2\pi$, so
$\mathbf{r}_u(u, v) = \cos u\cos v\,\mathbf{i} + \cos u\sin v\,\mathbf{j} + \mathbf{k}$ and $\mathbf{r}_v(u, v) = -\sin u\sin v\,\mathbf{i} + \sin u\cos v\,\mathbf{j}$. Thus,

$\mathbf{r}_u \times \mathbf{r}_v = \begin{vmatrix} \mathbf{i} & \mathbf{j} & \mathbf{k} \\ \cos u\cos v & \cos u\sin v & 1 \\ -\sin u\sin v & \sin u\cos v & 0 \end{vmatrix} = -\sin u\cos v\,\mathbf{i} - \sin u\sin v\,\mathbf{j} + \sin u\cos u\,\mathbf{k}$ and

$|\mathbf{r}_u \times \mathbf{r}_v| = \sqrt{\sin^2 u\cos^2 v + \sin^2 u\sin^2 v + \sin^2 u\cos^2 u} = \sin u\sqrt{1 + \cos^2 u}$, so the required area is

$A = \iint_D |\mathbf{r}_u \times \mathbf{r}_v|\, du\, dv = \int_0^{2\pi} \int_0^{\pi} \sin u\sqrt{1 + \cos^2 u}\, du\, dv = \int_0^{2\pi} \int_{-1}^{1} \sqrt{1 + x^2}\, dx\, dv$ (let $x = \cos u$)

$= \int_0^{2\pi} \left[\tfrac{1}{2}x\sqrt{1 + x^2} + \tfrac{1}{2}\ln\left(x + \sqrt{1 + x^2}\right)\right]_{-1}^{1} dv = \int_0^{2\pi} \left[\sqrt{2} + \tfrac{1}{2}\ln\left(3 + 2\sqrt{2}\right)\right] dv$

$= \left[2\sqrt{2} + \ln\left(3 + 2\sqrt{2}\right)\right]\pi$

41. a. Here $x = a \sin u \cos v$, $y = b \sin u \sin v$, and $z = c \cos u$, so

$$\left(\frac{x}{a}\right)^2 + \left(\frac{y}{b}\right)^2 + \left(\frac{z}{c}\right)^2 = \sin^2 u \cos^2 v + \sin^2 u \sin^2 v + \cos^2 u$$

$$= \sin^2 u + \cos^2 u = 1$$

This equation represents an ellipsoid.

b.

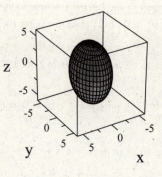

c. Here $\mathbf{r}(u, v) = 3 \sin u \cos v\,\mathbf{i} + 4 \sin u \sin v\,\mathbf{j} + 5 \cos u\,\mathbf{k}$, so

$\mathbf{r}_u(u, v) = 3 \cos u \cos v\,\mathbf{i} + 4 \cos u \sin v\,\mathbf{j} - 5 \sin u\,\mathbf{k}$ and $\mathbf{r}_v(u, v) = -3 \sin u \sin v\,\mathbf{i} + 4 \sin u \cos v\,\mathbf{j}$. Thus,

$$\mathbf{r}_u \times \mathbf{r}_v = \begin{vmatrix} \mathbf{i} & \mathbf{j} & \mathbf{k} \\ 3 \cos u \cos v & 4 \cos u \sin v & -5 \sin u \\ -3 \sin u \sin v & 4 \sin u \cos v & 0 \end{vmatrix} = 20 \sin^2 u \cos v\,\mathbf{i} + 15 \sin^2 u \sin v\,\mathbf{j} + 12 \cos u \sin u\,\mathbf{k}$$

and $|\mathbf{r}_u \times \mathbf{r}_v| = \sqrt{400 \sin^4 u \cos^2 v + 225 \sin^4 u \sin^2 v + 144 \cos^2 u \sin^2 u}$, so the required area is

$A = \iint_D |\mathbf{r}_u \times \mathbf{r}_v|\, du\, dv = 8 \int_0^{\pi/2} \int_0^{\pi/2} |\mathbf{r}_u \times \mathbf{r}_v|\, du\, dv \approx 199.455$.

43. Let $\mathbf{r}(u, v) = u\mathbf{i} + v\mathbf{j} + \sqrt{u^2 + v^2}\,\mathbf{k}$. Then $\mathbf{r}_u(u, v) = \mathbf{i} + \dfrac{u}{\sqrt{u^2 + v^2}}\mathbf{k}$ and

$\mathbf{r}_v(u, v) = \mathbf{j} + \dfrac{v}{\sqrt{u^2 + v^2}}\mathbf{k}$, so

$$\mathbf{r}_u \times \mathbf{r}_v = \begin{vmatrix} \mathbf{i} & \mathbf{j} & \mathbf{k} \\ 1 & 0 & \dfrac{u}{\sqrt{u^2 + v^2}} \\ 0 & 1 & \dfrac{v}{\sqrt{u^2 + v^2}} \end{vmatrix} = -\dfrac{u}{\sqrt{u^2 + v^2}}\mathbf{i} - \dfrac{v}{\sqrt{u^2 + v^2}}\mathbf{j} + \mathbf{k}$$ and

$|\mathbf{r}_u \times \mathbf{r}_v| = \sqrt{\dfrac{u^2}{u^2 + v^2} + \dfrac{v^2}{u^2 + v^2} + 1} = \sqrt{2}$. Thus, $A = \iint_R \sqrt{2}\, dA = \sqrt{2}\pi$.

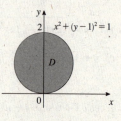

45. We use the vector representation of the surface of revolution: $\mathbf{r}(u, v) = u\mathbf{i} + f(u) \cos v\,\mathbf{j} + f(u) \sin v\,\mathbf{k}$

with parameter domain $D = \{(u, v) \mid a \le u \le b, 0 \le v \le 2\pi\}$ (see Equation 1 on page 1283/1279

in ET). $\mathbf{r}_u(u, v) = \mathbf{i} + f'(u) \cos v\,\mathbf{j} + f'(u) \sin v\,\mathbf{k}$ and $\mathbf{r}_v(u, v) = -f(u) \sin v\,\mathbf{j} + f(u) \cos v\,\mathbf{k}$,

so $\mathbf{r}_u \times \mathbf{r}_v = \begin{vmatrix} \mathbf{i} & \mathbf{j} & \mathbf{k} \\ 1 & f'(u) \cos v & f'(u) \sin v \\ 0 & -f(u) \sin v & f(u) \cos v \end{vmatrix} = f(u) f'(u)\,\mathbf{i} - f(u) \cos v\,\mathbf{j} - f(u) \sin v\,\mathbf{k}$. Thus,

$|\mathbf{r}_u \times \mathbf{r}_v| = \sqrt{[f(u)]^2 [f'(u)]^2 + [f(u)]^2 \cos^2 v + [f(u)]^2 \sin^2 v} = f(u) \sqrt{1 + [f'(u)]^2}$ since $f(u) \ge 0$. Then

$S = \iint_D |\mathbf{r}_u \times \mathbf{r}_v|\, du\, dv = \int_0^{2\pi} \int_a^b f(u) \sqrt{1 + [f'(u)]^2}\, du\, dv = \int_a^b f(u) \sqrt{1 + [f'(u)]^2}\, du \int_0^{2\pi} dv$

$= 2\pi \int_a^b f(u) \sqrt{1 + [f'(u)]^2}\, du = 2\pi \int_a^b f(x) \sqrt{1 + [f'(x)]^2}\, dx$

47. False. $\mathbf{r}_u(u, v) = \cos v\,\mathbf{i} + \sin v\,\mathbf{j} + \mathbf{k}$ and $\mathbf{r}_v(u, v) = -u\sin v\,\mathbf{i} + u\cos v\,\mathbf{j}$, so

$$\mathbf{r}_u \times \mathbf{r}_v = \begin{vmatrix} \mathbf{i} & \mathbf{j} & \mathbf{k} \\ \cos v & \sin v & 1 \\ -u\sin v & u\cos v & 0 \end{vmatrix} = -u\cos v\,\mathbf{i} - u\sin v\,\mathbf{j} + u\mathbf{k} = \mathbf{0} \text{ if } u = 0. \text{ Therefore, the surface described by } \mathbf{r}(u, v)$$

is not smooth.

15.7 Concept Questions   ET 14.7

1. a. See page 1292 (1288 in ET).   **b.** See page 1292 (1288 in ET).   **c.** See page 1294 (1290 in ET).

3. a. See page 1296 (1292 in ET).   **b.** See page 1297 (1293 in ET).   **c.** See page 1299 (1295 in ET).

15.7 Surface Integrals   ET 14.7

1. $3x + 2y + z = 6 \Rightarrow z = g(x, y) = 6 - 3x - 2y$, so

$$\iint_S f(x, y, z)\,dS = \iint_S (x + y)\,dS$$

$$= \iint_R (x + y)\sqrt{[g_x(x, y)]^2 + [g_y(x, y)]^2 + 1}\,dA$$

$$= \iint_R (x + y)\sqrt{(-3)^2 + (-2)^2 + 1}\,dA$$

$$= \sqrt{14}\int_0^2\int_0^{(6-3x)/2}(x + y)\,dy\,dx$$

$$= \sqrt{14}\int_0^2\left[xy + \tfrac{1}{2}y^2\right]_{y=0}^{y=(6-3x)/2}dx = -\frac{\sqrt{14}}{8}\int_0^2\left(3x^2 + 12x - 36\right)dx = -\frac{\sqrt{14}}{8}\left(x^3 + 6x^2 - 36x\right)\Big|_0^2$$

$$= 5\sqrt{14}$$

3. $z = g(x, y) = 2x + y^2 \Rightarrow g_x(x, y) = 2$ and $g_y(x, y) = 2y$, so

$$\iint_S f(x, y, z)\,dS = \iint_R y\sqrt{[g_x(x, y)]^2 + [g_y(x, y)]^2 + 1}\,dA = \iint_R y\sqrt{4y^2 + 5}\,dA$$

$$= \int_0^2\int_0^1 y\left(4y^2 + 5\right)^{1/2}dy\,dx = \int_0^2\left[\left(\tfrac{1}{8}\right)\left(\tfrac{2}{3}\right)\left(4y^2 + 5\right)^{3/2}\right]_0^1 dx$$

$$= \int_0^2 \tfrac{1}{12}\left(27 - 5\sqrt{5}\right)dx = \tfrac{1}{6}\left(27 - 5\sqrt{5}\right)$$

5. $y + z = 4 \Rightarrow z = g(x, y) = 4 - y$, so $g_x(x, y) = 0$ and $g_y(x, y) = -1$. Thus,

$$\iint_S f(x, y, z)\,dS = \iint_S (x + 2y + z)\,dS = \iint_S [x + 2y + (4 - y)]\,dS$$

$$= \iint_R (x + y + 4)\sqrt{[g_x(x, y)]^2 + [g_y(x, y)]^2 + 1}\,dA$$

$$= \sqrt{2}\int_0^{2\pi}\int_0^1 (r\cos\theta + r\sin\theta + 4)r\,dr\,d\theta$$

$$= \sqrt{2}\int_0^{2\pi}\left[(\cos\theta + \sin\theta)\left(\tfrac{1}{3}r^3\right) + 2r^2\right]_{r=0}^{r=1}d\theta$$

$$= \sqrt{2}\int_0^{2\pi}\left[\tfrac{1}{3}(\cos\theta + \sin\theta) + 2\right]d\theta = \sqrt{2}\left[\tfrac{1}{3}(\sin\theta - \cos\theta) + 2\theta\right]_0^{2\pi} = 4\sqrt{2}\pi$$

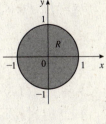

7. $z = g(x, y) = \sqrt{x^2 + y^2} \Rightarrow g_x(x, y) = \dfrac{x}{\sqrt{x^2 + y^2}}$ and

$g_y(x, y) = \dfrac{y}{\sqrt{x^2 + y^2}}$, so

$$\iint_S f(x, y, z)\, dS = \iint_S x^2 z\, dS = \iint_S x^2 \sqrt{x^2 + y^2}\, dS$$

$$= \iint_R x^2 \sqrt{x^2 + y^2} \sqrt{[g_x(x, y)]^2 + [g_y(x, y)]^2 + 1}\, dA$$

$$= \iint_R x^2 \sqrt{x^2 + y^2} \sqrt{\dfrac{x^2}{x^2 + y^2} + \dfrac{y^2}{x^2 + y^2} + 1}\, dA$$

$$= \sqrt{2} \int_0^{2\pi} \int_0^1 (r\cos\theta)^2 \sqrt{r^2}\, r\, dr\, d\theta = \sqrt{2} \int_0^{2\pi} \left[\left(\cos^2\theta\right)\left(\tfrac{1}{5}r^5\right) \right]_{r=0}^{r=1} d\theta = \dfrac{\sqrt{2}}{5} \int_0^{2\pi} \cos^2\theta\, d\theta$$

$$= \dfrac{\sqrt{2}}{5} \int_0^{2\pi} \tfrac{1}{2}(1 + \cos 2\theta)\, d\theta = \dfrac{\sqrt{2}\pi}{5}$$

9. We project the relevant part of the cylinder onto the yz-plane.

$x = g(y, z) = \sqrt{4 - y^2} \Rightarrow g_y(y, z) = -\dfrac{y}{\sqrt{4 - y^2}}$ and $g_z(y, z) = 0$, so

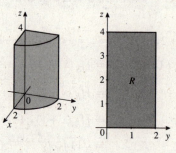

$$\iint_S f(x, y, z)\, dS = \iint_S xyz\, dS$$

$$= \iint_R yz\sqrt{4 - y^2} \sqrt{[g_y(y, z)]^2 + [g_z(y, z)]^2 + 1}\, dA$$

$$= \iint_R yz\sqrt{4 - y^2} \left(\dfrac{2}{\sqrt{4 - y^2}} \right) dA = 2 \int_0^2 \int_0^4 yz\, dz\, dy$$

$$= 2 \int_0^2 \left[\tfrac{1}{2} yz^2 \right]_{z=0}^{z=4} dy = 16 \int_0^2 y\, dy = 32$$

11. $\mathbf{r}(u, v) = u\mathbf{i} + v\mathbf{j} + \left(v^2 - 1\right)\mathbf{k} \Rightarrow \mathbf{r}_u(u, v) = \mathbf{i}$ and $\mathbf{r}_v(u, v) = \mathbf{j} + 2v\mathbf{k}$, so

$$\mathbf{r}_u \times \mathbf{r}_v = \begin{vmatrix} \mathbf{i} & \mathbf{j} & \mathbf{k} \\ 1 & 0 & 0 \\ 0 & 1 & 2v \end{vmatrix} = -2v\mathbf{j} + \mathbf{k} \Rightarrow |\mathbf{r}_u \times \mathbf{r}_v| = \sqrt{4v^2 + 1}, \text{ so}$$

$$\iint_S f(x, y, z)\, dS = \iint_D f(\mathbf{r}(u, v)) |\mathbf{r}_u \times \mathbf{r}_v|\, dA$$

$$= \iint_D \left(u + \dfrac{v}{\sqrt{4(v^2 - 1) + 5}} \right) \sqrt{4v^2 + 1}\, dA$$

$$= \int_{-1}^1 \int_0^1 \left(u\sqrt{4v^2 + 1} + v \right) du\, dv = \int_{-1}^1 \left[\tfrac{1}{2}u^2 \sqrt{4v^2 + 1} + uv \right]_{u=0}^{u=1} dv = \int_{-1}^1 \left(\tfrac{1}{2}\sqrt{4v^2 + 1} + v \right) dv$$

$$= \tfrac{1}{2} \left\{ \tfrac{1}{2} \left[v\sqrt{4v^2 + 1} + \tfrac{1}{2}\ln\left(2v + \sqrt{4v^2 + 1}\right) \right] + v^2 \right\} \bigg|_{-1}^1 = \tfrac{1}{4} \left(2\sqrt{5} + \tfrac{1}{2}\ln \dfrac{\sqrt{5} + 2}{\sqrt{5} - 2} \right)$$

$$= \tfrac{1}{4} \left[2\sqrt{5} + \ln\left(\sqrt{5} + 2\right) \right]$$

13. $\mathbf{r}(u, v) = u \cos v\, \mathbf{i} + u \sin v\, \mathbf{j} + v\mathbf{k} \Rightarrow \mathbf{r}_u(u, v) = \cos v\, \mathbf{i} + \sin v\, \mathbf{j}$ and $\mathbf{r}_v(u, v) = -u \sin v\, \mathbf{i} + u \cos v\, \mathbf{j} + \mathbf{k}$, so

$$\mathbf{r}_u \times \mathbf{r}_v = \begin{vmatrix} \mathbf{i} & \mathbf{j} & \mathbf{k} \\ \cos v & \sin v & 0 \\ -u \sin v & u \cos v & 1 \end{vmatrix} = \sin v\, \mathbf{i} - \cos v\, \mathbf{j} + u\mathbf{k} \Rightarrow |\mathbf{r}_u \times \mathbf{r}_v| = \sqrt{u^2 + 1}. \text{ Thus,}$$

$$\iint_S f(x, y, z)\, dS = \iint_D f(\mathbf{r}(u, v)) |\mathbf{r}_u \times \mathbf{r}_v|\, dA = \iint_D v\sqrt{1 + (u \cos v)^2 + (u \sin v)^2} \cdot \sqrt{u^2 + 1}\, dA$$

$$= \int_0^1 \int_0^{2\pi} v\left(u^2 + 1\right) dv\, du = \int_0^1 \left[\tfrac{1}{2} v^2 \left(u^2 + 1\right)\right]_{v=0}^{v=2\pi} du = 2\pi^2 \int_0^1 \left(u^2 + 1\right) du = 2\pi^2 \left(\tfrac{1}{3} u^3 + u\right)\Big|_0^1$$

$$= \tfrac{8}{3}\pi^2$$

15. We have $\rho(x, y, z) = kx^2$, where k is the constant of proportionality. Now $x + 2y + 3z = 6 \Rightarrow x = g(y, z) = 6 - 2y - 3z$. The projection of S onto the yz-plane is the region shown. Then

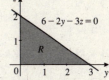

$$m = \iint_S \rho(x, y, z)\, dS = \iint_R kx^2 \sqrt{[g_y(y, z)]^2 + [g_z(y, z)]^2 + 1}\, dA$$

$$= k \int_0^3 \int_0^{(6-2y)/3} (6 - 2y - 3z)^2 \sqrt{(-2)^2 + (-3)^2 + 1}\, dz\, dy$$

$$= \sqrt{14}\, k \int_0^3 \int_0^{(6-2y)/3} (6 - 2y - 3z)^2\, dz\, dy = \sqrt{14}\, k \int_0^3 \left[\left(-\tfrac{1}{3}\right)\left(\tfrac{1}{3}\right)(6 - 2y - 3z)^3\right]_{z=0}^{z=(6-2y)/3} dy$$

$$= \tfrac{8\sqrt{14}}{9} k \int_0^3 (3 - y)^3\, dy = -\tfrac{8\sqrt{14}}{9} k \left(\tfrac{1}{4}\right)(3 - y)^4\Big|_0^3 = 18\sqrt{14}\, k$$

17. The density function is $\rho(x, y, z) = kz = k\sqrt{4 - x^2 - y^2}$, where k is the constant of proportionality. Furthermore, S has representation $z = g(x, y) = \sqrt{4 - x^2 - y^2} \Rightarrow g_x(x, y) = -\dfrac{x}{\sqrt{4 - x^2 - y^2}}$ and $g_y(x, y) = -\dfrac{y}{\sqrt{4 - x^2 - y^2}}$, so

$$m = \iint_S \rho(x, y, z)\, dS = \iint_R k\sqrt{4 - x^2 - y^2} \sqrt{\left(\dfrac{-x}{\sqrt{4 - x^2 - y^2}}\right)^2 + \left(\dfrac{-y}{\sqrt{4 - x^2 - y^2}}\right)^2 + 1}\, dA = \iint_R 2k\, dA$$

$$= 2k\pi(2)^2 = 8k\pi$$

19. $\mathbf{F} = P\mathbf{i} + Q\mathbf{j} + R\mathbf{k} = 2x\mathbf{i} + 2y\mathbf{j} + z\mathbf{k}$ and $z = g(x, y) = 4 - x^2 - y^2 \Rightarrow g_x(x, y) = -2x$ and $g_y(x, y) = -2y$. Then

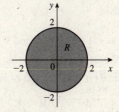

$$\iint_S \mathbf{F} \cdot d\mathbf{S} = \iint_R (-Pg_x - Qg_y + R)\, dA$$

$$= \iint_R \left[-(2x)(-2x) - (2y)(-2y) + \left(4 - x^2 - y^2\right)\right] dA$$

$$= \iint_R \left[3\left(x^2 + y^2\right) + 4\right] dA = \int_0^{2\pi} \int_0^2 \left(3r^2 + 4\right) r\, dr\, d\theta$$

$$= \int_0^{2\pi} \left[\tfrac{3}{4} r^4 + 2r^2\right]_0^2 d\theta = \int_0^{2\pi} 20\, d\theta = 40\pi$$

21. $\mathbf{F} = P\mathbf{i} + Q\mathbf{j} + R\mathbf{k} = x\mathbf{i} + y\mathbf{j} + z\mathbf{k}$ and $z = g(x, y) = 6 - 3x - 2y$, so

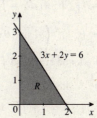

$$\iint_S \mathbf{F} \cdot d\mathbf{S} = \iint_R (-Pg_x - Qg_y + R)\, dA = \iint_R [-x(-3) - y(-2) + z]\, dA$$

$$= \iint_R [3x + 2y + (6 - 3x - 2y)]\, dA = \iint_R 6\, dA = 6\left(\tfrac{1}{2}\right)(3)(2)$$

$$= 18$$

23. $\mathbf{F} = P\mathbf{i} + Q\mathbf{j} + R\mathbf{k} = -y\mathbf{i} + x\mathbf{j} + 2z\mathbf{k}$. Also, $z = g(x, y) = \sqrt{4 - x^2 - y^2} \Rightarrow$

$g_x(x, y) = -\dfrac{x}{\sqrt{4 - x^2 - y^2}}$ and $g_y(x, y) = -\dfrac{y}{\sqrt{4 - x^2 - y^2}}$. Therefore,

$$\iint_S \mathbf{F} \cdot d\mathbf{S} = \iint_R (-Pg_x - Qg_y + R)\, dA$$

$$= \iint_R \left(-\frac{xy}{\sqrt{4 - x^2 - y^2}} + \frac{xy}{\sqrt{4 - x^2 - y^2}} + 2z \right) dA$$

$$= \iint_R 2\sqrt{4 - x^2 - y^2}\, dA = 2\int_0^{2\pi}\int_0^2 \sqrt{4 - r^2}\, r\, dr\, d\theta = 2\int_0^{2\pi}\left[\left(-\tfrac{1}{2}\right)\left(\tfrac{2}{3}\right)\left(4 - r^2\right)^{3/2} \right]_0^2 d\theta = \frac{16}{3}\int_0^{2\pi} d\theta = \frac{32\pi}{3}$$

25. $\mathbf{F} = P\mathbf{i} + Q\mathbf{j} + R\mathbf{k} = 2\mathbf{i} + 3\mathbf{j} + \mathbf{k}$. Also, $z = g(x, y) = \sqrt{x^2 + y^2} \Rightarrow$

$g_x(x, y) = \dfrac{x}{\sqrt{x^2 + y^2}}$ and $g_y(x, y) = \dfrac{y}{\sqrt{x^2 + y^2}}$. Therefore,

$$\iint_S \mathbf{F} \cdot d\mathbf{S} = \iint_R (-Pg_x - Qg_y + R)\, dA$$

$$= \iint_R \left(-\frac{2x}{\sqrt{x^2 + y^2}} - \frac{3y}{\sqrt{x^2 + y^2}} + 1 \right) dA$$

$$= \int_0^{2\pi}\int_0^1 \left(-\frac{2r\cos\theta + 3r\sin\theta}{r} + 1 \right) r\, dr\, d\theta = \int_0^{2\pi}\int_0^1 (-2\cos\theta - 3\sin\theta + 1)\, r\, dr\, d\theta$$

$$= \int_0^{2\pi}\left[(-2\cos\theta - 3\sin\theta + 1)\left(\tfrac{1}{2}r^2\right) \right]_{r=0}^{r=1} d\theta = \tfrac{1}{2}\int_0^{2\pi} (-2\cos\theta - 3\sin\theta + 1)\, d\theta$$

$$= \tfrac{1}{2}\left(-2\sin\theta + 3\cos\theta + \theta \right)\Big|_0^{2\pi} = \pi$$

27. $\mathbf{F}(x, y, z) = y^3\mathbf{i} + x^2\mathbf{j} + z\mathbf{k}$ and $S = S_1 \cup S_2 \cup S_3$. On S_1,

$\mathbf{r}(\theta, z) = 3\cos\theta\,\mathbf{i} + 3\sin\theta\,\mathbf{j} + z\mathbf{k}$, $0 \le \theta \le 2\pi$, $0 \le z \le 3 \Rightarrow$

$\mathbf{r}_\theta = -3\sin\theta\,\mathbf{i} + 3\cos\theta\,\mathbf{j}$ and $\mathbf{r}_z = \mathbf{k}$, so

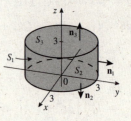

$$\mathbf{r}_\theta \times \mathbf{r}_z = \begin{vmatrix} \mathbf{i} & \mathbf{j} & \mathbf{k} \\ -3\sin\theta & 3\cos\theta & 0 \\ 0 & 0 & 1 \end{vmatrix} = 3\cos\theta\,\mathbf{i} + 3\sin\theta\,\mathbf{j}. \text{ Therefore,}$$

$$\iint_{S_1} \mathbf{F} \cdot d\mathbf{S} = \iint_{S_1} \mathbf{F} \cdot \mathbf{n}_1\, dS = \iint_{S_1} \left(27\sin^3\theta\,\mathbf{i} + 9\cos^2\theta\,\mathbf{j} + z\mathbf{k} \right) \cdot (3\cos\theta\,\mathbf{i} + 3\sin\theta\,\mathbf{j})\, dA$$

$$= \int_0^3\int_0^{2\pi} \left(81\sin^3\theta\cos\theta + 27\cos^2\theta\sin\theta \right) d\theta\, dz = \int_0^3 \left[\frac{81}{4}\sin^4\theta - 9\cos^3\theta \right]_0^{2\pi} dz = \int_0^3 0\, dz = 0$$

On S_2, $\mathbf{n}_2 = -\mathbf{k}$ and $z = 0$, so $\iint_{S_2} \mathbf{F} \cdot d\mathbf{S} = \iint_{S_2} \mathbf{F} \cdot \mathbf{n}_2\, dS = \iint_{S_2} \left(y^3\mathbf{i} + x^2\mathbf{j} + z\mathbf{k} \right) \cdot (-\mathbf{k})\, dS = \iint_{S_2} 0\, dS = 0$,

and on S_3, $\mathbf{n}_3 = \mathbf{k}$ and $z = 3$, so $\iint_{S_3} \mathbf{F} \cdot d\mathbf{S} = 3\iint_{S_3} dA = 3(\pi)(3)^2 = 27\pi$. Therefore,

$$\iint_S \mathbf{F} \cdot d\mathbf{S} = \iint_{S_1} \mathbf{F} \cdot d\mathbf{S} + \iint_{S_2} \mathbf{F} \cdot d\mathbf{S} + \iint_{S_3} \mathbf{F} \cdot d\mathbf{S} = 0 + 0 + 27\pi = 27\pi.$$

29. $m = \iint_S \rho \, dS = \iint_S k \, dS = k\left(\frac{1}{2}\right)\left(4\pi a^2\right) = 2k\pi a^2.$ By symmetry,

$\bar{x} = \bar{y} = 0.$

$\bar{z} = \frac{1}{m} \iint_S z\rho \, dS = \frac{k}{2k\pi a^2} \iint_R \sqrt{a^2 - x^2 - y^2} \sqrt{(g_x)^2 + (g_y)^2 + 1} \, dA,$ where

$z = g(x, y) = \sqrt{a^2 - x^2 - y^2} \Rightarrow g_x = -\dfrac{x}{\sqrt{a^2 - x^2 - y^2}}$ and

$g_y = -\dfrac{y}{\sqrt{a^2 - x^2 - y^2}},$ so

$\bar{z} = \dfrac{k}{2k\pi a^2} \iint_R \sqrt{a^2 - x^2 - y^2} \sqrt{\left(-\dfrac{x}{\sqrt{a^2 - x^2 - y^2}}\right)^2 + \left(-\dfrac{y}{\sqrt{a^2 - x^2 - y^2}}\right)^2 + 1} \, dA = \dfrac{1}{2\pi a^2} \iint_R a \, dA$

$= \dfrac{1}{2\pi a} \cdot \pi a^2 = \dfrac{a}{2}$

Therefore, the center of mass is $\left(0, 0, \frac{a}{2}\right).$

31. Because the spherical shell has uniform density, its moment of inertia about any diameter is the same, so we may place the shell so that its center lies at the origin. Then one of its diameters lies along the z-axis.

$z = g(x, y) = \sqrt{a^2 - x^2 - y^2} \Rightarrow g_x = -\dfrac{x}{\sqrt{a^2 - x^2 - y^2}}$ and

$g_y = -\dfrac{y}{\sqrt{a^2 - x^2 - y^2}},$ so

$I_z = 2 \iint_S \left(x^2 + y^2\right) \rho(x, y, z) \, dS = 2\rho \iint_R \left(x^2 + y^2\right) \sqrt{(g_x)^2 + (g_y)^2 + 1} \, dA$

$= 2\rho \iint_R \left(x^2 + y^2\right) \sqrt{\left(\dfrac{-x}{\sqrt{a^2 - x^2 - y^2}}\right)^2 + \left(\dfrac{-y}{\sqrt{a^2 - x^2 - y^2}}\right)^2 + 1} \, dA = 2\rho \iint_R \dfrac{ar^2}{\sqrt{a^2 - r^2}} \, dA$

$= 2a\rho \int_0^{2\pi} \left[\lim_{c \to a^-} \int_0^c \dfrac{r^2}{\sqrt{a^2 - r^2}} r \, dr\right] d\theta$

Using the substitution $u^2 = a^2 - r^2,$ so $du = -2r \, dr,$ we find that

$I_z = 2a\rho \int_0^{2\pi} \left[\lim_{c \to a^-} \left(\dfrac{r^2 + 2a^2}{3} \cdot \sqrt{a^2 - r^2}\right)\right]_c^0 d\theta = \dfrac{8}{3}\pi\rho a^4.$ But $m = 4\pi a^2 \rho,$ so $I_z = \frac{2}{3}ma^2.$

33. $2x + 3y + z = 6 \Rightarrow z = g(x, y) = 6 - 2x - 3y$ and

$\sigma(x, y, z) = kz^2 = k(6 - 2x - 3y)^2,$ where k is the constant of proportionality. If R is the projection of the relevant part of the plane $2x + 3y + z = 6$ onto the xy-plane, then

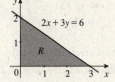

$Q = \iint_S \sigma(x, y, z) \, dS = k \iint_R (6 - 2x - 3y)^2 \sqrt{(g_x)^2 + (g_y)^2 + 1} \, dA$

$= k \iint_R (6 - 2x - 3y)^2 \sqrt{(-2)^2 + (-3)^2 + 1} \, dA$

$= \sqrt{14}\, k \int_0^3 \int_0^{(6-2x)/3} (6 - 2x - 3y)^2 \, dy \, dx = \sqrt{14}\, k \int_0^3 \left[\left(-\dfrac{1}{3}\right)\left(\dfrac{1}{3}\right)(6 - 2x - 3y)^3\right]_{y=0}^{y=(6-2x)/3} dx$

$= \dfrac{\sqrt{14}}{9} k \int_0^3 (6 - 2x)^3 \, dx = \dfrac{\sqrt{14}}{9} k \left(-\dfrac{1}{2}\right)\left(\dfrac{1}{4}\right)(6 - 2x)^4 \Big|_0^3 = 18\sqrt{14}\, k$

35. $z = g(x, y) = 6 - 2x - 3y \Rightarrow g_x = -2$ and $g_y = -3$. Also, $P = 2x$,

$Q = 2y$, and $R = 3z$, so f R is the projection of the relevant part of the plane

$2x + 3y + z = 6$ onto the xy-plane, then

$$\iint_S \mathbf{F} \cdot d\mathbf{S} = \iint_R (-Pg_x - Qg_y + R)\, dA$$

$$= \int_0^3 \int_0^{(6-2x)/3} \left[-(2x)(-2) - (2y)(-3) + 3(6 - 2x - 3y)\right] dy\, dx$$

$$= \int_0^3 \int_0^{(6-2x)/3} (18 - 2x - 3y)\, dy\, dx = \int_0^3 \left[18y - 2xy - \tfrac{3}{2}y^2\right]_{y=0}^{y=(6-2x)/3} dx$$

$$= \int_0^3 \left(\tfrac{2}{3}x^2 - 12x + 30\right) dx = \left(\tfrac{2}{9}x^3 - 6x^2 + 30x\right)\Big|_0^3 = 42$$

37. The heat flow is $\mathbf{q} = -k\nabla T = -5(2x\mathbf{i} + 2y\mathbf{j})$ and $x^2 + y^2 = 1 \Rightarrow y = g(x, z) = \sqrt{1 - x^2} \Rightarrow g_x(x, z) = -\dfrac{x}{\sqrt{1 - x^2}}$

and $g_z(x, z) = 0$. Then the rate of heat flow across the cylindrical surface $x^2 + y^2 = 1$ between $z = 0$ and $z = 1$ is

$$\iint_S \mathbf{q} \cdot \mathbf{n}\, dS = 2\iint_R (-Pg_x - Rg_z + Q)\, dA = 2\iint_R \left[-(-10x)\left(-\dfrac{x}{\sqrt{1 - x^2}}\right) - 10y\right] dA$$

$$= -20\int_0^1 \int_{-1}^1 \left(\dfrac{x^2}{\sqrt{1 - x^2}} + \sqrt{1 - x^2}\right) dx\, dz = -40\int_0^1 \dfrac{1}{\sqrt{1 - x^2}}\, dx = -40\sin^{-1} x \Big|_0^1 = -20\pi$$

39. a. Put $G(x, y, z) = y - g(x, z)$. Then the unit normal to S is

$$\mathbf{n} = \dfrac{\nabla G(x, y, z)}{|\nabla G(x, y, z)|} = \dfrac{-g_x(x, z)\mathbf{i} + \mathbf{j} - g_z(x, z)\mathbf{k}}{\sqrt{[g_x(x, z)]^2 + [g_z(x, z)]^2 + 1}}.$$ Also, $dS = \sqrt{[g_x(x, z)]^2 + [g_z(x, z)]^2 + 1}\, dA$, so

$$\iint_S \mathbf{F} \cdot d\mathbf{S} = \iint_S \mathbf{F} \cdot \mathbf{n}\, dS = \iint_R (P\mathbf{i} + Q\mathbf{j} + R\mathbf{k}) \cdot [-g_x(x, z)\mathbf{i} + \mathbf{j} - g_z(x, z)\mathbf{k}]\, dA = \iint_D (-Pg_x + Q - Rg_z)\, dA.$$

b. $x^2 + y^2 = 4 \Rightarrow y = g(x, z) = \sqrt{4 - x^2}$, so $g_x = -\dfrac{x}{\sqrt{4 - x^2}}$ and

$g_z = 0$. Also, $P = y$, $Q = z$, and $R = -3yz^2$, so

$$\iint_S \mathbf{F} \cdot d\mathbf{S} = \iint_D \left[-y\left(-\dfrac{x}{\sqrt{4 - x^2}}\right) + z - \left(-3yz^2\right)(0)\right] dA,\text{ where}$$

$D = \{(x, z) \mid 0 \le x \le 2, 0 \le z \le 3\}$. Thus,

$$\iint_S \mathbf{F} \cdot d\mathbf{S} = \iint_R \left(\dfrac{x\sqrt{4 - x^2}}{\sqrt{4 - x^2}} + z\right) dA = \int_0^2 \int_0^3 (x + z)\, dz\, dx = \int_0^2 \left[\left(xz + \tfrac{1}{2}z^2\right)\right]_{z=0}^{z=3} dx$$

$$= \int_0^2 \left(3x + \tfrac{9}{2}\right) dx = \left(\tfrac{3}{2}x^2 + \tfrac{9}{2}x\right)\Big|_0^2 = 15$$

41. We use the spherical representation of the sphere $\mathbf{r}(u, v) = R\sin u \cos v\, \mathbf{i} + R\sin u \sin v\, \mathbf{j} + R\cos u\, \mathbf{k}$ with parameter

domain $D = \{(u, v) \mid 0 \le u \le \pi, 0 \le v \le 2\pi\}$. We may assume that the fixed point is $(0, 0, R)$. Then the density at each

point of the thin spherical shell is $\rho = k\sqrt{(R\sin u \cos v)^2 + (R\sin u \sin v)^2 + (R\cos u - R)^2} = k\sqrt{2}R\sqrt{1 - \cos u}$, where

k is the constant of proportionality. Also, using the result of Example 15.6.10 (14.6.10 in ET), $|\mathbf{r}_u \times \mathbf{r}_v| = R^2\sin u$, so the

total mass is

$$m = \iint_S \rho\, dS = \iint_D k\sqrt{2}R\sqrt{1 - \cos u}\, R^2\sin u\, dA = k\int_0^{2\pi}\int_0^\pi \sqrt{2}R^3\sqrt{1 - \cos u}\sin u\, du\, dv$$

$$= \sqrt{2}kR^3 \int_0^{2\pi} \left[\left(\tfrac{2}{3}\right)(1 - \cos u)^{3/2}\right]_0^\pi dv = \sqrt{2}kR^3 \left(\tfrac{4\sqrt{2}}{3}\right)\int_0^{2\pi} dv = \tfrac{16}{3}\pi kR^3$$

43. A parametric representation of S is $r(u, v) = u\mathbf{i} + f(u)\cos v\,\mathbf{j} + f(u)\sin v\,\mathbf{k}$
with parameter domain $D = \{(u, v) \mid a \le u \le b, 0 \le v \le 2\pi\}$. Then

$$\mathbf{r}_u \times \mathbf{r}_v = \begin{vmatrix} \mathbf{i} & \mathbf{j} & \mathbf{k} \\ 1 & f'(u)\cos v & f'(u)\sin v \\ 0 & -f(u)\sin v & f(u)\cos v \end{vmatrix} = f(u)f'(u)\mathbf{i} - f(u)\cos v\,\mathbf{j} - f(u)\sin v\,\mathbf{k} \text{ and}$$

$|\mathbf{r}_u \times \mathbf{r}_v| = \sqrt{[f(u)f'(u)]^2 + [-f(u)\cos v]^2 + [-f(u)\sin v]^2} = f(u)\sqrt{1 + [f'(u)]^2}$, so the area of S is

$A = \iint_D |\mathbf{r}_u \times \mathbf{r}_v|\,dA = \int_0^{2\pi} \int_a^b f(u)\sqrt{1 + [f'(u)]^2}\,du\,dv = 2\pi \int_a^b f(x)\sqrt{1 + [f'(x)]^2}\,dx.$

45. a. We differentiate $F(x, y, z) = 0$ implicitly to obtain $\dfrac{\partial g}{\partial x} = -\dfrac{F_x(x, y, z)}{F_z(x, y, z)}$ and $\dfrac{\partial g}{\partial y} = -\dfrac{F_y(x, y, z)}{F_z(x, y, z)}$. Then

$$\iint_S f(x, y, z)\,dS = \iint_D f\sqrt{\left(\frac{\partial g}{\partial x}\right)^2 + \left(\frac{\partial g}{\partial y}\right)^2 + 1}\,dA = \iint_D f\sqrt{\left(-\frac{F_x}{F_z}\right)^2 + \left(-\frac{F_y}{F_z}\right)^2 + 1}\,dA$$

$$= \iint_D \frac{f\sqrt{F_x^2 + F_y^2 + F_z^2}}{|F_z|}\,dA$$

b. Here $f(x, y, z) = x$ and $F(x, y, z) = 2x + 3y + z - 6 = 0 \Rightarrow F_x = 2$, $F_y = 3$, and $F_z = 1$, so

$$\iint_S f(x, y, z)\,dS = \iint_D x \cdot \frac{\sqrt{2^2 + 3^2 + 1^2}}{1}\,dA = \sqrt{14} \iint_D x\,dA, \text{ where } D = R \text{ as in Example 1. Therefore,}$$

$\iint_S f(x, y, z)\,dS = 3\sqrt{14}$, as before.

47. Let S have parametric representation $\mathbf{r}(u, v) = x(u, v)\mathbf{i} + y(u, v)\mathbf{j} + z(u, v)\mathbf{k}$, where (u, v) lies in the
parameter domain D. Then $\iint_S (a\mathbf{F} + b\mathbf{G}) \cdot d\mathbf{S} = \iint_D [a\mathbf{F}(\mathbf{r}(u, v)) + b\mathbf{G}(\mathbf{r}(u, v))] \cdot (\mathbf{r}_u \times \mathbf{r}_v)\,dA$, where
$\mathbf{F}(\mathbf{r}(u, v)) = \mathbf{F}(x(u, v), y(u, v), z(u, v))$. Thus,

$$\iint_S (a\mathbf{F} + b\mathbf{G}) \cdot d\mathbf{S} = \iint_D a\mathbf{F}(\mathbf{r}(u, v)) \cdot (\mathbf{r}_u \times \mathbf{r}_v)\,dA + \iint_D b\mathbf{G}(\mathbf{r}(u, v)) \cdot (\mathbf{r}_u \times \mathbf{r}_v)\,dA$$

$$= a\iint_D \mathbf{F}(\mathbf{r}(u, v)) \cdot (\mathbf{r}_u \times \mathbf{r}_v)\,dA + b\iint_D \mathbf{G}(\mathbf{r}(u, v)) \cdot (\mathbf{r}_u \times \mathbf{r}_v)\,dA$$

$$= a\iint_S \mathbf{F} \cdot d\mathbf{S} + b\iint_S \mathbf{G} \cdot d\mathbf{S}$$

49. True. Since $\mathbf{F}$ is a constant vector field, there are planes that are perpendicular to the vector field. Pick one of those planes
that divides the sphere S into two hemispheres S_1 and S_2 so that $S = S_1 \cup S_2$. Since S is a closed surface, we have
$\iint_S \mathbf{F} \cdot d\mathbf{S} = \iint_{S_1} \mathbf{F} \cdot d\mathbf{S} + \iint_{S_2} \mathbf{F} \cdot d\mathbf{S}$, but $\iint_{S_1} \mathbf{F} \cdot d\mathbf{S} = -\iint_{S_2} \mathbf{F} \cdot d\mathbf{S}$ because S_1 and S_2 have opposite orientations. The
result follows.

15.8 Concept Questions   ET 14.8

1. See page 1304 (1300 in ET).

15.8 The Divergence Theorem   ET 14.8

1. If S is the surface of the cube, then $S = S_1 \cup S_2 \cup \cdots \cup S_6$. The corresponding outward unit normal vectors are $\mathbf{n}_1 = \mathbf{i}$, $\mathbf{n}_2 = \mathbf{j}$, $\mathbf{n}_3 = \mathbf{k}$, $\mathbf{n}_4 = -\mathbf{i}$, $\mathbf{n}_5 = -\mathbf{j}$, and $\mathbf{n}_6 = -\mathbf{k}$, so $\mathbf{F} \cdot \mathbf{n}_1 = (x\mathbf{i} + y\mathbf{j} + z\mathbf{k}) \cdot \mathbf{i} = x$ and similarly $\mathbf{F} \cdot \mathbf{n}_2 = y$, $\mathbf{F} \cdot \mathbf{n}_3 = z$, $\mathbf{F} \cdot \mathbf{n}_4 = -x$, $\mathbf{F} \cdot \mathbf{n}_5 = -y$, and $\mathbf{F} \cdot \mathbf{n}_6 = -z$. Then, because $x = 0$ on S_4, $y = 0$ on S_5, and $z = 0$ on S_6, we have

$$\iint_S \mathbf{F} \cdot d\mathbf{S} = \iint_{S_1 \cup S_2 \cup \cdots \cup S_6} \mathbf{F} \cdot d\mathbf{S}$$

$$= \iint_{S_1} 1\, dA + \iint_{S_2} 1\, dA + \iint_{S_3} 1\, dA + 0 = 1 + 1 + 1 = 3$$

On the other hand, $\operatorname{div} \mathbf{F} = \frac{\partial}{\partial x}(x) + \frac{\partial}{\partial y}(y) + \frac{\partial}{\partial z}(z) = 1 + 1 + 1 = 3$, so $\iiint_T \operatorname{div} \mathbf{F}\, dV = 3 \iiint_T dV = 3 \cdot 1^3 = 3$ and the Divergence Theorem is verified for this special case.

3. $x^2 + y^2 = 4 \Rightarrow y = g(x, z) = \sqrt{4 - x^2}$, giving a representation of S_1. Letting $G(x, y, z) = y - g(x, z) = y - \sqrt{4 - x^2}$, we find

$$\mathbf{n}_1 = \frac{\nabla G}{|\nabla G|} = \frac{\frac{x}{\sqrt{4-x^2}}\mathbf{i} + \mathbf{j}}{\sqrt{\frac{x^2}{4-x^2} + 1}} = \tfrac{1}{2}x\mathbf{i} + \tfrac{1}{2}\sqrt{4 - x^2}\,\mathbf{j}, \text{ so}$$

$$\mathbf{F} \cdot \mathbf{n}_1 = \left(y\mathbf{i} + z\mathbf{j} - 3yz^2\mathbf{k}\right) \cdot \left(\tfrac{1}{2}x\mathbf{i} + \tfrac{1}{2}\sqrt{4 - x^2}\,\mathbf{j}\right)$$

$$= \tfrac{1}{2}xy + \tfrac{1}{2}z\sqrt{4 - x^2} = \tfrac{1}{2}\sqrt{4 - x^2}\,(x + z),$$

$\mathbf{F} \cdot \mathbf{n}_2 = \left(y\mathbf{i} + z\mathbf{j} - 3yz^2\mathbf{k}\right) \cdot \mathbf{k} = -3yz^2$, $\mathbf{F} \cdot \mathbf{n}_3 = \left(y\mathbf{i} + z\mathbf{j} - 3yz^2\mathbf{k}\right) \cdot (-\mathbf{k}) = 3yz^2$,

$\mathbf{F} \cdot \mathbf{n}_4 = \left(y\mathbf{i} + z\mathbf{j} - 3yz^2\mathbf{k}\right) \cdot (-\mathbf{i}) = -y$, and $\mathbf{F} \cdot \mathbf{n}_5 = \left(y\mathbf{i} + z\mathbf{j} - 3yz^2\mathbf{k}\right) \cdot (-\mathbf{j}) = -z$. Therefore, because $z = 3$ on S_2 and $z = 0$ on S_3, we have

$$\iint_S \mathbf{F} \cdot d\mathbf{S} = \iint_{S_1 \cup S_2 \cup \cdots \cup S_5} \mathbf{F} \cdot d\mathbf{S} = \iint_{S_1} \mathbf{F} \cdot \mathbf{n}_1\, dS + \iint_{S_2} \mathbf{F} \cdot \mathbf{n}_2\, dS + \cdots + \iint_{S_5} \mathbf{F} \cdot \mathbf{n}_5\, dS$$

$$= \iint_{S_1} \tfrac{1}{2}\sqrt{4 - x^2}\,(x + z)\, dS - \iint_{S_2} 27y\, dS + \iint_{S_3} 0\, dS - \iint_{S_4} y\, dS - \iint_{S_5} z\, dS$$

$$= \frac{1}{2}\int_0^2 \int_0^3 \sqrt{4 - x^2}\,(x + z) \sqrt{\left(\frac{x}{\sqrt{4-x^2}}\right)^2 + 1^2 + 0^2}\, dz\, dx - 27\int_0^{\pi/2} \int_0^2 r\sin\theta\, r\, dr\, d\theta$$

$$+ 0 - \int_0^3 \int_0^2 y\, dy\, dz - \int_0^2 \int_0^3 z\, dz\, dx$$

$$= \int_0^2 \int_0^3 (x + z)\, dz\, dx - 27 \int_0^{\pi/2} \int_0^2 r^2 \sin\theta\, dr\, d\theta - \int_0^3 \int_0^2 y\, dy\, dz - \int_0^2 \int_0^3 z\, dz\, dx = 15 - 72 - 6 - 9$$

$$= -72$$

On the other hand, $\operatorname{div} \mathbf{F} = \frac{\partial}{\partial x}(y) + \frac{\partial}{\partial y}(z) + \frac{\partial}{\partial z}\left(-3yz^2\right) = -6yz$, so

$$\iiint_T \operatorname{div} \mathbf{F}\, dV = -6 \int_0^{\pi/2} \int_0^2 \int_0^3 (r\sin\theta)\, zr\, dz\, dr\, d\theta = -6 \int_0^{\pi/2} \int_0^2 \left[\left(r^2 \sin\theta\right)\left(\tfrac{1}{2}z^2\right)\right]_{z=0}^{z=3} dr\, d\theta$$

$$= -6 \int_0^{\pi/2} \int_0^2 \tfrac{9}{2} r^2 \sin\theta\, dr\, d\theta = -27 \int_0^{\pi/2} \left[\tfrac{1}{3}r^3 \sin\theta\right]_{r=0}^{r=2} d\theta = -72 \int_0^{\pi/2} \sin\theta\, d\theta = -72$$

verifying the Divergence Theorem for this special case.

5. $\mathbf{F}(x, y, z) = xy^2\mathbf{i} + 2yz\mathbf{j} - 3x^2y^3\mathbf{k} \Rightarrow \operatorname{div}\mathbf{F} = \frac{\partial}{\partial x}\left(xy^2\right) + \frac{\partial}{\partial y}(2yz) + \frac{\partial}{\partial z}\left(-3x^2y^3\right) = y^2 + 2z$, so

$$\iint_S \mathbf{F} \cdot \mathbf{n}\, dS = \iiint_T \operatorname{div}\mathbf{F}\, dV = \int_{-1}^{1}\int_{-1}^{1}\int_{-1}^{1}\left(y^2 + 2z\right) dx\, dy\, dz = \int_{-1}^{1}\int_{-1}^{1}\left[\left(y^2 + 2z\right)x\right]_{x=-1}^{x=1} dy\, dz$$

$$= \int_{-1}^{1}\int_{-1}^{1}\left(2y^2 + 4z\right) dy\, dz = \int_{-1}^{1}\left[\frac{2}{3}y^3 + 4yz\right]_{y=-1}^{y=1} dz = \int_{-1}^{1}\left(\frac{4}{3} + 8z\right) dz = \left(\frac{4}{3}z + 4z^2\right)\Big|_{-1}^{1} = \frac{8}{3}$$

7. $\mathbf{F}(x, y, z) = \left(x^3 + \cos y\right)\mathbf{i} + \left(y^3 + \sin xz\right)\mathbf{j} + \left(z^3 + 2e^{-x}\right)\mathbf{k} \Rightarrow$

$\operatorname{div}\mathbf{F} = \frac{\partial}{\partial x}\left(x^3 + \cos y\right) + \frac{\partial}{\partial y}\left(y^3 + \sin xz\right) + \frac{\partial}{\partial z}\left(z^3 + 2e^{-x}\right)$

$\quad = 3x^2 + 3y^2 + 3z^2$

so

$$\iint_S \mathbf{F} \cdot \mathbf{n}\, dS = \iiint_T \operatorname{div}\mathbf{F}\, dV = 3\iint_R \left(x^2 + y^2 + z^2\right) dA \text{ where}$$

$R = \left\{(y, z) \mid y^2 + z^2 \le 1\right\}$; that is,

$$\iint_S \mathbf{F} \cdot \mathbf{n}\, dS = 3\int_0^{2\pi}\int_0^1\int_0^3 \left(r^2 + x^2\right) dx\, r\, dr\, d\theta = 3\int_0^{2\pi}\int_0^1 \left[r^2 x + \frac{1}{3}x^3\right]_{x=0}^{x=3} r\, dr\, d\theta = 3\int_0^{2\pi}\int_0^1 \left(3r^3 + 9r\right) dr\, d\theta$$

$$= 3\int_0^{2\pi}\left[\frac{3}{4}r^4 + \frac{9}{2}r^2\right]_0^1 d\theta = \frac{63}{4}\int_0^{2\pi} d\theta = \frac{63\pi}{2}$$

9. $\mathbf{F}(x, y, z) = 2xy\mathbf{i} + y^2\mathbf{j} + \left(x^2 + yz\right)\mathbf{k} \Rightarrow$

$\operatorname{div}\mathbf{F} = \frac{\partial}{\partial x}(2xy) + \frac{\partial}{\partial y}\left(y^2\right) + \frac{\partial}{\partial z}\left(x^2 + yz\right) = 2y + 2y + y = 5y$, so

$$\iint_S \mathbf{F} \cdot \mathbf{n}\, dS = \iiint_T \operatorname{div}\mathbf{F}\, dV = \int_0^1\int_0^{1-x}\int_0^{1-x-y} 5y\, dz\, dy\, dx$$

$$= \int_0^1\int_0^{1-x} \left[5yz\right]_{z=0}^{z=1-x-y} dy\, dx = \int_0^1\int_0^{1-x} 5y\,(1 - x - y)\, dy\, dx$$

$$= 5\int_0^1\left[\frac{1}{2}y^2(1 - x) - \frac{1}{3}y^3\right]_{y=0}^{y=1-x} dx = \frac{5}{6}\int_0^1 (1 - x)^3\, dx$$

$$= \frac{5}{6}(-1)\left(\frac{1}{4}\right)(1 - x)^4\Big|_0^1 = \frac{5}{24}$$

11. $\mathbf{F}(x, y, z) = x\mathbf{i} + 2y\mathbf{j} + 3z\mathbf{k} \Rightarrow \operatorname{div}\mathbf{F} = \frac{\partial}{\partial x}(x) + \frac{\partial}{\partial y}(2y) + \frac{\partial}{\partial z}(3z) = 1 + 2 + 3 = 6$, so

$$\iint_S \mathbf{F} \cdot \mathbf{n}\, dS = \iiint_T \operatorname{div}\mathbf{F}\, dV = \iiint_T 6\, dV = 6\left(\frac{4}{3}\pi \cdot 3^3\right) = 216\pi.$$

13. $\mathbf{F}(x, y, z) = xz\mathbf{i} - yz\mathbf{j} + xy\mathbf{k} \Rightarrow \operatorname{div}\mathbf{F} = \frac{\partial}{\partial x}(xz) + \frac{\partial}{\partial y}(-yz) + \frac{\partial}{\partial z}(xy) = z - z = 0$, so

$$\iint_S \mathbf{F} \cdot \mathbf{n}\, dS = \iiint_T \operatorname{div}\mathbf{F}\, dV = \iiint_T 0\, dV = 0.$$

15. $\mathbf{F}(x, y, z) = \left(x^3 + 1\right)\mathbf{i} + \left(yz^2 + \cos xz\right)\mathbf{j} + \left(2y^2z + e^{\tan x}\right)\mathbf{k} \Rightarrow$

$\operatorname{div}\mathbf{F} = \frac{\partial}{\partial x}\left(x^3 + 1\right) + \frac{\partial}{\partial y}\left(yz^2 + \cos xz\right) + \frac{\partial}{\partial z}\left(2y^2z + e^{\tan x}\right) = 3x^2 + z^2 + 2y^2$, so

$$\iint_S \mathbf{F} \cdot \mathbf{n}\, dS = \iiint_T \operatorname{div}\mathbf{F}\, dV = \iiint_T \left(3x^2 + z^2 + 2y^2\right) dV$$

$$= \int_0^{2\pi}\int_0^{\pi}\int_0^1 \left(3\rho^2 \sin^2\phi\cos^2\theta + \rho^2\cos^2\phi + 2\rho^2\sin^2\phi\sin^2\theta\right)\rho^2\sin\phi\, d\rho\, d\phi\, d\theta$$

$$= \int_0^{2\pi}\int_0^{\pi}\left[\frac{1}{5}\rho^5\left(3\sin^3\phi\cos^2\theta + \cos^2\phi\sin\phi + 2\sin^3\phi\sin^2\theta\right)\right]_{\rho=0}^{\rho=1} d\phi\, d\theta$$

$$= \frac{1}{5}\int_0^{2\pi}\int_0^{\pi}\left(3\sin^3\phi\cos^2\theta + \cos^2\phi\sin\phi + 2\sin^3\phi\sin^2\theta\right) d\phi\, d\theta$$

$$= \frac{1}{5}\int_0^{2\pi}\left[\left(3\cos^2\theta + 2\sin^2\theta\right)\left(-\frac{1}{3}\right)\left(2 + \sin^2\phi\right)\cos\phi - \frac{1}{3}\cos^3\phi\right]_{\phi=0}^{\phi=\pi} d\theta$$

$$= \frac{1}{5}\int_0^{2\pi}\left[2(1 + \cos 2\theta) + \frac{4}{3}(1 - \cos 2\theta) + \frac{2}{3}\right] d\theta = \frac{1}{5}\int_0^{2\pi}\left(4 + \frac{2}{3}\cos 2\theta\right) d\theta = \frac{1}{5}\left(4\theta + \frac{1}{3}\sin 2\theta\right)\Big|_0^{2\pi} = \frac{8\pi}{5}$$

17. $\mathbf{F}(x, y, z) = xz\mathbf{i} + x^2 y\mathbf{j} + (y^2 z + 1)\mathbf{k} \Rightarrow$

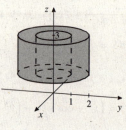

$\operatorname{div}\mathbf{F} = \frac{\partial}{\partial x}(xz) + \frac{\partial}{\partial y}(x^2 y) + \frac{\partial}{\partial z}(y^2 z + 1) = z + x^2 + y^2$, so

$$\iint_S \mathbf{F} \cdot \mathbf{n}\, dS = \iiint_T \operatorname{div}\mathbf{F}\, dV = \iiint_T (x^2 + y^2 + z)\, dV$$

$$= \int_0^{2\pi} \int_1^2 \int_1^3 (r^2 + z)\, r\, dz\, dr\, d\theta$$

$$= \int_0^{2\pi} \int_1^2 \left[r\left(r^2 z + \tfrac{1}{2} z^2\right) \right]_{z=1}^{z=3} dr\, d\theta = \int_0^{2\pi} \int_1^2 \left(2r^3 + 4r\right) dr\, d\theta$$

$$= \int_0^{2\pi} \left[\tfrac{1}{2} r^4 + 2r^2 \right]_1^2 d\theta = \tfrac{27}{2} \int_0^{2\pi} d\theta = 27\pi$$

19. Here $\mathbf{F}(x, y, z) = \mathbf{r} = x\mathbf{i} + y\mathbf{j} + z\mathbf{k}$, so $\operatorname{div}\mathbf{F} = \operatorname{div}\mathbf{r} = \frac{\partial}{\partial x}(x) + \frac{\partial}{\partial y}(y) + \frac{\partial}{\partial z}(z) = 3$, so

$\iint_S \mathbf{r} \cdot \mathbf{n}\, dS = \iiint_T \operatorname{div}\mathbf{r}\, dV = \iiint_T 3\, dV = 3\iiint_T dV = 3V(T) \Rightarrow V(T) = \frac{1}{3}\iint_S \mathbf{r} \cdot \mathbf{n}\, dS$.

21. Let $\mathbf{F}(x, y, z) = P(x, y, z)\mathbf{i} + Q(x, y, z)\mathbf{j} + R(x, y, z)\mathbf{k}$. Then

$$\operatorname{curl}\mathbf{F} = \begin{vmatrix} \mathbf{i} & \mathbf{j} & \mathbf{k} \\ \frac{\partial}{\partial x} & \frac{\partial}{\partial y} & \frac{\partial}{\partial z} \\ P & Q & R \end{vmatrix} = \left(\frac{\partial R}{\partial y} - \frac{\partial Q}{\partial z}\right)\mathbf{i} - \left(\frac{\partial R}{\partial x} - \frac{\partial P}{\partial z}\right)\mathbf{j} + \left(\frac{\partial Q}{\partial x} - \frac{\partial P}{\partial y}\right)\mathbf{k} \text{ and}$$

$$\operatorname{div}(\operatorname{curl}\mathbf{F}) = \frac{\partial}{\partial x}\left(\frac{\partial R}{\partial y} - \frac{\partial Q}{\partial z}\right) - \frac{\partial}{\partial y}\left(\frac{\partial R}{\partial x} - \frac{\partial P}{\partial z}\right) + \frac{\partial}{\partial z}\left(\frac{\partial Q}{\partial x} - \frac{\partial P}{\partial y}\right)$$

$$= \frac{\partial^2 R}{\partial x\, \partial y} - \frac{\partial^2 Q}{\partial x\, \partial z} - \frac{\partial^2 R}{\partial y\, \partial x} + \frac{\partial^2 P}{\partial y\, \partial z} + \frac{\partial^2 Q}{\partial z\, \partial x} - \frac{\partial^2 P}{\partial z\, \partial y} = 0$$

because the mixed partial derivatives are equal ($\mathbf{F}$ has continuous second order partial derivatives). Thus,

$\iint_S \operatorname{curl}\mathbf{F} \cdot \mathbf{n}\, dS = \iiint_T \operatorname{div}(\operatorname{curl}\mathbf{F})\, dV = \iiint_T 0\, dV = 0$.

Note that the result $\operatorname{div}(\operatorname{curl}\mathbf{F}) = 0$ was also proved in Example 9 on page 1239 (1235 in ET).

23. Let $\mathbf{c} = c_1\mathbf{i} + c_2\mathbf{j} + c_3\mathbf{k}$ be a constant vector, and define $\mathbf{F} = f\mathbf{c} = c_1 f\mathbf{i} + c_2 f\mathbf{j} + c_3 f\mathbf{k}$. Then

$\operatorname{div}\mathbf{F} = \frac{\partial}{\partial x}(c_1 f) + \frac{\partial}{\partial y}(c_2 f) + \frac{\partial}{\partial z}(c_3 f) = c_1\frac{\partial f}{\partial x} + c_2\frac{\partial f}{\partial y} + c_3\frac{\partial f}{\partial z} = \mathbf{c} \cdot \nabla f = \nabla f \cdot \mathbf{c}$. Using the Divergence

Theorem, we have $\iint_S \mathbf{F} \cdot \mathbf{n}\, dS = \iiint_T \operatorname{div}\mathbf{F}\, dV = \iiint_T \nabla f \cdot \mathbf{c}\, dV$. Since $\mathbf{c}$ is arbitrary, we can pick $\mathbf{c} = \mathbf{i}$. Then,

letting $\mathbf{n} = n_1\mathbf{i} + n_2\mathbf{j} + n_3\mathbf{k}$, we find $\iint_S f\mathbf{i} \cdot \mathbf{n}\, dS = \iiint_T \nabla f \cdot \mathbf{i}\, dV \Rightarrow \iint_S f n_1\, dS = \iiint_T \frac{\partial f}{\partial x}\, dV$. Similarly,

by taking $\mathbf{c} = \mathbf{j}$ and $\mathbf{c} = \mathbf{k}$ in succession, we have $\iint_S f n_2\, dS = \iiint_T \frac{\partial f}{\partial y}\, dV$ and $\iint_S f n_3\, dS = \iiint_T \frac{\partial f}{\partial z}\, dV$, so

$\iint_S f\mathbf{n}\, dS = \left(\iint_S f n_1\, dS\right)\mathbf{i} + \left(\iint_S f n_2\, dS\right)\mathbf{j} + \left(\iint_S f n_3\, dS\right)\mathbf{k} = \iiint_T \left(\frac{\partial f}{\partial x}\mathbf{i} + \frac{\partial f}{\partial y}\mathbf{j} + \frac{\partial f}{\partial z}\mathbf{k}\right) dV = \iiint_T \nabla f\, dV$, as

was to be shown.

25. First, we write $\iint_S (f\nabla g - g\nabla f) \cdot \mathbf{n}\, dS = \iint_S (f\nabla g) \cdot \mathbf{n}\, dS - \iint_S (g\nabla f) \cdot \mathbf{n}\, dS$. Then, applying the Divergence

Theorem to each term in the result, we have

$$\iint_S (f\nabla g - g\nabla f) \cdot \mathbf{n}\, dS = \iiint_T \operatorname{div}(f\nabla g)\, dV - \iiint_T \operatorname{div}(g\nabla f)\, dV$$

$$= \iiint_T \left[(\operatorname{div} f) \cdot \nabla g + f \operatorname{div}(\nabla g)\right] dV - \iiint_T \left[(\operatorname{div} g) \cdot \nabla f + g \operatorname{div}(\nabla f)\right] dV$$

$$= \iiint_T \left(\nabla f \cdot \nabla g + f\nabla^2 g\right) dV - \iiint_T \left(\nabla g \cdot \nabla f + g\nabla^2 f\right) dV = \iiint_T \left(f\nabla^2 g - g\nabla^2 f\right) dV$$

because $\nabla f \cdot \nabla g = \nabla g \cdot \nabla f$.

27. True. By the Divergence Theorem, $\iint_S \mathbf{F} \cdot d\mathbf{S} = \iint_S \mathbf{F} \cdot \mathbf{n}\, dS = \iiint_T \operatorname{div}\mathbf{F}\, dV = \iiint_T 0\, dV = 0$.

29. True. By the Divergence Theorem, $\iiint_T \operatorname{div}\mathbf{F}\, dV = \iint_S \mathbf{F} \cdot \mathbf{n}\, dS \leq \left|\iint_S \mathbf{F} \cdot \mathbf{n}\, dS\right| \leq \iint_S |\mathbf{F}|\, dS \leq \iint_S dS = A(S)$.

15.9 Concept Questions

ET 14.9

1. See page 1312 (1308 in ET).

15.9 Stokes' Theorem

ET 14.9

1. $\mathbf{F}(x, y, z) = 2z\mathbf{i} + 3x\mathbf{j} - 2y\mathbf{k}$, so $\operatorname{curl}\mathbf{F} = \begin{vmatrix} \mathbf{i} & \mathbf{j} & \mathbf{k} \\ \frac{\partial}{\partial x} & \frac{\partial}{\partial y} & \frac{\partial}{\partial z} \\ 2z & 3x & -2y \end{vmatrix} = -2\mathbf{i} + 2\mathbf{j} + 3\mathbf{k}.$

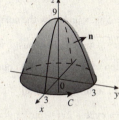

Let $z = g(x, y) = 9 - x^2 - y^2 \Rightarrow g_x = -2x$ and $g_y = -2y$. Then, using

Theorem 15.7.3 (14.7.3 in ET) with $R = \left\{ (x, y) \mid x^2 + y^2 \le 9 \right\}$, we have

$$\iint_S \operatorname{curl}\mathbf{F} \cdot d\mathbf{S} = \iint_R (-4x + 4y + 3)\, dA = \int_0^{2\pi} \int_0^3 (-4r\cos\theta + 4r\sin\theta + 3)\, r\, dr\, d\theta$$

$$= \int_0^{2\pi} \left[4(\sin\theta - \cos\theta)\left(\tfrac{1}{3}r^3\right) + \tfrac{3}{2}r^2 \right]_{r=0}^{r=3} d\theta = \int_0^{2\pi} \left[36(\sin\theta - \cos\theta) + \tfrac{27}{2} \right] d\theta$$

$$= \left[36(-\cos\theta - \sin\theta) + \tfrac{27}{2}\theta \right]_0^{2\pi} = 27\pi$$

Next, we see that the curve C has parametric representation $\mathbf{r}(t) = 3\cos t\, \mathbf{i} + 3\sin t\, \mathbf{j}, 0 \le t \le 2\pi$, and so

$$\oint_C \mathbf{F} \cdot d\mathbf{r} = \int_0^{2\pi} \mathbf{F}(\mathbf{r}(t)) \cdot \mathbf{r}'(t)\, dt = \int_0^{2\pi} (9\cos t\, \mathbf{j} - 6\sin t\, \mathbf{k}) \cdot (-3\sin t\, \mathbf{i} + 3\cos t\, \mathbf{j})\, dt = \int_0^{2\pi} 27\cos^2 t\, dt$$

$$= \tfrac{27}{2} \int_0^{2\pi} (1 + \cos 2t)\, dt = \tfrac{27}{2}\left(t + \tfrac{1}{2}\sin 2t \right)\Big|_0^{2\pi} = 27\pi$$

and Stokes' Theorem is verified for this special case.

3. $\mathbf{F}(x, y, z) = y\mathbf{i} + z\mathbf{j} + x\mathbf{k}$, so $\operatorname{curl}\mathbf{F} = \begin{vmatrix} \mathbf{i} & \mathbf{j} & \mathbf{k} \\ \frac{\partial}{\partial x} & \frac{\partial}{\partial y} & \frac{\partial}{\partial z} \\ y & z & x \end{vmatrix} = -\mathbf{i} - \mathbf{j} - \mathbf{k}.$ Let

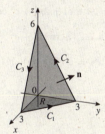

$z = g(x, y) = 6 - 2x - 2y \Rightarrow g_x = -2$ and $g_y = -2$. Then, using

Theorem 15.7.3 (14.7.3 in ET) with R the triangle shown in the diagram, we

have

$$\iint_S \operatorname{curl}\mathbf{F} \cdot d\mathbf{S} = \iint_R (-Pg_x - Qg_y + R)\, dA$$

$$= \iint_R [-(-1)(-2) - (-1)(-2) - 1]\, dA$$

$$= -5 \iint_R dA = -5\left(\tfrac{1}{2} \cdot 3 \cdot 3 \right) = -\tfrac{45}{2}$$

Next, observe that $C = C_1 \cup C_2 \cup C_3$ where $C_1 : \mathbf{r}_1(t) = (3 - 3t)\mathbf{i} + 3t\mathbf{j}, 0 \le t \le 1,$

$C_2 : \mathbf{r}_2(t) = (3 - 3t)\mathbf{j} + 6t\mathbf{k}, 0 \le t \le 1,$ and $C_3 : \mathbf{r}_3(t) = 3t\mathbf{i} + (6 - 6t)\mathbf{k}, 0 \le t \le 1.$ Thus,

$$\oint_C \mathbf{F} \cdot d\mathbf{r} = \int_C \mathbf{F}(\mathbf{r}(t)) \cdot \mathbf{r}'(t)\, dt$$

$$= \int_0^1 [3t\mathbf{i} + (3 - 3t)\mathbf{k}] \cdot (-3\mathbf{i} + 3\mathbf{j})\, dt + \int_0^1 [(3 - 3t)\mathbf{i} + 6t\mathbf{j}] \cdot (-3\mathbf{j} + 6\mathbf{k})\, dt + \int_0^1 [(6 - 6t)\mathbf{j} + 3t\mathbf{k}] \cdot (3\mathbf{i} - 6\mathbf{k})\, dt$$

$$= \int_0^1 (-45t)\, dt = -\tfrac{45}{2}t^2\Big|_0^1 = -\tfrac{45}{2}$$

and Stokes' Theorem is verified for this special case.

5. The surface S is bounded by the curve $x^2 + y^2 = 4$. A parametric
representation in a counterclockwise direction is
$C : \mathbf{r}(t) = 2\cos t\,\mathbf{i} + 2\sin t\,\mathbf{j}, 0 \le t \le 2\pi$. Using Stokes' Theorem with
$\mathbf{F}(x, y, z) = 2y\mathbf{i} + xz^2\mathbf{j} + x^2 y e^z\mathbf{k}$, we have

$$\iint_S \operatorname{curl}\mathbf{F} \cdot d\mathbf{S} = \oint_C \mathbf{F} \cdot d\mathbf{r} = \oint_C \mathbf{F}(\mathbf{r}(t)) \cdot \mathbf{r}'(t)\,dt$$

$$= \int_0^{2\pi} \left[4\sin t\,\mathbf{i} + (2\cos t)(0)\,\mathbf{j} + (2\cos t)^2(2\sin t)\,\mathbf{k}\right] \cdot (-2\sin t\,\mathbf{i} + 2\cos t\,\mathbf{j})\,dt$$

$$= \int_0^{2\pi} \left(-8\sin^2 t\right)dt = -4\int_0^{2\pi} (1 - \cos 2t)\,dt = -4\left(t - \tfrac{1}{2}\sin 2t\right)\Big|_0^{2\pi} = -8\pi$$

7. The surface S is bounded by the curve $x^2 + y^2 = 1$. Solving this equation and
$z = \sqrt{4 - x^2 - y^2}$ simultaneously, we find that $z = \sqrt{3}$, so a parametric
representation of C oriented in a counterclockwise direction is
$\mathbf{r}(t) = \cos t\,\mathbf{i} + \sin t\,\mathbf{j} + \sqrt{3}\mathbf{k}, 0 \le t \le 2\pi$. Using Stokes' Theorem with
$\mathbf{F}(x, y, z) = xyz\mathbf{i} + 2x\mathbf{j} + \tan^{-1} y^2\,\mathbf{k}$, we obtain

$$\iint_S \operatorname{curl}\mathbf{F} \cdot d\mathbf{S} = \oint_C \mathbf{F} \cdot d\mathbf{r} = \int_0^{2\pi} \mathbf{F}(\mathbf{r}(t)) \cdot \mathbf{r}'(t)\,dt$$

$$= \int_0^{2\pi} \left[(\cos t)(\sin t)\left(\sqrt{3}\right)\mathbf{i} + 2\cos t\,\mathbf{j} + \tan^{-1}\left(\sin^2 t\right)\mathbf{k}\right] \cdot (-\sin t\,\mathbf{i} + \cos t\,\mathbf{j})\,dt$$

$$= \int_0^{2\pi} \left(-\sqrt{3}\cos t\sin^2 t + 2\cos^2 t\right)dt = \int_0^{2\pi} \left(-\sqrt{3}\cos t\sin^2 t + 1 + \cos 2t\right)dt$$

$$= \left(-\tfrac{\sqrt{3}}{3}\sin^3 t + t + \tfrac{1}{2}\sin 2t\right)\Big|_0^{2\pi} = 2\pi$$

9. The boundary of S is the ellipse $9x^2 + 9y^2 = 36$; that is, the circle
$x^2 + y^2 = 4$. A parametric representation with counterclockwise orientation is
$C : \mathbf{r}(t) = 2\cos t\,\mathbf{i} + 2\sin t\,\mathbf{j}, 0 \le t \le 2\pi$, so
$\mathbf{F}(x, y, z) = z\sin x\,\mathbf{i} + 2x\mathbf{j} + e^x\cos z\,\mathbf{k} \Rightarrow$
$\mathbf{F}(\mathbf{r}(t)) = 0\mathbf{i} + 2(2\cos t)\mathbf{j} + e^{2\cos t}\mathbf{k}$. Using Stokes' Theorem, we have

$$\iint_S \operatorname{curl}\mathbf{F} \cdot d\mathbf{S} = \oint_C \mathbf{F}(\mathbf{r}(t)) \cdot \mathbf{r}'(t)\,dt = \int_0^{2\pi} \left(4\cos t\,\mathbf{j} + e^{2\cos t}\mathbf{k}\right) \cdot (-2\sin t\,\mathbf{i} + 2\cos t\,\mathbf{j})\,dt = \int_0^{2\pi} 8\cos^2 t\,dt$$

$$= 4\int_0^{2\pi} (1 + \cos 2t)\,dt = 4\left(t + \tfrac{1}{2}\sin 2t\right)\Big|_0^{2\pi} = 8\pi$$

11. $\mathbf{F}(x, y, z) = (y - z)\mathbf{i} + (z - x)\mathbf{j} + (x - y)\mathbf{k}$, so

$$\operatorname{curl}\mathbf{F} = \begin{vmatrix} \mathbf{i} & \mathbf{j} & \mathbf{k} \\ \frac{\partial}{\partial x} & \frac{\partial}{\partial y} & \frac{\partial}{\partial z} \\ y-z & z-x & x-y \end{vmatrix} = -2\mathbf{i} - 2\mathbf{j} - 2\mathbf{k}.$$ Writing $2x + 3y + z = 6$

in the form $z = g(x, y) = 6 - 2x - 3y$, we have $g_x = -2$ and $g_y = -3$. Using
Stokes' Theorem and Theorem 15.7.3 (14.7.3 in ET) with R as shown, we have

$$\oint_C \mathbf{F} \cdot d\mathbf{r} = \iint_S \operatorname{curl}\mathbf{F} \cdot d\mathbf{S} = \iint_R \left[-(-2)(-2) - (-2)(-3) - 2\right]dA$$

$$= -12\int_0^3 \int_0^{(6-2x)/3} dy\,dx = -12\int_0^3 \tfrac{1}{3}(6 - 2x)\,dx = -4\left(6x - x^2\right)\Big|_0^3 = -36$$

13. $\mathbf{F}(x, y, z) = 3xz\mathbf{i} + e^{xz}\mathbf{j} + 2xy\mathbf{k}$, so

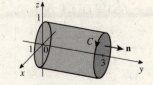

$$\text{curl }\mathbf{F} = \begin{vmatrix} \mathbf{i} & \mathbf{j} & \mathbf{k} \\ \frac{\partial}{\partial x} & \frac{\partial}{\partial y} & \frac{\partial}{\partial z} \\ 3xz & e^{xz} & 2xy \end{vmatrix} = (2x - xe^{xz})\mathbf{i} - (2y - 3x)\mathbf{j} + ze^{xz}\mathbf{k}. \text{ Since the}$$

circle is oriented counterclockwise when viewed from the right, we take $\mathbf{n} = \mathbf{j}$.

Then, using the fact that $y = 3$ on S,

$$\oint_C \mathbf{F} \cdot d\mathbf{r} = \iint_S \text{curl }\mathbf{F} \cdot \mathbf{n}\, dS = \iint_S (3x - 2y)\, dS = \iint_S (3x - 6)\, dS = \int_0^{2\pi} \int_0^1 (3r\cos\theta - 6)\, r\, dr\, d\theta$$

$$= \int_0^{2\pi} \left[r^3 \cos\theta - 3r^2 \right]_{r=0}^{r=1} d\theta = \int_0^{2\pi} (\cos\theta - 3)\, d\theta = (\sin\theta - 3\theta)\big|_0^{2\pi} = -6\pi$$

15. In the figure, C_1 joins the points $(0, 0, 0)$, $(0, 1, 0)$, $(2, 1, 0)$, $(2, 0, 0)$, and

$(0, 0, 0)$, while C_2 joins the points $(0, 0, 0)$, $(2, 0, 0)$, $(2, 0, 1)$, $(0, 0, 1)$, and

$(0, 0, 0)$. By Stokes' Theorem,

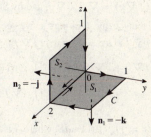

$$\oint_C \mathbf{F} \cdot d\mathbf{r} = \oint_{C_1 \cup C_2} \mathbf{F} \cdot d\mathbf{r} = \oint_{C_1} \mathbf{F} \cdot d\mathbf{r} + \oint_{C_2} \mathbf{F} \cdot d\mathbf{r}$$

$$= \iint_{S_1} \text{curl }\mathbf{F} \cdot d\mathbf{S} + \iint_{S_2} \text{curl }\mathbf{F} \cdot d\mathbf{S}$$

But $\text{curl }\mathbf{F} = \begin{vmatrix} \mathbf{i} & \mathbf{j} & \mathbf{k} \\ \frac{\partial}{\partial x} & \frac{\partial}{\partial y} & \frac{\partial}{\partial z} \\ xe^y & ye^x & xyz \end{vmatrix} = xz\mathbf{i} - yz\mathbf{j} + (ye^x - xe^y)\mathbf{k}$. Also, $z = 0$ on S_1 and $y = 0$ on S_2, so

$$\oint_C \mathbf{F} \cdot d\mathbf{r} = -\int_0^1 \int_0^2 (ye^x - xe^y)\, dx\, dy + \int_0^1 \int_0^2 yz\, dx\, dz = -\int_0^1 \int_0^2 (ye^x - xe^y)\, dx\, dy = -\int_0^1 \left[ye^x - \tfrac{1}{2}x^2 e^y \right]_{x=0}^{x=2} dy$$

$$= -\int_0^1 \left(ye^2 - y - 2e^y \right) dy = -\left(\tfrac{1}{2}y^2 e^2 - \tfrac{1}{2}y^2 - 2e^y \right)\Big|_0^1 = -\tfrac{1}{2}e^2 + 2e - \tfrac{3}{2}$$

17. The work done by $\mathbf{F}$ is $W = \oint_C \mathbf{F} \cdot d\mathbf{r}$. To evaluate this integral, we use Stokes'

Theorem: $\text{curl }\mathbf{F} = \begin{vmatrix} \mathbf{i} & \mathbf{j} & \mathbf{k} \\ \frac{\partial}{\partial x} & \frac{\partial}{\partial y} & \frac{\partial}{\partial z} \\ e^x + z & x^2 + \cosh y & y^2 + z^3 \end{vmatrix} = 2y\mathbf{i} + \mathbf{j} + 2x\mathbf{k}$. The

surface S bounded by the closed path C has equation

$z = g(x, y) = -2x - 2y + 2 \Rightarrow g_x = -2$ and $g_y = -2$, so by Stokes'

Theorem,

$$W = \oint_C \mathbf{F} \cdot d\mathbf{r} = \iint_S \text{curl }\mathbf{F} \cdot d\mathbf{S} = \iint_R [-(2y)(-2) - (1)(-2) + 2x]\, dA = \int_0^1 \int_0^{1-x} (2x + 4y + 2)\, dy\, dx$$

$$= \int_0^1 \left[2xy + 2y^2 + 2y \right]_{y=0}^{y=1-x} dx = \int_0^1 (4 - 4x)\, dx = \left(4x - 2x^2 \right)\big|_0^1 = 2$$

19. Let $A(C)$ denote the area of the surface bounded by C. By Stokes' Theorem,

$$\oint_C \mathbf{B} \cdot d\mathbf{r} = \iint_S \text{curl }\mathbf{B} \cdot \mathbf{n}\, dS = \iint_S \mu_0 \mathbf{J} \cdot \mathbf{n}\, dS = \mu_0 \iint_S \mathbf{J} \cdot \mathbf{n}\, dS = \mu_0 |\mathbf{J}|\, A(C) = \mu_0 I. \text{ See Exercise 15.3.42 (14.3.42 in}$$

ET).

21. a. $f\nabla g = fg_x\mathbf{i} + fg_y\mathbf{j} + fg_z\mathbf{k}$, so

$$\text{curl}\,(f\nabla g) = \begin{vmatrix} \mathbf{i} & \mathbf{j} & \mathbf{k} \\ \frac{\partial}{\partial x} & \frac{\partial}{\partial y} & \frac{\partial}{\partial z} \\ fg_x & fg_y & fg_z \end{vmatrix} = \left[\frac{\partial}{\partial y}\,(fg_z) - \frac{\partial}{\partial z}\,(fg_y)\right]\mathbf{i} - \left[\frac{\partial}{\partial x}\,(fg_z) - \frac{\partial}{\partial z}\,(fg_x)\right]\mathbf{j} + \left[\frac{\partial}{\partial x}\,(fg_y) - \frac{\partial}{\partial y}\,(fg_x)\right]\mathbf{k}$$

$$= (fg_{zy} + f_y g_z - fg_{yz} - f_z g_y)\mathbf{i} - (fg_{zx} + f_x g_z - fg_{xz} - f_z g_x)\mathbf{j} + (fg_{yx} + f_x g_y - fg_{xy} - f_y g_x)\mathbf{k}$$

$$= (f_y g_z - f_z g_y)\mathbf{i} - (f_x g_z - f_z g_x)\mathbf{j} + (f_x g_y - f_y g_x)\mathbf{k}$$

using the fact that the mixed derivatives are equal.

On the other hand, $\nabla f \times \nabla g = \begin{vmatrix} \mathbf{i} & \mathbf{j} & \mathbf{k} \\ f_x & f_y & f_z \\ g_x & g_y & g_z \end{vmatrix} = (f_y g_z - f_z g_y)\mathbf{i} - (f_x g_z - f_z g_x)\mathbf{j} + (f_x g_y - f_y g_x)\mathbf{k}$. Therefore,

$\text{curl}\,(f\nabla g) = \nabla f \times \nabla g$, and by Stokes' Theorem, $\oint_C (f\nabla g) \cdot d\mathbf{r} = \iint_S \text{curl}\,(f\nabla g) \cdot d\mathbf{S} = \iint_S (\nabla f \times \nabla g) \cdot d\mathbf{S}$.

b. Replacing g by f in part a, we see that $\nabla f \times \nabla f = \mathbf{0}$, so

$\oint_C (f\nabla f) \cdot d\mathbf{r} = \iint_S \text{curl}\,(f\nabla f) \cdot d\mathbf{S} = \iint_S \mathbf{0} \cdot d\mathbf{S} = \iint_S 0\,dS = 0$.

c. $\oint_C (f\nabla g + g\nabla f) \cdot d\mathbf{r} = \iint_S \text{curl}\,(f\nabla g + g\nabla f) \cdot d\mathbf{S} = \iint_S \text{curl}\,(f\nabla g) \cdot d\mathbf{S} + \iint_S \text{curl}\,(g\nabla f) \cdot d\mathbf{S}$

$$= \iint_S (\nabla f \times \nabla g) \cdot d\mathbf{S} + \iint_S (\nabla g \times \nabla f) \cdot d\mathbf{S} = 0 \text{ because } \nabla f \times \nabla g = -\nabla g \times \nabla f.$$

23. $\oint_C \mathbf{F} \cdot d\mathbf{r} = \iint_S \text{curl}\,\mathbf{F} \cdot d\mathbf{S}$, where S is the surface enclosed by C. But

$\mathbf{F}(x, y, z) = xf(r)\mathbf{i} + yf(r)\mathbf{j} + zf(r)\mathbf{k}$, where $r = \sqrt{x^2 + y^2 + z^2}$, so

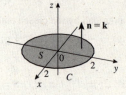

$$\text{curl}\,\mathbf{F} = \begin{vmatrix} \mathbf{i} & \mathbf{j} & \mathbf{k} \\ \frac{\partial}{\partial x} & \frac{\partial}{\partial y} & \frac{\partial}{\partial z} \\ xf(r) & yf(r) & zf(r) \end{vmatrix}$$

$$= \left\{\frac{\partial}{\partial y}\,[zf(r)] - \frac{\partial}{\partial z}\,[yf(r)]\right\}\mathbf{i} - \left\{\frac{\partial}{\partial x}\,[zf(r)] - \frac{\partial}{\partial z}\,[xf(r)]\right\}\mathbf{j}$$

$$+ \left\{\frac{\partial}{\partial x}\,[yf(r)] - \frac{\partial}{\partial y}\,[xf(r)]\right\}\mathbf{k}$$

$$= \left[zf'(r)\left(\tfrac{1}{2}r^{-1/2}\right)(2y) - yf'(r)\left(\tfrac{1}{2}r^{-1/2}\right)(2z)\right]\mathbf{i} - \left[zf'(r)\left(\tfrac{1}{2}r^{-1/2}\right)(2x) - xf'(r)\left(\tfrac{1}{2}r^{-1/2}\right)(2z)\right]\mathbf{j}$$

$$+ \left[yf'(r)\left(\tfrac{1}{2}r^{-1/2}\right)(2x) - xf'(r)\left(\tfrac{1}{2}r^{-1/2}\right)(2y)\right]\mathbf{k} = \mathbf{0}$$

Thus, $\oint_C \mathbf{F} \cdot d\mathbf{r} = \iint_S \text{curl}\,\mathbf{F} \cdot d\mathbf{S} = \iint_S 0\,dS = 0$.

25. The given integral is $\oint_C \mathbf{F} \cdot d\mathbf{r}$, where

$\mathbf{F}(x, y, z) = e^x \cos z\,\mathbf{i} + 2xy^2\mathbf{j} + \cot^{-1} y\,\mathbf{k}$. We can pick S lying in the

xy-plane (where $z = 0$) and bounded by $x^2 + y^2 = 4$. Observe that $\mathbf{n} = \mathbf{k}$.

Thus, $\text{curl}\,\mathbf{F} = \begin{vmatrix} \mathbf{i} & \mathbf{j} & \mathbf{k} \\ \frac{\partial}{\partial x} & \frac{\partial}{\partial y} & \frac{\partial}{\partial z} \\ e^x \cos z & 2xy^2 & \cot^{-1} y \end{vmatrix} = -\frac{1}{1 + y^2}\mathbf{i} - e^x \sin z\,\mathbf{j} + 2y^2\mathbf{k}$.

Using Stokes' Theorem, we find

$$\oint_C \mathbf{F} \cdot d\mathbf{r} = \iint_S \text{curl}\,\mathbf{F} \cdot d\mathbf{S} = \iint_S \left(-\frac{1}{1 + y^2}\mathbf{i} - e^x \sin z\,\mathbf{j} + 2y^2\mathbf{k}\right) \cdot \mathbf{k}\,dS = \iint_S 2y^2\,dS = \int_0^{2\pi}\int_0^2 2r^2 \sin^2\theta\,r\,dr\,d\theta$$

$$= \int_0^{2\pi}\left[\tfrac{1}{2}r^4 \sin^2\theta\right]_{r=0}^{r=2}d\theta = 8\int_0^{2\pi}\sin^2\theta\,d\theta = 4\int_0^{2\pi}(1 - \cos 2\theta)\,d\theta = 4\left(\theta - \tfrac{1}{2}\sin 2\theta\right)\Big|_0^{2\pi} = 8\pi$$

27. By the Divergence Theorem, $\iint_S \operatorname{curl} \mathbf{F} \cdot d\mathbf{S} = \iiint_T \operatorname{div} \operatorname{curl} \mathbf{F} \, dV$, but $\operatorname{div} \operatorname{curl} \mathbf{F} = 0$ (see Example 15.2.9/14.2.9 in ET), so $\iint_S \operatorname{curl} \mathbf{F} \cdot d\mathbf{S} = \iiint_T 0 \, dV = 0$.

29. The curve of intersection of the plane $x + z = 2$ and the cylinder $x^2 + y^2 = 1$ has vector representation $\mathbf{r}(t) = \cos t\,\mathbf{i} + \sin t\,\mathbf{j} + (2 - \cos t)\,\mathbf{k}, 0 \le t \le 2\pi$, so $\mathbf{F}(\mathbf{r}(t)) = \cos(2 - \cos t)\,\mathbf{i} + \cos^2 t\,\mathbf{j} + 2\sin t\,\mathbf{k}$, and since $\mathbf{r}'(t) = -\sin t\,\mathbf{i} + \cos t\,\mathbf{j} + \sin t\,\mathbf{k}$, we have $\oint_C \mathbf{F} \cdot d\mathbf{r} = \int_0^{2\pi} \left[(-\sin t)\cos(2 - 2\cos t) + \cos^3 t + 2\sin^2 t \right] dt$. Using a CAS, we find $\oint_C \mathbf{F} \cdot d\mathbf{r} = 6.28318530718\ldots = 2\pi$, as obtained in Example 2.

31. False. Referring to the figure, we have
$$\iint_{S_1} \operatorname{curl} \mathbf{F} \cdot d\mathbf{S} = \oint_C \mathbf{F} \cdot d\mathbf{r} = -\oint_{-C} \mathbf{F} \cdot d\mathbf{r} = -\iint_{S_2} \operatorname{curl} \mathbf{F} \cdot d\mathbf{S}.$$

Chapter 15   Review   ET 14

Concept Review

1. vector; vector

3. a. departs, accumulates

  b. $\dfrac{\partial P}{\partial x} + \dfrac{\partial Q}{\partial y} + \dfrac{\partial R}{\partial z}$; enters; departs; equals; departs; enters

5. a. $\lim_{n\to\infty} \sum_{k=1}^n f\left(x_k^*, y_k^*\right) \Delta s_k$

  b. $\int_a^b f(x(t), y(t)) \sqrt{[x'(t)]^2 + [y'(t)]^2} \, dt$

  c. $\lim_{n\to\infty} \sum_{k=1}^n f\left(x_k^*, y_k^*\right) \Delta x_k$; $\int_a^b f(x(t), y(t)) x'(t) \, dt$

  d. $\lim_{n\to\infty} \sum_{k=1}^n f\left(x_k^*, y_k^*\right) \Delta y_k$; $\int_a^b f(x(t), y(t)) y'(t) \, dt$

7. a. $\int_a^b \mathbf{F} \cdot \mathbf{T} \, ds = \int_a^b \mathbf{F}(\mathbf{r}(t)) \cdot \mathbf{r}'(t) \, dt$

  b. work

9. a. closed

  b. connected; conservative

11. $\operatorname{curl} \mathbf{F}$; $\dfrac{\partial Q}{\partial z}$; $\dfrac{\partial P}{\partial z}$; $\dfrac{\partial P}{\partial y}$

13. $\oint_C x \, dy$; $-\oint_C y \, dx$; $\dfrac{1}{2}\oint_C x \, dy - y \, dx$

15. a. $x(u, v)\,\mathbf{i} + y(u, v)\,\mathbf{j} + z(u, v)\,\mathbf{k}$; parameter; parametric equations

  b. $u\mathbf{i} + v\mathbf{j} + f(u, v)\,\mathbf{k}$

  c. $u\mathbf{i} + f(u)\cos v\,\mathbf{j} + f(u)\sin v\,\mathbf{k}$; $\{(u, v) \mid a \le u \le b, 0 \le v \le 2\pi\}$

17. a. $\iint_R F(x, y, f(x, y)) \sqrt{[f_x(x, y)]^2 + [f_y(x, y)]^2 + 1} \, dA$

  b. $\iint_D F(\mathbf{r}(u, v)) |\mathbf{r}_u \times \mathbf{r}_v| \, dA$

Review Exercises

1. a. $\operatorname{div} \mathbf{F} = \frac{\partial}{\partial x}\left(xy^2\right) + \frac{\partial}{\partial y}\left(yz^2\right) + \frac{\partial}{\partial z}\left(zx^2\right) = y^2 + z^2 + x^2$

b. $\operatorname{curl} \mathbf{F} = \begin{vmatrix} \mathbf{i} & \mathbf{j} & \mathbf{k} \\ \frac{\partial}{\partial x} & \frac{\partial}{\partial y} & \frac{\partial}{\partial z} \\ xy^2 & yz^2 & zx^2 \end{vmatrix} = -2yz\mathbf{i} - 2xz\mathbf{j} - 2xy\mathbf{k}$

3. a. $\operatorname{div} \mathbf{F} = \frac{\partial}{\partial x}\left(e^x \sin y\right) + \frac{\partial}{\partial y}\left(e^x \cos y\right) + \frac{\partial}{\partial z}\left(e^z\right) = e^x \sin y - e^x \sin y + e^z = e^z$

b. $\operatorname{curl} \mathbf{F} = \begin{vmatrix} \mathbf{i} & \mathbf{j} & \mathbf{k} \\ \frac{\partial}{\partial x} & \frac{\partial}{\partial y} & \frac{\partial}{\partial z} \\ e^x \sin y & e^x \cos y & e^z \end{vmatrix} = \left(e^x \cos y - e^x \cos y\right)\mathbf{k} = \mathbf{0}$

5. A parametric representation of C is $\mathbf{r}(t) = t\mathbf{i} + \sqrt{t}\mathbf{j}$, $1 \le t \le 4$. Then

$$\int_C y\, ds = \int_1^4 \sqrt{t}\sqrt{1 + \left(\tfrac{1}{2}t^{-1/2}\right)^2}\, dt = \int_1^4 \sqrt{t}\sqrt{\tfrac{4t+1}{4t}}\, dt = \tfrac{1}{2}\int_1^4 \sqrt{4t+1}\, dt = \tfrac{1}{2}\left(\tfrac{1}{4}\right)\tfrac{2}{3}(4t+1)^{3/2}\Big|_1^4$$

$$= \tfrac{1}{12}\left(17\sqrt{17} - 5\sqrt{5}\right)$$

7. $\int_C xy^2\, ds = \int_0^{\pi/2} \sin t \cos^2 t \sqrt{\cos^2 t + \sin^2 t + 1}\, dt = \sqrt{2}\int_0^{\pi/2} \cos^2 t \sin t\, dt = \sqrt{2}\left(-\tfrac{1}{3}\cos^3 t\right)\Big|_0^{\pi/2} = \tfrac{\sqrt{2}}{3}$

9. $C: \mathbf{r}(t) = t\mathbf{i} + t^{1/3}\mathbf{j}$, $1 \le t \le 8$, so

$$\int_C x^2 y\, dx + x^3 y\, dy = \int_1^8 \left(t^2 \cdot t^{1/3} dt + t^3 \cdot t^{1/3} \cdot \tfrac{1}{3}t^{-2/3} dt\right) = \int_1^8 \left(t^{7/3} + \tfrac{1}{3}t^{8/3}\right) dt = \left(\tfrac{3}{10}t^{10/3} + \tfrac{1}{11}t^{11/3}\right)\Big|_1^8$$

$$= \tfrac{54{,}229}{110}$$

11. $C: t\mathbf{i} + \cos t\mathbf{j} + \sin t\mathbf{k}$, $0 \le t \le \tfrac{\pi}{2}$, so

$$\int_C yz\, dx - y\cos x\, dy + y\, dz = \int_0^{\pi/2}\left[\cos t\,(\sin t\, dt) - \cos t\,(\cos t)(-\sin t\, dt) + \cos t\,(\cos t\, dt)\right]$$

$$= \int_0^{\pi/2}\left(\sin t \cos t + \cos^2 t \sin t + \cos^2 t\right) dt = \left[\tfrac{1}{2}\sin^2 t - \tfrac{1}{3}\cos^3 t + \tfrac{1}{2}\left(t + \tfrac{1}{2}\sin 2t\right)\right]_0^{\pi/2}$$

$$= \tfrac{5}{6} + \tfrac{\pi}{4} = \tfrac{3\pi + 10}{12}$$

13. A parametric representation of C is $\mathbf{r}(t) = t\mathbf{i} + t\mathbf{j} + 2t\mathbf{k}$, $0 \le t \le 1$, so

$$\int_C xy\, dx + e^{-y}\, dy + ze^x\, dz = \int_0^1 \left[t\,(t\, dt) + e^{-t}\, dt + 2te^t\,(2\, dt)\right] = \int_0^1 \left(t^2 + e^{-t} + 4te^t\right) dt$$

$$= \left[\tfrac{1}{3}t^3 - e^{-t} + 4(t-1)e^t\right]_0^1 = \tfrac{16}{3} - e^{-1}$$

15. A parametric representation of C is $\mathbf{r}(t) = (1+t)\mathbf{i} + (1+2t)\mathbf{j} + (1+4t)\mathbf{k}$, $0 \le t \le 1$, so

$$W = \int_C \mathbf{F} \cdot d\mathbf{r} = \int_0^1 \mathbf{F}(\mathbf{r}(t)) \cdot \mathbf{r}'(t)\, dt$$

$$= \int_0^1 \left\{(1+t)(1+2t)\mathbf{i} + [(1+2t)+(1+4t)]\mathbf{j} + (1+4t)^2\mathbf{k}\right\} \cdot (\mathbf{i} + 2\mathbf{j} + 4\mathbf{k})\, dt$$

$$= \int_0^1 \left[\left(1 + 3t + 2t^2\right) + 2(2+6t) + 4\left(1 + 8t + 16t^2\right)\right] dt = \int_0^1 \left(66t^2 + 47t + 9\right) dt = \left(22t^3 + \tfrac{47}{2}t^2 + 9t\right)\Big|_0^1 = \tfrac{109}{2}$$

17. Here $P(x, y) = 4xy + 3y^2$ and $Q(x, y) = 2x^2 + 6xy$, so $\frac{\partial P}{\partial y} = 4x + 6y = \frac{\partial Q}{\partial x}$ for all (x, y). Thus, $\mathbf{F}$ is a conservative vector field, and so there exists a function f such that $\frac{\partial f}{\partial x} = 4xy + 3y^2$ and $\frac{\partial f}{\partial y} = 2x^2 + 6xy$. Integrating the first equation with respect to x, we have $f(x, y) = 2x^2 y + 3xy^2 + g(y)$, and using the second equation, we get $\frac{\partial f}{\partial y} = 2x^2 + 6xy + g'(y) = 2x^2 + 6xy \Rightarrow g'(y) = 0 \Rightarrow g(y) = C$, a constant. The required function is thus $f(x, y) = 2x^2 y + 3xy^2 + C$.

19. $\mathbf{F}(x, y, z) = \left(2xy + y^3\right)\mathbf{i} + \left(x^2 + 3xy^2\right)\mathbf{j}$, so $\frac{\partial P}{\partial y} = \frac{\partial}{\partial y}\left(2xy + y^3\right) = 2x + 3y^2 = \frac{\partial Q}{\partial x}$, showing that $\mathbf{F}$ is a conservative

vector field. Therefore, $\int_C \mathbf{F} \cdot \mathbf{T}\, ds$ is independent of path and is equal to $\int_{C_1} \mathbf{F} \cdot \mathbf{T}\, ds$, where C_1 is the line segment from

$(-5, 0)$ to $(0, 3)$. C_1 has representation $\mathbf{r}(t) = (-5 + 5t)\mathbf{i} + 3t\mathbf{j} = -5(1 - t)\mathbf{i} + 3t\mathbf{j}, 0 \le t \le 1$, so

$\int_C \mathbf{F} \cdot \mathbf{T}\, ds = \int_{C_1} \mathbf{F} \cdot \mathbf{T}\, ds = \int_0^1 \mathbf{F}(\mathbf{r}(t)) \cdot \mathbf{r}'(t)\, dt$

$\qquad = \int_0^1 \left\{ \left[2(-5)(1-t)(3t) + (3t)^3 \right]\mathbf{i} + \left[25(1-t)^2 - 15(1-t)(3t)^2 \right]\mathbf{j} \right\} \cdot (5\mathbf{i} + 3\mathbf{j})\, dt$

$\qquad = \int_0^1 \left(540t^3 - 180t^2 - 300t + 75\right) dt = \left(135t^4 - 60t^3 - 150t^2 + 75t\right)\Big|_0^1 = 0$

21. $\oint_C \left(y^2 + \sec x\right) dx + \left(x^2 + y^5\right) dy$

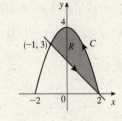

$\qquad = \iint_R \left[\frac{\partial}{\partial x}\left(x^2 + y^5\right) - \frac{\partial}{\partial y}\left(y^2 + \sec x\right) \right] dA = \iint_R (2x - 2y)\, dA$

$\qquad = 2\int_{-1}^2 \int_{-x+2}^{4-x^2} (x - y)\, dy\, dx = 2\int_{-1}^2 \left[xy - \tfrac{1}{2}y^2 \right]_{y=-x+2}^{y=4-x^2} dx$

$\qquad = 2\int_{-1}^2 \left(-\tfrac{1}{2}x^4 - x^3 + \tfrac{11}{2}x^2 - 6 \right) dx$

$\qquad = 2\left(-\tfrac{1}{10}x^5 - \tfrac{1}{4}x^4 + \tfrac{11}{6}x^3 - 6x \right)\Big|_{-1}^2 = -\tfrac{171}{10}$

23. $\oint_C \left(x^2 y + e^x\right) dx + \left(e^{-y} - xy^2\right) dy$

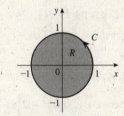

$\qquad = \iint_R \left[\frac{\partial}{\partial x}\left(e^{-y} - xy^2\right) - \frac{\partial}{\partial y}\left(x^2 y + e^x\right) \right] dA$

$\qquad = \iint_R \left(-y^2 - x^2\right) dA = -\int_0^{2\pi} \int_0^1 r^2 r\, dr\, d\theta$

$\qquad = -\int_0^{2\pi} \left[\tfrac{1}{4}r^4 \right]_0^1 d\theta = -\tfrac{1}{4}\int_0^{2\pi} d\theta = -\tfrac{\pi}{2}$

25. $2x + 2y + 3z = 6 \Rightarrow z = g(x, y) = \tfrac{1}{3}(6 - 2x - 2y) \Rightarrow g_x(x, y) = -\tfrac{2}{3}$ and

$g_y(x, y) = -\tfrac{2}{3}$, so

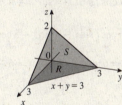

$\iint_S (y + xz)\, dS = \iint_R \left[y + x \cdot \tfrac{1}{3}(6 - 2x - 2y) \right] \sqrt{\left(-\tfrac{2}{3}\right)^2 + \left(-\tfrac{2}{3}\right)^2 + 1}\, dA$

$\qquad = \int_0^3 \int_0^{3-x} \tfrac{\sqrt{17}}{9}\left(3y + 6x - 2x^2 - 2xy \right) dy\, dx$

$\qquad = \tfrac{\sqrt{17}}{9}\int_0^3 \left[\tfrac{3}{2}y^2 + 6xy - 2x^2 y - xy^2 \right]_{y=0}^{y=3-x} dx$

$\qquad = \tfrac{\sqrt{17}}{9}\int_0^3 \tfrac{1}{2}\left(2x^3 - 9x^2 + 27 \right) dx = \tfrac{\sqrt{17}}{18}\left(\tfrac{1}{2}x^4 - 3x^3 + 27x \right)\Big|_0^3 = \tfrac{9\sqrt{17}}{4}$

27. $y = h(x, z) = 1 - x^2 - z^2 \Rightarrow h_x = -2x$ and $h_z = -2z$, so with

$\mathbf{F}(x, y, z) = x\mathbf{i} + y\mathbf{j} + z\mathbf{k}$ (so $P = x$, $Q = y$, and $R = z$) we have

$\iint_S \mathbf{F} \cdot \mathbf{n}\, dS = \iint_R \left[-x(-2x) + y - z(-2z) \right] dA$ where

$R = \left\{ (x, z) \mid x^2 + z^2 \le 1 \right\}$. Since $y = 1 - x^2 - z^2$, we get

$\iint_S \mathbf{F} \cdot \mathbf{n}\, dS = \iint_R \left[2x^2 + \left(1 - x^2 - z^2\right) + 2z^2 \right] dA$

$\qquad = \int_0^{2\pi} \int_0^1 \left(r^2 + 1\right) r\, dr\, d\theta = \int_0^{2\pi} \left[\tfrac{1}{4}r^4 + \tfrac{1}{2}r^2 \right]_0^1 d\theta = \tfrac{3}{4}\int_0^{2\pi} d\theta = \tfrac{3\pi}{2}$

29. $x + y + z = 1 \Rightarrow x = g(y, z) = 1 - y - z \Rightarrow g_y = -1$ and $g_z = -1$. We have

$\rho(x, y, z) = kx^2$, where k is the constant of proportionality, so

$$m = \iint_S \rho(x, y, z)\, dS = \iint_R kx^2 \sqrt{(-1)^2 + (-1)^2 + 1}\, dA$$

$$= k\sqrt{3} \int_0^1 \int_0^{1-y} (1 - y - z)^2\, dz\, dy$$

$$= k\sqrt{3} \int_0^1 \left[-\tfrac{1}{3}(1 - y - z)^3 \right]_{z=0}^{z=1-y} dy = k\sqrt{3} \int_0^1 \tfrac{1}{3}(1 - y)^3\, dy$$

$$= \tfrac{\sqrt{3}}{3} k \left[-\tfrac{1}{4}(1 - y)^4 \right]_0^1 = \tfrac{\sqrt{3}}{12} k$$

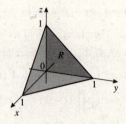

31. $\iint_S \mathbf{F} \cdot \mathbf{n}\, dS = \iiint_T \operatorname{div} \mathbf{F}\, dV = \iiint_T \left[\frac{\partial}{\partial x}(x) + \frac{\partial}{\partial y}(-y) + \frac{\partial}{\partial z}(z) \right] dV$

$$= \iiint_T (1 - 1 + 1)\, dV = \iiint_T dV = \pi(2)^2(3) = 12\pi$$

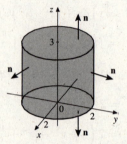

33. $\mathbf{F}(x, y, z) = \left(x + y^2 - 2 \right)\mathbf{i} - 2xy\mathbf{j} - \left(x^2 + yz^2 \right)\mathbf{k}$ and

$C : \mathbf{r}(t) = 2\cos t\,\mathbf{i} + 2\sin t\,\mathbf{j}$, $0 \le t \le 2\pi$. Thus,

$\iint_S \operatorname{curl} \mathbf{F} \cdot \mathbf{n}\, dS = \oint_C \mathbf{F} \cdot \mathbf{T}\, ds$

$= \oint_C \mathbf{F} \cdot d\mathbf{r}$

$= \oint_C \mathbf{F}(\mathbf{r}(t)) \cdot \mathbf{r}'(t)\, dt$

$= \int_0^{2\pi} \left\{ \left[2\cos t + (2\sin t)^2 - 2 \right]\mathbf{i} - 2(2\cos t)(2\sin t)\mathbf{j} - (2\cos t)^2\mathbf{k} \right\} \cdot (-2\sin t\,\mathbf{i} + 2\cos t\,\mathbf{j})\, dt$

$= \int_0^{2\pi} \left(-4\cos t\sin t - 8\sin^3 t + 4\sin t - 16\cos^2 t\sin t \right) dt$

$= \int_0^{2\pi} \left(-4\cos t\sin t - 4\sin t - 8\cos^2 t\sin t \right) dt = \left(2\cos^2 t + 4\cos t + \tfrac{8}{3}\cos^3 t \right)\Big|_0^{2\pi} = 0$

35. $2x + y + z = 6 \Rightarrow z = g(x, y) = 6 - 2x - y \Rightarrow g_x = -2$ and $g_y = -1$. Next,

$$\operatorname{curl} \mathbf{F} = \begin{vmatrix} \mathbf{i} & \mathbf{j} & \mathbf{k} \\ \frac{\partial}{\partial x} & \frac{\partial}{\partial y} & \frac{\partial}{\partial z} \\ 2x + y & -3x - z & y - z \end{vmatrix} = 2\mathbf{i} - 4\mathbf{k}.$$ Observe that the normal

vector $\mathbf{n}$ points downward, so

$$\oint_C \mathbf{F} \cdot \mathbf{T}\, ds = \iint_S \operatorname{curl} \mathbf{F} \cdot \mathbf{n}\, dS = -\iint_R [-2(-2) - (0)(-1) - 4]\, dA$$

$$= -\int_0^3 \int_0^{6-2x} 0\, dA = 0$$

37. $\mathbf{F}(x, y) = 2xy^3\mathbf{i} + 3x^2y^2\mathbf{j} \Rightarrow \frac{\partial}{\partial x}\left(3x^2y^2 \right) = 6xy^2 = \frac{\partial}{\partial y}\left(2xy^3 \right)$, so $\mathbf{F}$ is a conservative vector field.

Therefore, the line integral $\oint_C \mathbf{F} \cdot d\mathbf{r}$ is independent of path. By the Fundamental Theorem for Line

Integrals, we have $\oint_C \mathbf{F} \cdot d\mathbf{r} = f(2, 4) - f(0, 0)$ where f is a potential function of $\mathbf{F}$. To find f, note that

$\frac{\partial f}{\partial x} = 2xy^3$ (1) and $\frac{\partial f}{\partial y} = 3x^2y^2$ (2). From (1), we have $f(x, y) = x^2y^3 + g(y)$, and using (2) we obtain

$\frac{\partial f}{\partial y} = 3x^2y^2 + g'(y) = 3x^2y^2 \Rightarrow g'(y) = 0 \Rightarrow g(y) = C$, a constant, and so $f(x, y) = x^2y^3 + C$. Thus,

$\oint_C \mathbf{F} \cdot d\mathbf{r} = f(2, 4) - f(0, 0) = (2)^2(4)^3 - 0 = 256.$

39. $F(x, y) = \dfrac{x - y}{x^2 + y^2}\mathbf{i} + \dfrac{x + y}{x^2 + y^2}\mathbf{j}$. Since $\dfrac{\partial}{\partial x}\left(\dfrac{x + y}{x^2 + y^2}\right) = \dfrac{\left(x^2 + y^2\right) - (x + y)(2x)}{\left(x^2 + y^2\right)^2} = \dfrac{-x^2 - 2xy + y^2}{\left(x^2 + y^2\right)^2}$ and

$\dfrac{\partial}{\partial y}\left(\dfrac{x - y}{x^2 + y^2}\right) = \dfrac{-x^2 - 2xy + y^2}{\left(x^2 + y^2\right)^2}$ are equal, we see that $\mathbf{F}$ is a conservative vector field. Therefore, the line integral is

independent of path, and since C is a closed path, $\oint_C \mathbf{F} \cdot d\mathbf{r} = 0$.

41. True. curl $\mathbf{F}$ is a vector field on R, so curl (curl $\mathbf{F}$) is a vector field on R.

43. False. $\nabla \times (\nabla f)$ is a vector field, so $\nabla \cdot \left[\nabla \times (\nabla f)\right]$ is a scalar field. In fact, $\nabla \cdot \left[\nabla \times (\nabla f)\right] = 0$ everywhere.

45. True. Consider $\nabla f(x, y) = \mathbf{0}$ (the three-variable case is similar). Then $f_x = 0$ (1) and $f_y = 0$ (2). From (1), we have

$f(x, y) = g(y)$, and then (2) gives $g'(y) = 0 \Rightarrow g(y) = C$, a constant. Thus, $f(x, y) = C$.

47. True. By the Divergence Theorem, $\iint_S \mathbf{F} \cdot \mathbf{n}\, dS = \iiint_T \operatorname{div} \mathbf{F}\, dV$, where T is the solid region enclosed by S. Since

$\operatorname{div} \mathbf{F} = 0$, the result follows.

Challenge Problems

1. $F(x, y) = \dfrac{1}{\sqrt{4 - x^2 - 4y^2}}\mathbf{i} + \dfrac{1}{\sqrt{4x^2 + 4y^2 - 1}}\mathbf{j}$. We require that

$4 - x^2 - 4y^2 > 0$ and $4x^2 + 4y^2 - 1 > 0 \Rightarrow \dfrac{x^2}{4} + \dfrac{y^2}{1} < 1$ and

$x^2 + y^2 > \left(\frac{1}{2}\right)^2$. Thus, the domain of $\mathbf{F}$ is

$\left\{(x, y) \mid \frac{1}{4}x^2 + y^2 < 1 \text{ and } x^2 + y^2 > \frac{1}{4}\right\}$.

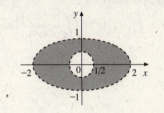

3. a. Let $t = \frac{y}{x}$, so $y = tx$. Substituting into $x^3 + y^3 = 3axy$ gives $x^3 + (tx)^3 = 3ax(tx) \Rightarrow x^3 + t^3x^3 = 3ax^2t \Rightarrow$

$x^3\left(1 + t^3\right) = 3ax^2t \Rightarrow x = \dfrac{3at}{1 + t^3}$ and $y = tx = \dfrac{3at^2}{1 + t^3}$. These are the required parametric equations, with $t \geq 0$.

b. The required area is

$A = \dfrac{1}{2}\oint_C x\, dy - y\, dx = \dfrac{1}{2}\lim_{b \to \infty}\int_0^b \left[\dfrac{3at}{1 + t^3} \cdot \dfrac{6at - 3at^4}{\left(1 + t^3\right)^2}\, dt - \dfrac{3at^2}{1 + t^3} \cdot \dfrac{3a - 6at^3}{\left(1 + t^3\right)^2}\, dt\right] = \dfrac{1}{2}\lim_{b \to \infty}\int_0^b \dfrac{9a^2t^2}{\left(1 + t^3\right)^2}\, dt$

$= \dfrac{1}{2}\lim_{b \to \infty}\left[-\dfrac{3a^2}{1 + t^3}\right]_0^b = \dfrac{1}{2}\lim_{b \to \infty}\left[-\dfrac{3a^2}{1 + b^3} + 3a^2\right] = \dfrac{3}{2}a^2$

5. Here $x = 1 + \cos t$, $y = 1 + \sin t$, and $z = 1 - \sin t - \cos t$, so

$z = 1 - (y - 1) - (x - 1) \Rightarrow x + y + z = 3$, and we see that the curve $\mathbf{r}$ lies in

the plane $x + y + z = 3$. To find the projection of C onto the xy-plane, we

observe that $\cos t = x - 1$ and $\sin t = y - 1$, so

$1 = \cos^2 t + \sin^2 t = (x - 1)^2 + (y - 1)^2$, showing that the projection is the

circle centered at $(1, 1, 0)$ with radius 1.

$\operatorname{curl} \mathbf{F} = \begin{vmatrix} \mathbf{i} & \mathbf{j} & \mathbf{k} \\ \dfrac{\partial}{\partial x} & \dfrac{\partial}{\partial y} & \dfrac{\partial}{\partial z} \\ y\cos x & x + \sin x & \cos z \end{vmatrix} = \mathbf{k}$. The surface S with boundary C is $x + y + z = 3 \Rightarrow z = g(x, y) = 3 - x - y$,

so $g_x = -1$ and $g_y = -1$. Using Stokes' Theorem with $R = \left\{(x, y) \mid (x - 1)^2 + (y - 1)^2 \leq 1\right\}$ we have

$\oint_C \mathbf{F} \cdot d\mathbf{r} = \iint_S \operatorname{curl} \mathbf{F} \cdot \mathbf{n}\, dS = \iint_R \left(-Pg_x - Qg_y + R\right) dA = \iint_R dA = \pi(1)^2 = \pi$.

7. Using Theorem 15.5.2 (14.5.2 in ET), we have $A(R) = \iint_R dx\, dy = \int_{C_2} x\, dy$, where C_2 is the boundary of R. But $x = g(u, v)$ and $y = h(u, v) \Rightarrow dy = \frac{\partial h}{\partial u}\, du + \frac{\partial h}{\partial v}\, dv$, so if the positive direction of the boundary C_1 of S is taken to correspond to that of the boundary C_2 of R under the mapping T, then $\int_{C_2} x\, dy = \int_{C_1} g(u, v)\left(\frac{\partial h}{\partial u}\, du + \frac{\partial h}{\partial v}\, dv\right) = \int_{C_1} g(u, v)\frac{\partial h}{\partial u}\, du + g(u, v)\frac{\partial h}{\partial v}\, dv$.

Using Green's Theorem with $P(u, v) = g(u, v)\dfrac{\partial h}{\partial u}$ and $Q(u, v) = g(u, v)\dfrac{\partial h}{\partial v}$, we have

$\int_{C_2} x\, dy = \pm \iint_S \left\{ \frac{\partial}{\partial u}\left[g(u, v)\frac{\partial h}{\partial v}\right] - \frac{\partial}{\partial v}\left[g(u, v)\frac{\partial h}{\partial u}\right] \right\} dA$. The sign depends on the orientation of C_2. Thus,

$$\int_{C_2} x\, dy = \pm \iint_S \left[\frac{\partial g}{\partial u}\frac{\partial h}{\partial v} + g(u, v)\frac{\partial^2 h}{\partial u\, \partial v} - \frac{\partial g}{\partial v}\frac{\partial h}{\partial u} - g(u, v)\frac{\partial^2 h}{\partial v\, \partial u} \right] dA = \pm \iint_S \left(\frac{\partial g}{\partial u}\frac{\partial h}{\partial v} - \frac{\partial g}{\partial v}\frac{\partial h}{\partial u} \right) dA$$

$$= \pm \iint_S \frac{\partial(x, y)}{\partial(u, v)}\, du\, dv$$

since $\dfrac{\partial^2 h}{\partial u\, \partial v} = \dfrac{\partial^2 h}{\partial v\, \partial u}$. Because $A(R)$ is positive, we choose the sign to match that of $\dfrac{\partial(x, y)}{\partial(u, v)}$. Then

$$A(R) = \iint_R dx\, dy = \iint_S \left| \frac{\partial(x, y)}{\partial(u, v)} \right| du\, dv.$$